NATURAL ENEMIES

Natural Enemies

*The Population Biology of Predators,
Parasites and Diseases*

EDITED BY

MICHAEL J. CRAWLEY

PhD, FLS, FIBiol
Department of Biology, Imperial College
Silwood Park, Ascot, Berks

OXFORD

BLACKWELL SCIENTIFIC PUBLICATIONS

LONDON EDINBURGH BOSTON
MELBOURNE PARIS BERLIN VIENNA

©1992 by
Blackwell Scientific Publications
Editorial offices:
Osney Mead, Oxford OX2 0EL
25 John Street, London WC1N 2BL
23 Ainslie Place, Edinburgh EH3 6AJ
238 Main Street, Cambridge
  Massachusetts 02142, USA
54 University Street, Carlton
  Victoria 3053, Australia

Other Editorial Offices:
Librairie Arnette SA
2, rue Casimir-Delavigne
75006 Paris
France

Blackwell Wissenschafts-Verlag
Meinekestrasse 4
D-1000 Berlin 15
Germany

Blackwell MZV
Feldgasse 13
A-1238 Wien
Austria

First published 1992

Set by Setrite Typesetters, Hong Kong
Printed and bound in Great Britain by
the University Press, Cambridge

DISTRIBUTORS

Marston Book Services Ltd
PO Box 87
Oxford OX2 0DT
(*Orders*: Tel: 0865 791155
           Fax: 0865 791927
           Telex: 837515)

USA
Blackwell Scientific Publications, Inc.
238 Main Street
Cambridge, MA 02142
(*Orders*: Tel: 800 759−6102
           617 225−0401)

Canada
Oxford University Press
70 Wynford Drive
Don Mills
Ontario M3C 1J9
(*Orders*: Tel: 416 441−2941)

Australia
Blackwell Scientific Publications
  (Australia) Pty Ltd
54 University Street
Carlton, Victoria 3053
(*Orders*: Tel: 03 347−0300)

A catalogue record for this book is
available from the British Library

ISBN 0−632−02698−7

Library of Congress
Cataloging in Publication Data

Natural enemies: the population biology
of predators, parasites, and diseases/
edited by M.J. Crawley.
       p.   cm.
   ISBN 0−632−02698−7 (limp)
   1. Predation (Biology) 2. Predatory
animals. 3. Population biology.
4. Parasitism. 5. Parasites.
6. Diseases. 7. Pests − Biological
control.
I. Crawley, Michael J.
QL758.N37 1992
591.53 − dc20

Contents

Contributors

JOHN BEDDINGTON *Renewable Resources Assessment Group, Centre for Environmental Technology, Imperial College of Science, Technology and Medicine, London SW7 1NA*

TIM CARO *Department of Wildlife and Fisheries Biology, University of California, Davis, CA 95616, USA*

MICHAEL J. CRAWLEY *Department of Biology, Imperial College at Silwood Park, Ascot, Berks SL5 7PY*

ANDREW P. DOBSON *Department of Ecology and Evolutionary Biology, Princeton University, Princeton, NJ 08540, USA*

CHRISTOPHER DYE *Department of Medical Parasitology, London School of Hygiene and Tropical Medicine, Keppel Street, London WC1E 7HT*

CLARE D. FITZGIBBON *Large Animal Research Group, Department of Zoology, University of Cambridge, Cambridge CB2 3EJ*

P.G. FAIRWEATHER *Graduate School of the Environment, Macquarie University, NSW 2109, Australia*

JOHN L. GITTLEMAN *Department of Zoology and Graduate Program in Ecology, University of Tennessee, Knoxville, TN 37996–0810, USA*

H. CHARLES J. GODFRAY *Centre of Population Biology, Imperial College, Silwood Park, Ascot, Berks SL5 7PY*

TIM GUILFORD *Animal Behaviour Research Group, Department of Zoology, University of Oxford, South Parks Road, Oxford OX1 3PS*

ILKKA HANSKI *Department of Zoology, University of Helsinki, P. Rautatiekatu 13, SF-00100, Helsinki, Finland*

PAUL H. HARVEY *Department of Zoology, University of Oxford, South Parks Road, Oxford OX1 3PS*

MICHAEL P. HASSELL *Department of Biology, Imperial College at Silwood Park, Ascot, Berks, SL5 7PY*

PETER J. HUDSON *Scottish Grouse Research Project Game Conservancy, Crubenmore Lodge, Newtonmore, Inverness-shire, Scotland*

JOHN R. KREBS *Department of Zoology, University of Oxford, South Parks Road, Oxford OX1 3PS*

ANNARIE M. LYLES *Ornithology Department, The Bronx Zoo, New York Zoological Society, Bronx, NY 10460, USA*

STEPHEN B. MALCOLM *Department of Biological Sciences, Western Michigan University, Kalamazoo, MI 49008, USA*

ROBERT M. MAY *Department of Zoology, University of Oxford, South Parks Road, Oxford OX1 3PS*

NICK J. MILLS *Division of Biological Control, College of Natural Resources, University of California, Albany, CA 94706, USA*

IAN NEWTON *Institute of Terrestrial Ecology, Monks Wood Experimental Station, Abbots Ripton, Cambridgeshire PE17 2LS*

D. JAMES NOKES *Department of Biology, Imperial College, London SW7 2BB*

SIMON NORTHRIDGE *Renewable Resources Assessment Group, Centre for Enviromental Technology, Imperial College of Science, Technology and Medicine, London SW7 1NA*

STUART L. PIMM *Department of Zoology and Graduate Program in Ecology, University of Tennessee, Knoxville, TN 37996–0810, USA*

ANDREW REDFEARN *Graduate Program in Ecology, University of Tennessee, Knoxville, TN 37996, USA*

SUSAN E. RIECHERT *Department of Zoology, University of Tennessee, Knoxville, TN 37996, USA*

MAURICE W. SABELIS *Department of Pure and Applied Ecology, University of Amsterdam, Kruislaan 302, 1098 SM Amsterdam, The Netherlands*

JON SEGER *Department of Biology, University of Utah, Salt Lake City, Utah 84112, USA*

A.J. UNDERWOOD *Institute of Marine Ecology, University of Sydney, NSW 2006, Australia*

JEFF K. WAAGE *CAB Institute of Biological Control, Silwood Park, Buckhurst Road, Ascot, Berks SL5 7TA*

CHARLOTTE H. WATTS *Department of Zoology, University of Oxford, South Parks Road, Oxford OX1 3PS*

Preface

This book is about death and disease. It is an ecologist's view of Darwin's vivid evocation of Nature, red in tooth and claw. As such, it is definitely not recommended for the squeamish. The aim is to look for broad patterns in the population biology of natural enemies, and to address general questions about the role of natural enemies in the population dynamics and evolution of their prey. In doing this, we aim to highlight the similarities and differences between the various taxonomic groups of natural enemies from lions to viruses.

A wide range of questions is addressed. For example, how do large natural enemies like wolves differ from small natural enemies like bacterial diseases in their effects on prey abundance? How do specialist natural enemies like insect parasitoids differ from generalist arthropod predators like spiders in their effects on prey evolution? How do gregarious natural enemies compare with solitary natural enemies in their success as biological control agents? How does harvesting of animal populations by man differ in its effects from attack by generalist natural enemies? How should predators behave in order to maximize their fitness? Is it better to chase after the prey, or sit and wait for them to come to you? Is it better to build a trap to catch the prey, or to rely upon grabbing them in jaws or talons? How should prey behave in order to minimize the risk of their being eaten by predators, e.g. habitat selection (the search for enemy-free space), adoption of nocturnal habits, stealth, alertness, group living. The book deals with the interplay of these ecological and evolutionary forces.

Views about the importance of natural enemies in prey population dynamics have tended to polarize at two ends of a spectrum. At one end, predators were thought of as mopping up a doomed surplus of prey individuals, and as having no important effect on long-term prey densities. At the other end, keystone predators were seen as controlling ecosystem structure and function by regulating the populations of potentially dominant prey species at such slow densities that resources were not depleted, and species richness was enhanced as a result of competitor release (e.g. Hairston *et al.*, 1960). As usual in these disputes, the truth lies somewhere in the middle.

Another difficulty in the study of predator–prey interactions has been a more than usually prevalent tendency to promulgate single-factor explanations. This has meant that the realizations that populations are not necessarily regulated all of the time, and that the factor responsible for regulation can change from place to place and from time to time, have come relatively late.

Views also differ about the evolutionary implications of natural enemy attack (e.g. its importance in the evolution of sex). Predation is almost always selective according to the age, size and sex of the prey. Sometimes, variation in susceptibility to natural enemy attack also has a genetic basis, and in these cases selective predation can act as an important agent of natural selection, making future generations of prey more difficult to catch. This, in turn, is likely to act as a force of natural selection on the natural enemies, increasing their efficiency in finding, catching and handling their prey. This evolutionary play forms a central theme of the book.

The book is arranged in three parts. The first covers background material on evolution, morphology and population dynamics. The second describes how the different taxonomic groups of natural enemies affect, and are affected by, their prey. The final section provides a synthesis of the material in the preceding sections, with particular emphasis on humans as predators (in harvesting natural resources) and as managers of natural enemies (in biological pest control).

M.J.C.

1

BACKGROUND

1: Evolution of Exploiter–Victim Relationships

JON SEGER

Everything is in flux (Τὰ πὰντα 'ρεî)
[Heraclitus]

God is in the details
[Mies van der Rohe]

INTRODUCTION

Almost every TV nature show has at least one scene in which the potential victim of predation pricks up its ears and looks about nervously for the predator. This behaviour and the evasive responses that go with it are often presented as Eternal Truths of Life: the fox and the hare (and every other predator and prey) have been doing this dance, in just this way, from The Beginning, and they always will. Leaving aside the implied value judgement, is the factual assertion likely to be true, even as a metaphor? Its logic is that predation is inevitable, therefore so must be the classic paranoid behaviour of the prey. At a suitably coarse-grained level, the argument seems entirely plausible; prey and predator seldom trade roles, and each typically belongs to a higher taxon that is deeply committed to a way of life that cements the relationship in its current form. However, at a more fine-grained level, the *details* of the interaction between prey and predator (or host and parasite) may be far from constant.

The main theme of this chapter is the idea that the evolutionary dynamics of asymmetrical antagonistic relationships (attacker versus attacked) are inherently unstable, which implies that rapid and erratic coevolution may be characteristic of such relationships. I will focus almost exclusively on microevolutionary processes that operate on ecological time scales. These processes seem likely to have macro-evolutionary consequences; for example, in some cases they could promote speciation, and I will point this out in passing. But I will otherwise not say anything about the description and analysis of macro-evolutionary patterns (e.g. co-speciation of hosts and parasites), which is an important and rapidly developing subject in its own right (for introduction to the literature see Mitter & Brooks, 1983; Brooks, 1988; Hafner & Nadler, 1988).

[3]

Interest in the evolutionary dynamics of genetically determined host–parasite interactions has increased sharply during the last decade (e.g. Jaenike, 1978; Bremermann, 1980; Hamilton, 1980, 1982; Lewis, 1981a,b; Levin & Lenski, 1983, 1985; May & Anderson, 1983a,b; Bremermann & Fiedler, 1985; Bell & Maynard Smith, 1987; Seger, 1988). This interest has been fuelled in large part by the proposition that an endless coevolutionary chase could give rise to persistent heritable fitness variation of the kind needed to overcome the 'cost of sex' in large, low-fecundity organisms without male parental investment (e.g. Williams, 1975; Maynard Smith, 1978b; Bell, 1982; various chapters in Michod & Levin, 1988). The surface of this very rich subject has hardly been scratched, even at a theoretical level, but volatile dynamics appear with such regularity that they have almost come to be expected, particularly in models where the interaction between attacker and attacked is suitably 'qualitative', in a sense which is developed further below. These models trace back to earlier work (e.g. Mode, 1958, 1961; Person, 1959, 1966; Leonard, 1969, 1977; Jayakar, 1970; Yu, 1972; Clarke, 1976) that was motivated in large part by empirical descriptions of 'gene-for-gene' interactions that had been described in some agricultural plant populations and their fungal pathogens (e.g. Flor, 1956; see Barrett, 1985; Newton & Crute, 1989).

There is also increasing interest in the coevolution of prey and predator species (e.g. various chapters in Futuyma & Slatkin, 1983; Taylor, 1984; various chapters in Feder & Lauder, 1986; Allen, 1988; Endler, 1988; Spencer, 1988). However, more of this work is focused on the details of particular *outcomes* (e.g. physiology, structure and behaviour) than on the evolutionary *dynamics* of the interaction (but see Stewart, 1971; Levin & Udovic, 1977; Rosenzweig & Schaffer, 1978; Schaffer & Rosenzweig, 1978; Slatkin & Maynard Smith, 1979; Levin & Segel, 1982; Roughgarden, 1983; Abrams, 1986; Gould, 1988). At least three reasons can be given for this difference of emphasis.

1 The physiological and behavioural adaptations of macroscopic prey and predator species are relatively easy to observe, and thus have long demanded explanation. In some cases (e.g. ungulates and carnivores) there is even a fossil record showing in detail how morphologies (and by inference, behaviours) coevolved over vast stretches of time (Bakker, 1983; Vermeij, 1987).

2 The mutually adapted traits of prey and predator species tend to be *quantitative*, involving sizes, shapes, colours, rates or other continuously adjustable variables with (presumably) continuous effects on performance. Models for the coevolution of quantitative characters can be more difficult to analyse than their qualitative counterparts, especially when population dynamics are also included, and in many cases they are less inclined to exhibit the perpetual evolutionary motion that has come to be expected of qualitative models.

3 There is a long history of interest in the *population* dynamics of predator–prey systems (also inherently unstable), going back to the early years of this century (see Scudo & Ziegler, 1978). Much of the work on predator–prey coevolution comes out of this tradition, and is focused more on the problem of community stability than on the problem of genetic variation within species; models in this tradition are often cast in terms of global quantitative variables such as mean density-dependent birth and death rates, and the aim is often to find equilibrium conditions.

For these and other reasons, there are fewer fully dynamical models of antagonistic coevolution for *quantitative* characters than there are for *qualitative* characters, and these models have less often been designed and analysed with the aim of finding permanent instability. This suggests that suitably designed models of quantitative predator–prey interactions might well exhibit the kind of dynamical richness found in genetic models of qualitative host–parasite interactions.

There are at least two levels on which a model may be said to be either qualitative or quantitative, and in principle these levels are independent of each other. Both the *mode of phenotypic interaction* between antagonists, and the *genetic systems* controlling variation in the expression of the phenotypes, may be either qualitative or quantitative. There is a tendency, in the minds of theorists and perhaps also in nature, for qualitative phenotypes to be controlled by one or a few major-gene loci and for quantitative phenotypes to be controlled by polygenes. But this association is not inevitable, either in theory or in fact. A qualitative phenotypic difference can be caused by developmental canalization of what would otherwise be continuous variation influenced by genes at many loci (and by environmental variation as well); such phenotypes are known in quantitative genetics as 'threshold characters'. And a quantitative phenotypic difference whose expression and consequences are both naturally measured on ordinal scales (e.g. a dimension or a rate) can be determined by alleles at a single locus; we will consider one such model below. Of course on both levels (mode of phenotypic interaction and underlying genetic system) the distinction between 'qualitative' and 'quantitative' is somewhat arbitrary. But the distinction is a useful one nonetheless, in part because it provides a framework on which the structures of the models can be related to their ranges of behaviour.

Many organisms respond *phenotypically* to the expected or actual presence of antagonist species, in ways that mimic certain aspects of the *evolutionary* responses considered in this chapter. For example, immune responses are often viewed as forms of simulated evolution that allow slow-growing hosts to keep up with fast-growing, fast-evolving pathogens. Seasonal and facultative variation in morphology, life history and behaviour is also found in prey species (particularly in aquatic invertebrates) subject to varying intensities and sources of predation (for review see Dodson, 1989a). These adaptive 'polyphenisms' reveal much about the kinds of selection pressures to which victim species are exposed, and about feasible responses to them. But there has been relatively little exploration of the important connection between phenotypic and evolutionary responses to antagonist species.

FROM STABILITY TO CHAOS

How did the modern 'dynamic' view of antagonistic coevolution arise in the first place? And why was the earlier view comparatively static? I will not attempt to answer these important and fascinating questions, except to point out that fragmentary hints of the modern view can easily be found in earlier writings. However, it is seldom clear to what extent these earlier commentaries directly stimulated subsequent developments, and to what extent they merely expressed ideas that were 'in the air' at the time.

For example, in Chapter III of *On the Origin of Species* (*Struggle for Existence*), Darwin (1859) offers an image that in retrospect seems to embody a very fluid view of competition:

> The face of Nature may be compared to a yielding surface, with ten thousand sharp wedges packed close together and driven inwards by incessant blows, sometimes one wedge being struck, and then another with greater force.

This seems to imply that as some wedges advance inwards, they do so at the expense of others, which are forced backwards. The image conveys a sense of dynamical instability, and hence of permanent, churning motion, of the kind now commonly referred to by means of the 'Red Queen' (Van Valen, 1973) and 'arms race' (Dawkins & Krebs, 1979) metaphors. However, even though Darwin's *images* tend to be full of movement, and he emphasizes repeatedly that the struggle is mainly with other species (a staggeringly complex 'action and reaction of . . . innumerable plants and animals'), his *analyses* tend to emphasize equilibrium.

> Battle within battle must ever be recurring with varying success; and yet in the long-run the forces are so nicely balanced, that the face of nature remains uniform for long periods of time, though assuredly the merest trifle would often give the victory to one organic being over another.

Early in this century Fisher (1930) discussed the idea that any well-adapted species will experience a 'constantly deteriorating' environment, owing to 'the evolutionary changes in progress in associated organisms'. Van Valen's (1973) Red Queen hypothesis ('Now here, you see, it takes all the running you can do, to keep in the same place' Lewis Carroll, *Through The Looking-Glass*) is an elaboration of the deteriorating-environment idea, for which Van Valen credits Lyell (1832) in addition to Darwin and Fisher. All of these authors seem to have been thinking more about competition than about asymmetrical antagonistic relationships of the host–parasite and predator–prey type, and they seem to have been thinking mainly about evolutionary changes taking place on geological, as opposed to ecological, time scales; the emphasis is much more on the extinction of entire species than on the microevolutionary changes taking place within individual populations.

Asymmetrical antagonism, genetic polymorphism and rapid evolution were apparently connected for the first time in a modern way by Haldane (1949), who argued that 'the struggle against [infectious] disease' has been an extremely important 'evolutionary agent', with consequences 'rather unlike those of the struggle against natural forces [e.g. cold], hunger, and predators, or with members of the same species'. The argument begins from the 'very elementary fact' that in 'all species investigated the genetical diversity as regards resistance to disease is vastly greater than that as regards resistance to predators'. (We will return shortly to Haldane's separation of host−parasite from predator−prey coevolution.) In support of this assertion Haldane mentions the variable resistance of wheat cultivars to different strains of the fungus *Puccinia graminis*, and various other examples. 'To put the matter rather figuratively, it is much easier for a mouse to get a set of genes which enable it to resist *Bacillus typhi murium* than a set which enable it to resist cats.' The reason for this difference is that 'a small biochemical change will give a host species a substantial degree of resistance to a highly adapted microorganism'.

However, a species is unlikely to evolve *lasting* immunity, because 'microscopic and submicroscopic parasites can evolve so much more rapidly than their hosts'. Therefore 'it is an advantage to the individual to possess a rare biochemical phenotype' which will, on account of its rarity, 'be resistant to diseases which attack the majority'. In addition, 'it is an advantage to a species to be biochemically diverse, and even to be mutable as regards genes concerned in disease resistance'. Although these key predictions (diversity and mutability) are derived from a verbal species-advantage argument that would be viewed with scepticism today, Haldane clearly sees how diversity can be favoured by frequency-dependent natural selection; for example, he invokes the analogy to self-sterility alleles in plants where rare alleles always enjoy higher fitness than do common alleles.

Haldane speculates that the 'surprising' serological diversity of bird and mammal species 'may play a part in disease resistance, a particular race of bacteria or virus being adapted to individuals of a certain range of biochemical constitution, while those of other constitutions are relatively resistant'. He also cites evidence consistent with the prediction of high mutability:

> Many pure lines of mice have split up into sublines which differ in their resistance to tumour implantation. This can only be due to mutation. The number of loci concerned is comparable, it would seem, with the number concerned with coat colour. But if so their mutation frequency must be markedly greater.

Finally, Haldane argues 'that the selection of rare biochemical genotypes has been an important agent not only in keeping species variable, but also in speciation'. His scenario is the standard allopatric model with hybrid inferiority following secondary contact (Mayr, 1963); here the main *cause* of hybrid inferiority is divergence at loci affecting resistance to coevolving pathogens.

> ... under the pressure of disease, every species will pursue a more or less random path of biochemical evolution. Antigens originally universal will disappear because a pathogen had become adapted to hosts carrying them, and be replaced by a new set, not intrinsically more valuable, but favouring resistance to that particular pathogen. Once a pair of races is geographically separated they will be exposed to different pathogens. Such races will tend to diverge antigenically, and some of this divergence may lower the fertility of crosses.

I have quoted at such length from this remarkable paper because it is not available in all university libraries, and because it articulates so many issues and beliefs that are still central to the subject. In particular, Haldane: (i) identifies biochemical diversity as a problem; (ii) suggests that the fitness of a given phenotype may depend mainly on the relative frequencies of complementary phenotypes in antagonist species; and (iii) notes the all-important evolutionary implication, which is endless evasion and pursuit through the space of feasible biochemical defences and attacks, along trajectories that are, at least in certain respects, 'random'.

With the benefit of hindsight we are naturally led to wonder why Haldane never pursued the fascinating dynamical possibilities he had so clearly identified. No doubt there are many reasons, among them the extreme difficulty and expense involved in carrying out extensive computations in the early 1950s, which tended to encourage the view that interesting dynamical systems were those whose behaviour could be described analytically (Stewart, 1989). Yet ecologists had been well aware of oscillatory *population* dynamics (in theory, in experiments and in nature) for the previous two decades (Kingsland, 1985; Kareiva, 1989). Haldane's (1949) paper even includes a discussion of host–parasitoid population dynamics, complete with a Nicholson–Bailey model, equations for the equilibrium densities of host and parasitoid and the observation that 'there is perpetual oscillation round this equilibrium'. (Since the time of its publication the paper has been cited regularly for its discussion of the way ecological factors interact to limit population sizes; only in recent years has it come to be cited as often for its discussion of the way pathogens contribute to the maintenance of genetic variation in their hosts.)

A few years later Haldane (1954) again discussed host–pathogen coevolution, in a paper called 'The statics of evolution'. As its title would suggest, this paper is mainly about the 'approximate equilibrium' that renders evolution 'almost unobservable'. An extensive list of factors responsible for maintaining genetic variation includes (as number 7) 'Selection by Labile Pathogens'. Haldane very briefly summarizes the logic laid out in the 1949 paper, and concludes that 'both host and pathogen will constantly alter their prevailing genotype, in so far as it affects the host–parasite relation'. This process will favour genetic diversity. 'It is, perhaps, too rapid and reversible to be regarded as evolution, but cannot be without effect on evolution, particularly by favouring divergence of separated populations.' Haldane goes so far as to suggest that this process might even be considered an instance of 'non-Darwinian selection', because 'the relative fitness of two genotypes may depend on their frequency'; other instances would include selection at the Rh (Rhesus blood group) locus in the human species and some cases of frequency-dependent mimetic polymorphism. It seems both amazing and sobering that a mere 40 years ago, frequency dependence could be viewed as a non-Darwinian side-show, while today it dominates our thinking about many kinds of ecological and evolutionary phenomena.

Antagonistic coevolution was firmly connected to population regulation by Pimentel (1961b, 1968). As the population size of a parasite or predator species increases, it causes greater mortality in its host. If this increased mortality in turn causes stronger selection for effective defences, then the host may become so resistant that its parasite or predator declines in numbers, thereby lowering the intensity of selection on the host. Pimentel originally conceived this 'genetic feedback mechanism' as a process that would stabilize the ecological interactions of antagonists; this view was supported by long-term experimental studies of insect host–parasitoid systems, by the mutual adjustment of rabbits and myxoma virus in Australia, and by a great deal of circumstantial evidence derived from agricultural experience with arthropod and fungal enemies of crops. However, mathematical models of the process soon revealed that it could lead to instability as well as to stability, depending on many details of the underlying biology and genetics (see, for example, Levin, 1972).

This theme, that the god of dynamics is in the details, reappears over and over again during the next two decades of work on a wide range of coevolutionary models. It is perhaps most obvious in the context of host–parasite and predator–prey evolution, but it also appears in other settings, for example, in models of frequency-dependent competition *within* species; one such model is mentioned below, in the section on quantitative interactions.

QUALITATIVE INTERACTIONS — HOST AND PARASITE

By the late 1950s a few people had begun to pursue the dynamical possibilities so clearly identified by Haldane, but the pattern of citations seems to imply that they were generally unaware of his 1949 paper. As was mentioned above, most of the early genetic models are based at least loosely on the 'gene-for-gene' pattern of interaction that had been described for a few agricultural systems, most notably for flax (*Linum usitatissimum*) and the rust *Melampsora lini*, where multiple alleles at corresponding loci in each species determined the host's resistance and the parasite's virulence (Flor, 1956). As the genetics of both species had been characterized, it was natural to include both host and parasite explicitly (though always in a highly simplified and abstracted form) in these models, and it soon became clear that a wide range of outcomes was possible, depending on many details of the genetic systems, life histories and patterns of costs and benefits assumed for each species; the range includes fixation of alleles for maximum resistance in the host and maximum virulence in the parasite, stable polymorphic equilibria and various forms of permanent cycling, including chaos (see, for example, Mode, 1958, 1961; Leonard, 1969, 1977; Jayakar, 1970; Levin, 1972, 1983; Yu, 1972; Rocklin & Oster, 1976; Levin & Udovic, 1977; Auslander *et al.*, 1978; Hamilton, 1980; Rapport & Person, 1980; Lewis, 1981a,b; May & Anderson, 1983a,b; Bremermann & Fiedler, 1985; Barrett, 1988; Seger, 1988).

The simplest models almost always cycle, and they show at once why cycling is at least an underlying tendency wherever antagonists interact through complementary phenotypes (Eshel & Akin, 1983). Consider a haploid host with genotypes H_1 and H_2, and a parasite with complementary genotypes P_1 and P_2. Here 'complementary' means that parasite P_1 is fitter on host H_1 than on H_2, because it is better able to penetrate the defences of H_1; as a consequence, an attack by P_1 lowers the fitness of host H_1 more than it does the fitness of H_2. The symmetrically opposite pattern holds for parasite P_2, which is better able to attack host H_2 than H_1. Thus for each species, the relative *fitness* of each of its two genotypes is determined by the relative *frequencies* of the genotypes of the *other* species. The population sizes of the two species are assumed to be regulated by density-dependent factors at least partly extrinsic to the interaction modulated by these genotypes, and are therefore not specified.

If most parasites are P_1, then H_2 will be fitter than H_1 and will tend to increase in frequency; but as H_2 becomes the predominant host genotype, P_2 becomes the fitter parasite and begins to increase in frequency; then as P_2 increases in frequency, the relative fitness of H_2 declines, eventually dropping below that of H_1, which begins to increase in frequency; the ascendancy of H_1 makes P_1 the fitter parasite, and P_1 again increases to predominance, making H_2 again the fitter host; and so on *ad infinitum*. Each genotype plants the seeds of its own undoing, through the effect its success has on the relative fitnesses (and thus eventually the relative frequency) of the genotypes of the other species; the high are indeed made low, and the low are made high.

This sounds like negative frequency-dependence ('I'm my own worst enemy'), and it is, but the mechanism is somewhat unusual in being both *indirect* and *time delayed*. The frequency-dependence is indirect because, strictly speaking, the fitness of a given genotype (say, H_1) is independent of its *own* frequency; instead, its fitness depends on the frequencies of genotypes (say P_1 and P_2) of *other* species. Of course the current frequencies of P_1 and P_2 reflect the recent frequencies of H_1 and H_2, such that being common today tends to make H_1 less fit tomorrow, so in a larger sense (and in the long run) H_1 experiences a *statistical* or *virtual* form of negative frequency-dependence, with a time delay (because current fitness is best correlated with frequency a number of generations in the past).

If the genetics are as simple as those just described (two alleles at one haploid locus in each of two species) then the dynamics tend towards limit cycles whose exact character depends on various details of the model such as mutation and migration rates (if any), frequency-dependent interactions *within* each species (if any) and whether the model is cast in continuous (differential equation) or discrete (difference equation) time. For example, in a model with discrete time, with no frequency-dependence and no mutation or migration, the populations spiral outward to the boundaries (Fig. 1.1).

If the genetics are made even slightly more complicated than this (e.g. by introducing diploidy in one or both species, or more than one locus, or more than two alleles at each locus), then the dynamical possibilities become extremely rich. Depending on the biological details and parameter values, such models may exhibit the full range of

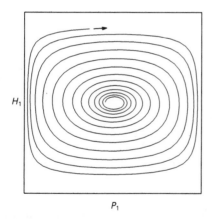

H_1

P_1

Fig. 1.1 Dynamics of a haploid one-locus, two-allele model of host–parasite coevolution. This phase diagram shows the simultaneous frequencies of allele H_1 in the host and allele P_1 in the parasite. The trajectory begins near the centre of the gene frequency plane ($P_1 = H_1 = 0.45$) and follows the evolution of the system for 1000 generations, by which time the gene frequencies are becoming 'stuck' near the boundaries (corresponding to a state of near-fixation of one allele in one of the two species). The selection coefficients are $s = 0.1$ (hosts infected by the parasite able to overcome their defences have a relative fitness of 0.9) and $t = 0.3$ (parasites on the host which is able to resist them have a relative fitness of 0.7). Hosts and parasites encounter each other at random. There is no mutation in either species. Small amounts of mutation are sufficient to keep the gene-frequency trajectories away from the boundaries, in which case the system eventually settles into a stable limit cycle (see Fig. 1.3a). Additional details of this model are described by Seger (1988).

behaviours from point equilibria through simple cycles, to complex cycles and chaos. The papers cited earlier in this section give some examples, but taken together they represent no more than an initial foray into this vast and essentially uncharted territory. The models studied to date are only a small fraction of those that could be created by reassembling their own biological and genetic assumptions in new combinations; in addition, many interesting biological and genetic assumptions have yet to be incorporated into any models at all. As a consequence, it is still easy to concoct simple new models, many of which exhibit dazzling and apparently novel behaviour. For example, Fig. 1.2 shows a diploid generalization of the haploid model of Fig. 1.1, and Fig. 1.3 shows a two-host, two-parasite version of the same model. The difficult task that remains is to systematically explore the space of relevant models, in a way that will eventually yield biologically meaningful principles, or at least some useful generalizations.

Most host−parasite models are qualitative in both of the senses

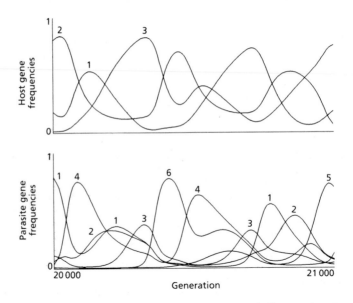

Fig. 1.2 A multi-allele diploid generalization of the model illustrated in Fig. 1.1. Here the host has three alleles (1−3) at a diploid locus, giving six diploid genotypes, while the parasite has six alleles (1−6) at a haploid locus, each allele corresponding to one of the six diploid genotypes of the host. For example, parasites of genotype P_1 are able to overcome the defences of the first host genotype (H_1H_1). With so many genotypes in each species it is not possible to display the trajectories as a phase portrait (Fig. 1.1), so here the gene frequencies are shown as functions of time. The behaviour of this system appears to be chaotic. Even though there is again no mutation, the trajectory shows no tendency to become stuck at the boundaries (although individual parasite alleles may have very low frequencies for several hundred generations at a time). This is probably a consequence of the host's diploidy, which allows the alleles that make up a temporarily unfit genotype to 'escape' into other, relatively more fit genotypes, thereby weakening the correlations between genotypic and genic fitness.

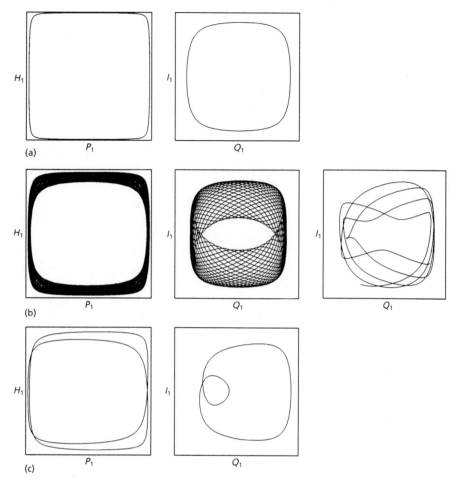

Fig. 1.3 A two-host, two-parasite generalization of the model illustrated in Fig. 1.1. Two species of host are attacked by two species of parasite, but each of the two parasites has a different favoured host: parasite P is fittest on host H, and parasite Q is fittest on host I. The phase diagrams are as in Fig. 1.1, with the $P–H$ pair illustrated on the left, and the $Q–I$ pair on the right. The parasite has a mutation rate of 10^4 per locus per generation (forward and back, between the two alleles). In all cases studied, some form of apparently stable limit cycle was reached after many generations. The cases illustrated here were all run for 30 000 or more generations before these plots were made, so as to show only the limit cycles. (a) Parasites P_1 and P_2 both have zero fitness on hosts I_1 and I_2, while Q_1 and Q_2 have zero fitness on H_1 and H_2; thus the two host–parasite pairs are completely uncoupled, and the two phase portraits represent models that are equivalent to the one illustrated in Fig. 1.1. The pair on the left interact with each other more strongly than do the pair on the right (i.e. s and t are larger for P and H than they are for Q and I). (b) The parasites have low but finite fitness on their 'alternative' hosts. For Q on H, the *sign* of the interaction between genotypes is reversed (e.g. Q_1 is fitter on H_2 than on H_1). (c) The fitness parameters are similar to those illustrated in (b), except that the sign of the genotypic interaction on the alternative host is reversed for *both* parasites.

mentioned earlier and seen in the examples (i.e. distinct, complementary phenotypes correspond directly to simple genotypes). Many of them also assume that phenotypes are intrinsically interchangeable or *symmetric*, i.e. there is no real difference between H_1 and H_2 except that they are vulnerable to P_1 and P_2, respectively. Thus differences in fitness among the phenotypes within a population arise purely as a result of interactions with the antagonist species; in particular, no phenotype is assumed to be intrinsically more costly to develop than another. Such phenotypes may be said to be *strategic* or *informational*. They correspond to Haldane's view of antigenic diversity or, to give a more recent example, to the DNA sequence specificities of bacterial restriction enzymes, which could also generate fitness differences that existed solely because of the biological context established by another species (in this case, different phage genotypes).

The assumption that all phenotypes have exactly the same intrinsic cost is of course just a special case of the assumption that they have slightly different cost. Not surprisingly, the behaviour of most models seems not to change qualitatively when differential costs are imposed, as long as high costs do not render some phenotypes unconditionally worse than others. If the pattern of costs and benefits is further modified to make the more *costly* phenotypes the more *effective* against *all* the alternatives presented by the antagonist species, then the model becomes *quantitative* at the phenotypic level, while remaining qualitative at the level of the underlying genetic system. It is not hard to imagine biochemical scenarios that would have this property (e.g. different levels of expression of a metabolically costly gene product, or of one that becomes otherwise harmful at too high a concentration), but this model can also be viewed as the simplest genetic representation of the 'arms race' that is often taken to characterize the relationship between prey and predator.

QUANTITATIVE INTERACTIONS — PREY AND PREDATOR

Why should it be 'much easier for a mouse to get a set of genes which enable it to resist *Bacillus typhi murium* than a set which enable it to resist cats'? One reason is that the necessary defences may be more like bulwarks than passwords. Such defences are forms of armour that must be paid for in currencies such as energy and time — resources that might otherwise be spent on reproduction. The key assumption is that larger investments in defence (e.g. a thicker shell) always lower the prey's risk of being eaten, while larger investments in mechanisms of attack (e.g. heavier jaws and teeth) increase the predator's rate of success against such defences. Each side faces an unavoidable tradeoff, in which the net marginal value of increasing or decreasing the current investment in defence or attack depends on the investments currently being made by the other side.

The 'arms race' metaphor provides a natural way to think about such interactions (e.g. Dawkins & Krebs, 1979), but its heuristic value has been called into question (e.g. Abrams, 1986), in part because it blurs the distinctions between several different aspects of predator–prey coevolution, only some of which have inescapable economic tradeoffs at their core. For example, both prey and predator may have opportunities to incorporate *absolutely* beneficial new mutations that improve their respective skills (and hence their fitness) under all conditions. If so, interest is likely to centre on questions concerning the two species' relative rates of evolution, and the effects this process will have on their population sizes. This interpretation of the metaphor leads to models that are cast in very general terms, such that the variables of interest are equivalent to population mean fitness, rate of population growth, carrying capacity, and the like (e.g. Levin & Udovic, 1977; Schaffer & Rosenzweig, 1978); such models do not usually incorporate specific economic and genetic constraints that limit improvement in each species. (Note that these constraints might equally well be qualitative or quantitative, in any of the senses developed above.) In effect, the evolutionary process being considered is one in which the constraints themselves are overcome, *expanding* the boundary of the set of feasible phenotypes, rather than one in which the species evolve *along* the boundary defined by the existing constraints. By contrast, the majority of host–parasite models define a fixed set of genetic and ecological constraints, and ask what happens within that framework.

Other interpretations of the arms race metaphor emphasize quantitative interactions between prey and predator, and thus lead to models that focus on cost–benefit tradeoffs. There has been much work on the behavioural ecology of these tradeoffs, in the tradition of optimal foraging (see Chapter 4), and the interface between individual behaviour and population dynamics has recently become an especially active focus of research (e.g. Holt, 1977, 1985; various chapters in Sibly & Smith, 1985; Holt & Kotler, 1987; Ives & Dobson, 1987). There have been remarkably few fully coevolutionary analyses of mutual antagonisms specifically modelled on a quantitative interpretation of the predator–prey relationship (e.g. Abrams, 1986), and even fewer with explicit genetic models for the expression of the phenotypes in question. Instead, interest in the dynamics of predator–prey systems has been concentrated overwhelmingly on their *population* dynamics, for which there is a large body of work in the tradition of community ecology (e.g. May, 1974; Hassell, 1978; various chapters in Diamond & Case, 1986 and Roughgarden *et al.*, 1989; Yodzis, 1989).

Quantitative counterparts of the qualitative models considered in the previous section can easily be made to cycle. The simplest model of this kind would again have a haploid locus in each species, with alleles H_1 and H_2 in the prey, and alleles P_1 and P_2 in the predator. The prey genotypes differ with respect to defence (H_2 being better defended than H_1, and thereby suffering a fixed cost s), while the predator genotypes differ with respect to attack (P_2 being stronger than P_1, and thereby suffering a fixed cost t). Let the weakly defended prey genotype be equally vulnerable to both predator genotypes (thereby suffering a cost of predation u), but let the more resistant prey be vulnerable only to predator individuals with the stronger attack; this implies that the weaker (P_1) predators experience reduced success (at a cost of failure v) in the presence of the more resistant (H_2) prey. These assumptions and definitions give rise to the following sets of fitnesses for the prey and predator genotypes, conditioned on the presence of the genotypes of the other species.

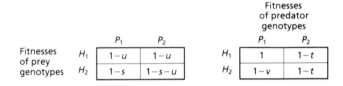

		P_1	P_2			P_1	P_2
Fitnesses of prey genotypes	H_1	$1-u$	$1-u$		H_1	1	$1-t$
	H_2	$1-s$	$1-s-u$		H_2	$1-v$	$1-t$

For example, if $s = t = 0.1$ and $u = v = 0.15$, then the fitnesses are:

		P_1	P_2			P_1	P_2
Fitnesses of prey genotypes	H_1	0.85	0.85		H_1	1	0.9
	H_2	0.9	0.75		H_2	0.85	0.9

These fitnesses do not exhibit the symmetries seen in the corresponding qualitative model, but they retain the same essential property that gives rise to cycling in that model: the more common genotype within each species establishes, in the *other*, selection for the genotype that lowers its own relative fitness. For example, when the stronger predator is the more common, the *weaker* prey is actually the more fit, because the stronger predator successfully attacks *both* prey types, rendering the more costly defence unprofitable. Thus it is not surprising that the behaviour of this model is exactly like that of simplest 'host−parasite' model illustrated in Fig. 1.1, except that the unstable interior equilibrium point and the gene-frequency trajectories around it are not centred in the gene-frequency plane.

The model can easily be extended in various ways. Perhaps the most obvious extension is that to several levels of defence in the prey, with several corresponding levels of attack in the predator. Each step in each sequence costs a little more than the last. If each phenotype is determined by a unique haploid genotype, as in the simple model just described, then some rather complex and beautiful cycles can arise; but not surprisingly, the cycles tend to be more orderly than in comparable host−parasite models, moving sequentially upward through the sequence of graded steps before returning to the weak end of the spectrum. And in keeping with a principle that seems to emerge from all the work done to date on models of antagonistic coevolution, *different underlying genetic systems can produce very different evolutionary dynamics*, in otherwise similar models. Figure 1.4 shows the three-step version of the model just described, with additive diploid genetics in each species (e.g. H_1H_1, H_1H_2 and H_2H_2 giving rise to weak, medium and strong prey defences, respectively). The behaviour of this model differs qualitatively, in several respects, from that of the corresponding three-allele haploid model.

What happens, then, when we infinitely subdivide the ranges of possible defence and attack phenotypes, and place their expression under the control of quantitative genetic systems, where many loci make individually small contributions to the final phenotype? Extrapolating from the one- and two-locus models, where increasing the complexity of the genetic systems almost always increases the complexity of the dynamics, we might expect to find the greatest complexity of all in the quantitative-genetic case. Instead, what happens is the opposite; at least in the simplest cases, there is no inherent tendency to cycle. The reason for this is fairly subtle, but easy to understand

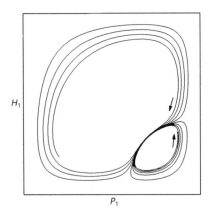

Fig. 1.4 A diploid model with quantitative phenotypic interactions between host and parasite (or prey and predator). Both host and parasite are diploid. They interact according to a three-genotype extension of the two-genotype quantitative model described in the text. Thus predator genotype 3 (the P_2P_2 homozygote) pays $2t$ fitness units for the development of its attack phenotype, and always succeeds in taking prey regardless of the prey's genotype; predator genotype 2 (the P_1P_2 heterozygote) pays t units and succeeds except when it meets host genotype 3 (H_2H_2), in which case it also loses v units owing to the prey's successful resistance; and so on. For the case illustrated, the fitness parameters are $s = t = 0.05$, $u = v = 0.115$, and the system always finds the (counterclockwise) limit cycle in the lower right-hand corner. Other parameter sets can give qualitatively different patterns. As with the model illustrated in Fig. 1.2 (diploid host, haploid parasite), and unlike the models illustrated in Figs 1.1 and 1.3 (haploid host, haploid parasite), no mutation is needed to keep the populations away from the boundaries of the gene-frequency plane.

once it has been seen. The problem is that most phenotypes in the population must always be near the *mean* phenotype. As a consequence, rare phenotypes far removed from the mean can increase in frequency only if the mean begins to move in their direction. In other words, the relative frequencies of the different phenotypes are *coupled* (constrained to co-vary) under quantitative inheritance, in a way they are not under major-gene inheritance. This has profound effects on the dynamical possibilities because populations easily become 'trapped' on local fitness maxima, when higher global maxima exist, that if attainable would lead to another round of coevolution (Fig. 1.5).

Maynard Smith and Brown (1986) found exactly this effect of the genetic system on the propensity to cycle, in a model that on its surface appears to have nothing to do with predator–prey or host–parasite coevolution. The model is called the 'size game', and it concerns

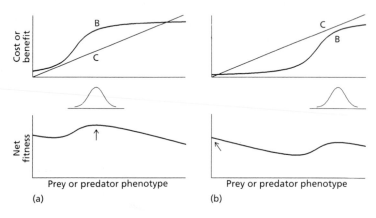

Fig. 1.5 Illustration of the process by which a prey or predator may become 'trapped' on a local fitness maximum, in a model with quantitative inheritance of quantitative phenotypes. Points along the horizontal axis represent states of development of a costly quantitative defence or attack phenotype. Thus cost (curve C) increases for phenotypes further to the right. Benefit (curve B) also increases, but not linearly, and its inflection point can move, because the benefit realized from a given investment depends on the current distribution of corresponding phenotypes in the antagonist species. For example, the rate of gain per unit invested decreases rapidly for phenotypes strong enough to resist (or overcome) most corresponding phenotypes in the other species. Net fitness, shown in the lower panels, is proportional to the difference between benefit and cost ($W = 1 + B - C$). (a) At time 1, a hypothetical population is distributed around a local maximum of fitness, which also happens to be the global maximum (arrow). But if coevolution with the antagonist takes both species to higher and higher levels of investment in defence and attack, then the inflection point of the benefit curve will also move to the right, reflecting the changed distribution of the antagonist's defence or attack phenotypes. (b) At time 2, the population is still centred at a local fitness maximum, and therefore cannot evolve very far in either direction, but the global maximum is now at the opposite (low) end of the range of possible investments (b). The phenotypes have become so costly that individuals would have higher net fitness without them, but there is 'no way to get there from here'.

a form of frequency-dependent intraspecific competition that is closely related to the 'war of attrition' (Maynard Smith, 1982; see also Parker, G.A., 1983, 1985). The essential idea is that large size is costly, but leads to victory in fitness-related encounters with smaller individuals; thus it is always better to be slightly larger than the current mean, and the population gradually increases in mean body size. There comes a point, however, where small individuals who lose all fights nonetheless have greater than average fitness, because unlike everyone else in the population, they do not pay the high costs involved in being very large. With asexual inheritance this model readily cycles, because a sub-population of small, rare individuals can persist and grow, even as mean body size in the rest of the population is continuing to increase. With sexual polygenic inheritance, however, there is either a stable equilibrium or a 'runaway' toward infinite size, because the smallest size classes cannot establish themselves while the rest of the population is large. Environmental variation in size can also stabilize what would otherwise be a dynamically unstable version of the model, because it too tends to normalize the distributions of cost and benefit, and lowers the correlation between the phenotypes of parent and offspring.

This interpretation seems to leave open the interesting theoretical possibility that a model with continuous underlying genetic variation could nonetheless be made to cycle, if the phenotypic alternatives controlled by that variation took the form of distinct, canalized threshold characters (e.g. behaviour or colour morphs). At least in the two-phenotype case, the less costly, less competitive phenotype could increase in frequency without increasing the frequency of seriously disadvantageous intermediate phenotypes (because such phenotypes simply do not occur under the assumed developmental canalization). However, there seems to have been no work on models of this kind, and it is not clear that they would apply to very many natural situations.

For the fully continuous case there is a radical but effective way to solve this problem (that an ecological opportunity may exist, e.g. for a small morph, but cannot be exploited owing to the normal distribution of phenotypes produced by sexual reproduction). The solution is to speciate. A small-bodied daughter *species* could efficiently exploit the opportunity created by the increasing gigantism of its otherwise ident-ical parent, because it would not produce any offspring that were worst-of-both-worlds intermediates. Maynard Smith and Brown (1986) suggest a similar possibility in connection with the 'size game' model:

> ... it is worth remembering that the fossil record shows that
> some 90% of mammalian lineages increase in size. Since,
> averaged over species, mammals have not continued to increase
> in size, extinction must have been commoner in large species,
> and speciation among smaller ones.

This mechanism provides another means (in addition to the one suggested by Haldane) by which the microevolution of host–parasite or predator–prey interactions could promote speciation.

The dynamical consequences of mutual antagonism that emerge from this brief review of theoretical results seem to be: (i) that persistent evolution is much more likely to arise where phenotypes correspond directly to genotypes, than it is where the genetics are quantitative; and (ii) that the *mode* of phenotypic interaction makes no essential difference. In other words, phenotypically qualitative 'host–parasite' models seem no more inclined to cycle than their phenotypically quantitative 'predator–prey' counterparts, given similarly 'qualitative' systems of genetic control of the phenotypic alternatives. Permanently dynamical behaviour also turns up often in evolutionary models of many *intraspecific* evolutionary 'games' that bear little superficial resemblance to host–parasite and predator–prey coevolution, such as the 'size game' mentioned above (for another example see Maynard Smith & Hofbauer, 1987). The key features shared by all these cases are: (i) that fitnesses are directly or indirectly frequency-dependent; and (ii) that phenotypes map onto genotypes in a simple way.

Does this mean that we should always expect to find dynamical instability in antagonistic systems where the phenotypic alternatives (and underlying genetics) are discrete, but never where they are continuous? The answer is almost certainly that we should *not* expect the world to be so orderly, even though the models might well be correct in suggesting that cycling will be more common where the alternatives are qualitative. Even if we had no qualms about the highly abstract, metaphorical nature of the models studied to date, they would tell us almost nothing about what to expect when pairwise antagonisms are embedded in networks of multispecies interactions. Multispecies interactions are bound to be important, and in principle they could either restrict or enhance the possibilities for rapid coevolution (e.g. Atsatt & O'Dowd, 1976). For example, one mechanism with potentially powerful effects is predator 'switching' among alternative prey species, on the basis of their current *relative* profitability.

The population dynamics of multispecies communities can be highly volatile (e.g. May, 1974; May & Leonard, 1975; Hassell, 1978; Diamond & Case, 1986; Roughgarden *et al.*, 1989; Yodzis, 1989), but there has been relatively little work on models of their evolutionary dynamics, particularly where the interactions are antagonistic and explicitly constrained by ecologically motivated patterns of costs and benefits. It would be surprising if multispecies evolutionary models did not turn out to exhibit a wide range of dynamical behaviours not seen in their two-species counterparts. Such models are inherently very complex and therefore difficult to analyse by traditional means, but the rapidly decreasing cost of fast computation is now making the numerical study of large multispecies models truly feasible for the first time.

Although the phenotypic variation that modulates predator–prey interactions may more often be continuous (hence 'quantitative' in the sense developed here) than the variation that modulates host–parasite interactions, this association is by no means a strong one. For example, qualitative colour-pattern polymorphisms occur in many species subject to visual predation (e.g. Lepidoptera, land snails), which implies that intermediate phenotypes may often suffer reduced fitness (for an introduction to the huge literature on this subject see Chapters 16 and 20; Turner, 1984; Allen, 1988; Endler, 1988). If these polymorphisms are typically maintained by frequency-dependent selection arising from the *learned* behaviour of predators, then cycling is possible (Levin & Segel, 1982) but relatively unlikely (see Parker, G.A., 1985). However, if 'hard-wired' neural mechanisms of perception play an important role in modulating the predator's choice of prey, or if more than one predator species is involved, then such systems can easily be destabilized (Endler, 1988).

More generally, it seems likely that many behavioural tactics employed by prey and predator species take the form of distinct alternatives. This possibility is illustrated (if only hypothetically) in one of the few fully genetic 'predator–prey' models analysed to date. This model treats a *qualitative* interaction between 'cats' that adopt one of two hunting strategies (sit-and-wait or stalk), and 'mice' that adopt one of two defensive strategies (run or freeze) when they scent a cat (Stewart, 1971). If the model is cast in continuous time, with diploid genetics in both species, then there is a stable polymorphic equilibrium; other assumptions may lead to instability, and presumably to cycling. In this model the mice *do* get genes that enable them to resist cats, at least some of the time. (Stewart's paper does not cite Haldane (1949).)

In short, while there are important distinctions to be made between qualitative and quantitative phenotypes, and between qualitative (major-gene, or else asexual) and quantitative (polygenic, necessarily sexual) genetic systems, *all* are potentially susceptible to evolutionary instability, given suitable forms of ecological interaction between two (or more) antagonists. These possibilities are best appreciated for qualitative phenotypes controlled by qualitative genetic systems (as motivated by host–parasite relationships), but this is due in large part to the relative lack of work on comparable quantitative models. It is not clear how often host–parasite relations may turn out to be quantitative (e.g. grades of resistance to grades of virulence), and how often predator–prey relations may turn out to be qualitative (e.g. distinct defensive strategies versus distinct hunting strategies); presumably, these kinds of relationships are fairly common. And for no case do we have more than hints as to what may happen in communities where several parasites (or predators) attack several hosts (or prey).

DOES LIFE IMITATE ART?

Roughgarden (1983) distinguishes between *simplifying* models that are intended to give insight into general processes, and *summarizing* models that are intended to represent what is known about the details of a particular situation. The models used in fisheries management are instances of the latter, and the models considered here are instances of the former. Although simplifying models do not yield specific quantitative predictions about particular systems (as summarizing models must do, to be of any use), they can nonetheless point to broad patterns that ought to exist if their assumptions are approximately correct. Some general predictions that seem to follow from the work described here are:

1 that mutual antagonisms should often give rise to cyclical or erratic patterns of persistent, *rapid evolutionary change* in phenotypic attributes that modulate the antagonistic interaction;

2 that as a spatial consequence of these temporal patterns, we should often observe *local coadaptation* of the antagonists; and

3 that at least in some cases we should expect to find considerable genetic polymorphism (and in particular, heritable fitness variation) associated with these traits.

The first prediction is the most direct of the three, but the hardest to test, because the time scales involved are expected to be long relative to those of research grants and human lifetimes, even though they are very short relative to species' lifetimes. For example, the period of the simple two-allele haploid model illustrated in Fig. 1.1 is roughly $4\pi/(st)^{\frac{1}{2}}$ generations, where s and t are the selective disadvantages suffered by hosts and parasites, respectively, when they find themselves in association with the antagonist genotype to which they are least adapted (Seger, 1988). Even when the fitness differences among genotypes are as large as one-half, each cycle takes more than 25 generations, and for fitness differences more modest than these, the cycles become much longer. Diploidy, multiple alleles and many other realistic complications tend to further increase the system's natural period. This seems to imply that many familiar species could be involved in fast-moving evolutionary chases, without our having yet noticed any obvious evidence of change.

Other than being very patient, what can we do? One promising approach is to study table-top systems with extremely short generation times, such as bacteria and their viruses (e.g. Levin & Lenski, 1983, 1985; Lenski, 1984a, 1988a,b). Like all such models, these systems necessarily exist under artificial, highly simplified conditions, but they may provide a half-way house between mathematics and nature that will prove very useful (e.g. Lenski & Levin, 1985).

There is one area in which we know more than we might wish to about rapid evolution. The quick responses made by pathogens and 'pests' to human alterations of their environment (e.g. antibiotics, insecticides, new strains of crop plants) clearly demonstrate that many species contain (or rapidly manufacture) genetic variation that provides useful defences even against novel kinds of attack. Our subsequent counterattacks are cultural inventions, however, not evolved ones, so these examples cannot be taken as strong support for the idea that similar kinds of events occur frequently in nature. The coevolution that has occurred between rabbits and the myxoma virus in Australia (since the introduction of the virus in 1950) remains the best-documented example of complementary genetic changes occurring in two species under semi-natural conditions (Fenner & Ratcliffe, 1965; see also May & Anderson, 1983b; Pimentel, 1984). The usual interpretation of this remarkable history implies that an equilibrium has been reached, with increased resistance in the rabbits and decreased virulence in the virus. Whether or not this apparent phenotypic equilibrium corresponds to an underlying genetic equilibrium is apparently not known (see Barrett, 1984).

Since we can travel in space much more easily than in time, localized coadaptation is easier to study than its temporal counterpart, and it seems to be widespread. For example, Lively (1989) has recently shown that a trematode parasite (*Microphallus* sp.) is locally adapted to its freshwater snail host (*Potamopyrgus antipodarum*). Lively conducted a series of reciprocal cross-infection experiments between populations of snails and trematodes from four lakes. The distances between lakes ranged from less than 10 km to nearly 100 km. In every case examined, the parasite more readily infected hosts from its own lake than those from other lakes. Lively also discusses evidence of local coadaptation in other snail–trematode systems.

In a similar kind of study, M.A. Parker (1985) demonstrated strong local adaptation of a specialized fungal pathogen (*Synchytrium decipiens*) to a leguminous annual (*Amphicarpaea bracteata*). Dramatic differences in susceptibility to particular strains of the fungus were found between host populations separated by only 1 km. At even smaller distances, Edmunds and Alstad (1978, 1981) (see also Alstad & Edmunds, 1983, 1989; Alstad & Corbin, 1990) found significant adaptation of a sedentary homopteran (the black pine leaf scale, *Nuculaspis californica*) to individual trees of its coniferous host (*Pinus ponderosa*). Whether this last example qualifies as an instance of *spatial* differentiation depends on whether the interaction is viewed from the trees' or the scales' point of view. Neighbouring trees can have distinct defensive phenotypes and correspondingly differentiated populations of scales, so from their point of view this might better be considered an instance of phenotypic (and presumably genetic) polymorphism.

Antonovics and Ellstrand (1984), Kelley *et al.* (1988), and others have shown that genetically variable (sexual) progenies of the grass *Anthoxanthum odoratum* tend to be fitter than equivalent clonal (asexual) progenies. These results imply that there is significant genetic variation in the grass, to which one or more parasites can quickly adapt on microgeographic spatial scales, but not that there is necessarily strong local differentiation of the grass, on the scale of its genetic neighbourhoods.

These examples all concern host–parasite systems. Do prey and predator species show similar patterns of local coadaptation? It seems likely that many potential examples of size adjustment could be extracted from the literature on prey selection. For example, the bill sizes of Darwin's finches vary in relation to the available distributions of seed size and hardness (Grant, 1986). However, in this case as in most others, the variation has been studied mainly from the point of view of the predator, often with the aim of understanding how several competing predators (consumers) divide up a range of prey (resources).

The third prediction, that such polymorphisms should be widespread, is of course abundantly satisfied (e.g. Day, 1974; Holmes, 1983; Barrett, 1984; Rollinson & Anderson, 1985; Day & Jellis, 1987). Indeed, it was in large part the *observation* of polymorphism (in general, and for resistance to disease in particular) that led Haldane and others to ask whether parasites might be its cause. However, the fact that antagonists often present highly diverse faces to each other does not prove that dynamic instability and rapid evolution are the rule. Large amounts of variation could be maintained within local populations by parasites or predators that generated frequency-dependent selection of a kind that tended to accumulate such variation in a dynamically stable way (e.g. Clarke, 1979). One such mechanism has already been mentioned — 'apostatic' selection by visually hunting predators — and it seems entirely possible that realistic extensions of the models described earlier (e.g. to include spatial population structure) could stabilize their dynamics in many cases (but see Frank, 1991).

An interesting and apparently open question is whether dynamically unstable situations might be distinguished from dynamically stable ones on the basis of the kinds of gene-frequency distributions to be expected. Nadeau *et al.* (1988) surveyed H-2 and allozyme polymorphisms in two subspecies of house mice (*Mus musculus domesticus* and *M. m. musculus*) from several sites in Europe, North Africa and South America. (H-2 is homologous to the human HLA complex; its numerous loci are involved in the regulation of immune responses to foreign antigens and they tend to be extremely polymorphic.) Genetic distances derived from the distributions of alleles at the allozyme loci clearly reflected geographic distance and other known barriers to gene flow; for example, the two subspecies were much more distant from each other than were any two populations within either subspecies. But genetic distances derived from the distribution of alleles at the H2-K and H2-D loci showed no such simple pattern, and varied far less between closely and distantly related populations. A hierarchical analysis of diversity indicated that variation within populations accounted for nearly 83% of the total H-2 diversity, but only 48% of the allozyme diversity. From these and other analyses of the allele-frequency data, the authors conclude that variation at the H-2K and H2-D loci is more uniformly distributed than that at enzyme loci, and may therefore be protected by a form of balancing selection that does not act on allozymes, at least not as strongly. (The authors of *this* paper about mice *do* cite Haldane (1949).)

At first glance these findings seem to imply a fairly static selective regime, in which negative frequency-dependence always acts to pull up the frequencies of rare alleles. This interpretation may well be correct, but it is not obvious what kinds of allele-frequency (or gamete-frequency) distributions we should *expect* to find at highly polymorphic loci subject to various different patterns of ongoing antagonistic coevolution. If 'snapshots' of dynamically volatile systems were expected to look very different from those of their relatively placid counterparts, then it might in principle be possible, with large amounts of data, to distinguish among at least the more extreme alternatives.

In summary, we do not yet know whether the evolutionary battles between natural enemies are often as fluid and fast-paced as theory seems to suggest they could be. But if the right sorts of conditions are common (and there is not yet any compelling reason to suppose that they are not), then the face of nature may eventually come to seem even less uniform than it does today.

ACKNOWLEDGEMENTS

For helpful suggestions and comments on the manuscript I thank J.J. Bull, A.P. Dobson, R.D. Holt and N.A. Moran.

2: Correlates of Carnivory: Approaches and Answers

PAUL H. HARVEY & JOHN L. GITTLEMAN

INTRODUCTION

How do carnivores differ from non-carnivores? That is a question for comparative biologists, and is also the topic of this chapter. We cannot review the enormous literature pertaining to this topic in a single chapter. Instead, we shall describe why our question must be approached comparatively, how our comparisons should be made, and what sorts of conclusions can be drawn. As we shall see, cross-taxonomic comparisons are best used as a complement to the findings of detailed observational studies and controlled manipulative experiments on particular test cases. In this way, we can both explain general patterns and account for exceptions to those patterns. Throughout this chapter we shall use descriptive vignettes to illustrate some of the correlates of carnivory, and the observations and experiments which demonstrate why they have come to be.

EXPLAINING SPECIES DIFFERENCES

Evolutionary biologists seek general principles to explain general patterns. As with other historical sciences, experiments are not always possible. It would not be feasible to rerun evolution under controlled conditions, even if we had the time. Instead, cross-taxonomic data are treated as the results of a series of natural experiments from which the consequences of the presence and absence of carnivory are traced. Use of the comparative method allows data from all the available species to be incorporated into a single study, and thereby claims generality. However, if we want to test our ideas about correlates of carnivory using the comparative method, we must be very careful about how we proceed.

The comparative method

Closely related species contain many characteristics that are identical by descent through common ancestry. If we wish to seek the evolutionary consequences of carnivory, we must be sure that each evolutionary origin of carnivory is treated only once in our sample. Otherwise we should be inflating our sample size and adding additional unwarranted degrees of freedom to our statistical tests. For example, McNab (1986a,b) pointed out that mammals which feed on vertebrate prey have higher metabolic rates for their body weights than those feeding on invertebrates. This is true when species are used for the comparisons, largely because all those species belonging to the orders Cetacea and Pinnipedia for which we have data both eat vertebrates and have high metabolic rates, while all those species belonging to the Monotremata, Tublidentata and Pholidota eat invertebrates and have low metabolic rates. In contrast, when we look *within* the only two orders of mammals containing both vertebrate and invertebrate eaters for which we know the metabolic rates, the chiropterans and marsupials, those species which feed on invertebrates have *higher* metabolic rates than those feeding on vertebrates (Elgar & Harvey, 1987)! A good experiment controls for the influence of extraneous factors and is based on independent replicates. Similarly, a good comparative study can control for many third variables by making comparisons between closely related taxa, and can be based on independent replicates by making comparisons within different phylogenetic lineages (Pagel & Harvey, 1988; Harvey & Pagel, 1991). In fact, a phylogenetically based analysis reveals no association between metabolic rate and diet in mammals when body weight is held constant (Harvey *et al.*, 1990).

Unfortunately, comparative tests can reveal only correlated differences. They cannot usually distinguish case from effect, because we do not usually know which of two correlated evolutionary events preceded the other. For example, we might observe that many snake species use venom to kill their prey, while others use venom to defend themselves against would-be predators. The question arises: did venom evolve as a predator-defence mechanism or as a means of becoming a more effective predator? We should need to know the circumstances under which the use of venom evolved. If it evolved in a group of predatory snakes to help them subdue their prey, then in Gould and Vrba's (1982) terminology, the use of venom as a predatory device would be an *adaptation*, but its use as a predator-defence mechanism would be an *exaptation*.

A nice example of historical analysis leading us to pose different questions about carnivory comes from Coddington's study of web structure in spiders. An accepted dogma in arachnology was that the orb web evolved from a primitive cob web as an adaptation for catching flies more efficiently. However, careful cladistic analysis indicates that orb webs were probably ancestral to cob webs (Coddington, 1986a,b), so the proper question is: what were cob webs an adaptation for? If an analysis of their relative efficiencies at catching prey shows that orb webs are indeed better than cob webs, how should that variation be interpreted? We should need to search for some hitherto unsuspected advantage of cob webs over orb webs.

Optimality theory and comparative studies

Many comparative studies can be set firmly within the framework of optimality theory (Harvey & Pagel, 1991), and comparative studies of the correlates of carnivory are no exception. In this section we shall explore some of the links between the two approaches. This is an important exercise because optimality theory frequently forms the basis for many experimental studies, and its use can help illustrate the links between the two approaches to problem solving in evolutionary biology.

What exactly is an optimality model and how is it applied? The procedure is to specify an optimality criterion, to define alternative strategy sets, and to estimate the payoffs for each strategy, thus determining which is optimal under the conditions specified. The optimal strategy is the one expected to occur. By way of example, consider a predatory insect, say a scorpion, with a potential diet consisting of two species of grasshopper, one of which is much larger than the other. Under the Darwinian scheme of things, scorpions are selected to leave the maximum number of offspring. Assume that grasshopper biomass can be converted to scorpion offspring biomass using a simple multiple (<1). Then, scorpions will have been selected to maximize the weight of grasshoppers eaten per unit time (optimization criterion). This might be achieved by adopting one of three foraging strategies: eat all grasshoppers encountered, eat only large grasshoppers, eat only small grasshoppers (alternative strategies). Different foraging strategies will result in different weights of grasshopper consumed per unit time, depending upon the handling time for each type of prey, and the weight of large versus small grasshoppers (different payoffs). Scorpions would be expected to adopt that feeding strategy with the highest payoff.

Shore crabs (*Carcinus maenas*) feeding on mussels (*Mytilus edulis*) seem to forage more or less according to the optimization principles enunciated for our would-be scorpion. Very large mussels take a long time to crack, whereas very small mussels contain so little flesh that they have very little food value. Elner and Hughes (1978) showed that crabs prefer to eat the size of mussel giving the highest rate of energy return (Fig. 2.1). Why they occasionally tackle both very small and very large mussels is not explained by the study, though it could be that either (i) the crabs can detect meaty small mussels and cracked large ones; or (ii) the crabs utilize such prey after long unprofitable search periods.

The optimization criterion adopted by the scorpions in the previous paragraphs was to maximize the weight of prey eaten per unit time. The same might be true for other predators feeding on alternative prey types, say a bird feeding on worms. However, other factors frequently come into play. For example, a bird feeding on worms is spending time away from its nest, leaving its offspring unprotected. It may be that the value of additional worm biomass to the offspring decreases beyond

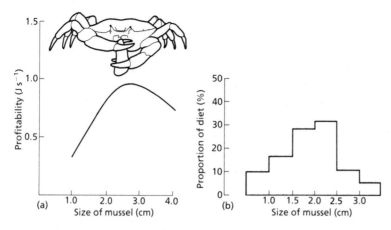

Fig. 2.1 The size of mussel *Mytilus edulis* which provides the highest rate of energy return for foraging shore crabs *Carcinus maenas* (a) is the one they prefer to eat in aquaria (b). From Krebs & Davies (1987).

some critical value. For example, the young can grow only so fast, irrespective of how much food they get beyond a critical amount. Now the optimality criterion might be different: minimize the amount of time spent acquiring a set weight of food items. If the bird does that, it will spend the maximum time at the nest protecting its offspring. Different optimization criteria and a host of alternative strategy sets have been examined in the literature, and numerous experiments have been performed to help understand the behaviour of feeding animals (see Chapter 4).

HOME RANGE AND TERRITORY SIZE

How do such optimality studies relate to comparative tests? A brief historical account of the development of our understanding of differences in home range and territory size between animals with different diets helps demonstrate the link, and shows how experimental tests can complement comparative generalities. Preliminary comparative studies from the 1950s to the 1970s examined the ways in which species territory and home range sizes change with body size and diet (Hutchinson & MacArthur, 1959; McNab, 1963; Schoener, 1968; Milton & May, 1976). A reasonable question to ask at that time was whether larger animals need larger territories to satisfy their greater metabolic needs. As predicted, territory size did increase with body size and, furthermore, species living on more sparsely distributed food resources also had larger territories. In particular, carnivorous mammals had larger home ranges for their body sizes than did herbivorous mammals (McNab, 1963), and the size of a bird's territory could be predicted in large part from the proportion of animal matter in its diet (Schoener, 1968).

The next step was to ask whether territory size increased with body size in a quantitatively sensible way. Since metabolic rate (energy used per unit time) increases with body weight raised to the power 0.75, presumably the minimum-size, continuously productive territory (energy produced per unit time) for animals with similar diets living in similar habitats would also be expected to increase with the 0.75 power of body mass (Kleiber, 1932, 1961; contra Lindstedt *et al.*, 1986). The data did not accord with that expectation: territory or home range size in a variety of taxa increased with body size with an exponent appreciably greater than the metabolic exponent (e.g. Harvey & Clutton-Brock, 1981; Gittleman & Harvey, 1982; Mace & Harvey, 1983; Lindstedt *et al.*, 1986). Several possible causes for this discrepancy seemed likely: (i) suitable habitat is not continuous, and larger-bodied species must take in a disproportionate area of unsuitable habitat; and (ii) the acceptable food spectrum may change with body size (Schoener, 1983). For example, birds supplying food to a nest in the centre of a territory might have evolved an optimal foraging strategy resulting in the selection of only the larger food items at an increased foraging distance from their nest. Larger-bodied species might become increasingly selective at greater distances from the nest, possibly because of intruder pressure at the nest.

It is both unnecessary and impractical to perform detailed field observations and food manipulation studies on more than a few species in order to determine the likely reasons for the comparative relationship. A series of optimality models have been developed, based on a variety of optimization criteria, which predict different relationships between territory size and body size (Schoener, 1983). Also, the likely optimization criteria are becoming better known through carefully controlled field and laboratory experiments (Davies & Houston, 1984; Stephens & Krebs, 1986). For example, Davies and Lundberg (1984) were able to show that female dunnocks (*Prunella modularis*) change the size of their territories so that they will provide enough food supplies for themselves and their young, whereas males defend an area which contains as many females' ranges as possible. The optimization criterion for females (smallest territory with enough food supplies) is different from that for males (territory containing as many mates as possible).

The integration of foraging theory with comparative studies has been a long time coming, in part because it has proved difficult to identify the cost and benefit curves for foraging animals.

SOME CHARACTERISTICS OF CARNIVORES

Carnivores have to catch their prey, kill it, ingest it and digest it. Most biologists are familiar with adaptations for each aspect of carnivory: the spider's web or the cheetah's speed, the scorpion's sting and the boa's constriction, the raptor's beak and the lion's canines, the sundew's enzymes and the viper's venom are cases in point. With such an astonishing diversity of adaptations, is it possible to detect pattern? How are the various adaptations distributed taxonomically, which of them have additional functions, and which should properly be viewed as exaptations? In the following sections we consider some morphological and behavioural correlates of carnivory, bearing in mind the different questions posed in the previous sentences. In each case we concentrate on the different comparative, behavioural and experimental studies that have been used, each to reciprocally illuminate the results of the others. (At this point it is useful to bear in mind differences between animals as prey and plants as food. Shipman and Walker (1989) mention that, compared with plants, animal prey are more wary and mobile, they are more likely to possess active defence mechanisms and, in dietary terms, they contain less fibre and more calories per unit weight of food. Finally, fermentation does not help digestion of animal matter as it does certain plant materials.)

Body size

Larger predators from many taxonomic groups tend to feed on larger prey. Examples from insectivorous birds, mammals, raptors and other animals are referenced in Peters (1983), and many of the data are brought together in Fig. 2.2. However, there are many exceptions to this general rule. For example, some whales feed on invertebrates, while many parasites feed on and ultimately kill hosts that are very much larger than themselves. Nevertheless, when comparisons are made among closely related species (Gittleman, 1985) from the same ecological guild (Van Valkenburgh, 1988), prey size is generally positively related to predator body size.

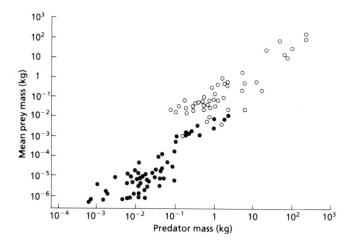

Fig. 2.2 Prey mass increases with predator mass for mammals and birds of prey (large-prey eaters) (○) and for lizards, amphibians, seabirds, and insectivorous birds (small-prey eaters) (●). From Calow (1981).

It is often considered that large size is an advantage to predators, because larger predators are more likely to be able to overpower larger prey. It has even been suggested that there has been an arms race between carnivores and ungulates, with each getting larger over evolutionary time — the carnivores to overpower the ungulates, and the ungulates to escape from the carnivores — although there are other possible interpretations for this particular manifestation of Cope's rule (Cope, 1885) of phyletic size increase (C.M. Janis, in preparation), if, indeed, it actually exists (Radinsky, 1978).

Although large size can be an advantage to predators, this is not always the case. We have already mentioned parasites, but many conventional predators seem to do rather well as a consequence of being small (King, 1989). For example, the astonishing manoeuvrability of small raptors probably helps them catch flying prey (see Chapter 6).

A particularly nice example of small size favouring predators is provided by Simms' (1979) study of the stoat *Mustela erminea*. Small stoats can chase prey down their burrows, and Simms demonstrates a correlation between prey burrow size and the size of female stoats from different areas. It is not known what the advantages to large size are in female stoats, but there must be some, otherwise stoats would merely need to be large enough to overcome their prey. One possible advantage to large size in female stoats concerns energy reserves. Larger-bodied animals often have enough spare reserves of energy to more than compensate for their increased metabolic rate (energy reserves scale with the 1.0 power of body weight, whereas energy needs scale with the 0.75 power of body weight), so that larger animals can survive longer without food or, in the case of females, may have more energy resources available during the critical period of lactation. One advantage of large size to male stoats not enjoyed by large females is that they are more successful in competition for mates (Erlinge, 1977). Sandell (1989) recently applied an energetically based optimality model which suggests that male body sizes are near to the survival optimum, whereas females seem to be away from the survival optimum, possibly as a result of additional selection pressures operating during the breeding season.

Gut morphology

Carnivorous animals feed on food that should be relatively easily assimilated. Indeed, Calow (1981) refers to the 'broad generalisation which is widely quoted is that detritus (rich in lignin and cellulose residues of plants) will be less easily and less well digested than live plant material and this in turn less well digested than meat'. However, comparative work across invertebrates does not indicate that their guts have inordinately high assimilation efficiencies (Table 2.1). Perhaps more fine-grained analyses taking into account subtle dietary differences would tell a somewhat different story. At the level of enzyme secretion, clear and expected differences are well described and associated with

Table 2.1 Absorption (assimilation) efficiencies for a variety of invertebrates. From Calow (1981)

				Efficiency (%)						
0	10	20	30	40	50	60	70	80	90	100

⟵   Filter-feeding herbivores   ⟶

⟵   Detrivores   ⟶

⟵   Invertebrate carnivores   ⟶

⟵   Gastropod macroherbivores   ⟶

⟵   Insect macroherbivores   ⟶

⟵   Microalgal herbivores   ⟶

⟵   Bacterivores   ⟶

the need to digest particular food items. For example, across a wide range of invertebrate taxa, the carnivores concentrate on protease production, while the herbivores produce amylases, glycosidases and cellulases (Mansour-Bek, 1954).

Returning to gut morphology, the vertebrates have been relatively well studied with respect to diet. Broadly speaking, the vertebrate gastrointestinal tract can be divided into four main compartments: stomach, small intestine, caecum and colon. Following initial work by Chivers and Hladik (1980), Martin et al. (1985) analysed data on the dimensions of each gut component for 168 specimens from 73 species. Since, as we have seen, energy needs scale with approximately the 0.75 power of body weight, it was not surprising that within dietetic groups the surface area of each gut component scaled with a similar exponent. Folivorous species from different mammalian orders have either an expanded stomach or an expanded midgut (caecum and proximal part of the colon), never both. The caecum, a compartment that was probably present in the common ancestor of mammals (Hill & Rewell, 1954) and which is often used to house the symbiotic bacteria that aid the digestion of plant material, has been lost in many insectivores, carnivores, edentates and other mammals with a diet which contains a predominance of animal matter (Martin et al., 1985). Several species with carnivorous ancestors that have secondarily evolved a herbivorous diet, such as the giant and red pandas, do not have a caecum and show both slow development and have poor reproductive records (Schaller et al., 1985; Gittleman & Oftedal, 1987).

Two additional related and interesting points arise from comparisons of gut morphology among vertebrates. First, Martin et al.'s (1985) study examined the relative placement of different mammal species in multidimensional space, with the various axes representing size-corrected gut component dimensions. Humans cluster with a group containing species from the order Carnivora, two meat-eating cetaceans, and the carnivorous African water shrew (Potomogale). This finding suggests that the modern human gut is adapted for a diet containing a fair proportion of animal matter. Lest we draw hasty conclusions, we must consider the second point: diets received by captive animals may be quite unnatural, and guts adapt their dimensions to an animal's diet during its lifetime. For example, wild-caught squirrel monkeys had colons that were on average 48% longer than those of captive specimens, the stomachs were 25% longer and the small intestines 16% longer. As Martin and his colleagues emphasize, humans from different cultures living at different times could be exposed to quite different diets, and the response of guts to different diets in humans is not well described.

Catching prey

Many carnivorous species outrun their prey. What do comparative studies tell us about the adaptations carnivores are likely to possess if they have been selected to run fast? Here we find comparisons, observations and experiments used together in several different ways. Obviously, a cheetah can run faster than a mongoose. Is this simply a consequence of their size difference?

One advantage of larger body size is that strides are longer, so it takes fewer of them to cover the same distance. How, then, does stride length change with body size? Alexander *et al.* (1979) found that the length of limbs (*l*) of mammals varying in size from shrews (*Sorex*) to elephants (*Loxodonta*) increased with the 0.35 power of body weight (*w*). If all mammals were the same shape, linear dimensions like limb length would increase with the one-third power of body weight. The implication of the 0.35 exponent is that mammals of different size *are* roughly the same shape. Now, limb length should be linearly related to stride length, so the frequency of strides for two animals running at the same speed should differ according to their difference in $w^{0.35}$, with the larger animal taking fewer strides.

What about the energetic cost of running? In a subsequent paper, Alexander *et al.* (1981) found that, across mammals as a whole, leg muscle mass (a good measure of the force of contraction) increases roughly in direct proportion to body mass. As a consequence, the work per step is proportional to w^1 and, since the number of steps per unit distance is proportional to $w^{-0.35}$, the work needed to travel over unit distance is proportional to

$$w^1 \times w^{-0.35} = w^{0.67}$$

However, larger animals are heavier and we need to divide this expression by body mass to estimate the work per unit distance per unit body weight, and

$$w^{0.67} \times w^{-1.0} = w^{-0.33}$$

In summary, larger animals take larger strides but the cost per stride is directly proportional to body weight, so it costs smaller animals more per unit body weight to move the same distance (Schmidt-Nielsen, 1984). In a series of treadmill experiments with different mammal species, Taylor *et al.* (1970) found that the amount of oxygen consumed per unit weight per unit distance run decreases with body weight according to $w^{-0.4}$, an exponent roughly in line with the prediction made from Alexander's two sets of morphological comparisons described earlier (Fig. 2.3). So it costs smaller animals more per unit weight to run the same distance, but it does not, of course, cost them less in absolute terms: total costs of running increase roughly with $w^{0.67}$ as we saw above.

However, success as a predator depends not only on the cost of running but also on how fast you can run. Two studies point to roughly the same conclusion, but both should be considered preliminary. Pennycuick (1975) found that stepping frequency for cantering East African mammals, ranging in weight from 15 g to about 3000 g, decreased with increased body weight according to $w^{-0.17}$. We have seen that stride length increases with $w^{0.33}$, which means that speed should increase according to

$$w^{-0.17} \times w^{0.33} = w^{0.16}$$

A study by Heglund *et al.* (1974) which recorded the stride frequency at the transition point between trotting and galloping suggests that speed should increase with $w^{0.20}$, a value roughly in line with Penny-cuick's result. Calculations by Schmidt-Nielsen (1984) suggests that the maximum power animals are likely to have available for running should increase with a similar exponent.

All these calculations and observations are for animals running on the flat. Going uphill is a different matter because the product of weight and vertical distance moved is a measure of potential energy gained, which is energy that must be used to get the animal uphill. It could cost a lot more to get a heavier animal up a hill, which accords

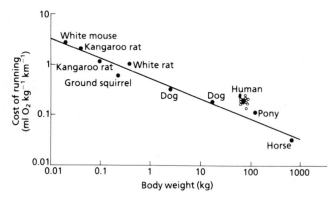

Fig. 2.3 The cost of running, expressed as oxygen needed to transport 1 kg of body weight over 1 km, for animals of different body weights. From Schmidt-Nielsen (1984).

with observations by Taylor *et al.* (1972) that the rate of oxygen consumption for a mouse running up a 15% incline is much the same as on the flat, but for a chimpanzee the rate was almost doubled! Small prey running from large predators should do better if they go up hills rather than along the flat.

Group living

There are many advantages and disadvantages to group living in animals. The various costs and benefits differ with the ecological habitats of each particular species (Bertram, 1978a; Harvey & Greene, 1981; Krebs & Davies, 1987; Gittleman, 1989). The benefits can derive from locating and catching food, detecting predators and avoiding predation, winning a mate, learning from others through cultural transmission, and rearing offspring. The costs can include some of the same factors as well as others. For example, both prey and predators are more likely to detect groups of animals than singletons, males living in groups must invest more heavily in avoiding cuckoldry, while disease is more likely to be transmitted among the members of a group.

An interesting question which we can use to motivate our discussion of group living is: who gets the benefits? Is group hunting coordinated for the good of the group or the good of the individual? Of course, individuals would not hunt together if each did not benefit from the association. But there is a more subtle distinction. Consider a group of four male chimpanzees attempting to catch a young baboon which runs up a tree. One chimpanzee follows it up the tree, while others climb adjacent trees thus cutting off the escape routes for the young baboon. Is each chimpanzee attempting to maximize the chances that it will catch the baboon, or are the chimpanzees which guard the less likely escape routes acting to maximize the chances that the baboon will be caught by some chimpanzee in the group? Assume that the probability of the baboon trying to escape along the most favoured route is 40%, compared with 5% along the fourth least likely route. Further assume that a guarded route guarantees that the baboon trying to escape along it will be caught, and that the three most likely escape routes are guarded. In that case, the last chimpanzee on the scene would go for the most favoured route if it was acting in its own direct interests (thus expecting 20% of a baboon meal), but would guard the least likely escape route if it was acting in the group's interest (increasing the chances of 5% that the group would catch the baboon). Of course, the decision in this case would be influenced by whether the meat from the captured baboon was shared among the group or not.

An example of unequal benefits to animals living in groups involves the predatory jack (*Caranx ignobilis*) feeding on schooling anchovy (*Stolephorus purpureus*). The jack lives in small groups which separate an anchovy from its school and then chase it until one of them catches it and eats it. Major (1978) was able to show under experimental conditions that jacks living in groups of three to five fish typically got twice as many anchovies per individual per experiment than solitary jacks. Furthermore, the leading jack in the group caught about four times as many prey, as the second or third in line. Under the conditions of the experiment, it was more profitable to feed alone than occupy the fourth or fifth position in a hunting group.

Do the same individuals always benefit from feeding in groups? One explanation for group living would suggest that the answer should be no. Ward and Zahavi (1975) suggest that communal roosts and breeding colonies in many birds exist primarily because they enable more efficient exploitation of unpredictable food sources by acting as 'information centres'. The argument is that individuals learn the location of rich food sources from other, more successful birds, and hence reduce the time spent searching for food. However, if an information centre is to evolve, we should expect the relative success of individuals to change, otherwise it would not pay a consistently successful individual to join a colony.

Particularly convincing evidence for the information centre hypothesis comes from Brown's (1986) study of cliff swallows (*Hirundo pyrrhonata*), which breed in colonies of up to 3000 nesting pairs in western North America. The birds build gourd-shaped mud nests which they fasten on concrete bridges or, more naturally, underneath overhanging rock ledges and on the sides of cliffs and canyons. Cliff swallows feed exclusively on insects that are typically found in localized swarms lasting about 25 minutes that can support about 500 swallows. However much food they have captured, foraging swallows return to their colony after an absence of about 5 minutes.

Brown found that swallows typically depart from the colony in groups, and birds that leave together also feed together. In accord with the information centre hypothesis, the tendency of an individual to be followed or to follow is influenced by its previous foraging success; successful foragers tend to be followed but unsuccessful foragers follow. Unsuccessful foragers can locate good food sources by simply following those individuals that recently arrived back at the colony with a beak full of insects. Overall, Brown found that almost all individuals were likely to follow others on about 40% of foraging trips from the nest, suggesting that different colony members benefited about equally from opportunities to receive information. This passive recruitment highlights an important comparative prediction — information centres should evolve only in those species like the cliff swallow where there are no obvious disadvantages to the successful birds of foraging in flocks. Otherwise birds would be selected to disguise their foraging success. At the other extreme, black-billed gulls actually seem to recruit others into a foraging session by calling, perhaps because they benefit from foraging in flocks (Evans, 1982).

Unfortunately for the cliff swallows, information about food sources is not the only thing that is transferred in large colonies. Fumigation experiments by Brown and Brown (1986) show that a haematophagous cimicid swallow bug (*Oeciacus vicarius*), which is much more common in large colonies, reduces nestling survivorship in large but not small colonies by up to 50%. It is always necessary to identify both the costs and benefits of living in groups.

We have already discussed the general relationship between body size and prey size. It has been suggested that group hunting can often result in predators being able to deal with larger prey than would be expected for their body size. Indeed, this seems to be the case for species in some families of the mammalian order Carnivora: in both the canids and hyenids, Kruuk (1972) found a positive association between group living and prey size. However, the felids reveal a different picture: several solitary cats, such as cougars and leopards, prefer to hunt relatively large prey.

Even if the prey size hypothesis does not claim generality across the felids as a whole, surely cooperative hunting is the primary benefit to lion sociality? For many years this seemed to be the case, but it now seems that Kruuk's comparative message may have been correct all along. Caraco and Wolf (1975) analysed data from Schaller's (1972) monograph *The Serengeti Lion* in an attempt to identify the foraging group size that maximized per capita food intake rates, and their calculations suggested an optimal group size of two for all prey sizes. Subsequently, Packer (1986) reanalysed Schaller's data (as well as data from other studies) and, using more realistic assumptions about the way multiple-kill rates change with group size, found that solitary lions were predicted to have the highest rate of food intake.

So why do lions live in groups? Packer suggests that the high frequency of scavenging from conspecifics provides the key. His argument is based on three aspects of lion ecology: (i) lions eat large prey, which allows several individuals to feed from a single carcass; (ii) lions live in open habitats which means that carcasses are conspicuous to scavengers, including other lions, and (iii) lions live at unusually high densities for large cats. It follows that a kill made by a single lion would be likely to attract others because it is conspicuous, available for a long time, and there are likely to be several lions in the vicinity. In support of Packer's argument are the findings that the most common context in which pridemates come together is after a kill, the number of lions feeding from a carcass increases with carcass size, and most inter-pride encounters occur at kills.

To explain how sociality in lions evolved, Packer (1986) considers an ancestral lioness who hunts alone to feed herself and her cubs. The large prey that she captures feed herself, her cubs and other nearby lions that scavenge from the carcass. As the cubs grow, they depress their mother's food intake but do not disperse because it would be difficult to find their own hunting ranges in such a high density population. Instead, they remain with their mother whose inclusive fitness is thereby raised — if her own food intake is decreased, that of her daughters is increased. Males who gain mating access to this primordial group might attempt to kill unweaned offspring (infanticide is very common in lions; Packer & Pusey, 1984), thereby bringing the mother into oestrus. The mother and daughters would then be selected to rear their cubs communally in order to defend them against such attacks (Packer *et al.*, 1988).

Packer's scenario takes into account important aspects of both the behaviour of felids in general (infanticidal males) and the ecology of lions (high population densities, open terrain, and a preference for large prey). At the same time it explains a lot about lion social behaviour — matrilineal groups, intragroup sharing of prey, and defence of home ranges. Packer's conclusion, then, is that much of female cooperative foraging behaviour in lions is a consequence and not a cause of group living. The family that stays together preys together.

CONCLUSION

Carnivory is a lifestyle that has evolved on many occasions during the history of life. Animals with carnivorous diets from different taxa frequently face the same selective pressures and have produced similar adaptations for dealing with them. The comparative method, supported by appropriate theories, observations and experimental tests helps us to identify and make sense of the morphological and behavioural correlates of carnivory.

3: Population Dynamics of Natural Enemies and their Prey

MICHAEL J. CRAWLEY

INTRODUCTION

This chapter is about the population dynamics of natural enemies and their prey. The aim is to make general points that apply equally well to predators, parasites and diseases, but to keep things simple I shall refer to predators and prey (rather than host–parasitoid, host–disease, and so on). The chapter is focused on four interrelated questions: (i) how do natural enemies affect the abundance of their prey?; (ii) can natural enemies regulate prey population density?; (iii) what determines the abundance of natural enemies?; and (iv) what factors influence the pattern of dynamics exhibited by a given predator–prey interaction? These questions are tantalizingly simple at first glance. Surely, if predators kill prey, they must make prey less abundant? Again, if prey numbers decline, then surely predators must either switch to other prey or decline in abundance themselves? In order to see some of the subtleties involved in these questions, it is necessary to work methodically through the consequences of some simple assumptions, before looking at the interactions that produce the rich variety of dynamic behaviour.

THE COMPONENTS OF DYNAMICS

In order to understand any given predator–prey interaction, we shall need to consider the following set of questions.
1 How abundant is the predator?
2 Is the predator monophagous or polyphagous?
3 Is the prey a preferred food species?
4 How abundant is the prey, compared with more preferred prey species?
5 To what extent do the spatial ranges of the predator and prey overlap?
6 Is the abundance of the predator limited by the availability of the prey species in question?
7 Is the prey species sufficiently abundant as to be subject to density-dependence in its birth, death or dispersal rates?
8 To what extent do the predators and prey disperse from patch to patch, both within and between generations?

9 What is the timing of predation in the life cycle of the prey?

10 Does predation relax the impact of density-dependence, or does it occur after density-dependence has already reduced prey numbers?

11 Is the predator capable of rapid numerical responses to changes in prey density by dispersal and aggregation?

12 What are the relative magnitudes of λ_{max} for the prey and the predator?

13 To what extent is the current rate of predation determined by previous predation at various times in the past?

14 How important are time-lags in determining the patterns of predator and prey population?

15 Are there invulnerable age or size classes of the prey, and how important are age-structure effects in general?

It should be immediately clear that these questions cannot be tackled simultaneously, either by experiment or theory. Even if there were only two states for each of these categories, this would give an experiment with $2^{15} = 32\,768$ different treatment combinations. As a starting point, we shall look at each of these questions in isolation, but bearing in mind that interaction effects will be commonplace, and that trends may be reversed by other circumstances.

We need to be especially clear about the scale at which we are attempting to describe the dynamics of the predator–prey interaction. At the smallest spatial scale, every death caused by predation is a local extinction. At a large spatial scale, the interaction may be stable despite local instability, or because of local regulating mechanisms. There may be metapopulation dynamics that reflect differential rates of dispersal and occupancy of predator-free patches by the prey, with local extinction of the prey inevitable following discovery of a patch by the predators (see Chapter 10). At the other extreme, density-dependent attack by natural enemies may regulate prey population density in all patches, with mean density varying from patch to patch as a result of differences in resource productivity (see Chapter 11). Alternatively, the large-scale dynamics may owe their persistence to the existence of refuges in which a fixed or varying proportion of the prey are protected from predation. The refuges may be physical 'safe places' (see Box 3.2), or they may be statistical, in the sense that some of the prey find themselves in predator-free places by chance alone (Chesson, 1985; Pacala & Crawley, 1992).

PATTERNS OF DYNAMICS

The term 'patterns of dynamics' refers to the ways in which numbers change from year to year, or from generation to generation. Under field conditions, there will always be more or less pronounced random fluctuations in abundance, caused mainly by changes in the weather. From time to time, there will also be catastrophes that inflict massive mortality on predator and prey alike. These sources of variation constitute the *noise* in the system. Underlying this background noise, and sometimes obscured by it, is a *signal* which reflects the nature of the dynamical properties of the predator–prey interaction itself.

The signal represents the underlying deterministic dynamics, and can be classified according to the following simple scheme, which might apply to either predator or prey numbers:

1 exponential increase,
2 exponential decrease,
3 exponential stability,
4 damped oscillations,
5 increasing oscillations,
6 cycles,
7 chaos,
8 random walk.

These are illustrated in Fig. 3.1. One of the aims of any study into the population dynamics of a predator–prey interaction is to tease apart the noise and the signal, and to understand the roles of random and deterministic processes in population regulation.

We begin by considering the impact of generalist predators, and then look at systems involving specialist natural enemies.

GENERALIST NATURAL ENEMIES

A generalist natural enemy eats a sufficiently large number of prey species such that its abundance is unaffected by changes in the population density of any one prey species. This allows us to assume that the numbers of the predator, P, are constant (or vary randomly), and that predator and prey dynamics are effectively *uncoupled*, e.g. the predator might be regulated by the availability of suitable breeding sites. There are four different ways in which such a generalist enemy might affect the abundance of a prey species (i) by taking a fixed number of prey individuals each time period; (ii) by taking a fixed proportion of the prey population; (iii) by taking a random but variable proportion of the prey each year; or (iv) by taking a variable proportion of the prey, where the proportion killed increases as a function of prey density, i.e. density-dependent predation, see Box 3.1 (Table 3.1 shows these models for generalist predators). We shall analyse each of these cases in turn.

Fixed number predation

Suppose that each of the P predators eats a fixed number of prey per unit time, c, irrespective of the size of the prey population. In the absence of predation the prey population, N_t, would increase exponentially until it ran out of resources, because we assume that the net multiplication rate $\lambda > 1$ for prey persistence (this is known as the *invasion criterion*). For the time being, we assume that the prey popu-

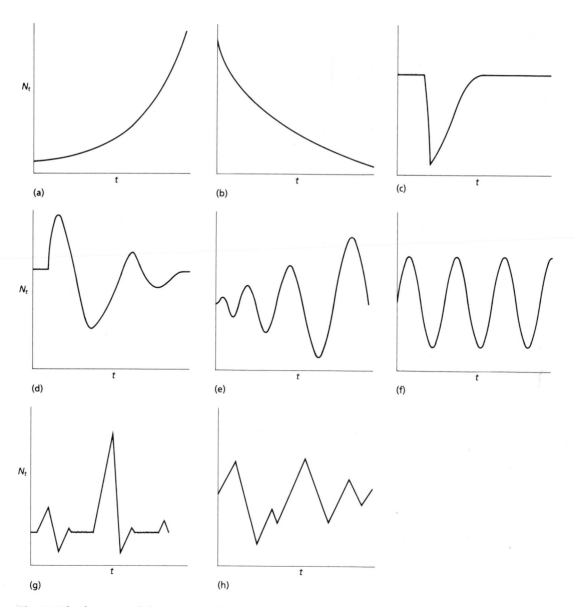

Fig. 3.1 The dynamics of theoretical model populations of predators and prey. The graphs show prey numbers at time t (N_t) against time (generations, t). If predator numbers were food-limited, then predator numbers would follow a similar pattern after a more or less protracted time-lag (see text for details). (a) Exponential increase. (b) Exponential decline. (c) Exponential stability, showing a smooth return to equilibrium following perturbation. (d) Damped stability, showing a return to equilibrium following perturbation through a series of damped oscillations. (e) Instability, showing a series of oscillations of increasing amplitude (eventually, one of the 'lows' would cause local extinction). (f) Cycles (these would be stable limit cycles if prey numbers returned to the same cyclical trajectory following perturbations; the neutral cycles of Lotka–Volterra dynamics do not do this; see text). (g) Chaos (deterministic dynamics which appear to be random fluctuations). (h) Random walk (a series of ups and downs in prey numbers caused by random changes in environmental conditions from year to year (e.g. weather, numbers of generalist predators)).

Table 3.1 Models for generalist predators, and the conditions for prey persistence. The equation for equilibrium prey population with density-dependent prey and fixed number predation is transcendental, and has to be solved numerically. With $\lambda = 2$, $\alpha = 0.01$ and $cP = 1.1931$ a hand calculator can be used to demonstrate that $N^* = 50$

Prey dynamics	Conditions for prey population growth	Comments
$N_{t+1} = \lambda N_t$	$\lambda > 1$	Invasion criterion
$N_{t+1} = \lambda N_t - cP$	$(\lambda - 1)N_t > cP$	Fixed number predation
$N_{t+1} = \lambda N_t - cPN_t$	$(\lambda - 1) > cP$	Fixed proportion predation
$N_{t+1} = \lambda N_t(1 - x_t)$	$(\lambda - 1) > \bar{x}, \ (0 \le x_t \le 1)$	Random predation
$N_{t+1} = \lambda N_t - cP[1 - e^{-bN_t}]$	$(\lambda - 1) > cP\,[1 - \exp(-bN_t)]$	Saturating functional response
$N_{t+1} = \lambda N_t - cPN_t^{1+b}$	$N^* = \left(\dfrac{\lambda - 1}{cP}\right)^{\frac{1}{b}}$	Density-dependent generalist predation
$N_{t+1} = \lambda N_t e^{-\alpha N_t} - cP$	$1 = N^*(\lambda e^{-\alpha N^*} - cP)$	Density-dependent prey with fixed number predation
$N_{t+1} = \lambda N_t e^{-\alpha N_t} - cPN_t$	$N^* = \dfrac{\ln\left(\dfrac{\lambda}{cP}\right)}{\alpha}$	Density-dependent prey with fixed proportion predation

Box 3.1 Use, suitability, availability and preference

One of the most difficult aspects of foraging ecology concerns the definition and measurement of use, suitability, availability and preference. Part of the problem is the great variation in the confidence with which these different concepts can be addressed.

1 It is relatively straightforward to measure the *use* of different prey species either by direct observation of predator behaviour and the careful analysis of their gut contents, or in the case of parasitoids by determining the percentage of prey that are parasitized.

2 We can make reasonable estimates of the prey that are *available* to the predator by sampling prey abundance in those parts of the habitat that are frequented by the predators (although it must be admitted that we can never know how the predators themselves perceive availability).

3 We can get some idea about *suitability* of different prey by carrying out studies on the profitability of different prey types under controlled conditions, but field estimates of suitability (or *food quality* as it is sometimes known) are notoriously difficult to obtain.

4 It is extremely difficult, however, to obtain unbiased estimates of *preference* under field conditions.

The notion of preference has evolved from studies of animals with diets made up of complex mixtures of species (Ivlev, 1961; Crawley, 1983). It is an essentially statistical notion which aims to predict the proportion of the diet made up by prey species-i as a function of the proportional availability of prey species-i in the environment.

The existence of preference is typically based on the observation that use is not perfectly correlated with availability. When items appear in the diet more frequently than in the environment, they are said to be preferred, and when they occur less frequently in the diet they are said to be avoided (Fig. B3.1). Direct, experimental assessments of preference are dogged by the difficulties of making availabilities equal, e.g. in cafeteria trials, where preference is measured by the relative amounts of equally available resources consumed. But there are great difficulties in carrying out experimental feeding trials sufficiently thoroughly to factor out effects due to the *order* in which foods are eaten, the *combinations* of prey offered, the *range of different relative availabilities* assessed, and the *condition* of the predators (their age, sex, hunger level, reproductive condition, and so forth). A further difficulty is presented by the fact that, in the case of parasitoids, we almost never know the population densities of the host-searching, adult insects, and so *use* (i.e. per cent parasitism) is the only thing we can measure.

Quantitative indexes of preference (Ivlev, 1961; Chesson, 1978; Greenwood & Elton, 1979) are heavily dependent upon decisions about whether or not to include abundant, low preference prey species in the spectrum of available foods. A simple numerical example makes this point.

Suppose that the percentages of three prey species (A, B and C) in the habitat and in the diet are as shown in Table B3.1.

continued on p. 46

Box 3.1 *contd*

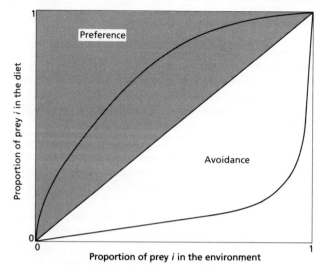

Fig. B3.1 The relationship between predator preference and prey relative abundance. When the predator diet contains a higher proportion of a given prey than that prey makes of all prey species in the environment, then that prey is said to be preferred (shaded region). When the proportion in the predator diet is less than the proportion in the environment the prey species is said to be avoided. Equality between the proportion in the predator's diet and the environment demonstrates neutral, non-selective predation.

Table B3.1 Proportion of three potential prey species in habitat and predator diet

Species	A	B	C
% in the habitat (availability)	1	9	90
% in the predator's diet	5	95	Trace

When prey species C is included as available food, then A and B both appear to be strongly preferred (preference ratios of 5.0 and 10.6). However, if species C is omitted from the analysis as not representing an available prey species, then the habitat contains 10% of prey species A and 90% of species B. In this case, B is still preferred (slightly), but now species A is quite strongly *avoided* (ratios of 1.1 and 0.5 respectively). Thus, decisions about which unpalatable prey species to include have strong effects on the interpretation of preference measures of the more preferred species.

Optimal foraging arguments suggest that the structure of the diet should be influenced by the abundance of the most rewarding prey species, but should not be affected by the abundance of unprofitable prey (see Chapter 4). In particular, reductions in the abundance of the most rewarding prey should lead to increases in the number of inferior prey eaten, but increases in the number of inferior prey should not lead to their being included in the diet (unless, of course, the increase

continued on p. 47

Box 3.1 *contd*

was associated with a decline in the abundance of superior prey, as a result of interspecific competition between the prey).

To understand the consequences of selective predation by generalist natural enemies, it is important to distinguish between the following.

1 *Preference*, which is the probability of taking species i when the predator is offered a *simultaneous* choice between species i and one or more other species.

2 *The zero-one rule* or *prey rule*, which states that a prey species is either taken every time it is encountered, or is ignored every time it is encountered.

In terms of foraging theory, a prey is either worth taking or it is not. If it is worth taking it should be taken every time the predator finds one. The reason that a predator might turn down a prey species is that by feeding on an inferior prey item it would lose the opportunity to find and feed from a superior prey (see Chapter 4 for details).

lation is sufficiently scarce that it does not suffer from any density-dependent constraints on its growth (e.g. competition for resources). Thus, without predators:

$$N_{t+1} = \lambda N_t \tag{3.1}$$

which gives unbounded exponential growth for the prey population because $\lambda > 1$. Now each of the P predators eats c prey per time period, so the population dynamics of the prey are described by:

$$N_{t+1} = \lambda N_t - cP \tag{3.2}$$

This means that the prey will continue to increase if

$$N_t(\lambda - 1) > cP \tag{3.3}$$

and will decline to extinction if predation is more intense than this. The important point is that prey populations with higher net rate of increase λ are capable of persisting in habitats with higher densities of generalist predators. The equilibrium prey population

$$N^* = \frac{cP}{\lambda - 1} \tag{3.4}$$

is artificial and highly unstable. It is a knife-edge, so that any relaxation of predation would lead to unbounded growth of the prey population, while any increase in predation would drive the prey inexorably to extinction.

Fixed proportion predation

It is unrealistic to assume that each individual predator kills the same number of prey, irrespective of prey density. It might be more reasonable to expect that the number of prey killed would fall as prey density declines, because prey will be more difficult to find. The simplest way to model this is to assume that the predators kill a constant proportion of the prey population in each time period. Thus, for example, if the predation rate were 10%, the predators would eat 10 animals from a population of 100 prey, but only one from a population of 10.

When the generalists take a more or less constant proportion $s = cP$ $(0 < s < 1)$ of the prey (e.g. they search only a fraction of the prey's habitat) then the dynamics are described by:

$$N_{t+1} = \lambda N_t - sN_t = N_t(\lambda - cP) \tag{3.5}$$

from which it is clear that the prey population will increase so long as $(\lambda - 1) > cP$ and decline to extinction otherwise. The harvestable excess is $\lambda - 1$, and any proportional predation greater than this will cause the prey population to decline.

As before, the equilibrium is a trivial one in which the predation rate cP exactly matches the production rate of prey $(\lambda - 1)$. At equilibrium, the predators are eating up the precise reproductive surplus, and the prey population would remain at whatever density it began, e.g. starting with 100 prey would leave 100 prey forever, whereas starting with 1000 prey would leave 1000 prey forever. This, again, seems most implausible, and suggests that one or more of the assumptions is wrong.

Random proportion predation

In many real world habitats, the mortality inflicted by generalist predators will be a random variable, unrelated to prey density and unpredictable in its magnitude:

$$N_{t+1} = \lambda N_t(1 - x_t) \tag{3.6}$$

where x_t is a random predation pressure $(0 \leq x_t \leq 1)$. Rather little work has been done on the dynamics of this system, e.g. specifying different probability distributions for x_t, investigating the importance of the relationship between λ and the variance in x_t, and so on. The average behaviour of this simple system produces a counter-intuitive result of the kind that is quite frequent in stochastic modelling. If the average proportion taken, $\overline{x}$, is less than $(\lambda - 1)$ then the system will be expected to increase exponentially. On the other hand, since x_t can be as high as 1, then if we wait long enough, the population is bound to go extinct. Thus, at the same time, we have an expected population size of infinity, yet the probability that the population will be extinct is equal to 1.

Over shorter periods of time, random predation could easily keep prey numbers within narrow bounds, and give a completely false impression of density-dependent regulation. It is ironic that in order to obtain convincing evidence of population regulation, you need to observe substantial departures from equilibrium density. This is because density-dependence is inferred by observing that large negative changes in log density tend to occur when population density is high, and large positive changes when population density is low. If there are no changes in density (e.g. because the population is perfectly regulated) then there is no possibility of detecting population regulation in this way. The alternative, of course, is to carry out manipulative experiments to see if the prey population returns to equilibrium following perturbation.

Functional response

Apparently, neither fixed number predation, fixed proportion predation nor random predation by a generalist enemy can regulate prey population density in a biologically sensible way, i.e. these processes can produce equilibria, but the equilibria are trivial and unstable. Perhaps the most obvious flaw in the present model lies in the relationship between prey abundance and predator feeding. Neither fixed number nor fixed proportion predation are satisfactory assumptions; the first assumes that each predator would eat c prey, no matter how scarce the prey became, while the second assumes that the predators have a limitless appetite, and eat a fraction s of the prey, no matter how abundant the prey become. We need to combine the realistic elements from both these models: (i) limited feeding per predator at high prey densities (predator satiation) from the fixed number model; and (ii) reduced predation at low prey densities (the functional response) from the fixed proportion model.

Both of these more realistic assumptions are built into the *functional response curve* (Box 3.2). A convenient mathematical form was provided by the Russian fisheries ecologist Ivlev (1961), and the generalist predator model with an Ivlev functional response is:

$$N_{t+1} = \lambda N_t - cP(1 - e^{-bN_t}) \qquad (3.7)$$

where c is now the maximum number of prey that can be eaten by each predator (the asymptote of the functional response curve) and b is the parameter which describes the steepness of the curve. Prey numbers will decline to extinction if the kill exceeds $N_t(\lambda - 1)$, and increase without bounds otherwise.

The predation rate always declines with prey density (Fig. 3.2d) so that generalist predators with this kind of functional response cannot regulate prey density. Their limited appetite means that they become *satiated* as prey density increases, at which point the prey population escapes control. Only with *sigmoid functional responses* (Box 3.1) in which predation is density-dependent at low prey densities, can generalist predators regulate prey abundance through functional responses alone (see below).

Density-dependent generalist

Generalist predators are capable of inflicting density-dependent mortality on their prey. For example, by *switching* to prey species that

Box 3.2 Functional responses

The term functional response was coined by Solomon (1949) to describe the way in which the number of prey eaten by a predator changed with prey density. In the simplest case, each predator eats a fixed number of prey per unit time, irrespective of prey density. This model may work reasonably well at high prey densities, but when prey are scarce the model breaks down, because predators require more prey than are present. In any event, such *fixed number predation* leads to a reduction in the percentage of the prey population eaten as prey density rises.

A simple alternative, which may be preferable when prey are scarce is the linear functional response, which is based on the assumption that a *fixed proportion* of the prey population is eaten by each predator (Fig. 3.2a). Thus, while the number of prey eaten per predator increases indefinitely as prey density rises, the fraction of the prey population taken is density-independent (Fig. 3.2b).

Neither the fixed number nor the fixed proportion functional responses is a satisfactory description over the full range of prey densities. A model is required that allows prey consumption to increase with prey density when prey are scarce, but also to allow that the number of prey that an individual predator can handle in a given time is limited by the capacity of the predator's gut, or by the time available in which the predator can handle prey. These two requirements are met by the so-called Type II functional response (Fig. 3.2c), described independently by Ivlev (1961) for fish and Holling (1959a) for shrews. Ivlev's equation for the number of prey encountered per predator (before allowance is made or the effects of exploitation) is:

$$N_e = c_{max}(1 - e^{-bN})$$

where c_{max} is the maximum number of prey that an individual predator can consume per unit time, and b describes the steepness of the ascending limb of the curve. Holling's disc equation (so called because the original experiments involved blindfold human subjects searching for sandpaper discs by touch) is:

$$N_e = \frac{aT_tN}{1 + aT_hN}$$

where a is the searching rate, T_h is the handling time, and T_t is total time available for search. When N is very large, the predator spends all its time handling prey, and none of it searching, and the asymptotic number of prey encountered is given by T_t/T_h. In practice, it would not be possible to decide between the two equations on the basis of their fit to the data. The decision as to which equation to use should be based on the particular predator and prey involved. If handling time can be measured independently, and hunting is clearly time-limited, then Holling's equation is most suitable. If predator intake is limited by gut capacity, then Ivlev's equation is preferable. By plotting the

continued on p. 51

Box 3.2 *contd*

percentage of prey consumed against prey density (Fig. 3.2d) it is clear that both these equations lead to inversely density-dependent predation.

The third example is the sigmoid functional response (or Type III; see Fig. 3.2e). This is of considerable theoretical interest because it is the only functional response that allows density-dependent predation. Note how in Fig. 3.4f the percentage predation rises with prey density to a maximum and then tails off. There are many equations to describe S-shaped functional responses, but one of the most appealing is based on the assumption that Holling's attack rate, *a*, varies asymptomatically with prey density:

$$a = \frac{bN}{1 + cN}$$

This means that predators are ineffective at finding prey when prey are scarce, but become more effective as prey density rises, up to a maximum efficiency (b/c) when prey are abundant. Substituting for *a* in Holling's functional response equation, gives

$$N_e = \frac{bT_tN^2}{1 + dN + bT_hN^2}$$

Sigmoid functional responses are described in detail by Murdoch and Oaten (1975) and Hassell (1978).

The least stable form of the functional response is the domed, or Type IV curve. This comes about as a result of a loss in hunting efficiency at high prey densities. This could result from confusion of the predators by mass scattering of the prey, or from 'defence in numbers' by the prey (see Fig. 3.2g). For a general discussion of herding behaviour and prey defence, see Hamilton (1971).

become temporarily abundant, or by moving to areas and *aggregating* in regions of high prey density, the percentage of prey killed by the predator can increase with prey density. Another way in which the attack of a generalist natural enemy might be density-dependent, is when the prey have a *fixed number refuge* from attack (Box 3.3). In this case, the prey population suffers little if any predation at low densities, because all the prey population can live safely within the protection of the refuge. At high densities, some prey are forced to live outside the refuge where they are vulnerable to predation. The result is that the predation rate is higher at high prey densities than at low. This, in turn, allows the possibility of prey population regulation. Note that fixed proportion refuges cannot stabilize prey numbers with a generalist predator (see Box 3.3).

A simple way to model density-dependent predation is to make the rate of predation explicitly density-dependent, so that instead of re-moving a constant fraction of the prey population, *cP*, in each time

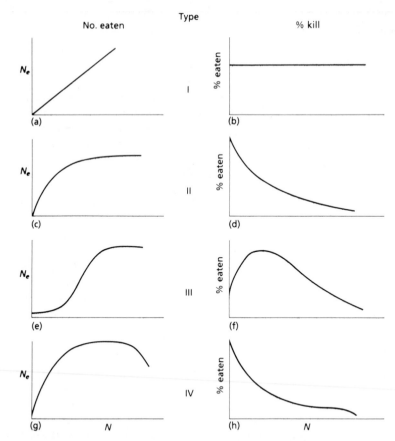

Fig. 3.2 The functional response of predator feeding to changes in prey density. N_e is the number of prey encountered, and N is the number of prey available per unit area. Type I: (a) a linear relationship between prey density and the number of prey eaten per predator per unit time; this produces (b) a constant percentage kill. Type II: (c) an asymptotic curve, with a limit to the number of prey eaten per predator which is set either by limited gut capacity, or by restrictions on handling time; this produces (d) a declining (inverse density-dependent) percentage kill. Type III: (e) a sigmoid functional response where foraging is inefficient at low prey densities; this produces (f) direct density-dependence in percentage kill at low prey densities, but inverse density-dependence at high. Type IV: (g) a humped functional response where the number of prey killed declines at very high prey densities (e.g. group defence effects); this produces (h) a non-linear, but inverse density-dependence in percentage kill. See text for details.

period, the predators remove an increasing proportion $cP.f(N_t)$ where $f(.)$ increases with N_t. For example,

$$N_{t+1} = \lambda N_t - cPN_t^{1+b} \qquad (3.8)$$

where $b > 0$, which means that the predation rate increases with prey density. This gives a stable equilibrium prey population of:

$$N^* = \left(\frac{\lambda - 1}{cP}\right)^{\frac{1}{b}} \qquad (3.9)$$

Box 3.3 Prey refuges

Fixed number refuge

The habitat may provide a fixed number of hiding places in which individual prey are safe from attack by natural enemies. These may be holes in the ground, inaccessible ledges on cliffs, or crevices in the bark of trees. Consider the dynamics of a prey population that is subject to attack by a fixed number generalist predator. Let the size of the refuge population of prey be N_r and the prey requirement of the predator be cP (see equation 3.2). Since the predator cannot gain access to the refuge, we can assume that the refuge is full of prey (even if it were not full, it would fill up through reproduction at a rate of λ per generation). Now individuals in the refuge breed to produce $\lambda.N_r$ individuals in the following generation; N_r of these replace the population in the refuge, while $(\lambda - 1).N_r$ must venture into the perilous habitat beyond the refuge. If the refuge surplus $(\lambda - 1).N_r$ is greater than the predation pressure cP, then the prey population will escape control and grow exponentially. If the predation pressure is greater than the refuge surplus, then all prey outside the refuge are killed, and prey population size is constant at N_r. This is the model that has traditionally been invoked to explain the population dynamics of red grouse on heather moorlands. In that example, territory holders were in refugia, relatively safe from predation, while those excluded from territory ownership were eaten up by generalist predators (see Chapter 14 for an alternative view of red grouse population dynamics).

Next, consider a generalist predator that takes a fixed proportion of the prey available rather than a fixed number. As before, $N-N_r$ prey are available to the predator, and of these, a fraction e is eaten. Thus, the prey equation becomes:

$$N_{t+1} = \lambda[N_t - e(N_t - N_r)]$$

This gives an equilibrium prey population of

$$N^* = \frac{e\lambda N_r}{1 - \lambda + e\lambda}$$

which is stable, so long as the predation rate $e > (\lambda - 1)/\lambda$. If the predation rate is lower than this, the prey population escapes control and increases exponentially.

Fixed proportional refuge

In other circumstances, it may be appropriate to consider the refuge as harbouring a constant fraction of the prey population rather than a constant number of individuals. For instance, if the natural enemy did not forage over 20% of the range occupied by the prey species, then the fraction of the prey population in that part of the range would be safe. Again, if a parasitoid wasp laid eggs into the larvae of stored products pests living at different depths below the surface of the grain,

continued on p. 54

Box 3.3 *contd*

then larvae that were deeper than the length of the ovipositor would be secure from attack. Does a constant proportional refuge stabilize prey numbers in the same way as the fixed number refuge discussed above? The answer is no.

Consider the N hosts as being divided into a safe fraction, s, and a vulnerable fraction, $(1 - s)$. For a fixed number predator, requiring cP hosts, we have:

$$N_{t+1} = \lambda N_t - \min[cP, \lambda(1 - s)N_t]$$

which leads to one of two things. Either the predators can take all the prey they need (cP), or they take all the surplus living outside the refuge $(\lambda(1 - s)N_t)$. In the first case, the prey population is bound to increase exponentially, because $(\lambda - 1)N_t > cP$. In the second case, the predators eat all the prey outside the refuge, so the outcome depends upon the relative magnitudes of λ and s: $\lambda s > 1$, exponential prey increase; $\lambda s < 1$, prey extinction. A proportional refuge cannot stabilize prey numbers in the face of fixed number predation from an uncoupled generalist predator.

Would it make a difference if the predation pressure were fractional, rather than fixed number? Again, the answer is no. The prey equation in this case would be

$$N_{t+1} = \lambda N_t - cP(1 - s)\lambda N_t$$

When the predation rate (the constant proportion cP) is such that:

$$cP = \frac{1 - \dfrac{1}{\lambda}}{1 - s}$$

the host population does not change (it remains at whatever initial value it had). However, for predation rates slightly lower than this the host population will increase without limit, while for predation rates slightly higher, the prey population will plummet inexorably to extinction, despite the proportional refuge.

Thus, for this particular kind of uncoupled generalist predator, a fixed number refuge will stabilize prey numbers, but a fixed proportion refuge will not. Neither kind of refuge can provide global stability, because in both cases if the predation rate is too low (or the refuge too large) the prey population will escape control. In coupled systems, involving specialist natural enemies whose numbers are linked to prey abundance, both fixed and proportional refuges can be stabilizing. The simplest way of modelling a proportional refuge in a coupled system is to assume that the proportion of prey escaping predator attack is described by the zero term of a negative binomial distribution (see p. 61). As the aggregation parameter k gets smaller, the refuge gets bigger. Population stability is enhanced, but equilibrium prey densities are increased in a tradeoff that is at the heart of biological control theory (see Chapter 18).

which makes good intuitive sense; it says that the size of the equilibrium prey population increases with prey reproductive rate λ, and declines with the number of predators (P) and the appetite of each predator (c). As the density-dependence coefficient $(b > 0)$ increases, so the equilibrium prey population declines. For example, if $b = 1$ then N^* is just $(\lambda - 1)/cP$ while if $b = 2$ then the equilibrium is reduced to the square root of this density.

Because the number of predators is a constant in equation 3.8, it means that the form of the prey equation is exactly the same as in a single-species, density-dependent prey model *without predation.*

$$N_{t+1} = \lambda N_t(1 - eN_t^b) \qquad (3.10)$$

This makes the important general point that it is impossible to say *what* is regulating a population from the observation that a population *is* regulated (although this conclusion, itself, is usually difficult enough to reach; see Crawley, 1990). With a generalist predator, there would be no obvious correlation between prey numbers and predator numbers that might lead us to suspect that predation was the regulating process in prey population dynamics.

Sigmoid functional response

Combining density-dependent predation with predator satiation leads to the definition of sigmoid (Type III) functional responses. With a sigmoid functional response and no prey density-dependence, the prey equation becomes:

$$N_{t+1} = \lambda N_t - cP \left(\frac{bN_t^2 T}{1 + dN_t + bT_hN_t^2} \right) \qquad (3.11)$$

The model has two equilibria (Fig. 3.3): (i) a stable low density equilibrium regulated by predation (recall that there is no other density-dependence in the model); and (ii) a higher, unstable equilibrium (at prey densities above this threshold, the prey escape control as a result of predator satiation).

Density-dependence in the prey population

It is not realistic to assume that the prey population would increase indefinitely in the absence of predation. At some point, intraspecific competition for food, epidemic disease, or some other density-dependent process would slow the rate of growth, eventually placing an upper limit on prey abundance. The simplest way to include density-dependence in the prey equation is to assume that instead of increasing at a constant rate, λ per generation, the population increases at a declining rate, $\lambda \exp(-\alpha N_t)$.

Prey population density need not be particularly high in absolute terms for this to be important, because competition may be for resources

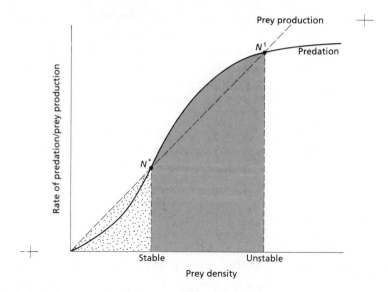

Fig. 3.3 The dynamics of the sigmoid functional response. If the prey production graph is such that it intersects the S-shaped predation graph in the density-dependent part of the curve, then predators are capable of maintaining a stable, low density prey equilibrium ▯, prey increase; ▧ prey decline. If prey numbers increase above N^1 then the prey population escapes from control by the predator. N^*, stable equilibrium; N^1, unstable equilibrium.

that are themselves at low absolute densities. Thus, for example, birds of prey may compete for safe nest sites on cliffs, and such cliffs may be several kilometres apart.

If there is density-dependence in the prey, then the question of whether generalist predators can regulate host density is not a major issue. The questions now become: (i) how far below the equilibrium set by competition can predators reduce prey density; and (ii) is there a limit to the stability of the prey population, e.g. can the predators drive the prey to extinction? The answers depend upon the kind of predation and the kind of competition. If a fixed number of prey are killed, we have:

$$N_{t+1} = \lambda N_t e^{-\alpha N_t} - cP \tag{3.12}$$

The result of the density-dependent term $\exp(-\alpha N_t)$ is that, in the absence of predation, the population will rise to an equilibrium at the point where $\lambda \exp(-\alpha N^*) = 1$, so that the equilibrium population density $N^* = \ln(\lambda)/\alpha$. The steeper the slope of the prey density-dependence function (α), the lower the prey equilibrium population will be.

Now predation reduces the prey population below this level (Fig. 3.4). So long as the predators take less than cP prey the possibility of a stable equilibrium remains. However, if prey numbers fall below N_T, then the prey will be driven to extinction.

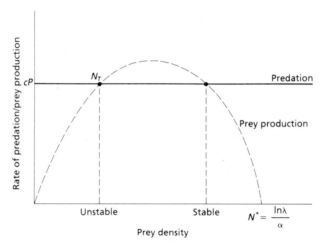

Fig. 3.4 The dynamics of constant-harvest predation. If a fixed number of prey cP is removed per unit time then there is a stable equilibrium prey density that is lower than the equilibrium caused by competition for resources (N^*, see text for details). If prey numbers are perturbed downwards, so that they fall below the unstable equilibrium N_T, then constant-harvest predation will drive the prey to local extinction.

Figure 3.5 shows a density-dependent prey population subject to constant proportional predation. This differs from the last case, because the predation graph is a straight line passing through the origin, whose slope is the proportion of prey eaten cP

$$N_{t+1} = \lambda N_t e^{-\alpha N_t} - cPN_t \qquad (3.13)$$

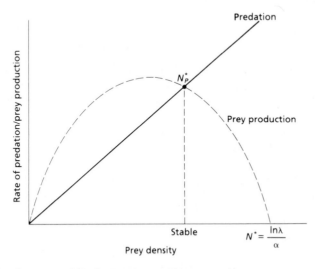

Fig. 3.5 The dynamics of fixed-proportion predation. In contrast to Fig. 3.4, the amount of predation increases linearly with prey availability. Given the kind of density dependence in the prey population represented by the n-shaped production curve, fixed-proportion predation will always produce a stable equilibrium N_P^* at the point where production and consumption are equal.

which has a stable equilibrium prey population at

$$N_P^\star = \frac{\ln\left(\dfrac{\lambda}{cP}\right)}{\alpha} \tag{3.14}$$

The equilibrium exists so long as $\lambda > cP$, because the prey must be able to increase when rare at a greater rate than the predators are exploiting them. Notice that with proportional predation there is no equivalent of the threshold population density that led to prey extinction in the case of constant numerical predation (compare Figs 3.4 and 3.5). As before, the greater the rate of predation, the lower the equilibrium prey population (but see Box 3.4 for an example where this is not the case).

Random search and exploitation

In all the models up to this point, we have assumed that if the predators required cP prey, and the prey population contained at least this many animals, then all cP prey would be killed. This simple procedure hides a number of implicit assumptions: (i) that the predators will not run out of hunting time before they catch the required number of prey; (ii) that the predator population searches in a systematic way, so that areas cleared of prey are not re-searched and prey are not inadvertently passed over; and (iii) that there are no behavioural interactions between the predators leading to inefficiencies in prey gathering at high predator densities. Each of these assumptions can be relaxed, and their effects investigated in turn.

Box 3.4 Prey density-dependence: contest or scramble?

Density-dependence in the prey population has profound effects on the dynamics of the predator–prey interaction. Both the nature of the density-dependence, and the timing of predation relative to the action of density-dependence are vitally important.

Consider a prey species that exhibited contest competition (Fig. B3.2a). This means that after attaining a certain high population density, say N_H, the prey population would not change any further following immigration of prey from other areas. Recruitment at high densities is constant, and this is depicted by the asymptote of Fig. B3.2a. Now suppose that predators kill cP prey at this high density. The figure shows that prey numbers in the next generation will not be depressed at all. Contest competition among the prey means that *compensatory reductions in other mortality factors* can ameliorate the effects of predation. Now suppose that the same predation pressure was applied at low prey densities (Fig. B3.2a). In this case, the number of prey in the next generation *is* depressed by

continued on p. 59

Box 3.4 *contd*

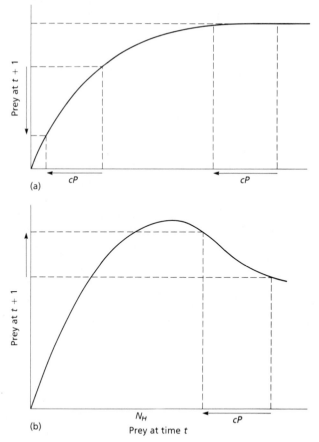

Fig. B3.2 Prey population dynamics with (a) contest and (b) scramble competition amongst the prey. With contest competition (a) removal of *cP* prey from a dense prey population causes no change in prey numbers in the next generation. With scramble competition (b) removal of *cP* prey from a dense population can actually lead to increased numbers of prey in the next generation because of competitor release.

predation. The general point is that compensatory changes in mortality are not possible when the prey are scarce.

Now consider a prey population that exhibits over-exploitation of its resources leading to scramble competition (Fig. B3.2b). This means that when prey numbers rise above a threshold density N_H, prey numbers in the subsequent generation actually decline because the consequences of competition are so severe. This opens up the intriguing possibility that predation in one generation could actually lead to higher prey densities in the next generation. As the predators take *cP* prey at high densities, the number of prey in the next generation goes up because of *predator-induced competitor release*. As with contest competition, any predation at low prey densities is bound to lead to lower prey numbers in the next generation.

One way to allow for time limitations and non-systematic search is to assume *random search* by the predators. In random search the predator is assumed to encounter hosts in a manner described by the Poisson distribution. This specifies the mean number of encounters per prey, $\mu = cP/N$, and predicts the proportion of the prey population that would be encountered 0, 1, 2, 3, ..., n times. The proportion of prey surviving is given by P_0, the *zero term* of the Poisson distribution (i.e. the animals that were never encountered).

$$P_0 = e^{-\mu} \tag{3.15}$$

The proportion of prey eaten is simply $1 - P_0$. Predators and parasitoids differ in their response to *prey exploitation*, because insect parasitoids can rediscover parasitized hosts during subsequent foraging. Time may be wasted in handling parasitized hosts that are later rejected, or superparasitism may result (the fitness implications of this are discussed in Chapter 11). With predators, of course, attack and consumption can only happen once, so no time is wasted re-encountering or rehandling prey. This means that more time is available for searching in predators than in parasitoids (Rogers, 1972): $T_s = T - cT_h$ where T_s is searching time, T is total time, c is number of prey eaten per predator, and T_h is handling time per prey. Thus, with P predators and assuming random search the *random predator equation* is:

$$cP = N\{1 - \exp[-aP(T - cT_h)]\} \tag{3.16}$$

where cP is the number of prey consumed by the population of P predators. There is no analytical solution to equation (3.16), and the value of c must be determined numerically, in order to find the total prey consumption at a given prey density. In practice, the predator model predicts fewer prey killed than the disc equation (because the latter assumed systematic search), but more than the *random parasite equation* (which allows for time-wasting during re-encounters with parasitized hosts):

$$cP = N\left[1 - \exp\left(-\frac{aTP}{1 + aT_hN}\right)\right] \tag{3.17}$$

(see Rogers, 1972 for details). The difference between these models is only important if foraging bouts are long, handling time is protracted, and attack rates are high.

Aggregated search

Random search is often assumed for mathematical convenience rather than as an accurate description of events. Many field studies have shown that predator foraging is aggregated, with the result that a larger proportion of the prey population escapes detection than is predicted by the Poisson distribution (above). The *negative binomial* distribution has proved to be an excellent description of the way that attacks by

natural enemies are distributed, both over prey and through space. The
zero term of the negative binomial is:

$$P_0 = \left(1 + \frac{\mu}{k}\right)^{-k} \qquad (3.18)$$

where μ is the average number of predator encounters per prey (as
before) and k is the aggregation parameter. As k gets large (greater than
about 10), the negative binomial approaches the Poisson distribution.
For small values of k (less than 1) the distribution of predator attacks
is highly aggregated. As the degree of aggregation increases, the value
of k declines and the zero term gets bigger. This means that a larger
proportion of the prey escapes predator attack, so the likelihood of
over-exploitation leading to local extinction of the prey is reduced (see
Fig. 3.6).

Aggregated attack is typically stabilizing in coupled predator–prey
dynamics (see Chapter 11), because it produces a refuge for the prey.
However, because the zero term of the negative binomial produces a
proportional refuge, it cannot regulate prey population density on its
own in a system with generalist predators and no other form of density-
dependence. While fixed number refuges are stabilizing with both
uncoupled generalists and coupled specialist predators, proportional
refuges are only stabilizing with coupled specialist predators (see
Box 3.3).

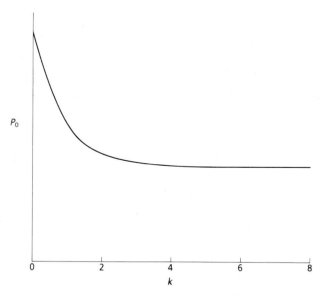

Fig. 3.6 The relationship between the proportion of prey escaping predation (P_0) and
the value of the aggregation parameter k of the negative binomial. As k is reduced
below 1 (high aggregation) so the proportion of the prey population escaping attack
increases rapidly. For large values of k (approaching random attack), the proportion
of prey escaping predation approaches a constant, minimal value (the zero term of a
Poisson distribution).

THE DYNAMICS — OF
COUPLED PREDATOR–PREY SYSTEMS

The reason that prey populations so often escaped control in the models described in the previous sections was that the predator population was fixed in size, so that its killing power was strictly limited. When a particular prey species is abundant, however, the predators may exhibit a *numerical response to prey density*; they may reproduce more frequently, their young may show increased survival or attain sexual maturity earlier, or there may be immigration from adjacent areas which boosts local predator populations. This means that P is really a variable, with its size determined (at least in part) by the number of prey available.

The simplest system of coupled specialist predators involves a strictly monophagous predator whose numbers are determined by prey availability, and a prey species whose density is determined entirely by predation. This system was analysed in the 1920s by Lotka and Volterra, and is described in detail in the next section. The difficulty is that without any density-dependence, the system exhibits *neutral stability* in continuous time, an artificial property which means that the amplitude of fluctuations depends entirely upon initial conditions, and that once perturbed, the system stays perturbed (see p. 66).

The components of a symmetric predator–prey system are shown in Table 3.2. In the absence of predation the prey will increase exponentially at a rate λ, while in the absence of prey the predators decline exponentially at a rate d. For simplicity, we consider the case

Table 3.2 The components of a symmetrical predator–prey interaction can be modelled with two equations: one for the prey population and one for the predator. These equations may be continuous-time differential equations, like this: $dN/dt = A - B$, $dP/dt = C - D$; or they may be discrete-time difference equations like this: $N_{t+1} = N_t f(A - B)$, $P_{t+1} = P_t g(C - D)$. In either case, the precise form of the equation for each of the four components may take any number of different forms, depending upon the system in question and the precise questions being investigated. The formulations in the table show some of the simplest forms. More detail can be found in Crawley (1983)

Component	Equation	References
A Prey increase	λN	Deevey (1947), Birch (1948), Cole (1954)
B Prey decrease	$cP[1 - \exp(-aN)]$	Nicholson & Bailey (1935), Hassell (1978)
C Predator increase	$gP[1 - \exp(-aN)]$	Crawley (1975)
D Predator decrease	dP	May (1974)

in which density-dependence is confined to the functional and numerical responses, as shown below:

$$N_{t+1} = \lambda N_t - cP_t f(N_t) \tag{3.19}$$

$$P_{t+1} = P_t g[f(N_t)] - dP_t \tag{3.20}$$

where the function $f(.)$ describes the functional response of the predator to changes in prey density and the function $g(.)$ is the numerical response (it converts the number of prey captured by each predator into a per capita birth rate).

The dynamics depend upon the functional forms of $f(.)$ and $g(.)$. With linear or saturating functional responses (see equations 3.5 and 3.7) the term $f(.)$ cannot be stabilizing because the rate of predation is either density-independent or declines with increasing prey density. Only when $f(.)$ increases with prey density over all or part of its range can this term contribute towards the stability of the interaction (e.g. with sigmoid functional responses; see Fig. 3.3). The predator's numerical response can stabilize the interaction so long as R_0 for the predator is high enough and $g(.)$ declines at high predator density so that over-exploitation does not drive the prey to extinction (such stabilizing mechanisms include *mutual interference* (Hassell, 1978), and the *pseudo-interference* that comes about as a result of predators aggregating in high density patches of prey (Free *et al.*, 1977)).

The general point is that without density-dependence in at least one of its components, the coupled interaction is unstable. Predator and prey would be unlikely to coexist for long in any one place in the absence of immigration. The second point is that it does not matter greatly where the density-dependence comes from. We included it in the functional and numerical responses, but it could equally well have entered in the prey recruitment term (as in equation 3.12) or in the predator death rate term. The kind of dynamics that are produced (exponential stability, damped oscillations, increasing oscillations, neutral cycles, stable limit cycles or chaos) depend upon the timing of predation relative to prey reproduction and prey density-dependence (see Box 3.5), and upon the precise parameter values (see below and Chapter 19), but usually *not* on the particular term in which the density-dependence appears.

Phase-plane analysis

If the analysis of predator–prey isoclines is new to you, the best plan is to work painstakingly through the following examples on paper, tracing out the changes in predator and prey abundance from one time step to the next. If you are familiar with phase-plane analysis, you should turn directly to p. 66.

In the simplest case, prey numbers are determined only by predation, and predator numbers are determined only by prey availability. Instead

Box 3.5 The timing of predator attack

We consider two important cases: (i) the timing of predation relative to prey reproduction; and (ii) the timing of predation relative to the action of prey density-dependence.

In the first case, timing is important because it determines whether or not predation eats into the reproductive capital of the prey population, or whether it takes some of the annual productivity of young. In terms of our models, a generalist taking cP prey before reproduction would have the following effect on prey population:

$$N_{t+1} = \lambda(N_t - cP)$$

Compare this with equation 3.2 where the predators attack after reproduction. Prey dynamics will only be affected by the timing of predator attack if this influences the subsequent impact of prey or predator density-dependence.

The timing of prey density-dependence in relation to the phenology of predator attack is important because of the potential that exists for compensatory reductions in other mortality factors. When predation happens after density-dependence, as in equation 3.12, then there is no scope for compensation, and prey numbers in the next generation are bound to be reduced. If, on the other hand, predation occurs before prey density-dependence, then predation relaxes the intensity of competition. Then, depending upon whether the competition is scramble or contest in type (see Box 3.4), prey numbers in the next generation may not be reduced at all (in some extreme cases, they may even by higher). The model for pre-density-dependent predation is:

$$N_{t+1} = \lambda(N_t - cP)e^{-\alpha(N_t - cP)}$$

assuming, of course, that $N_t > cP$.

Thus, the order of events within each generation is important, and doubly important if the prey population suffers scramble competition. If predator exploitation is continuous within each prey generation, then it may be preferable to model the interaction with continuous-time differential equations rather than the discrete difference equations employed here (see Chapter 19).

of writing out a pair of difference equations, as we would do in extending the models used in the previous section, consider a graph that has prey abundance on the x axis and predator abundance on the y axis. The aim is to depict population dynamics by tracing a path through the space defined by these axes (so-called phase-plane analysis).

1 *Prey dynamics.* Prey increase when predators are scarce and decrease when predators are abundant. To show this we draw a horizontal line (the prey isocline) halfway up the predator axis (Fig. 3.7a). Above the line, predators are abundant so prey decrease. The arrows for prey dynamics all point *to the left* in this zone. Below the line, predators are scarce, so prey increase. Thus all prey arrows point *to the right* below the predator isocline.

2 *Predator dynamics.* Predators increase when prey are abundant and decline when prey are scarce. To show this we draw a vertical line halfway along the prey axis (the predator isocline). To the left of the line, prey are scarce, so predator numbers decline. The arrows for predator abundance point *downwards* in this zone. To the right of the

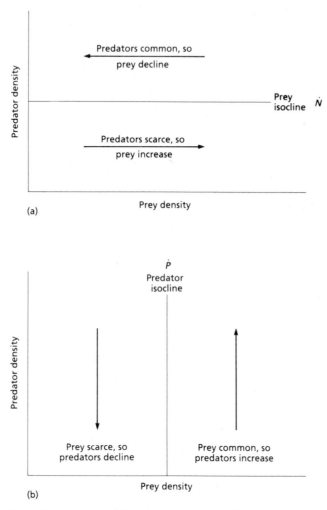

Fig. 3.7 Phase-plane analysis of the Lotka–Volterra predator–prey model. (a) The prey isocline. Prey increase when predators are scarce (arrow towards the right) and decrease when predators are abundant (arrow towards the left) $\dot{N}$ is $dN/dt = 0$; the point where change in prey numbers is zero. (b) The predator isocline. Predators increase when prey are abundant (arrow upwards), and decrease when prey are scarce (arrow downwards). $\dot{P}$ is $dP/dt = 0$; the point where change in predator numbers is zero. Predators are assumed to be the only force acting on the prey and prey availability is assumed to be the only factor limiting the increase of the predators. These assumptions are embodied in the fact that both isoclines are straight lines at 90° to the axes.

line, prey are abundant, so predators increase. Thus, all predator arrows point *upwards* in this zone (Fig. 3.7b).

By combining the predator and prey vectors, we obtain the resultant motion in each of the four quadrants (Fig. 3.8). In the top right, predators and prey are both abundant, so the motion is to the north-west (prey decline but predators increase); in the top left, prey are scarce but predators are abundant, so the motion is to the south-west (both prey and predators decline); in the bottom left, prey and predators are both scarce, so the motion is to the south-east (prey increase but predators decrease); in the bottom right, prey are abundant, but predators are scarce, so the motion is to the north-east (both predators and prey increase).

These simple rules apply throughout the following studies of predator–prey dynamics, and using only these four vectors, we can deduce a great deal about the dynamics of predator–prey interactions under a variety of ecological circumstances.

Linear predator and prey isoclines

The first predator–prey model was developed independently in the 1920s by Lotka (1925) and Volterra (1926). In the Lotka–Volterra model, the predator and prey isoclines are at their simplest; the predator isocline is a straight vertical line, and the prey isocline is a straight

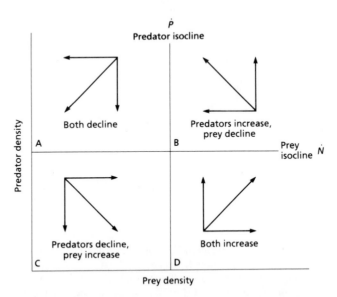

Fig. 3.8 The resultant motions of predator and prey numbers in the four quarters of the Lotka–Volterra phase space. A, predators and prey both decline because prey are scarce but predators are abundant. C, predators decline but prey increase because predators are scarce. D, predators and prey both increase because prey are abundant but predators are scarce. B, predators increase because prey are abundant, but prey decline because predators are abundant.

horizontal line. To understand the dynamics of the system we proceed as follows. Take a starting point anywhere on the phase plane (say in the middle of the lower right hand box). Then look up the motion appropriate to this zone (it is towards the north-east; Fig. 3.8). Draw a line north-eastwards from the starting point until it intersects an isocline (the prey isocline in this case). Stop at this point. Now consider the motion in the quadrant that has just been entered (in this case, the top right box, where subsequent motion is to the north-west). Now draw a line north-westwards until the next isocline is intersected. Repeat this procedure for six or seven more steps (Fig. 3.9a). If the y axis is intersected, the prey population has been driven to extinction. If the x axis is intersected, the predators have gone extinct.

The first thing you will notice is that if you have drawn the picture carefully, you begin to retrace your path exactly on returning to the initial quadrant. Try starting in a different initial position. You will then go round and round on a new closed loop. Thus, the initial conditions determine the trajectory. In order to see what is happening, it is often helpful to plot numbers against time. Assume that there are just three density categories for each population: high, medium and low. Prey numbers are high when they are to the right of the predator isocline, medium when they are at the predator isocline, and low when they are to the left of it. Similarly, predator numbers are high when they are above the prey isocline, medium when they are on the prey isocline, and low when they are below it.

We began in the lower right hand box, with high prey density and low predator density. Plotting prey numbers first, we see that at the end of the first step prey numbers were to the right of the predator isocline (high); after the second, they were 'on' the predator isocline

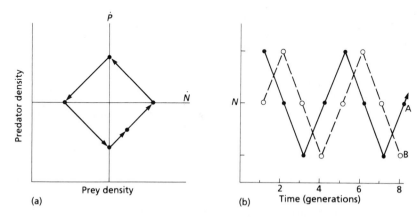

Fig. 3.9 The dynamics of the Lotka−Volterra predator−prey model. (a) Phase-space analysis shows that numbers cycle forever on a fixed trajectory determined by initial predator and prey numbers. (b) Time series analysis shows that predator (○) and prey (●) numbers cycle through time, that predator numbers lag behind changes in prey numbers, and that the size of the time-lag is one-quarter of a cycle period.

(moderate); after the third they were to the left of the predator isocline (low); after the fourth, they were on the predator isocline again (moderate); and so on. We can plot these densities against time (Fig. 3.9b), then repeat the exercise for predator densities. They go medium, high, medium, low for the first four steps, then repeat this pattern indefinitely.

Using this exceptionally simple technique, we can discover a great deal about the dynamics of this system.

1 Numbers of predators and prey cycle indefinitely.

2 The cycles are out of phase.

3 The predator curve lags behind the prey curve by one-quarter of a cycle.

4 The trajectory is determined uniquely by the initial densities of predators and prey.

And all this without any algebra!

Density-dependence in the prey

It is clear that the Lotka—Volterra model is hopelessly unrealistic. In the following sections we shall investigate the effects of making progressively more realistic assumptions about predator and prey biology, one step at a time. We do this by *changing the shape of the isoclines*. First, we consider the effects of density-dependence in the prey. The prey population would not increase indefinitely in the absence of predators, so the prey isocline must eventually bend down at the right, until it touches the x axis. The point at which it touches this axis is K, the prey equilibrium in the absence of predators. It might be determined by food availability, or by the number of safe nest sights. If prey numbers were to rise above this point (e.g. as a result of immigration), then the prey population would subsequently decline. The arrow for prey motion would point to the left, despite the fact that predators were scarce (see Fig. 3.10).

We can investigate three different kinds of prey density-dependence. Linear, continuous prey density-dependence means that prey populations continue to increase at higher predator densities when prey are scarce, than when prey are common. This is modelled by rotating the prey isocline about its mid-point clockwise until it intersects the x axis. The effects of this kind of linear prey density-dependence on population dynamics are shown in Fig. 3.10. Most starting conditions lead to a stable equilibrium, e.g. trajectory A, but trajectory B leads to extinction of the predator during step 1, while trajectory C leads to extinction of the prey in step 2. Thus prey density-dependence can produce a stable equilibrium, but the equilibrium is not globally stable, and some initial conditions (and hence, some perturbations) can lead to extinctions.

A second possibility is that the prey isocline is abruptly non-linear, as in Fig. 3.11. This might happen when prey numbers increase ex-

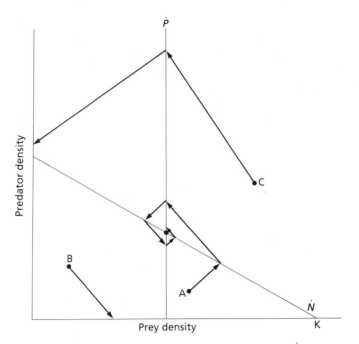

Fig. 3.10 Prey density-dependence. By rotating the prey isocline ($\dot{N}$) clockwise, we can model the effects of density-dependence acting on the prey population. The intersection with the predator isocline is now a stable equilibrium reached from initial conditions like A. The model is not globally stable, however, because initial condition B leads to extinction of the predator after one generation, and condition C leads to extinction of the prey after two generations.

ponentially up to some fixed limit as might be imposed by the number of safe nesting sites. In this case, the dynamics depend critically upon the position of the predator isocline. In Fig. 3.11a, where the predator isocline cuts the density-independent part of the prey curve, the dynamics are cyclic. *Neutral cycles*, as produced by the Lotka–Volterra model, are observed for starting conditions close to the intersection of the isoclines, while *stable limit cycles* arise when starting conditions are further away, i.e. trajectories converge on the *same* cycle no matter where, outside the cycle, they originate. If the predator isocline intersects the density-dependent, descending portion of the prey isocline (Fig. 3.11b), then the dynamics are quite different, and show damped oscillations to a stable equilibrium. Note that many more density combinations lead to extinction when there is abrupt density-dependence than when there is continuous density-dependence.

A third possibility arises if the prey population exhibits inverse density-dependence (the so-called *Allee effect*). In this case the prey population grows more slowly when it is scarce (e.g. as a result of difficulties in finding mates, or in lone individuals exploiting food less effectively; see Crawley, 1983, for examples). To model this we, draw the prey isocline as descending to the left. If there were a threshold prey density below which the prey could not persist, the curve would

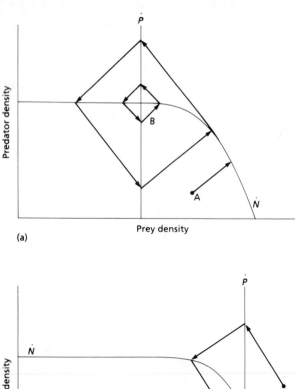

(a)

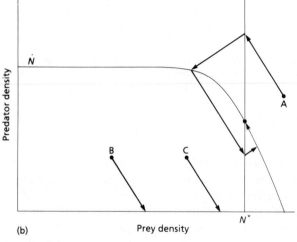

(b)

Fig. 3.11 Non-linear prey density-dependence. (a) If the predator isocline intersects the horizontal (density-independent) part of the prey isocline then we can obtain stable limit cycles (A) or neutral cycles (B) depending upon initial conditions. (b) If the predator isocline intersects the declining (density-dependent) part of the prey isocline, the resulting equilibrium is stable (A). Neither case is globally stable, however, and certain starting conditions (B, C) lead to extinction of the predators.

intersect the *x* axis above zero (Fig. 3.12). When the predator isocline intersects this ascending part of the curve, the prey population is inevitably driven to extinction.

As before, the graphical technique has enabled us to discover a great deal about dynamics.

1 Density-dependence acting on the prey is stabilizing.
2 This stability is local, not global.
3 Predator–prey dynamics need not be cyclic.
4 Predators can produce a stable prey equilibrium but only if the

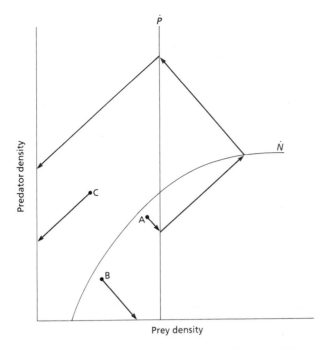

Fig. 3.12 Allee effects (inverse density-dependence) on the prey. If the predator isocline intersects an increasing prey isocline, then the resulting equilibrium is unstable, and leads to increasing oscillations with (A) prey extinction after four generations. Initial conditions B lead directly to predator extinction and C to prey extinction in generation one.

predator isocline cuts the descending (density-dependent) part of the prey isocline.

5 Predator and prey densities at equilibrium are independent of initial densities.

6 Equilibrium population size is determined by the *interaction* between the density-independent and density-dependent components, not by either one or the other alone, i.e. the equilibrium changes when *either* of the isoclines is moved.

7 The shape of the density-dependent prey isocline is important (if it is smooth and continuous, there is a stable equilibrium and no prospect of limit cycles, but if it is abrupt, then stable limit cycles are possible when the predator isocline does not intersect the descending part of the curve).

8 Allee effects (inverse density-dependence) in the prey are destabilizing, and are associated with a high probability of prey extinction.

Density-dependence in the predator

There may be an upper limit to predator abundance that is set by its intolerance of crowding or by a shortage of safe breeding sites. In order to make predator dynamics density-dependent, we alter the shape of the predator isocline. A linear density-dependent predator isocline might come about as a result of intraspecific competition, and would mean that the decline in predator abundance should begin at higher prey densities when predators are common than when predators are scarce. This is achieved by rotating the predator isocline clockwise by about 20°. In order to see the dynamics that result from this, take the simplest possible density-independent prey isocline (a horizontal straight line). There are three possibilities for the dynamics (Fig. 3.13). From most sets of initial conditions (A), the trajectory spirals rapidly inwards to a stable equilibrium. For lower initial prey abundances (B), the predator would go extinct during time step 1, while for higher predator densities (C), the prey would be consumed to extinction during step 1.

Alternatively, the predator population may experience abrupt density-dependence (Fig. 3.14), as might occur if the upper limit on predator abundance were set by the number of secure roosting sites. In this case, the predator isocline bends abruptly to the right, so that predators can increase when prey are common, but only up to a limit. If predator numbers were to be increased above this level (e.g. by immigration), then predator numbers would decline, despite the abundance of prey. Starting at high prey densities (A), we intercept the horizontal part of the predator isocline. In the zone above this, we want to travel south-westwards, but we are prevented from doing so by

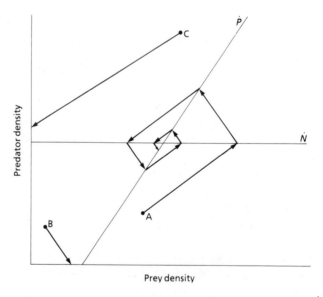

Fig. 3.13 Predator density-dependence. By rotating the predator isocline ($\dot{P}$) clockwise we model the effects of predator density-dependence. With density-independent prey (a horizontal prey isocline) this produces a locally stable equilibrium at the point of intersection for initial conditions like A. Starting condition B leads directly to the extinction of the predator and C to extinction of the prey in generation one, so the model is not globally stable.

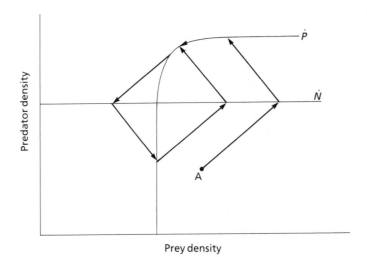

Fig. 3.14 Non-linear predator density-dependence. If there was a fixed upper limit to predator abundance (e.g. predator territoriality), then the predator isocline would bend over to become horizontal. This produces stable limit cycle behaviour for initial conditions like A. There is no stable point equilibrium with density-independent prey, and many initial conditions lead to predator or prey extinction.

the predator isocline itself. When this happens, the rule is to *travel along the isocline* towards the left (prey abundance declining all the time as a result of high predator numbers). When we come to the 'corner' of the isocline we can move off south-westwards. What happens next depends upon the position of the predator isocline. If it is high, we intercept the *y* axis and the prey are driven to extinction. If the isocline is lower we enter a stable limit cycle ending up on the same cyclical trajectory, no matter where we begin. Starting at intermediate prey densities close to the predator isocline we would obtain neutral cycles like those in the Lotka–Volterra model.

Inverse density-dependence in predator dynamics (an Allee effect might involve low-density problems like difficulty in finding a mate, or genetic problems due to inbreeding depression) is modelled by having the predator isocline bend to the right at low predator densities (Fig. 3.15). The consequence of this is that the predators go can extinct if they reduce prey availability too far (A), but also if there is a catastrophic reduction in predator numbers even when prey are abundant (B).

The dynamics of a density-dependent specialist predator feeding on a density-independent prey species can be summarized as follows.
1 A linear declining predator isocline can give a stable equilibrium.
2 This equilibrium is not globally stable, and perturbation can lead to prey extinction.
3 With a non-linear predator isocline, the maximum predator density may be so high that the prey will be driven to extinction, i.e. predator density-dependence is not inevitably stabilizing.

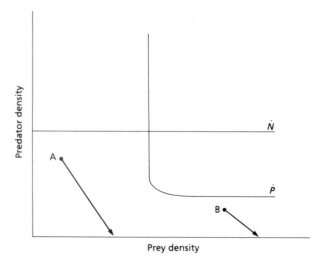

Fig. 3.15 Allee effects on the predators. If predators have difficulty finding mates at low predator densities, then the predator isocline may become horizontal parallel with the prey axis. This means that predators can go extinct from certain starting conditions (B) even when prey are abundant. This system can exhibit only neutral cycles or local extinctions (A).

4 Abruptly non-linear predator isoclines tend to cause stable limit cycles (but initial densities close to the intersection of predator and prey isoclines could produce neutral cycles).
5 Predator populations that exhibit Allee effects are prone to extinction.

Prey refuges

A prey refuge exists when the predators are unable to drive the prey to extinction (Table 3.3; Box 3.3). This is modelled by turning the prey isocline upwards through 90° at the lowest density to which the prey population can be reduced by predation. There are two possibilities for the dynamics of a system exhibiting a prey refuge, but without any other form of density-dependence (Fig. 3.16). The outcome depends upon the *interaction between the refuge and the predator isocline*: (i) if the refuge is small and the predator isocline is well to the right, the predator goes extinct and the prey population escapes control; (ii) if the refuge is large or the predator isocline lies further to the left, then the system produces stable limit cycles because the refuge is sufficiently productive to ensure that the predator is not driven to extinction before the prey population begins to recover.

Prey refuges can be combined with other forms of density-dependence. For example, combining a prey refuge with prey density-dependence (as in Fig. 3.10) may increase the stability. Note, however, that the model is not globally stable, because if the refuge is too small,

Table 3.3 Refuges from predation

Kind of refuge	Example	Reference
Hiding places	*Tribolium* in glass tubes	Crombie (1945), Gause (1934)
	Covered oranges	Huffaker et al. (1963)
Inaccessible microhabitat	Ovipositor length in parasitoids	Askew (1961), Arthur (1962)
Habitat selection	Incomplete spatial overlap	Vinson (1981) Huffaker & Kennet (1959)
Microhabitat selection	Differential use of host plant parts	Gardner & Dixon (1985)
Predator behaviour	Aggregation in high-density prey patches	Hassell & May (1974)
Predator selectivity	Switching behaviour	Murdoch & Oaten (1975)
Prey behaviour	Predator avoidance	Hamilton (1971)
Prey morphology	Crypsis	Guilford (Chapter 16)
Prey polymorphism	Shell banding in snails	Jones et al. (1977)
Multilayed egg batches	Central eggs inaccessible	Braune (1982)
Parental care	Limited number of young protectable	Callan (1944)
Inaccessible reserves	Protected parts of colonial animals	Noy-Meir (1975)
Phenology	Differential susceptibility of different cohorts	Crawley & Akhteruzzaman (1988)
Differential mobility	Migratory prey with sedentary predators	Sinclair (1979)
Other trophic levels	Predators of the predators	Davidson et al. (1985)
Host plant effects (i)	Lack of predator attractants	Wesloh (1976)
Host plant effects (ii)	Presence of predator deterrents	Woets & van Lenteren (1976)
Lack of other similar prey	Low predator density	Messenger (1975)

or the predator isocline is too far to the right, then the trajectory intersects the x axis before intersecting the predator isocline, so the predator goes extinct. Complete, global stability is depicted in Fig. 3.17a. Here, both predator and prey have a refuge to protect them from extinction (i.e. we assume the existence of some external source of predator immigration), while in Fig. 3.17b both populations also exhibit linear density-dependence. The general conclusions are the following.

1 Stable equilibria are only possible when the predator isocline inter-

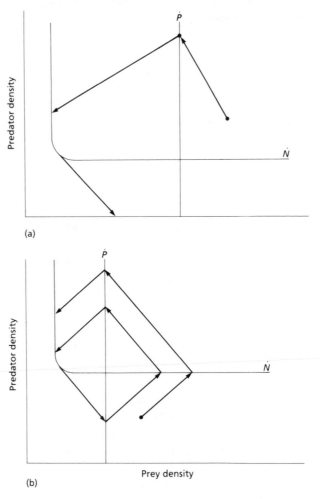

Fig. 3.16 Prey refuge. A refuge exists when the prey population cannot be driven to zero, no matter how abundant the pedators become. This is modelled by bending the prey isocline to the vertical, parallel with the predator axis. The dynamics depend upon the position of the vertical part of the prey isocline (the size of the refuge), relative to the location of the predator isocline. (a) Unstable dynamics: the refuge is too small to stabilize the interaction. (b) Stable limit cycle: the refuge is sufficiently large that the trajectory from the base of the refuge intersects the prey isocline. Neither of these cases is globally stable.

sects a part of the prey isocline that is not inversely density-dependent.
2 In order to be stabilizing, a prey refuge must be sufficiently large that its production of prey is sufficient to see the predator population though the lean period.
3 Refuges for both the predator *and* the prey are necessary to guarantee global stability.

These graphical models do have a number of shortcomings, e.g. it is difficult to deal with time-lags, or with quantitative differences in

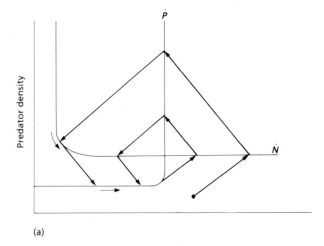

(a)

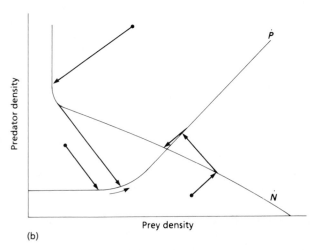

(b)

Prey density

Fig. 3.17 Globally stable predator–prey dynamics. (a) A globally stable system exhibiting stable limit cycle dynamics in which the only density-dependence consists of a refuge for both the predators and their prey. (b) A globally stable point equilibrium with density-dependent isoclines for both predators and prey, and refuge for each population. Global stability requires that both populations have a refuge.

parameter values, like changes in prey reproductive rate. Nevertheless, their simplicity, and the ease with which they allow insights into predator–prey dynamics easily compensate for these deficiencies. In subsequent chapters, we shall revisit many of the concepts introduced in this section. Sometimes the models will be written as coupled difference equations (e.g. host–parasitoid interactions in Chapter 11) and sometimes as coupled differential equations (e.g. human disease models in Chapter 15). In either case, it will be useful to refer back to the phase-plane diagrams introduced here.

All the preceding models were based on the assumption that both the prey and their predators were uniformly distributed throughout suitable habitats. In reality, of course, it is more likely that the prey would be distributed in patches of differing density, and that predators would be non-randomly distributed over these patches. Patchiness in prey distribution might result from underlying heterogeneity in the distribution of their food resources, from clustering of progeny around parents, or from physical limitations on dispersal.

From an evolutionary perspective, we would expect the predators to be distributed over prey patches in the way that maximized their fitness. Other things being equal, therefore, the entire predator population should be found in the single patch supporting the highest prey density, because this is where each individual predator would encounter prey at the maximum rate. Lower density patches should only be exploited if the number of predators in the best prey patch became so high that the cost of foraging in a poorer patch was outweighed by the benefit of experiencing reduced competition from other predators. Alternatively, exploitation of the best patches may be so rapid that they quickly stop being 'the best'. The stepwise occupation of progressively poorer prey patches as predator density increases is known as the *ideal free distribution*, and is discussed in detail in the next chapter (see p. 99). Since the low density prey patches are under-exploited, they form a *refuge* for the prey. Thus, foraging behaviour that leads to predator aggregation in the high density patches, tends to stabilize predator–prey dynamics, because it prevents over-exploitation of the prey (see Fig. 3.16).

When predators aggregate for other reasons (e.g. around roost sites, shade, water holes or mating arenas) then they tend to have less impact in depressing overall prey density. Nevertheless, density-independent predator aggregation is still stabilizing because a refuge is provided for the prey in the predator-free patches away from their aggregation zones.

The maximum degree of prey population depression comes about in the unlikely event that there is a perfect correlation between prey density and predator density (leading to a *fine-grained* pattern of predation; see Crawley, 1983). As the correlation between prey and predator densities becomes more and more imperfect as a result of predator aggregation, so the degree of prey population depression is reduced (because predation pressure is 'wasted' either in searching empty patches, or in competition with other predators in high density patches). Stability of the predator–prey interaction increases as predators increase their degree of aggregation, but only up to a point, because stability will eventually decrease as the refuge becomes too large (see Box 3.3).

The important point about refuges is that there must be heterogeneity in the probability of attack on different prey individuals; this can come about in numerous ways. It might result from density-dependent, inverse density-dependent or density-independent changes in the distribution of predation pressure, or it might be an entirely stochastic process.

INDIVIDUAL VARIATION

Most predator–prey models assume that all predators are alike, and all prey are alike. This will almost always be untrue. Individuals will differ in age and in sex. The proportion of the population made up by individuals of different ages (its *age structure*) and sexes (its *sex ratio*) will change from year to year. Individuals of the same age and sex will differ in their phenotypes as a result of differences in their histories, the environmental conditions they experience, and a host of other causes. There will often be differences between the genotypes of individuals of the same age and sex, and different genotypes may perform differently in different environments. And so on.

All this would be mere detail were it not for the fact that individual differences can have extremely important effects on dynamical behaviour. In general, individual differences tend to be stabilizing, because invulnerable classes of individuals create a refuge from predation, and, as we have seen, refuges tend to be stabilizing. Models that allow for vulnerable and invulnerable age classes (Smith & Mead, 1974), susceptible and immune individuals (Anderson & May, 1991) or differential exposure to predation (Bailey *et al.*, 1962), all show pronounced increases in stability, compared with similar models that assume all individuals to be homogeneous. It will become clear in the following chapters that the existence of individual differences is universal, and that individual differences capable of influencing population dynamics are to be found in all facets of predator and prey behaviour, morphology and life history.

DYNAMICS OF REAL NATURAL ENEMIES AND THEIR PREY

The numbers of animals never remain constant for very long, and usually fluctuate considerably and often rather regularly. The primary cause of these fluctuations is usually the unstable nature of the animals' environment, as is shown by the effects of periodic bad winters on the numbers of certain small birds. These fluctuations in numbers affect other animals dependent upon them as prey, and the final effects upon animal communities are very complex. [Elton, 1927]

Population cycles in arctic small mammals

Charles Elton's (1927) classic book *Animal Ecology* brought the existence of two long-term sets of data on the abundance of predators and their prey to the attention of a wide readership: (i) arctic lemmings and arctic fox (Fig. 3.18); and (ii) the snowshoe hare and its predators (chiefly lynx and red fox, but also hawks, owls, weasels, ermine, martens, wolverine, skunks and mink; Fig. 3.19). The lemming cycles show an average of 3.75 years between peaks over a 60-year period, while the snowshoe hare show an average of 10 years between their peaks over a similar period of time. Two important questions arise from these data: (i) what determines the duration of the cycles (why 4

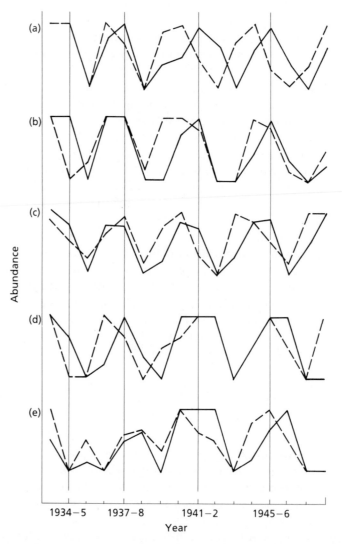

Fig. 3.18 Cyclic populations of arctic lemmings (– – –) and foxes (——).
Population cycles over the period 1933–1948 in five geographic locations:
(a) northern Baffin Island; (b) southern Baffin Island; (c) northern Quebec; (d) eastern Victoria Island; (e) western Victoria Island. Note the synchrony between peaks and troughs in different locations, and the roughly equal cycle lengths. Vertical lines show the major migrations of snowy owls. After Chitty (1950).

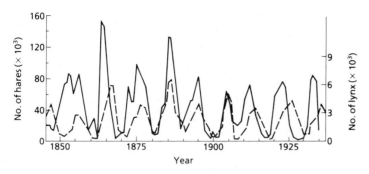

Fig. 3.19 Cyclic populations of snowshoe hares (———) and lynx (– – –). Note the wide amplitude, but relatively constant period of the cycles. After MacLuick, (1937).

years for lemming and 10 for snowshoe hare); and (ii) what drives the cycles (do predators cause the cycles, or are the cycles generated in some other way, with predator numbers simply driven along by cycles in their prey)?

There are two competing explanations for the cause of the cycles, and these can be caricatured by two hypothetical extreme views as the *pro-predator* and the *anti-predator* arguments. The pro-predator argument goes like this:

Year 1. Prey numbers are at rock bottom, and all predators have died or emigrated; prey numbers increase exponentially.

Year 2. Prey numbers increase exponentially in the absence of natural enemies.

Year 3. Prey numbers reach very high densities; predators immigrate in high numbers; predator breeding success is very high.

Year 4. Massive numbers of predators are now confronted with rather low prey numbers, leading to virtual extermination of preferred prey; many predators starve, others emigrate.

The anti-predator argument goes like this:

Year 1. Prey numbers are at rock bottom, and all predators have died or emigrated; food plants begin to recover but prey numbers increase slowly.

Year 2. Prey numbers increase exponentially in the absence of natural enemies following full recovery of food plants.

Year 3. Prey reach very high densities; preferred, high quality food plants are in short supply; prey show low reproductive success and poor body condition; predators immigrate in high numbers; predator breeding success very high.

Year 4. Vegetation is severely depleted; massive numbers of predators are now confronted with reduced prey numbers, leading to virtual extermination of preferred prey; many predators starve, others emigrate.

The two explanations are very similar, and, in principle, it should be straightforward to distinguish between them. The pro-predator argument says that natural enemies are responsible for halting prey increase and initiating prey decline. If this is so, then a predator exclusion experiment should show that prey populations continue to increase longer and reach higher peak densities in exclosures than on matched plots subject to predation. The anti-predator argument says that other factors cause the prey population to stop growing and to begin to decline (over-exploitation of preferred, high quality food is the favourite alternative). If this is so, then a food supplementation experiment should show that prey populations continue to increase longer, and reach higher peak densities, than on matched plots that receive no extra food resources.

Unfortunately, the spatial scale over which these high-latitude predators move is of the order of thousands of square kilometres, which makes realistic predator exclusion experiments completely impractical. Food supplementation studies also need to be carried out over large areas and they, too, need to be protected from immigration of extra natural enemies. This is because prey will be scarce outside the food supplementation areas, but natural enemies will be abundant (and hungry), and immigration of predators onto the food-supplemented plots might obscure the experimental result, by reducing prey densities to background levels. The experiments of Krebs and his colleagues (Krebs *et al.*, 1986), on the relative importance of food and predators in driving the snowshoe hare cycle, should be consulted as a model of the kind of practical difficulties that arise in attempting to test what look like quite straightforward alternative hypotheses under field conditions.

It is interesting to note that Elton (1927) considered that the crash in herbivore numbers in both lemming and hare cycles was brought about by an 'epidemic which kills them off when a certain density of population is reached'. He was clearly a subscriber to the anti-predator school of thought, although it is not clear whether he had any evidence for his belief that the population declines were the result of disease epidemics. It is typical of the difficulties involved in unravelling complex dynamics that these high-latitude cyclic populations involve what turn out to be the most unwieldy kinds of natural enemies: generalist predators that can show strong numerical responses to the abundance of a single prey species (see above).

Tawny owls and woodland rodents

Mick Southern studied the tawny owl *Strix aluco* population in Wytham Wood near Oxford for 13 years between 1947 and 1959. Parallel studies were made of the distribution and abundance of the owl's main prey species, the woodmouse *Apodemus sylvaticus* and the bank vole *Clethrionomys glareolus*. The owl's vocal defence of rigid territories enabled accurate census of population density to be made each spring. The fact that fledged young remained for a long time in their parents' territories made it possible to determine the number of young produced per year.

The number of fledged young fell far short of what was possible. Some owl pairs refrained from breeding at all, others laid eggs but failed to hatch them, yet others hatched young but failed to rear them to fledgling. By and large, this reproductive failure was associated with the availability of rodent prey. Above a threshold rodent abundance, fledging was high and unaffected by mouse and vole numbers. When rodents were exceptionally scarce, no owls even attempted to breed. Over a range of low prey densities there was a positive relationship between rodent density and breeding success (Fig. 3.20). Despite this considerable variation in recruitment, the population of adult owls remained remarkably constant, changing only between a low of 17 pairs in 1947 and a high of 30 pairs in 1955.

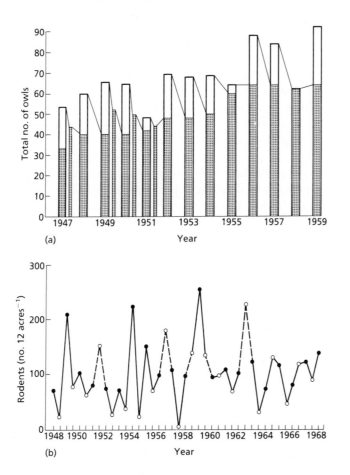

Fig. 3.20 Tawny owls and rodents. A relatively constant breeding population of owls over a 13-year period (a), despite wide fluctuations in the abundance of their prey (b). The stippled bars in (a) show breeding birds and the open bars show juvenile recruitment. Note that breeding failed, or was not even attempted, in years when prey were scarce (e.g. 1955 and 1958). From Southern (1970).

It appears that reproduction in tawny owls was food-limited, and fluctuated markedly with variations in rodent density. Adult population size, however, appeared to be limited by the number of territories, and not by recruitment of juveniles (adult population was stable despite erratic recruitment). It is clear that young which failed to find a territory either starved or moved outside the area (where, presumably, most of them starved). Territory area varied with habitat (e.g. 12 ha in closed woodland and 20 ha in mixed, open woodland), but did not change from year to year with rodent abundance. Possession of a territory was vital, because a tawny owl without a territory could not feed. It is possible that mean rodent abundance might influence mean owl density in the long term, if there was a relationship between mean territory size and mean rodent numbers in different habitats.

What of the effect of tawny owl predation on rodent numbers? The second lowest rodent numbers were observed in the spring of 1955 after myxomatosis had virtually eliminated rabbits from the woodland, and habitual rabbit predators like the fox *Vulpes vulpes* had to turn to feeding on small mammals. Thus, it is clear that vertebrate predators are capable of exerting at least a short-term, depressive effect on prey abundance. In other years, however, the combined abundance of mice and voles fluctuated rather little about the long-term average density of $20\,\text{ha}^{-1}$, and there is no evidence that owl predation had any appreciable impact on the separate or combined densities of mice and voles.

This example highlights the asymmetry that is often observed in studies of population dynamics. The rodents influenced the owls in the short term by determining their annual breeding success (the more rodents, the higher the number of young owls fledged). In the long run, they may have influenced owl breeding densities by determining average territory size. But there is no evidence of any influence of owl predation on rodent abundance in most years. The owls appear to be prey-limited, but the rodents are not predator-limited (see Southern, 1970 for details).

The wolves and moose of Isle Royale

David Mech (1966) began work on the moose and wolves on Isle Royale, an island 30 km from the shore of Lake Superior, in June 1958. Since then the populations have been studied by a progression of ecologists (Jordan *et al.*, 1967; Peterson & Page, 1983). Mech's data suggested that the large wolf pack killed one moose every 3 days or so, and that the smaller wolf packs killed about one-third this many, for a total annual kill of 163 moose (adults and calves were killed in approximately equal numbers). An estimated 564 moose were present in late May when calves were born. Calves comprised about 25% of the summer population, suggesting that at least 188 calves were produced. More detailed figures suggested that about 83 adult moose were killed each year while about 85 adults were recruited through the survival of yearlings. This simple arithmetic suggested that the moose population was predator-limited and would remain stable, or would increase slightly as long as these figures applied.

What the arithmetic cannot tell us, however, is the degree to which the predation rate was dependent on the density of moose, nor does it throw any light on the regulation of wolf numbers. In fact, the moose population increased substantially following Mech's study (Taylor, 1984), suggesting that the wolves were *not* regulating moose numbers. The important question is whether wolves kill moose in a density-dependent manner, or whether wolf predation merely substitutes for other mortality factors. The data are equivocal, but circumstantial evidence in favour of moose limitation by wolf predation is: (i) moose were much more abundant on Isle Royale before the introduction of wolves; (ii) moose severely overgrazed many of the browse species prior to the introduction of wolves, suggesting food-limitation; and (iii) browse species have increased substantially in abundance following the introduction of wolves.

Recent studies in south-western Quebec lend further support to the notion that moose populations can be limited by wolf predation. Messier and Crete (1985) measured moose densities, wolf densities and predation losses at three sites. Year-long predation rates were density-dependent, varying between 6.1% at a site with 0.17 moose km^{-2} to 19.3% at a site where there were 0.37 moose km^{-2}, consistent with predation playing a regulating role. The relationship between wolf predation rate and moose density does not appear to increase over the full range of densities, however, and there may be an n-shaped relationship, with lower predation rates at peak moose densities (where moose populations would increase until they were food-limited). Of course these studies lack any statistical replication, and differences in moose density are confounded by countless other differences between sites. Convincing demonstrations of density-dependent predation are only likely to come from manipulative experiments, but these are immensely difficult to do (consider, for example, the large spatial scale involved with moose dispersal, and the complex social system involved in wolf pack behaviour).

The implicit assumption in the original Isle Royale study was that the population density of the wolves was regulated by the abundance of moose. Data on this question are provided by Messier (1985) who found that wolf populations in areas where moose were at low densities suffered higher mortality as a result of malnutrition and intraspecific strife, and had reduced success in producing pups, compared with wolf packs in areas of higher moose density. Messier suggests that in the absence of alternative prey, 0.2 moose km^{-2} would be the minimum population necessary to support a viable wolf pack. Social factors of wolf pack organization also appear to be important, since the wolf population on Isle Royale did not show a numerical response to in-

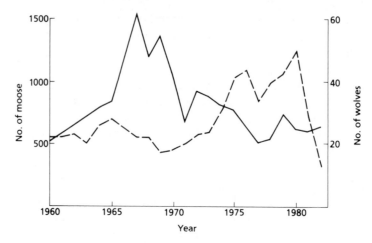

Fig. 3.21 Wolves (– – –) and moose (——) on Isle Royale. The decline in moose numbers is attributed to wolf predation. From Taylor (1984).

creased moose abundance throughout the 1960s (Fig. 3.21) until the Big Pack split up in 1972 and 1974, after which wolf numbers doubled (Taylor, 1984).

Rats and cats on oceanic islands

Alien vertebrate predators have had a profound impact on the faunas of islands into which they have been introduced, and extinctions of native island animals as a result of introduced predators are numerous. The most serious losses are amongst land birds, and the main culprits have been rats *Rattus norvegicus* and feral cats *Felis catus*. The importance of 'buffering prey species' is shown by the example of two birds of Macquarie Island in Antarctica. Banded rail and parakeet survived for 70 years following the introduction of feral cats and wekas *Gallirallus australis*, but both birds declined to extinction following the introduction of rabbits. The reason appears to have been that cats increased greatly in numbers by feeding on the rabbits, and this led to increased predation on the endemic birds, which eventually drove them to extinction (Taylor, 1979). Not all the examples are so gloomy. Nesting seabird numbers have increased dramatically following the removal of feral cats from islands to the north of New Zealand (Moors & Atkinson, 1984).

Ladybirds and aphids

The population dynamics of arthropod predator–prey systems in temperate and boreal environments are often determined chiefly by year to year differences in weather conditions. This is because with temperature-conforming ('cold-blooded') species, both the time at which the individuals become active in spring, and their potential rates of population increase, are strongly temperature-dependent. Furthermore, the temperature-dependence of these demographic parameters is often different for predators and prey.

Ladybirds are beetles (family Coccinellidae) that are specialist predators of aphids both as adults and as larvae. Adult ladybirds are mobile and can fly many kilometres, laying their eggs on a range of plant species, wherever aphids happen to be abundant. Aphid populations are typically parthenogenetic in the summer time, consisting entirely of females, and capable of exhibiting prodigious powers of population increase. Two questions about the predator–prey interaction arise immediately: (i) how important is ladybird predation in keeping down aphid numbers; and (ii) what determines ladybird abundance?

Frazer and Gilbert (1976) studied the ladybird *Coccinella trifasciata* in alfalfa fields where it fed on the pea aphid *Acyrthosiphon pisum*. Aphid survival rate was significantly affected by factors in addition to predator and prey densities (e.g. ambient temperature and the age structure of the aphid population both influenced the rate of aphid kill per predator). They also showed that the predation process differed qualitatively (not just quantitatively) when they studied it in the field rather than the laboratory. For example, ladybirds routinely eliminated entire aphid colonies in laboratory feeding trials, but never eliminated prey populations in the field. Some aphids always survived, no matter how low aphid densities were initially, because the ladybirds simply *did not have enough time* to find sufficient prey. The functional response of field populations, therefore, was unlike any of the textbook examples because it did not pass through the origin (see Fig. 3.22).

Ladybird numbers are determined by a complex of factors operating over an enormous spatial scale. The previous generation of beetles may have exploited many different species of aphid, feeding on a range of host plants, in a variety of different habitats, with different individuals experiencing markedly different conditions. Thus, the adult ladybirds in a given alfalfa field on a particular sampling occasion may belong to several different cohorts and have come from many kilometres away. At present, we lack the resources to contemplate studying population dynamics on this sort of scale. Probably the best that we can do is to treat ladybird predation pressure as a random variable determined partially by weather conditions. Thus, we would predict that the impact of coccinellid predation on aphid population dynamics would be greatest in hot dry summers and least in years with long cool springs.

In terms of pest control within alfalfa fields, it may be that the less conspicuous, generalist predators like spiders and anthocorid bugs are more important than the immigrant ladybirds. These generalists are active early in the year, and can nip potential aphid outbreaks in the bud, either by killing the founder insects or by exterminating small colonies. In this way, the generalists might have more impact in reducing summer aphid numbers than the specialized aphidophagous predators like ladybirds, hoverflies and lacewings that immigrate into the crop once aphid numbers have built up.

Perhaps the most important lesson to be learned from this example is that the importance of invertebrate predators depends upon the weather. In cool summers, the aphids become active well before the ladybirds, and this head start allows their numbers to build up to high densities before ladybird feeding begins. Under these circumstances, ladybird predation has no appreciable effect in checking aphid population increase. In hot dry summers, however, the ladybirds are active from the outset, and small founder colonies of aphids are discovered early in the season and eradicated. Thus, ladybirds *can* keep aphid populations down, but only if the weather is right. It is likely that ladybird numbers are affected by aphid abundance, but the spatial scales involved are so

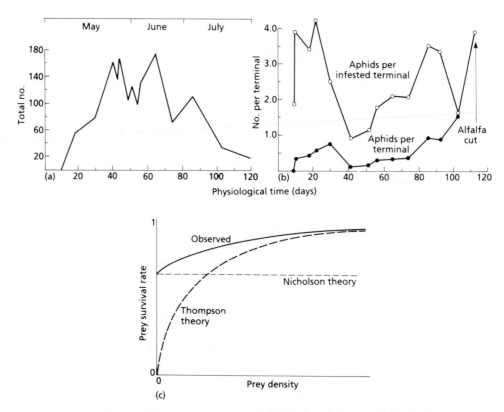

Fig. 3.22 Predator–prey systems driven by immigration and dispersal. (a) The number of coccinellids over the period May–July in alfalfa fields in North America. (b) Aphid (prey) numbers per infested shoot tip show two peaks in early and late season. Predation by coccinellids may contribute towards reduced aphid numbers in June, but they do not prevent the second peak, nor the inexorable rise in the number of aphids per terminal (Frazer & Gilbert, 1976). (c) The functional response of coccinellids to aphids demonstrates elements of both Nicholson and Thompson models; there is rapid predator satiation at high prey densities, but at low prey densities, the predators have too little time to find all the prey, and so prey survivorship does not fall to zero (Gilbert et al., 1976).

large, and the numbers of aphid species so great, that there is no realistic prospect of predicting ladybird abundance from one year to the next. As Neil Gilbert is fond of saying, the most important aspect of insect predator–prey dynamics is the relative temperature thresholds of the predator and the prey. When the predator's temperature threshold is substantially greater than the prey's, it is most unlikely to have a profound impact on prey population dynamics, because the predator will always arrive on the scene too late to prevent prey population build-up.

CONCLUSION

In the introductory sections, we considered a variety of important ecological processes in a stepwise fashion. Each of the components, taken on its own, was a simple idea, but put together, in order to explain the dynamics of a real predator–prey interaction, the processes can form systems of Byzantine complexity. It is salutary, also, to recall that even the simplest mathematical models (e.g. a two-parameter non-linear difference equation) can exhibit chaotic dynamics of almost unimaginable intricacy.

Consider the minimum amount of information that would be necessary to understand the dynamics of just one specialist predator and its prey.

1 The intrinsic rate of increase of the prey.
2 The functional response of the predator.
3 The predator's spatial foraging behaviour.
4 The nature of prey density-dependence.
5 The nature of predator density-dependence.
6 The densities at which the predator and prey isoclines intersect.
7 The relative slopes of the two isoclines at the intersection.
8 The size of the prey refuge.
9 The rate of predator immigration.
10 The rate of between-patch prey dispersal.

A bare minimum description of this set of processes would require at least one parameter for each component, and so a full model of a specialist predator–prey interaction would need to have at least 10 parameters. At this point it is worth recalling the physicist Linus Pauling's famous remark about parameter-rich models: 'Give me four parameters and I'll draw you an elephant: give me five and I'll waggle its trunk'.

4: Foraging Theory

MICHAEL J. CRAWLEY & JOHN R. KREBS

INTRODUCTION

Foraging theory is a branch of evolutionary biology that deals with the behavioural ecology of food-gathering. When an animal harvests food it has to make decisions: choices about where to hunt for food, which kinds of prey to eat, when to move to a new patch, and so on. This chapter discusses how animals make these choices, not from the point of view of detailed behavioural mechanisms, but by considering general strategic rules that might apply to a wide range of natural enemies. The rationale is that, as a result of evolutionary selection pressures, animals will tend to hunt for their food efficiently, so that if we can work out in theory the decision rules which would maximize an animal's efficiency, these rules ought to predict how a predator makes its choices.

Optimality modelling is an explicit way of thinking about the features of predators, parasites and diseases that might arise through natural selection. One of the key figures responsible for introducing selectionist thinking to ecological theory was David Lack. First, in his work on Darwin's Finches, he introduced the idea that ecologically similar species might have evolved differences in their feeding niche and morphology because of selection to avoid competition. Second, he proposed the first experimentally testable optimality model in ecology — his theory of clutch size in birds (Lack, 1954). This stated that birds which feed their young lay the number of eggs that maximizes the number of young fledged; they could lay more eggs, but if they did so they would fledge fewer progeny because of competition for food between the nestlings.

The theory of optimal foraging has a well-defined beginning in 1966 with the simultaneous publication of two papers, one by J.M. Emlen and another by R.H. MacArthur and E.R. Pianka (for a complete review of the history of these ideas, replete with amusing anecdotes, see Schoener, 1987). The background justification for optimal foraging theory, as for biological optimality models of any sort, is that natural selection influences many of the features of organisms that biologists study, and that these features are *adaptations*. Natural selection, being a cumulative, competitive and iterative process, tends to maximize the utility of adaptations in promoting survival and reproduction.

Optimality models are simply tools for asking what adaptations would look like if they had been favoured by selection to maximize survival and reproduction in a particular way. They provide a benchmark against which observed features can be compared. For example, one could ask what group size a bird might live in, if group size had been selected to achieve an optimal balance between the benefits related to predator defence, and the costs arising from competition for food. Notice that the assertion is that natural selection *tends to maximize,* and there is no implication that it necessarily produces perfect outcomes.

MODELS

Foraging models typically have three components: (i) *decision assumptions* (which of the predator's choices is to be analysed); (ii) *currency assumptions* (how are the various choices to be evaluated); and (iii) *constraint assumptions* (what limits the animal's feasible choices, and what limits the payoff that may be obtained). These components are expanded in Box 4.1, and this breakdown of the formal structure of a foraging model demonstrates what kinds of ideas an optimality approach might allow us to test; assumptions about the decision, the currency and the constraints. When an optimality model produces predictions that agree with observations, confidence in these assumptions increases slightly. When a model fails to account for what happens in nature, then confidence in the assumptions diminishes. The general background rationale of optimality modelling is not itself under direct test in any one study, although, in the long term, repeated success or repeated failure of individual models would either sustain or undermine belief in the general assumptions surrounding individual models (Maynard Smith, 1978a; Schoener, 1987).

Perhaps the most important assumptions of what are referred to as the 'classical' foraging models are: (i) *rate maximization* (or long-term average-rate maximization, in full); and (ii) the *exclusivity of searching*

Box 4.1 Elements of foraging models

Decision assumptions

All optimality models consider the 'best' way to make a particular decision. Foraging models have studied two basic problems: (i) which prey items to consume; and (ii) when to leave a patch. In the first case, most models of diet choice solve for the optimal probability that the forager will consume a given prey type after encountering it. This is the decision variable. It assumes that the predator can recognize different categories (prey types) on encountering them, and decide whether or not to pursue them at that stage. In the case of patch models, the decision variable is the *time spent per patch* (also called

continued on p. 92

Box 4.1 *contd*

the patch residence time). This is both general and mathematically tractable, time being a convenient and continuous variable. Whether or not real predators measure time spent in a patch as such, or use simple rules of thumb in their decision making, is explained in the text.

Currency assumptions

The currency of a model is the criterion used to compare alternative values of the decision variable. Currencies are as diverse as the adaptations they are used to study. For example, in designing the optimal caddisfly net, the currency is the *number of particles filtered per minute* and the choice principle is *maximization*. Conventional foraging models use the *net rate of energy gain while foraging* as their currency, and optimal foragers are assumed to maximize this quantity. More energy is assumed to be better, because a forager with more energy will be more likely to meet is metabolic requirements, and will be able to spend spare energy on important non-feeding activities like fighting, fleeing, nest-making and reproducing. An alternative currency for foraging models is time spent foraging; an optimal forager might be a *time minimizer*. For most practical purposes, however, both currencies are equivalent to rate maximization (see Pyke *et al.*, 1977).

Constraint assumptions

Constraints are those factors that limit and define the relationship between the currency and the decision variable. Limitations are of two biologically different types, intrinsic and extrinsic. Intrinsic constraints are: (i) limitations in the abilities of animals (e.g. honeybees cannot distinguish red from grey); and (ii) limitations in the tolerances within which animals must live (e.g. the animal can survive only 2 hours without food). Extrinsic constraints are placed on the animal by the environment. For example, stream velocity limits the number of particles a caddisfly net can filter per hour, and a forager cannot eat more prey than it is able to encounter in the time available. Intrinsic and extrinsic constraints are not independent. Thus, foraging by a lizard is constrained by the interaction between its muscle physiology (an intrinsic constraint) and the air temperature (an extrinsic constraint) which jointly determine its running speed.

Conventional foraging models make three constraint assumptions.
1 *Exclusivity of search and exploitation*: a predator cannot search for new prey while handling a captured prey.
2 *Sequential random encounters*: prey itims are encountered one at a time, and the probability of encountering each prey type is constant.
3 *Complete information*: the forager knows, or behaves as if it knows, all the rules of the model. A completely informed forager is like a gambler who knows the odds involved in roulette, but cannot predict what number will come up on the next spin of the wheel.

and exploiting (an animal can search for prey or handle captured prey, but not both at once). Taken together, these two assumptions lead to what Stephens and Krebs (1986) call *the principle of lost opportunity.* Briefly, decisions about exploiting a food item can be assessed by comparing the potential gain from exploiting the prey in question, with the potential loss of opportunity to do better. For example, if the prey is of the best possible type, then no opportunity is lost by eating it. The best outcome that might result from not eating it would be to happen immediately upon another prey item of exactly the same type. Conversely, if an inferior prey is eaten, then the forager loses some opportunity of discovering a superior prey item. We consider rate maximization models first before going on to examine other currencies.

The classical prey model

The prey model asks whether a foraging animal should attack the prey individual it has just encountered, or pass it over. Therefore, the prey question is: *attack or continue searching?*

A particular type of prey item provides a fixed mean amount of energy, and a fixed mean amount of time is required to pursue, capture and consume it. Taken together, these activities constitute the *handling time.* Assume that searching costs s per time unit. Let there be a set of n possible types of prey. Each type of prey, i, is characterized by four variables: h_i, the expected handling time spent with an individual prey item of type i, if it is attacked upon encounter; e_i, the expected net energy gained from an individual prey item of type i, if it is attacked upon encounter; λ_i, the rate at which the forager encounters items of type i while searching; p_i, the probability that items of type i will be attacked upon encounter; this is the decision variable.

Also assume that the time taken to handle an item encountered but not attacked is zero, and that the net energy gained from an item encountered but not attacked is also zero. This means that the net rate of energy gain is

$$R = \frac{\sum_{i=1}^{n} p_i \lambda_i e_i}{1 + \sum_{i=1}^{n} p_i \lambda_i h_i} \tag{4.1}$$

The value of p_i that maximizes R must be either the largest feasible value ($p_i = 1$) or the smallest ($p_i = 0$). The algebra behind this result is presented in full by Stephens and Krebs (1986). What this means is that a prey is either always attacked on encounter, or always ignored (the *zero–one rule*). Now the important question is what determines whether p_i is equal to zero or one? In order to maximize the average rate of intake, p_i is set to zero if

$$e_i c_i - h_i k_i < 0 \quad \text{or} \quad \frac{e_i}{h_i} < \frac{k_i}{c_i} \tag{4.2}$$

and to one in all other cases. Here k_i is the sum of all terms not involving p_i in the top line of equation 4.1 and c_i is the sum of all terms not involving p_i in the top line. Thus, both k_i and c_i are constant with respect to p_i.

The *prey algorithm* can then be used to determine which prey types should be included in the diet: (i) rank the n prey types from most to least profitable $(e_1/h_1 > e_2/h_2 > \ldots > e_n/h_n)$; (ii) add prey to the diet in order of increasing rank until:

$$\frac{\sum_{i=1}^{j} \lambda_i e_i}{1 + \sum_{i=1}^{j} \lambda_i h_i} > \frac{e_{j+1}}{h_{j+1}} \tag{4.3}$$

(iii) the highest value of j that satisfies equation 4.3 defines the breadth of the diet; (iv) if there is no value of $j < n$ that satisfies this equation, then all prey types should be taken when encountered.

This chain of reasoning leads to three principal predictions of the prey model.

1 *The zero-one rule*: prey types are either always taken upon encounter $(p_i = 1)$ or always ignored $(p_i = 0)$.

2 *Ranking by prey profitability*: prey types are ranked by the ratio of energy gained per attack to the handling time per attack (e_i/h_i), this ratio being called the profitability of a prey type; prey types are added to the diet stepwise, in the order of their ranks according to the prey algorithm (equation 4.3).

3 *Independence of encounter rate*: prey types are included on the basis of their profitability (above), and on the characteristics of prey of higher rank; it may appear odd that the inclusion of a prey type does not depend upon its own encounter rate.

Recall that the prey model asks whether a prey type should be attacked after it has been encountered. No opportunity can be lost by attacking an item of the highest possible rank. For a low-ranking item, however, the lost opportunity is the expected gain from searching for and eating a higher ranked item. If this opportunity loss exceeds the immediate gain from the attack, then it never pays to eat the low-ranking type, no matter how frequently it is encountered.

The prey model is sometimes used to claim that if the profitability of prey type X is greater than the profitability of prey type Y, then type X should be preferred to type Y. What this means, however, is that X should be preferred to Y when both types are offered *simultaneously*. In the prey model, types are encountered *sequentially*; they are not alternatives, and this notion of preference does not apply. Item Y should either be taken every time it is encountered, or not taken at all.

There have been several experimental tests of the prey model (see Box 4.2), but the prediction that has attracted the most controversy is the zero−one rule (Gray, 1987). The problem arises because, even under controlled laboratory conditions, animals always take a small

Box 4.2 Assumptions and predictions of classical foraging models

The assumptions and predictions of the prey model, the patch model and the central-place foraging model have been subjected to a variety of tests, of differing degrees of rigour. The results of these tests are summarized by Stephens and Krebs (1986). The literature is dominated by qualitative tests (64% of cases) rather than by quantitative predictions. Only 13% of tests, however, clearly contradict the predictions. Prey and patch models are about equally well supported, although an exact match to the zero−one rule has never been observed (animals always take a few inferior prey types, now matter how frequent the superior prey). The assumptions least often tested are random sequential encounter in prey models and the nature of the gain function in patch models. This is important, because failing to verify the assumptions renders many of the supposed tests ambiguous.

Table B4.1 Assumptions and predictions of classical foraging models

	Assumptions	Predictions
Prey model	Exclusivity of search and handling Sequential encounters Finite handling time	Preference for more profitable prey Increased selectivity at higher encounter rates Selectivity independent of abundance of lower ranking prey Threshold for dropping prey types from diet
Patch model	Patch quality recognizable Gain function known	Longer patch times in poorer environments Longer patch times with longer travel times All patches reduced to similar marginal value More resources extracted from better patches
Single-prey loader Central-place model	Availability of prey of different sizes similar at all sites	Bigger prey are brought from greater distances Prey classes should be dropped from diet at threshold distances

proportion of inferior prey (types that the zero−one rule predicts they should have ignored). When the encounter rate with profitable prey types λ is greater than some threshold rate $\hat{\lambda}$, then the rule predicts that none of the inferior types should be taken (Fig. 4.1a). What happens under experimental conditions (e.g. Krebs *et al.*, 1977) is that instead of an abrupt step function, a smoother sigmoid function is observed

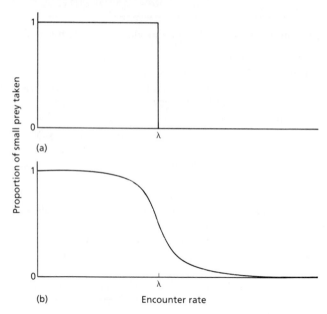

Fig. 4.1 Proportion of small prey taken as a function of encounter rate (λ).
(a) Inferior prey should either be taken each time they are encountered or not at all,
depending upon λ, the encounter rate with more profitable prey. (b) In practice,
some inferior prey are taken even when the encounter rate with more profitable
prey is relatively high, so the step function is blurred.

(Fig. 4.1b), with some inferior prey taken even when the encounter rate
with superior prey is greater than $\hat{\lambda}$. Stephens and Krebs (1986) argue
that the reason for the sigmoid rather than step function is that the
threshold value $\hat{\lambda}$ is not constant, but fluctuates in ways beyond the
experimenter's control (examples of why the threshold might fluctuate
are given by Krebs & McCleery, 1984). There are a number of other
explanations (for review see McNamara & Houston, 1987).

The patch model

The patch model asks how long a forager should hunt in the patch it
has just encountered. Therefore, the patch question is: *how long to
stay before leaving?*

The model presented here is the *marginal value theorem* (Charnov,
1976). Assume that there are n patch types, and that travelling between
patches costs s per unit time. Each patch is characterized by three
quantities:

λ_i, the encounter rate with patches of type i;

t_i, the time spent hunting in patches of type i, known as the *patch
residence time* (this is the decision variable);

$g_i(t_i)$, the *gain function* for patches of type i, specifying the expected
net energy gain from a patch of type i if t_i units of time are spent
hunting within it.

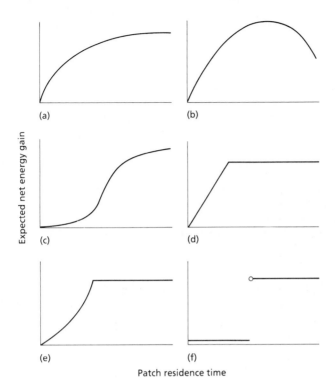

Fig. 4.2 The gain function shows the expected net energy gain from a patch of a given type for a given patch residence time. (a) Smooth asymptotic gain function; (b) humped gain function; (c) sigmoid gain function; (d) linear gain function with asymptotic cut-off; (e) accelerating gain function with asymptotic cut-off; (f) stepped gain function. Humped gain functions like (b) are more likely if patch depletion is important.

The gain function is assumed to be a well-defined, continuous, deterministic function (Fig. 4.2 shows six hypothetical gain functions). We assume further that: (i) the net energy gain when zero time is spent in a patch is zero; (ii) the function increases, at least initially; and (iii) the slope of the function eventually declines. If patch depletion is important, then humped functions like Fig. 4.2b may be more realistic than asymptotic functions like Fig. 4.2a, because, as the patch becomes depleted, more energy will be expended than gained.

Charnov's marginal value theorem states that a rate-maximizing forager will choose the patch residence time for each patch type so that the marginal rate of gain at the time of leaving is equal to the long-term average rate of energy intake in the habitat (Fig. 4.3). The phrase 'marginal rate' comes from economics and translates as *slope* or *derivative*.

There are three results of this.

1 The marginal rate at leaving must be the same in all patches visited.

2 If the habitat becomes poorer, in such a way that the average rate of

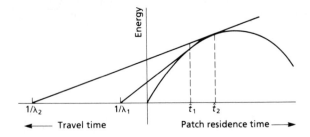

Fig. 4.3 Charnov's marginal value theorem. A rate-maximizing forager will choose a patch residence time (t) for each patch such that the marginal rate of gain at the time of leaving is equal to the long-term average rate of energy intake for the habitat as a whole. Patch residence time increases with increasing travel time between patches. The slope of the line is the long-term average rate of energy intake, because $1/\lambda$, the reciprocal of the encounter rate, is the average time required to travel between patches.

energy intake decreases without affecting the gain function of patches of type i, then a rate-maximizer will stay longer in patches of type i.
3 If the habitat becomes richer, the forager will leave patches earlier. Habitats may become poorer without affecting the gain function of a given type of patch, either because travel costs between patches increase (e.g. they become further apart), or because the set of patch types visited is reduced.

A considerable literature has built up on the question of how natural enemies might use simple rules of thumb in order to deal with decisions on when to leave patches (Green, 1980). The most frequently investigated rules are: (i) the *number rule* (leave after catching n prey); (ii) the *time rule* (leave after t seconds); (iii) the *giving-up-time* rule (leave after g seconds of unsuccessful search have elapsed). An important question, therefore, is how closely these rules of thumb can approximate the *rate rule* (leave when the instantaneous intake rate drops to a critical value). The rule which works best depends upon both the environment and upon the gain function. With a single patch type and a gain function that rises steeply then levels off, the number rule is best, because the other rules involve staying in the patch during the less profitable part of the gain function. With gradual resource depression, however, the giving-up-time rule works best.

Patches will often vary in quality in ways which cannot be recognized before visiting them. Under these circumstances, the forager may indulge in *patch sampling* (Iwasa, 1981; Green, 1984). When the number of prey per patch has a high variance, then the giving-up-time rule works best. But the best solution depends upon the distribution of patch quality, because this influences the information value of a prey capture. With a fixed number of prey per patch, each capture 'tells' the forager that the patch is getting worse, and it should leave. But if the variance is high, a capture 'tells' the forager it may have hit the jackpot, and should stay around. If prey are randomly distributed over patches, then a capture gives no information at all about relative patch quality. For an experimental test of patch sampling see Lima (1983).

Exploiting non-patchy environments

A patch is an abstraction for the convenience of modelling. In some natural situations it is easy to visualize the way in which a predator or parasitoid might encounter patchy prey, but in others the nature of patchiness is less obvious. Arditi and d'Acorogna (1988) analyse the rate-maximizing foraging strategy of an animal in an environment in which there are spatial variations in prey density but no discrete patches. They show that in this kind of environment the animal should travel through areas of low prey density and exploit only high density places. Thus, even when the world is not made up of discrete patches, a rate-maximizing forager should divide its behaviour into travel and exploitation as though the world was patchy.

The ideal free distribution

The ideal free distribution describes the way in which animals compete for patchily distributed resources (Fretwell & Lucas, 1970; Parker, 1970); these resources may be food, nest sites or mates. Fretwell and Lucas assumed that animals were *free* to go wherever they wanted to, and were *ideal* in that they had a perfect knowledge of the quality of all the different habitats. Their theorem states that where there are several habitats, the distribution of individuals over habitats should be such that each individual has the same fitness prospects. Thus, in the absence of competition, all individuals should occupy the single habitat in which fitness is highest (the 'best' habitat). As the number of individuals increases to the point where exploitation competition within the best habitat begins to reduce the fitness of animals foraging there, then the next best habitat should be occupied. The number of animals occupying the second best habitat should be just sufficient that the fitness of individuals is equal in the two habitats, and no animal can *improve* its fitness by moving to another habitat. If density continues to increase, then the third best habitat should be occupied, and so on.

In the context of predators foraging for patchily distributed prey, the predictions of the simplest version of the ideal free distribution are as follows.

1 The number of habitats occupied increases with population density.
2 When the ideal free distribution has been attained, there is no increase in fitness to be gained by changing habitats.
3 That mean fitness of animals in the best patch should decline progressively as the number of inferior patches occupied increases.

A number of experimental tests of the ideal free distribution have confirmed some of these predictions (for review, see Milinski & Parker, 1991). For example, Milinski (1979) fed six stickleback fish at two ends of a tank. When one end of the tank received five times the resource supply rate of the other, then, on average, one fish was observed at the poor end of the tank and five at the good end. When the resource input was changed, so that the good end of the tank received twice the resource input of the poor end, the fish redistributed accordingly, with four feeding at the good end, and two at the poor end. The theorem assumes that all animals have the *same competitive ability*, and while average feeding rates were the same at both ends of the tank, it was clear that certain individual fish did consistently better than others.

Sutherland and Parker (1985) developed a model in which animals were assumed to differ in their ability to compete for food. The probability that a given animal obtains food in a given patch is equal to its competitive ability, divided by the sum of the competitive abilities of all the other animals in the patch. At equilibrium, it is assumed that no animal should be able to improve its fitness by moving to another patch. Sutherland and Parker found that with unequal competitive abilities, there was no longer a unique distribution (as in the ideal free), but a variety of possible distributions. For example, assume that there are two patches, a good patch providing 12 prey per unit time, and a poor patch providing six prey per unit time. Now take 12 animals, six good competitors and six poor competitors, and allow that the good competitors are twice as good as the poor competitors. The set of possible equilibrium distributions is shown in Fig. 4.4; in one case, all the good competitors might be in the good patch and all the poor competitors in the poor patch (no individual can improve its fitness by moving). Note that distribution (c), with four good competitors and four poor competitors in the good patch, and two good competitors and two poor competitors in the poor patch, matches the predictions of the ideal free distributions (i.e. twice as many animals in the patch with twice the resource supply). But in this case, the animals exhibit strikingly different competitive abilities. Sutherland and Parker make the intriguing suggestion that this kind of apparent conformity to the ideal free distribution may be only a matter of chance. Since there is only one way that distribution (a) can come about, but 225 ways that distribution (c) might arise, then distributions that look like the ideal free would result, even though its central assumption of equal competitive ability was wrong. Further work, however, has shown that this result does not hold when the number of individuals is large. Houston *et al.* (1988) found that with unequal competitive abilities, the most likely distribution would have a smaller proportion of animals in the best patch than predicted on the basis of resource supply.

Studies of animals in simple laboratory environments show that the ideal free distribution (or the modification for unequal competitors) is often observed. This raises two kinds of question: (i) how do individual decisions produce a population distribution in which fitness is equalized across patches; and (ii) what are the consequences for density-dependence of their prey if predators form ideal free distributions?

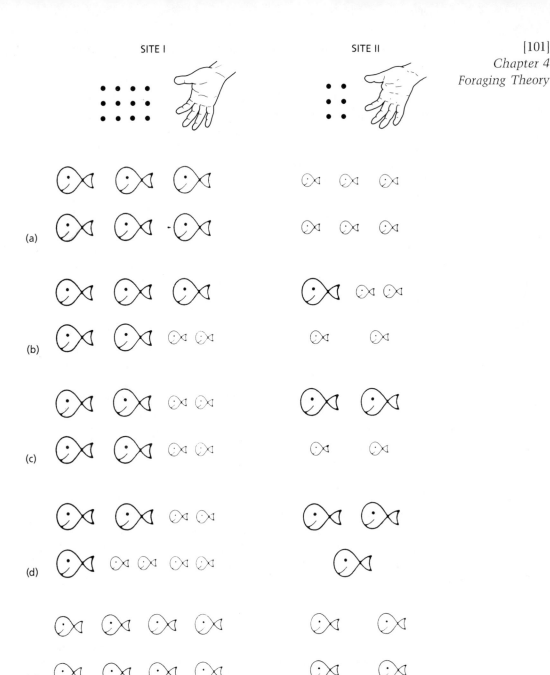

Fig. 4.4 The range of possible solutions for individuals differing in competitive ability. For this example food is added at rates of 6 and 12 items per unit time in the two sites. In (a), (b), (c) and (d) the 12 fish differ in their competitive abilities such that each of the six big fish always capture twice as much as each of the six small fish; in (e) all 12 fish are equally competitive. Note that (i) in none of the cases could any individual gain by moving; (ii) the sums of competitive abilities always equals 12 : 6; (iii) the number of individuals and the mean intake in each site differs considerably between the different solutions. From Sutherland & Parker (1985).

Bernstein *et al.* (1988, 1991) examined the first of these questions in a simulation model in which individual predators or parasitoids searched for prey (hosts) in a patchy environment. The model showed that if individuals use a rule akin to that of the marginal value theorem (i.e. leave the current patch when expected gain is higher from moving), and if the expected gain is estimated from experience, then the population can end up in an ideal free distribution. Three factors influenced the likelihood of observing an ideal free equilibrium.

1 *The rate of depletion of prey in relation to speed of learning.* If prey deplete rapidly, the predators' estimate of expected gain lags behind the true value and this results in a distribution far from the ideal free.

2 *The cost of travel between patches.* If the cost of travel is high (as it might be for many relatively immobile parasitoids), individuals do not usually benefit from leaving their current patch, since the travel cost is subtracted from the expected gain resulting from moving. Therefore the predators do not move around much and fail to equilibrate near the ideal free.

3 *Environmental 'grain'.* Bernstein *et al.* (1991) showed that the 'grain' of the environment can have an important effect. If the grain, by which is meant the spatial change in average prey density per patch, is coarse relative to the distance moved by individual predators, then individuals do not experience a representative sample of patch qualities and hence do not adjust their distribution to the ideal free.

The general point here is that the relationship between behavioural properties (such as search speed and learning) of searching predators and the structure of the environment may play a key role in determining how populations of parasitoids/predators distribute themselves across patches. This may account for the great variety of patterns observed in nature (Lessells, 1985; Stiling, 1987; Walde & Murdoch, 1988).

The second question referred to above, whether or not ideal free distributed consumers exert spatial density-dependence on their prey, has been considered by Sutherland (1983) and Bernstein *et al.* (1991). Two points are worth noting. First, the simple laboratory experiments such as those of Milinski do not show density-dependence because all prey are consumed as soon as they appear in a patch, regardless of the number of consumers. Second, and more generally, whether or not ideal free distributed predators show density-dependence is largely related to the way in which they interfere with one another. If the interference coefficient, a measure of the extent to which an additional searching individual causes a decrease in average search efficiency, is less than 1, the ideal free distribution will be associated with density-dependence (Sutherland, 1983).

McNamara and Houston (1990c) develop a rather different model of the ideal free distribution. They assume that consumers maximize fitness rather than average net rate of intake, a fitness surrogate. They also allow for the possibility that patches vary in danger of predation as well as in terms of rate of intake. Finally, they make the realistic assumption that choice of feeding patch depends upon the consumer's state (e.g. hunger). This model has some interesting consequences. For example, imagine that there are two patches, one safe but with a low rate of intake and one dangerous from the point of view of predation, but with a high intake rate. Only animals that are close to starvation go to the dangerous patch, as a consequence of which most of the mortality occurs in this patch but individuals have a higher rate of intake there, even though the patch may be relatively rarely used.

This line of reasoning may have important consequences for conservation. Patches of habitat that are rarely used by a species may nevertheless be of critical importance for survival if they are places where animals only go when close to starvation, in order to benefit from a high intake rate. Thus the usual conservationist's practice of assuming that the only important habitat patches are those where most of the individuals are seen for most of the time may be quite wrong.

Central place foraging

Consider the problem faced by a bird feeding its young by foraging from a nest. The parent must return to the nest time and again, and must make decisions on how patches should be used, and at what distances from the *central place* different prey types should be attacked. Orians and Pearson (1979) investigated two types of system: the single-prey loader and the multiple-prey loader. The *single-prey loader* exploits patches with discontinuous gain functions, and it must search within the patch for a single prey item. If it takes a mean time to encounter of 5 seconds for prey type A and 10 seconds for prey type B, then the expected time to the first encounter *regardless of type* is 3.3 seconds (note that it is *not* the average of the two times, 7.5 seconds). Because of this, the patch-use tactic 'be unselective' will often be the rate-maximizing choice. Nevertheless, size selectivity increases with distance, and large prey are likely to be preferred in distant patches because travel costs are greater. In general, for short travel times, superiority of prey hinges on energy per unit handling time, but for long travel times, superior prey are those of higher energy, regardless of handling time (Fig. 4.5) (Krebs & Avery, 1985).

Some birds can carry many prey items at once. Examples of multiple-prey loaders include swallows, which catch numerous insects in a single flight, and starlings, which may line up several leatherjackets in their beaks. The marginal loading rate is bound to decrease, becase the load itself hinders further loading. Thus, a beak full of prey reduces a starling's ability to find and collect soil invertebrates, and this effect is cumulative with load size (Tinbergen, 1981). For multiple-prey loaders, patch residence time determines load size, and the marginal value theorem can be applied. The chief prediction is that increasing average distance from the central place should be matched by increasing load size.

Kacelnik (1984) tested this prediction quantitatively by training

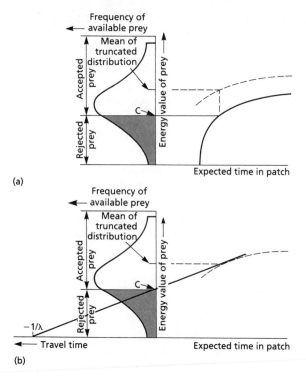

Fig. 4.5 Size selectivity in single-prey loading predators. Size selectivity increases with distance and large prey are likely to be preferred in distant patches because travel costs are greater. (a) The solid curve shows the relationship between the minimum acceptable energy value (C) and the mean patch residence time. The dashed curve shows the mean energy value. (b) Note that the tangent intercepts the energy axis at C.

starlings in the field to collect food (mealworms) from a feeder and deliver the items to their young. By controlling the time interval between successive mealworms, Kacelnik was able to present the birds with an identical loading curve at different distances from the nest. As predicted, load size per visit increased with round-trip travel time. As a further refinement, Kacelnik was able to compare the quantitative predictions of several different versions of the marginal value theorem based on slightly different currencies. He found that the quantitative fit to the model based on maximizing net gain to the family was better than the fit to versions based on maximizing gross delivery rate or ratio of gain:cost.

In further experiments, Cuthill and Kacelnik (1991) pointed out that even apparently convincing 'test' of the marginal value model may be subject to other interpretations. When starlings were presented with a linear loading curve instead of one with gradual depletion, they still showed a load-size–distance effect, even though the simplest interpretation of the marginal value model would predict no such effect (Fig. 4.6). These new results do not necessarily invalidate Kacelnik's earlier conclusion, because the quantitative pattern he observed differs for linear and gradually depleting loading curves, but they do show how the interpretation of qualitative trends must be treated with caution.

TESTS OF THE MODELS

Up to this point, we have illustrated the models with a small number of examples, but a wide range of studies, both in the laboratory and in the field, have been inspired by the conceptual framework of classical optimal foraging theory. These are reviewed in Stephens and Krebs (1986), Commons *et al.* (1987), Kamil *et al.* (1987), Shettleworth (1988) and Hughes (1990). The overall picture that emerges is one in which the qualitative predictions of rate-maximization models are frequently substantiated (see Box 4.2) although it is equally clear that there are situations in which total maximization does not apply (see below).

The situation with regard to quantitative predictions is more difficult to evaluate, partly because rather few studies have carefully evaluated all the assumptions of the models and ensured that they are met, and partly because there are disagreements about what constitutes a quantitative fit. For example, Gray (1987) argues that the failure to observe rigid adherence to the zero−one rule of prey choice shows that there is no quantitative support for the prey model. On the other hand, Krebs and Avery (1985) suggested that the hypothesis should be analysed by an appropriate statistical technique such as probit analysis.

Rather than continue the debate on whether or not 'the evidence supports classical foraging theory', we prefer to stress two points.

1 The widespread qualitative fit to rate maximization shows that even those very simple models can be a useful device for organizing data and predicting behaviour. The task at hand now is to place this

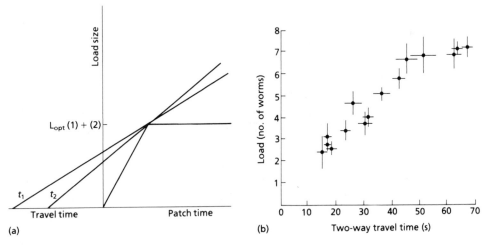

Fig. 4.6 Load size in multiple-prey loaders as a function of two-way travel time: (a) in theory; and (b) in an experiment with starlings. The predicted asymptotic load size in (a) was not observed in the experiment (b). See text for details.

behavioural information into an ecological context by analysing the consequences of rate-maximizing behaviour by individuals for populations and communities.

2 This point reflects an alternative line of development in studies of foraging. The failure to observe quantitative fits to simple foraging models is in itself a powerful heuristic, since it leads to an investigation of behavioural mechanisms that underlie prey or patch decisions. Often a better understanding of the mechanisms of choice may help to explain deviations (see, for example, Rechten *et al.*, 1983). Thus we see foraging theory giving rise on the one hand to ecological insights at the population and community levels, and on the other, as a tool for the investigation of mechanisms of behaviour.

State-dependent foraging models

The classical foraging models simplify the analysis of decision making by including a number of unrealistic assumptions. They do not consider how an animal might trade off alternative activities, such as foraging and vigilance; they do not analyse how the animal's internal state (e.g. hunger in the case of a predator, or egg reserves in the case of a parasitoid) might influence decision making; and they assume that the environment is represented by mean rates, rather than by means and variances. None of these omissions represent flaws in optimal foraging theory in general, since alternative models based on the logic of optimality are available to encompass each (or all) of the complications listed above.

The advantage of the simple rate-maximizing models is their analytical tractability and testability. The price they pay for this is to make what may be unrealistic simplifications.

RISK SENSITIVITY

One approach to incorporate internal state into foraging models is that suggested by Caraco *et al.* (1980). These authors offered yellow-eyed juncos (a small, seed-feeding passerine bird) choices between two sources of food: one offered a fixed amount (say two seeds) while the other offered a variable amount with the same average (say 0 or four seeds, each with probability 0.5). They observed the following striking result: when the birds were deprived of food for 1 hour before testing, they chose the fixed reward, but after a 4-hour deprivation, the variable reward was chosen. Two points are worth noting: (i) if the birds responded only to the mean rate, they should have been indifferent; and (ii) foraging decisions in this experiment *were* influenced by state.

The generally accepted interpretation of Caraco's result is as follows. When the birds were starved for 1 hour they were in *positive expected energy budget*; that is to say, by choosing the certainty of two seeds at each decision, they could meet their daily energy requirements. After the 4-hour deprivation, the birds were in *negative expected energy budget*, meaning that a fixed reward of two seeds per trial was not enough to meet their daily need. In this situation, the overall probability of survival is maximized by choosing the uncertain option (in the hope of getting lucky). If this yields the larger amount (four seeds) the bird is much more likely to survive than it would if it chose the fixed option. If the variable option yields no seeds, the bird's survival probability is not much lower than it would have been with the fixed option. So overall survival probability is higher as a result of choosing the variable outcome. When the bird is in positive expected energy budget, the fixed option offers near certainty of survival while the variable option may result in starvation. To summarize, 'risk-prone' behaviour (choosing the uncertain outcome) pays when 'risk-averse' behaviour would not meet daily requirements.

Caraco *et al.*'s (1980) original experiment had certain methodological problems (Krebs & Kacelnik, 1991) but subsequent studies have confirmed that it is possible to induce a switch from risk-averse to risk-prone behaviour by manipulating the animal's state. Two particularly striking results were those of Caraco *et al.* (1990) who manipulated expected energy budgets of juncos by altering the ambient temperature; the birds were risk-prone at 1°C and risk-averse at 19°C. Cartar and Dill (1990) caused worker bumble bees to shift from risk-averse to risk-prone behaviour by depleting the sugar reserves of the nest.

STOCHASTIC DYNAMIC PROGRAMMING

Models of risk sensitivity differ from classical rate maximization in terms of the currency (minimize risk of starvation instead of maximize long-term average rate of food intake), in terms of incorporating stochasticity and in terms of the fact that choices are state-dependent. In dealing with state-dependence, however, the risk sensitivity approach does not allow for feedback between choices, outcomes and state. Suppose, for example, that an animal in negative expected energy budget chooses a risky option, as a result of which it gains a high payoff and moves into positive energy budget. Its behaviour should now switch back to risk-averse. This illustrates a principle that is true of any model of foraging (or other) decisions; behaviour is affected by state, and in turn influences state, in a feedback loop. One technique that has grown in popularity over recent years is stochastic dynamic programming (SDP) (Houston *et al.*, 1988; Mangel & Clark, 1988). SDP models enable us to analyse state-dependent decisions in terms of their fitness consequences in a way that allows for dynamic changes in state. In brief, the technique is as follows.

1 The consequences for fitness are modelled over a fixed time period, e.g. a day, a season or a year. The relationship between state at the end of this period, and lifetime fitness is specified by a *terminal reward function*.

2   The time period is divided into discrete intervals. In each interval, the animal chooses between its behavioural options, and each option has fitness consequences, either directly (e.g. through danger of predation) or indirectly through affecting state. State could be represented by more than one variable, although the computations become cumbersome with more than two state variables.

3   The SDP algorithm allows one to calculate the choices as a function of state, through a series of time intervals, to the end of the time period that will maximize the terminal reward and hence lifetime fitness.

4   This trajectory is referred to as the *optimal policy*. In order to translate the optimal policy into predictions about decisions, it is necessary to calculate from the model which state the animal will be in at each time stage, if it follows the optimal policy.

APPLICATIONS OF SDP

Stochastic dynamic programming is a relatively new theoretical development and has not yet attracted a large body of experimental work in the way that simpler, rate-maximizing models have done. This reflects part of the choice of strategy of modelling; sometimes simpler 'incorrect' models may be a more powerful stimulus for empirical studies than structurally more complex 'realistic' models. Nevertheless, SDP models are able to account for many phenomena not explained by rate maximization, and empirical work based on this approach is expected to develop further over the coming years.

Two examples of the application of SDP models are to daily routines of small birds, and oviposition decisions of insect parasitoids.

Daily routines

SDP models have been successful in generating predictions and/or functional explanations of daily routines. One example is the work of McNamara *et al.* (1987) who modelled the dawn chorus of songbirds and produced a novel explanation of this widespread phenomenon. The observation to be explained is that many songbirds show a peak of singing activity in the early morning. At first sight this seems paradoxical, because it might be thought that birds would wake up with a deficit in body reserves, and need to forage intensively rather than sing just after dawn. Previous functional explanations for the dawn chorus in terms of ecological factors have suggested daily variations in sound transmission (Henwood & Fabrik, 1979), food availability (Kacelnik & Krebs, 1983) or intruder pressure in the case of territorial song (Kacelnik & Krebs, 1983) as being important. McNamara *et al.* (1987), in their SDP model, show that the dawn chorus could arise without *any* circadian variation in any ecological factors other than the day–night cycle, simply as a consequence of the way in which birds regulate their body reserves.

The essence of their argument is that small birds, rather than maximizing the body reserves they carry, optimize their reserves to reflect a tradeoff between the advantage of high reserves to counter starvation and the cost of carrying high energy reserves around (King, 1972; Lima, 1986; McNamara & Houston, 1990a,b). The cost of carrying body reserves may arise as follows. If metabolism is mass-dependent, extra reserves result in additional energy expenditure and therefore additional time spent foraging to replace lost energy and, if foraging is more dangerous in terms of predation risk than, say, resting, then maintaining extra reserves will result in a higher death rate from predation. A second way in which extra reserves might be disadvantageous is if they actually slow the animal down, so that it is more vulnerable to predation because its escape speed is reduced (this was not included in the model by McNamara *et al.*). Evidence that birds do not maximize their body reserves comes from observations of both seasonal and daily patterns of change in body mass. On a seasonal basis, many birds reach their highest mass on days when food availability might be expected to be lowest (e.g. extremely cold days in winter). This means that on other days, when food is more plentiful, the birds cannot be maximizing their body reserves. With regard to daily patterns, most small birds lose about 10−15% of their body weight overnight and do not fully regain these reserves until late in the afternoon, even if supplied with food *ad libitum*, implying that for much of the day they maintain reserves below the maximum level.

How does regulation of body reserves explain the dawn chorus? McNamara *et al.* suggest the explanation as follows. If overnight expenditure of energy is entirely predictable, birds should put on just enough reserves each afternoon to survive the night. If, however, overnight expenditure is unpredictable, the bird will have to put on more reserves than the average needed to survive the night, in order to survive on nights when expenditure is abnormally high. This leads to the paradoxical situation that on most mornings the birds start the day with an excess of body reserves. In their model, singing was a major alternative to foraging, so that when the bird had more than the optimal level of reserves, it would stop foraging and sing instead. Thus the consequence of optimal regulation of body reserves and unpredictable overnight expenditure is a peak of singing activity in the early morning. Consistent with this hypothesis, a number of studies (Garson & Hunter, 1979; Reid, 1987) have shown that the amount of singing in the dawn chorus is, needed, related to overnight temperature. Whilst McNamara *et al.* would not claim that their model is a complete explanation of the dawn chorus, it does make the important general point that daily routines in behaviour, such as the dawn chorus, could be generated without any variation through the course of the day in ecological factors that might influence the costs and benefits of different activities.

Oviposition decisions of insect parasitoids

A second example of the application of SDP refers to the oviposition behaviour of insect parasitoids (Mangel & Clark, 1988). In many insect parasitoids, the female lays her eggs in a host and the larvae develop inside the host, eating it as they grow (see Chapter 11). The female exerts control over the number of eggs laid on a single host ('clutch size'). It is easy to see why the fitness gain to a female, measured as number of surviving offspring to emerge from the host, might increase with clutch size at first, then level off, and finally decrease with increasing clutch size. In the increasing part of the fitness curve, resources inside the host are under-exploited so a larger clutch results in more emerging offspring but, as resources inside the host become limiting, a further increase in clutch size first results in no increase in number of surviving offspring, and finally results in a decrease in the number of surviving offspring as all resources are used up but no offspring obtain enough food to survive. This kind of *scramble competition* between larval parasitoids is discussed in detail in Chapter 11.

This means that, within a single host, there is a clutch size c^* that maximizes the number of offspring surviving. The value of c^* depends upon host size, and is bigger for larger hosts. Charnov and Skinner (1984) and Skinner (1985) noted that in the field the parasitoid *Nasonia vitripennis* lays clutches that are smaller than c^*. They suggest that this result can be explained by the marginal value theorem. Referring back to Fig. 4.3, if the gain curve can be taken to represent the left hand (increase) part of the relationship between clutch size (on the y axis) and fitness (x axis) within a host, then the rate of maximizing clutch size can be found by the tangent method. It will always be smaller than c^*. Thus, if a female maximizes her rate of production of offspring, rather than production per host, she will always lay a clutch that is smaller than the single host maximum.

Mangel and Clark (1988) show how the same result can be explained in terms of a SDP model. This model also makes additional predictions and explains additional features of the data. In the model, each female has a limited number of eggs (the egg reserve), a certain probability of survival between leaving one host and finding another, and a fixed time horizon in which to work (say the end of the day, or the end of the period during which susceptible hosts are present). This model makes three predictions.

1 *Time should affect choice.* As the female approaches the end of the fixed time period, she should be both less selective about hosts and lay larger clutches within each host. Take, for example, the last host to be encountered. The female parasitoid should lay as many eggs as possible, but not more than c^*, on the last host because there is no cost in terms of further opportunities to increase fitness. Similarly, a host which might not be worth parasitizing early in the time period is always worth parasitizing if it is the last host encountered (Lucas, 1987, presents a similar argument about prey choice is foraging).

2 *Optimal clutch size depends upon the number of eggs left in the female.* The larger the egg reserve, the smaller the opportunity cost of laying a large clutch in the present host. Therefore, there should be a relationship between clutch size and egg reserves for a given aged female on a given host individual.

3 *Future oviposition gains have to be discounted by the probability of survival.* If the probability of dying before finding the next host is high, then females should lay larger clutches for a given host size and number of egg reserves.

These predictions have not been tested, but they could contribute to an explanation of the variability in clutch size observed by Charnov and Skinner (1984). Other subtleties of female parasitoid behaviour, as they relate to determining the sex ratio of eggs laid, and to the outcome of competition between the larvae emerging from different clutches within the same host, are discussed in Chapter 11.

Tradeoffs between foraging and danger

There is ample evidence that animals may sacrifice rate of food intake to avoid the danger of predation. This may take the form of avoiding patches where danger is high (see, for example, Newman & Caraco, 1987) or altering behaviour within a patch to increase vigilance at the expense of intake rate when perceived danger of predation is high (see, for example, Krebs, 1980; Newman *et al.*, 1988). In some studies, the tradeoff has been shown to be state-dependent. For example, Milinski and Heller (1978) demonstrated that sticklebacks (*Gasterosteus aculeatus*) will choose to forage in a high density patch of *Daphnia* when hungry and in lower density patches when less hungry. In high density patches the intake rate is high, but so is the danger of predation because the fish, in order to feed successfully on high density swarms of *Daphnia*, must reduce their vigilance.

Attempts to quantify the tradeoff between foraging and danger have been of two sorts: descriptive and predictive (normative). Abrahams and Dill (1989) adopted a descriptive approach, by measuring the amount of food that the forager will sacrifice in order to avoid a predator. This is an estimate of the 'energy equivalent' of danger. Abrahams and Dill then used this value to predict how much extra food would have to be provided in a dangerous place to persuade the foragers to hunt there. Gilliam and Fraser (1987) used a predictive approach. Their model assumed that the forager had to meet a certain daily energy requirement, and that it did so by choosing patches that minimized mortality due to predation whilst meeting the minimum requirement. In experimental work with juvenile creek chub (*Semotilus atromaculatus*), the fish were presented with a choice of two ends of a stream tank with different combinations of food and danger (predators of the juvenile chub). In all six experimental treatments, the chub chose the patch predicted by the model.

LEARNING AND FORAGING

Foraging animals are frequently observed to adjust their strategy in relation to past experience. For example, in Kacelnik's (1984) study (see p. 104), starlings adjusted their patch residence time in relation to travel time. Similarly, redshank (*Tringa totanus*) feeding on the worm *Nereis diversicolor* adjust their diet in relation to encounter rate with more profitable size classes (Goss-Custard, 1977). This kind of observation suggests that foraging decisions might be based in part on learning. Some studies have attempted to investigate exactly which aspects of past experience influence foraging choices (reviewed in Commons *et al.*, 1987). For example, Cuthill *et al.* (1991) carried out an experiment in which starlings were exposed to different travel times between patches within a single foraging bout. Within a patch, the birds experienced gradual depletion, so the environment was one which could be modelled by the marginal value theorem. Cuthill and colleagues used patch residence time as the dependent variable and asked how it was influenced by recently experienced travel time. They concluded that the starling's response was only to the last one or two travel experiences, and that patch residence times were adjusted as if the birds had a very short-term memory.

The problem can be turned the other way round by asking how much an animal ought to remember to maximize its intake rate. As Cuthill *et al.* show, the answer to this question for their experiment is complex, and it depends upon detailed assumptions about the way in which the environment changes with time.

CRITIQUE

In addition to serving as a general background against which to organize observations about individual behaviour, foraging models have been used to ask a variety of questions.
1 How good are organisms at doing their jobs?
2 What are animals designed to do?
3 What constraint assumptions account for behavioural mechanisms?
4 How does behaviour affect the organization of populations and communities?
In principle, any model can be investigated by examining its assumptions, its predictions, or both. Foraging models make assumptions of two kinds. Some are part of the general background, e.g. that net rate of energy gain is related to fitness, or that natural selection optimizes design. Others are specific to the model under consideration, e.g. the

incompatibility of search and handling, or the importance of within-patch resource depression. Background assumptions are not usually tested, and may often be untestable (they may gain or lose credibility with the successes or failures of models based upon them). Model-specific assumptions can usually be tested directly by observation or experiment. However, when a model fails to fit real data one must know whether specific assumptions have been met, in order to determine what may have caused the discrepancy between predicted and observed outcomes. Too often, results that appear to run counter to the predictions of optimal foraging models may have done so simply because the assumptions of the model were not met.

Stephens and Krebs (1986) suggest a series of questions that should be asked before testing a given foraging model.

1 *Are the foragers playing the same game as the model?* For example, if the model assumes successive encounters, patches with resource depletion and incompatibility of search and handling, then make sure that these conditions apply to the foragers chosen to test the model.

2 *Are the assumptions of the model met?* For example, when testing a central-place foraging model, check that prey availability does not vary systematically with distance from the nest.

3 *Are the right variables being measured?* For example, check that the important variables like handling time and the gain function can be measured for the particular system under study.

4 *Is the test merely consistent with the model, or does it rule out alternative possibilities?* For example, is the result consistent with both an optimal foraging model and a null hypothesis of random choice (see Aronson & Givnish, 1983)?

There are a number of widespread misconceptions about testing foraging models. First, optimality is not the hypothesis under test. It is a technique used to work out the testable implications of specific hypotheses about design and constraint. Second, practitioners of foraging theory are accused of *ad-hoc*-ism, and of patching up their models when their predictions fail to be borne out in experimental trials. But *ad-hoc*-ism is not a specific vice of optimality models; science does not progress by the stepwise replacement of bad ideas by good ideas, but by the gradual improvement of inadequate ideas. Third, the notion that optimality hypotheses must be completely abandoned if they fail to match experimental data is absurd. The alternative, of assuming that behavioural traits are selectively neutral, would be worse, since if anything is untestable it is the notion that a trait has absolutely no selective value.

The question of testability does raise real problems, however. Optimality models are composite hypotheses, composed of assumptions about decision, constraint and currency. When an optimality model fails, we do not know which elements of the model are wrong. Again, as Oster and Wilson (1978) point out, 'the essentially innovative nature of the evolutionary process precludes an exhaustive list of strategies'.

Furthermore, design features of animals are not independent. Many of the models covered in this chapter have been concerned with maximizing the rate of food intake, and even the more complex, state-dependent tradeoff models ignore many potential design criteria, such as finding mates, building a nest, keeping dry or finding enough to drink. While it is convenient to assume that these traits are independent, this may often be wrong. On the other hand, there does not appear to be a sensible alternative. The piecemeal approach offers a way to start, and it has not done too badly in practice (Box 4.2). The holistic alternative, of trying to take everything into account, even if it were experimentally practical, would only lead to the 'hopeless position as seeing the whole organism as being adapted to the whole environment' (Lewontin, 1984).

In summary, optimization models are a way of studying the products of selection, namely the design features of animals. By formulating design hypotheses in a quantitative and rigorous way, they help to circumvent many of the criticisms levelled at the adaptationist approach. There are many criticisms of optimization modelling, including its lack of holism and its lack of attention to phylogenetic constraints. These criticisms amount to reasons why optimization models might be wrong, but not why they are bound to be wrong. Design hypotheses are essential features of most biological research, and optimization models seem to be the most explicit and powerful approach to the study of adaptation currently available.

2

POPULATION BIOLOGY OF NATURAL ENEMIES

5: Large Carnivores and their Prey: the Quick and the Dead

T.M. CARO & CLARE D. FITZGIBBON

Moving forward slowly, flashing a torch, I found two eyes glinting back at me. And there in all its splendour sat a leopard, feasting on a female chital ... I realised the rarity of what we had seen, and was mesmerized, sitting deathly still, not even daring to breathe as any movement might make the animal suspicious.

[Thapar, 1986]

INTRODUCTION

Large carnivores are fascinating, dangerous, beautiful and exciting to watch. They come into conflict with people because they often prey upon the large mammals which humans hunt for consumption and recreation. These two concerns have generated an enormous number of studies of carnivores and their prey. This chapter focuses on the most exciting and controversial issues surrounding these species and is divided into four parts. First, a background to the concepts and methods of studying large carnivores is given; then the effects that large terrestrial carnivores have on prey populations are considered. Next we discuss the evolution of cooperative hunting, and finally review antipredator strategies of ungulates, the principal prey of large carnivores.

The review is necessarily selective, and many of the examples are taken from behavioural studies. Broadly, ecological studies yield information on the effects that particular populations have on each other, whereas behavioural studies show how individuals mediate these effects by feeding or avoiding being eaten. Observation of predators and prey is the best available method for determining the direct effects of predation in terms of actual numbers and ages of prey animals removed. In addition, predators may also have far reaching indirect effects on prey populations, because reducing the risk of predation frequently conflicts with maximizing other components of fitness (Sih, 1987). For example, presence of predators may force prey individuals to stay away from areas of high food quality, or maintain high vigilance levels which reduce food intake and may thereby lower growth and fecundity. Detailed observation of predatory events also sheds light on the way predators and prey have shaped each other's ecology and behaviour during evolution (Dawkins & Krebs, 1979). Most of the examples are taken from large terrestrial carnivores, the dogs (Canidae), hyaenas (Hyaenidae), cats (Felidae) and bears (Ursidae).

Hunting techniques and prey

Different families of large terrestrial carnivores have typical hunting techniques. Dogs are adapted to swift and prolonged running that demands great stamina. These coursing predators kill a broad range of prey sizes relative to their own body size, but some species rely extensively on vegetable food as well. The hyaenas are also coursers, killing and scavenging a wide range of different sized prey; they also have extremely strong jaw muscles for crushing bones. Cats are strictly carnivorous. Their hunting method normally involves a slow concealed approach (stalk) followed by a quick dash. Bears are primarily omnivorous feeding on fruit, nuts, tubers and insects but will supplement the diet with items such as fish, ungulate fawns and carrion (Ewer, 1973).

Although these stereotypes may be useful for general comparative purposes, they blur differences in hunting styles that can vary according to conditions (see Taylor, 1989). For example, red foxes *Vulpes vulpes* approach black-headed gulls *Larus rudibundus* very fast on clear nights but stalk them on overcast nights (Kruuk, 1964). The same predator may hunt for different prey species using different methods: thus, red foxes pounce on rodents but they chase rabbits *Oryctolagus cuniculus* (Kruuk, 1964; for review, see Gittleman, 1985).

Moreover, our understanding of hunting techniques is continually being revised in the light of new data. Spotted hyaenas *Crocuta crocuta* were originally thought to be scavengers until, by watching them at night, Kruuk (1972) showed that they obtained most of their kills through hunting (93% in Ngorongoro Crater and 68% in the Serengeti National Park, Tanzania). Grizzly bears, *Ursus arctos*, often considered scavengers, actively hunt moose *Alces alces* calves in east central Alaska and obtain four times as much meat through hunting as through scavenging (Boertje *et al.*, 1988). In general, the most useful descriptions of hunting are quantitative, cover the whole 24-hour period, are reported separately by prey type, hunting group size and situation, and are repeated in different seasons (see, for an unusual example, Van Orsdol, 1984).

What is a hunt?

Casual observers can often agree on when a predator starts to hunt, but such conviction can be misleading. For example, cheetahs *Acinonyx jubatus* usually hunt visible prey such as Thomson's gazelles *Gazella thomsoni* by stalking, trotting, crouching, rushing at or chasing them (Caro 1987a), mothers spend an average of 3% of the day involved in these hunting activities. As a prelude to many hunts, however, mothers locate visible prey by sitting up for long periods to observe their surroundings (Caro, 1987b). Moreover, prey hidden in vegetation, such as hares *Lepus crayshawi* and *L. capensis* and newborn gazelles, are not hunted using a concealed approach but are stumbled upon while walking, and then briefly pursued. So there would be strong justification for using time spent searching (hunting, observing and moving) as an estimate of effort used to acquire prey (in this case an average of 32% of the day). In addition, it is sometimes difficult to decide when a hunt has ended (e.g. if predators rest between bouts of stalking).

Problems have arisen because it is not always clear which aspects of hunting or searching are being reported. Thus, the magnitude of 'hunt' durations, frequencies and success rates cannot be assessed with confidence. Issues such as these become acute when we want to make quantitative comparisons of hunting characteristics across species (especially between families) in order to understand differences in aspects of feeding behaviour such as prey selection (see, for example, Bertram, 1979). Thus, until definitions become more rigorous, it is only sensible to compare quantitative details of hunts between different classes of individuals of the same species.

Estimation of diet

Questions that concern carnivores' effects on prey populations and prey behaviour require knowledge of what prey species are eaten, and their importance in the diet. Different methods are used to determine carnivores' diets depending on habitat and ease of observation. In the tropical rainforest of Manu National Park, Peru where observation is difficult, Emmons (1987) studied diets of jaguars *Panthera onca*, cougars *Felis concolor* and ocelots *Felis pardalis* by analysing their scats. This painstaking work uses a microscope to compare scales and coloration pattern of hairs found in faeces to a reference collection of mammalian hairs; hard parts like bones, teeth and nails are compared with museum specimens. Scat analysis is subject to a number of errors: (i) scats of different predators may be confused (jaguars and cougars here); (ii) appearance of items may be affected by meal size or meal frequency; and (iii) less digestible food is likely to be over-represented. When it is difficult to know how many individuals of one prey species have been consumed, it is virtually impossible to estimate the relative importance of different prey species in the diet as a whole. Though relatively quick and inexpensive, scat analysis has low resolution.

Stomach contents of shot or trapped predators have often been used to obtain diet profiles in management-based studies. Results may be biased because certain individual predators (perhaps poor hunters) are more likely to be shot or trapped, and large prey may be over-represented because of slow gut passage time. Killing predators solely to obtain information on diet is now unacceptable in most countries. Neither scat nor stomach analyses can distinguish between killed or scavenged prey.

A third method of estimating diet is to examine carcasses. Careful detective work can lead to cause of death being assigned to a particular predator. In Serengeti, for example, vulture droppings around a carcass, or an eye removed, suggest death by natural causes; fresh grass in a herbivore's mouth indicates sudden death. Handfuls of plucked-out fur imply a leopard's *Panthera pardus* kill; missing limbs, the presence of spotted hyaenas; haemorrhaged marks around the muzzle, strangulation by lions *Panthera leo* (Bertram, 1978b). Such observations are not fool-proof, of course, and biases can develop because small carcasses are difficult to find, and disappear more quickly than large carcasses (Prins & Iason, 1989). Other methods include tracking predators, which can only be performed on certain substrates, or finding predators opportunistically or using radiotelemetry, but these are often heavily biased towards large prey (Mills, 1991).

Following a terrestrial predator continuously day and night is probably the best method for determining what it eats (see, for example, Packer *et al.*, 1990) but can only be attempted with habituated predators and still has drawbacks: (i) in the presence of observers, prey may become more vigilant and so reduce predators' hunting success and feeding rates; and (ii) prey selection could be influenced if certain age classes or species are made nervous. These effects are very difficult to document. Also, night-time observations are easiest to make when the moon is full, but hunting success in lions was found to be lower in moonlit hours than in moonless hours (Van Orsdol, 1984). Though least biased direct observation is labour intensive and impossible in some habitats.

Predators selecting prey

Classical optimal foraging models assume that animals maximize energy intake per unit time (Chapter 4). Accordingly, three factors are critical in determining selection between two kinds of prey: (i) the relative energy expected from the two sorts of prey; (ii) the time to find them; and (iii) their handling times (the time taken to hunt, subdue and eat them). This can be represented as follows.

$$\frac{1}{\lambda_1} < \frac{E_1 h_2}{E_2} - h_1 \tag{5.1}$$

where the expected time to find prey type 1 $(1/\lambda_1)$ must be less than the ratio of calories in prey 1 (E_1) to those in prey 2 (E_2) multiplied by the handling time of prey 2 minus the handling time of prey 1 (see p. 93). Much work has been directed at determining whether predators do select prey by comparing the relative frequencies of prey types in the diet with their relative frequencies in the environment, and many different measures of preference have been proposed (Chesson, 1978). For example, by recording the relative frequencies of prey species in all recorded kills made by lions, leopards, cheetahs and wild dogs in the

Table 5.1 Prey preference ratings (percentage of kills/percentage abundance) of four species of carnivores in Kruger National Park feeding on the five most common prey species. Scores greater than 1 denote prey taken in greater proportion than its relative abundance. From Pienaar (1969)

	Impala	Buffalo	Zebra	Wildebeest	Kudu
% relative abundance	53.4	8.7	8.1	7.8	2.9
Lion	0.37	1.06	1.98	3.06	3.82
Leopard	1.45	0.01	0.15	0.17	1.00
Cheetah	1.27	0.01	0.23	0.65	2.35
Wild dog	1.63	—	0.02	0.05	1.50

Kruger National Park, South Africa, and the relative frequencies of prey that live there, Pienaar (1969) was able to derive a preference rating, a 'forage ratio', of these two measures (Table 5.1). In this case, the optimal foraging model is somewhat limited because differences in preferences between predator species are likely to be due not only to differences in handling time, but also to the risks of injury in capturing prey. Obviously, it is both difficult and dangerous for the smaller predators such as cheetahs to bring down large buffalo *Syncerus caffer* and zebras *Equus burchelli*.

Attempts have hardly begun to examine selection in terms of maximizing energy intake among large carnivores (see Caro, 1989) because: (i) the body weights of different age classes of prey are poorly documented; (ii) relative abundances of different prey species are not always known; and (iii) species' differences in vulnerability to predation are restricted to anecdotes, e.g. wolves *Canis lupus* switch from moose *Alces alces* to much smaller caribou *Rangifer tarandus* when they are common because they are easier to kill (Bergerud, 1988).

Studies examining prey selectivity need to be conducted over a range of prey abundances because prey availability often changes with season. Table 5.2 shows that wild dogs took wildebeests *Connochaetes taurinus* and zebras chiefly during the first half of the year when these migratory prey were present on Serengeti's short grass plains. Also, some predators can kill immature prey individuals but not adults (so-

Table 5.2 Numbers of individuals of different prey species killed by wild dogs in the Serengeti. From Schaller (1972)

Prey type	January–June	July–December
Wildebeest	75	1
Zebra	11	0
Thomson's gazelle	32	52
Grant's gazelle	9	0
Other	5	13

called 'secondary predators'); they take newborn animals only during the prey's breeding season. Moreover, long-term studies show that proportions of species in the diet may change over a longer time scale. On Isle Royale, Michigan, USA, wolf predation on beaver *Castor canadensis* tripled over a 13-year period corresponding with a decline in moose numbers (Peterson, 1979).

The individual predator as the unit of analysis

Many studies assume that all members of a predator population select a particular profile of prey (see Gittleman, 1985, for references) but diet may differ between individual predators for a variety of reasons.

1 *Sex.* Many large carnivores are sexually dimorphic, and in some species the larger males can take bigger prey than females. Kills made

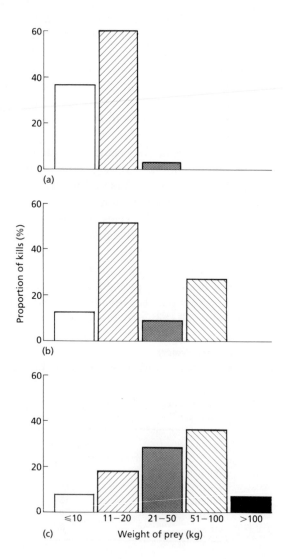

Fig. 5.1 Percentage of kills of different body size made by (a) single cheetah mothers (*n* = 71 kills); (b) single males (*n* = 33 kills); (c) males living in permanent groups of two and three (see Caro and Collins, 1987) (*n* = 39 kills). Prey that was stumbled upon (hares, Thomson's and Grant's gazelle fawns) rather than actively chosen by cheetahs were excluded from the figure.

Table 5.3 Percentage occurrence of major prey in bobcat stomachs from New Hampshire, USA, 1951–1962. White-tailed deer were found significantly less often in the stomachs of juveniles than in those of both other age classes. From Litvaitis et al. (1984)

	Juvenile bobcats (n = 63)		Yearling bobcats (n = 30)		Adult bobcats (n = 36)	
	M	F	M	F	M	F
Average weight of bobcat males (M) & females (F) (kg)	5.2	4.4	11.2	6.6	13.3	7.4
Prey						
White-tailed deer (18 kg)	9.5		13.3		11.1	
Snowshoe hares (1.5–5 kg)	34.9		20.0		19.4	
Cotton-tail rabbits (0.4–2.3 kg)	15.9		23.3		16.7	
Grey squirrels (0.4–0.7 kg)	9.5		13.3		11.1	
Small mammals (<0.03 kg)	22.2		13.3		5.6	

by single male cheetahs were larger than those made by cheetah mothers, whose offspring helped little in hunting (Fig. 5.1a,b).

2 *Reproductive status.* Energetic demands of lactating mothers are great, and in several species anecdotal accounts suggest mothers with dependent cubs hunt for different sorts of prey and at different rates than females without cubs. In cheetahs, pregnant females, lactating females and mothers with cubs of different age each kill adult Thomson's gazelles to differing extents (T.M. Caro, unpublished data; Laurenson, unpublished data).

3 *Age.* Many carnivores need considerable experience to capture prey effectively (Caro, 1980), and independent adolescents rely on different prey than do adults. For instance, Table 5.3 shows that juvenile bobcats consumed significantly less white-tailed deer *Odocoileus virginianus*, that were far larger than other prey, than did yearling or adult bobcats.

4 *Group size.* A number of carnivores hunt in groups (see below), and in many instances groups attempt different species and age–sex classes of prey than do single animals (see Fig. 5.1b,c).

These examples all show that predator population structure may be an important, but as yet unexplored, factor affecting the impact of predators on prey populations.

EFFECTS OF PREDATION ON PREY POPULATIONS

The impact of carnivores on prey

Less than 20 years ago biologists thought that most mammalian herbivore populations were primarily limited by food in the sense that food slowed down the rate of population increase more than any other factor (see, for example, Cowan, 1947). For example, after the disappearance of the exotic disease rinderpest, the Serengeti wildebeest population increased fourfold, but then stabilized at around 1.3 million animals, as a consequence of density-dependent mortality occurring in the dry season when food is least available (Sinclair *et al.*, 1985). However, Keith (1974), questioned the generality of this conclusion by reviewing studies showing that the impact of predation might be of sufficient magnitude to limit prey populations. Since then, several sources of evidence have confirmed that much of the mortality in prey populations can be attributed to predation, and that some prey populations may be limited in this way. In only a very few cases have carnivores been thought to regulate prey populations in the sense that the intensity of predation is positively correlated with prey population density (Murray, 1982, and see Chapter 2).

In northern latitudes of North America, evidence suggests that predators used to keep ungulate populations at low densities before predators were persecuted following the arrival of European people (Bergerud, 1974a). Predators often consume entire carcasses, making it difficult to assess their feeding habits. Using radiotransmitters that respond differently when prey is motionless (dead) however, it has been possible to determine the full extent of wolf, grizzly bear and black bear (*Ursus americanus*) predation. In some areas, approximately 70–85% of caribou, 94% of elk (*Cervus canadensis*) 85–94% of moose and 75% of deer (*Odocoileus heminus*) calf mortality is due to predation (see Bergerud & Elliot, 1986 for references). In eight studies of caribou where adults were radio-tagged, predation was the chief source of adult mortality in all of them. North of the tree line in Canada, wolf numbers have been reduced through mobilized hunting by man, and caribou numbers are increasing. In boreal forests, however, wolves are more difficult to shoot and can additionally take moose; here caribou populations are now in decline (Bergerud, 1988).

Considering moose, predation appears to be a primary factor limiting their density (Van Ballenberghe, 1987). Moose occur at high densities only where they are minor prey of bears and wolves in multiprey ecosystems, where they are preyed upon by just one prey species, or where wolves are absent (Gasaway *et al.*, 1991). Non-insular wolf populations can only sustain themselves when moose outnumber wolves by more than 30 : 1 (Messier & Crete, 1985). In simple Palaearctic ecosystems with few prey and few predators, predation has an enormous impact on prey.

In contrast, in Africa, where predator suites are larger and ungulate prey species more numerous, the impact of predation is tempered though it can still be high (Table 5.4). In Serengeti, just 23–28% of mortality in buffalo and 13–21% in wildebeest was due to predation (see Prins & Iason, 1989). This may be because the large herds of migratory herbivores can move beyond the foraging radius of sedentary carnivores. Indeed, it has been argued that predators could in part be responsible for the 10-fold differences in abundance between resident

Table 5.4 Mortality of large herbivores in natural ecosystems in Africa. From Prins and Iason (1989)

Prey	Main predator	% of mortality accounted for by predation	Region
Buffalo	Lion	89	Manyara
Springbok	Leopard	87	Kalahari
Wildebeest	Lion	96	Timbavati
Zebra	Lion	59–74	Serengeti

and migratory ungulates, because sedentary predators may be unable to show numerical responses to migratory prey but might with resident ungulates (Fryxell *et al.*, 1988). The extensive foraging trips made by predators and the large number of migratory Thomson's gazelles killed by them in this ecosystem argue against this hypothesis however (T.M. Caro, unpublished data).

In northern latitudes, microtine rodent populations show regular 3–4-year cycles with as much as a 40-fold difference between years of maximum and minimum abundance. Many hypotheses have been put forward to explain the causes of cycles (Krebs & Myers, 1974). In southern Sweden, populations of field voles *Microtus agrestis* and wood mice *Apodemus sylvaticus* do not cycle however. Erlinge *et al.* (1983) found that five terrestrial and three avian predators accounted for most of the annual mortality of both rodent species (Table 5.5). Much of the predation was due to generalist predators, primarily red foxes, domestic cats *Felis catus* and common buzzards *Buteo buteo*, that switched to rabbits when rodent populations fell, and thus maintained higher population densities than if they relied on rodents alone. This alternative prey was unavailable in the north. By examining vole predation in different months, the intensity of predation was found to increase in early summer as vole densities increased, in a typical

Table 5.5 Estimated numbers (and percentage) of small rodents taken per year by predators. From Erlinge *et al.* (1983a)

Species	No. produced year^{-1}	No. consumed by facultative predators (fox, cat, badger, tawny owl, common buzzard) year^{-1}	No. consumed by specialist predators (stoat, long-eared owl, kestrel) year^{-1}
Field voles	171 400	120 700 (70.4%)	36 165 (21.1%)
Wood mice[*]	20 100	17 180 (85.5%)	4 366 (21.7%)

[*]Number taken by predators needs to be reduced somewhat to exclude yellow-necked woodmice *Apodemus flavicollis*.

delayed density-dependent fashion. In this study, there is reasonably strong support for the hypothesis that a suite of generalist predators can regulate some of their prey species. As in the example of wolves in boreal forests, predators have a substantial impact on one preferred species, if there is an alternative prey species to buffer them against the impact of a declining prey base.

The classic example of predator–prey dynamics is the snowshoe hare *Lepus americanus* cycle. From trapping returns, it was discovered that both hares and lynx *Lynx canadensis* undergo marked 10-year cycles: hare densities can increase nearly 30-fold between years of least and most abundance. Several hypotheses concerning diet have been put forward to explain the causes of the hare cycles but none have withstood close inspection (Sinclair *et al.* 1988). In the Kluane Lake region of Yukon, Canada, Boutin *et al.* (1986) found that 58% of winter losses in the peak year were due to predation by locating the whereabouts of radio-tagged hares. In Rochester, 92% of winter losses following peak densities were caused by predation (Keith *et al.*, 1984). Predation appears to be the most important source of mortality for hares, and plays a strong role in halting the population increase and driving numbers down. It is not clear, however, whether healthy or starving hares (that were going to die soon anyway) were usually preyed upon. In these cycles, low hare densities eventually have an effect on lynx; their reproduction virtually ceases due to low survival of kittens, but adult lynx can survive until hare densities begin their recovery (Keith & Windberg, 1978). In summary, predation probably limits hare populations but its role in regulating them is not proven.

In conclusion, many studies have shown that predators kill large numbers of prey individuals, and that predation may be the most important source of mortality in many of these cases. Convincing evidence that large carnivores regulate their prey, however, is still some way off. Newsome (1990) has argued that predators can regulate prey populations if the latter is already low to begin with, for example as a result of drought. However, most of his examples are of predators such as feral cats or dingoes controlling rodents, rabbits, red kangaroos, in the sense of holding their numbers down, rather than removing them in a density-dependent way. To obtain a more accurate estimate of the impact of predation on populations, and the strength of predation as a selective force on individuals, it is necessary to know whether predators kill those individuals that would contribute to the population's growth, or mostly those that would contribute little to its increase (compensatory predation).

Do carnivores select sick prey?

Attempts have been made to determine whether predators select individuals in poor condition, i.e. those that are less likely to reproduce than healthy animals. Early studies simply noted the proportion of deformed skulls found without knowing whether these resulted from predation (Murie, 1944). Later studies recorded proportion of predator-induced deaths that were debilitated (e.g. seven out of 45 moose killed by wolves, Fuller & Keith, 1980) but could not compare these to the proportion of diseased individuals in the population. Still others have compared the health of predators' kills to the health of a 'random' sample of individuals in the population, but such control groups may themselves be biased. For example, kills by hunters are suspect because sick animals may be more approachable and easier to shoot. Again, 90% of a 'random' sample of deer that were killed by automobiles were in poor condition (O'Gara & Harris, 1988)! In addition, kidney fat and dry weight of bone marrow fat, the usual measures of body condition, change seasonally and according to reproductive state (Dunham & Murray, 1982) creating biases if control and prey subjects are sampled at different times of year.

Coursing predators which chase prey over long distances often select an individual only after a chase has been initiated (Estes & Goddard, 1967), which means that they might be able to survey fleeing prey and pick out a vulnerable victim. Stalkers, in contrast, which rely on surprise and short pursuits, have less time for prey selection during the chase and might be expected to kill a random sample of individuals. The data, however, are equivocal. When compared with cheetahs, wild dogs were found to take gazelles that were in worse condition, but wild dogs also selected a preponderance of male prey and these have lower fat reserves than females in any case (FitzGibbon & Fanshawe, 1989). Thus it is not clear whether selectivity was based on the sex of the prey, its health or both. Predators probably differ as to whether they select sick individuals according to their hunting technique, availability of alternative prey, and their own health and hunting abilities.

Do carnivores select old or young prey?

It has been argued that predation on old individuals will affect populations little because these animals may reproduce slowly and will die soon. Through examination of skull morphology and tooth wear, several studies have shown disproportionately high predation on old animals (see, for example, Hornocker, 1970). It is not clear, however, that old females in wild prey populations necessarily have lower reproductive rates. For example, reproductive output of red deer *Cervus elaphus* hinds remains high until death (Clutton-Brock *et al.*, 1988).

Predation on young animals may have less impact on prey populations than predation on mature individuals because the younger segment of the population can be replaced quickly. While most studies have reported that young animals suffer heavy predation (e.g. lynx taking caribou calves, Bergerud, 1971; spotted hyaenas taking wildebeest calves, Kruuk, 1972), fewer have provided the necessary data on prey population age structure to confirm that this predation is selective.

Table 5.6 Age classes, as determined from tooth wear, of Serengeti Thomson's gazelles killed by cheetahs and wild dogs compared with the overall population. From FitzGibbon & Fanshawe (1989)

	Killed by cheetahs		Killed by wild dogs		Proportion in population
	No.	%	No.	%	(%)
Fawns	44	40.7	13	20.3	3.0
Half-growns	14	13.0	6	9.4	2.9
Adolescents	13	12.0	4	6.3	9.2
Sub-adults	5	4.6	4	6.3	10.7
Adults	32	29.6	37	57.9	74.2

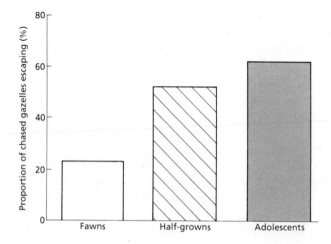

Fig. 5.2 The influence of age on the probability of escaping once Thomson's gazelles were chased by cheetahs. Fawns, $n = 47$; half-growns, $n = 25$; adolescents, $n = 16$. From FitzGibbon (1990a).

One study with cheetahs and wild dogs did find that they selected Thomson's gazelle fawns disproportionately from the population (Table 5.6). Young animals are more likely to be chosen by predators because: (i) they are easier to catch (Fig. 5.2); (ii) they have lower stamina and running speeds; (iii) they may not be able to outmanoeuvre predators in the same way as can adults (Fig. 5.3); and (iv) they may even fail to recognize predators!

Do carnivores select male prey?

Most mammals are polygynous and certain males may never get the opportunity to breed. Consequently, if predators concentrated on adult males in these species, their impact on the prey population would be lessened. As a rule, males are more likely to be preyed upon than females (see, for example, Table 5.7 and the wild dog example above),

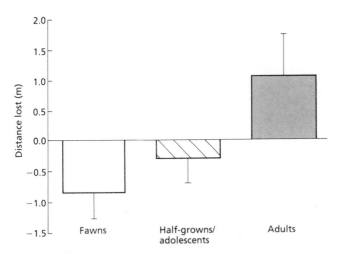

Fig. 5.3 The effect of Thomson's gazelle age on the mean distance lost by cheetahs during zigzagging of prey. Vertical lines denote standard errors. Fawns, $n = 18$; half-growns adolescents, $n = 14$; adults, $n = 26$. From FitzGibbon (1990a).

Table 5.7 Ungulate sex ratios in lion kills and in their populations in Nairobi National Park. From Rudnai (1974)

	Kills		Population		Significantly different?
Species	Male (%)	Female (%)	Male (%)	Female (%)	
Hartebeest	77.0	33.0	37.4	62.6	Yes
Zebra	50.0	50.0	27.1	72.9	Yes
Eland	88.9	11.1	46.6	53.4	Yes
Wildebeest	27.3	72.7	35.1	64.9	No

although this is not always the case (spotted hyaenas took more female zebras in Serengeti and Ngorongoro Crater, Kruuk, 1972). FitzGibbon (1990b) investigated the reasons that cheetahs hunted more adult male Thomson's gazelles (69%) than females (31%) despite males being outnumbered by 0.4 : 1 in the population. She found that males were: (i) less vigilant; (ii) were more often found alone; (iii) were usually on the periphery of groups; and (iv) had greater nearest-neighbour distances than did females. In general, males may seek less safe environments in order to maximize growth and so enhance competitive ability, and may be more vulnerable because they have become exhausted through competing for access to females.

Do carnivores select odd prey?

If predators select individuals that are conspicuous, uniformity will tend to be maintained among prey individuals. There are several anecdotal accounts which suggest that 'strange' animals are selected. Wildebeest that are infected by *Gedoelstia* flies can have an uncoordinated gait and are quickly picked off by predators. Lamprey (in Kruuk, 1972) stated that wildebeest whose horns he had painted white quickly disappeared from the population, while Jonsingh (1983) described dholes *Cuon alpinus* selecting chital *Axis axis* with particularly large antlers. It may pay to choose odd prey either because a predator is less confused by escaping prey if it follows a conspicuous individual, or because there are, as yet unproven, links between obvious characteristics that the predator can see, and reduced ability to escape that only become apparent once the prey has started to flee.

COOPERATIVE HUNTING IN LARGE CARNIVORES

Several species of terrestrial carnivores, for example dingoes *Canis dingo* (Corbett & Newsome, 1987) or brown hyaenas *Hyaena brunnea* (Mills, 1989), not only live in groups but also hunt in groups. Cooperative hunting involves hunting simultaneously, although it can be much more elaborate, and will only be favoured by natural selection when the foraging returns per individual (weight of food eaten per unit time) are greater as a result of their collaboration than if individuals hunt alone. Striking data showed that 74% of attempts at wildebeest calves were successful when two or more spotted hyaenas were involved in the hunt, but only 15% involving single hyaenas (Kruuk, 1972), and that golden jackals *Canis aureus* always caught Thomson's gazelle fawns when hunting in pairs but only rarely when hunting alone (Lamprecht, 1978). Such observations suggest carnivores might live in groups because of the benefits gained by individuals through hunting cooperatively.

It is worth remembering that individuals may hunt in groups for a variety of reasons, e.g. they do not want to lose contact with other residential group members who leave to begin a hunt. This means that all individuals present during a hunt may not actually be involved in hunting, and it will be very difficult for observers to discriminate 'laziness' from ineptitude. To determine the benefits of cooperative hunting, individual return rates of foraging need to be measured across groups of different size in terms of average weight of flesh ingested per group member per unit time. Variance in food intake, and the proportion of time that minimum food requirements are exceeded, are other important variables. Often it is difficult to measure these components as hunting groups may be smaller than foraging groups and because many carnivores scavenge from each other as well as make their own kills (e.g. lions, Packer, 1986).

In a series of game-theoretical models, Packer and Ruttan (1988) explored some of the consequences that individual selfish behaviour might have on hunting success when predators were hunting particular sorts of prey. Their models found that cooperative hunting is rarely favoured if groups hunt for single prey, because prey must be divided amongst group members. Individuals that rely on their companions to

Table 5.8 Summary of the predictions from Packer and Ruttan's (1988) models. An evolutionary stable strategy (ESS) denotes a strategy that cannot be invaded by the other stated alternatives

Prey type	Probability that cooperative hunting leads to gregariousness	Possible ESSs if constrained to live in groups	Effect of increase in predators' group size
Single large	Low	Cooperation, cheating, scavenging	Increase cheating, scavenging and solitary hunting
Multiple large	High	Cooperation, cheating, scavenging	Increase cheating and scavenging
Single small	Lowest	Cooperation	Increase solitary hunting
Multiple small	Highest	Cooperation	None

hunt (termed cheaters), and scavengers, will prosper in these circumstances (see Table 5.8). When several prey are captured in a hunt, and kills do not have to be shared, then cooperative hunting is more likely to be favoured.

When individual hunting success is relatively high, individuals will gain little benefit from hunting together, particularly when hunting single prey. Again, cheating will be favoured when prey is large enough to be scavenged, and cheating will increase as predator group size increases. Packer and Ruttan (1988) reasoned that if cooperative hunting was an important force in promoting group living, hunting success of individuals in group-living species that hunt alone should be much lower than in individuals of solitary species (when predators take single prey). Comparisons across taxa fail to support this prediction, suggesting that residential grouping patterns are not the result of cooperative hunting in many of these species.

The importance of this modelling is that it specifies the measurements that need to be made to determine whether group living has evolved in response to cooperative hunting.

1 The amount of food that individuals obtain from foraging in different sized groups needs to be quantified. This is difficult because: (i) weights of different age and sex classes of prey are unavailable for most species; (ii) weighing prey at death is intrusive and dangerous; (iii) weighing the amount of flesh left uneaten can be problematic if scavengers quickly remove remains; and (iv) observation conditions usually preclude recording the amount of time each group member spends feeding. Typically, the carcass weight is simply divided by the number of individuals feeding but individuals differ in their access to food because of competition, especially when prey is small.

2 Extensive data are needed because prey preferences may differ between different sized hunting groups, and individual differences in feeding habits are notorious amongst carnivores. For instance, zebras comprised 18%, 28% and 51% of the diet of three spotted hyaena clans in Ngorongoro Crater (Kruuk, 1972).

3 It is important to collect data on the frequency with which multiple kills are made according to predator group size. This requires observations across seasons because multiple kills often occur when migratory prey pass through predators' ranges, e.g. wildebeest and lions (Schaller, 1972), and instances of surplus killing can occur when vulnerable prey appear, e.g. wolves killing caribou calves (Miller *et al.*, 1985).

4 The costs of predation need to be specified. In the present context, time spent hunting different prey and energy expended (measured by different gaits for example), need to be determined as a function of predator group size. The risk of death or serious injury to the predator also needs to be systematically recorded, e.g. wolves killed by moose (Ballard *et al.*, 1987) or cougars killed by mule deer *Odocoileus hemionus* (Gashwiler & Robinette, 1957).

5 There may be hidden costs and benefits to group foraging that are not uncovered by estimates of the weight of flesh ingested per predator per hour. These include: (i) the constancy with which individuals eat meals, because variance in food intake might adversely affect growth; (ii) access to organs like livers and brains that contain special nutrients; (iii) enhanced ability of groups to open up a large carcass quickly; and (iv) possibilities that diseased or injured predators can survive temporary disabilities by relying on other group members.

Given these hurdles it is not surprising that very few studies have related per capita foraging returns to hunting group size. T.M. Caro (unpublished data) found that individual male cheetahs living and hunting in groups of two and especially three ingested more kilograms per unit time than those living alone despite having to share carcasses (Fig. 5.4). This was because groups chose to hunt larger prey, and not because they hunted cooperatively. Hunts in which coalition partners stalked prey simultaneously, chased it together or even tried to pull it down at the same time did not result in greater hunting success than hunts in which such coordination was lacking. In contrast, Packer *et al.* (1990) found that single lionesses had higher foraging returns than individuals belonging to any larger group sizes. An exception was that lionesses in groups of five and six matched singletons' foraging returns during the dry season because they hunted buffalo. In neither species did case did individuals experience increase per capita food intake as a result of hunting in groups, a result repeated in the handful of other studies that have measured this. At present there is insufficient widespread evidence to show that carnivores hunt cooperatively or that group hunting is responsible for group living in these species (T.M. Caro, unpublished data). In both of the examples above, enhanced ability to exclude conspecific competitors from territories was thought to be the most important factor maintaining group living, not foraging returns.

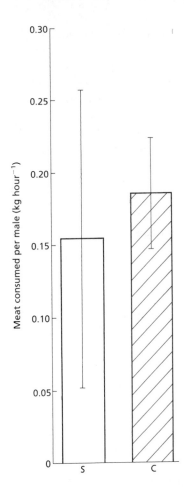

Fig. 5.4 Per capita number of kilograms of meat eaten per daylight hour by male cheetahs living alone (S), in coalitions of 2 and in coalitions of 3 (C); bars show standard errors. Males were observed for a minimum of 12 daylight hours. Numbers of individuals/coalitions watched: S, *n* = 9; C, *n* = 12.

UNGULATE ANTI-PREDATOR DEFENCES

Anti-predator defences are defined as any traits that confer a benefit of reduced risk of predation, whether or not the traits evolved primarily for anti-predator purposes. Prey animals can lower their vulnerability to predation by reducing encounters with predators, by minimizing their probability of being detected by any predators that they enounter, by reducing the probability that a predator's attack will be successful, and finally by diminishing the probability that they will be selected as the focus of any attack (see also Endler, 1986b and Box 5.1).

Reducing encounters with predators

Whereas predators concentrate their hunting activities in areas of high prey density, prey can, in turn, attempt to reduce encounters with predators by avoiding the areas frequented by predators. If predator movements are restricted for any reason, prey may congregate in areas of reduced predator densities. For example, Mech (1977) noted that white-tailed deer were concentrated in the overlapping areas between wolf-pack territories, where the wolves rarely hunted because of the dangers of encountering neighbouring packs. Barren-ground caribou exploit the fact that the movements of wolves are restricted to the availability of denning sites near the tree line and their need to remain near relatively immobile young (Kuyt, 1972). Female caribou migrate to traditional calving grounds that are a long way from wolf denning areas (Bergerud, 1974b). In some cases, individuals may sacrifice a high quality food source in order to remain in areas with low predator densities. For example, moose cows with calves on Isle Royale remained on small islands where both food quality and wolf density were low, rather than moving to the main island where food quality was higher but wolves were more common (Edwards, 1983). It has been shown in a number of models that the use of such refuges by prey enhances the stability of predator–prey systems.

Reducing the probability of being detected

Most ungulates rely on crypsis to reduce the probability that a predator will distinguish them from the background. Even the stripes of the zebra, which appear so conspicuous during the day, may be effective camouflage at night. Similarly, the black flank stripe of Thomson's gazelles may break up their outline at a distance, while their dark backs and light bellies may act as counter shading. Crypsis is best developed, however, in the small forest living species like duikers. Here, both visibility and flight possibilities are often restricted (Jarman, 1974; Table 5.9). Forest species often have dappled or striped coats that blend with the vegetation. On detecting a predator, they freeze or even lie down, remaining motionless until the predator approaches to within a few metres, and only then do they bolt. The effectiveness of crypsis can be further increased by selecting the appropriate background: dull coloured caribou mothers avoid areas with snow when they have calves (Bergerud & Page, 1987). Alternatively, the coat colour may be changed to match background, as species occupying Palaearctic habitats often develop white coats in the winter (e.g. reindeer and mountain hares *Lepus timidus*). In general, crypsis is thought to be effective only at low prey densities (Table 5.9), although the degree to which it affords protection has not been thoroughly investigated.

Fight or flight? Reducing the success rate of predator attacks

Large ungulate species like buffalo can depend on direct physical defence to reduce their vulnerability to predation. Smaller species rely on flight (Table 5.9). Some predators, such as spotted hyaenas and wild dogs, chase prey over long distances, relying on stamina rather than great speed to outrun their prey, whereas others such as cheetahs run down their prey in very short fast chases (see above). Ungulates exposed to both types of predators face a problem, because the structural and

Table 5.9 Anti-predator strategies of several ungulate species

Species	Weight (kg)	Normal group size	Habitat	Main anti-predator strategy	Hider/follower	Maternal defence
Kirk's dik-dik	4–5	2	Woodland	Crypsis	Hider?	No
Yellow-backed duiker	45–64	1	Forest	Crypsis	Hider	No
White-tailed deer	80–130	3–5	Woodland	Flight	Hider	Yes
Thomson's gazelle	18–25	5–6	Woodland/grassland	Flight	Hider	Yes
Caribou	75–125	1–20	Woodland/grassland	Flight	Follower	Yes
Himalayan tahr	55–120	2–23	Thick forest/grassland	Flight	Follower	?
Wildebeest	160–270	10–150	Woodland/grassland	Flight	Follower	Yes
Bison	550–800	4–30	Forest/grassland	Defence	Follower	Yes
Eland	500–900	15–25	Woodland/grassland	Defence	Hider	Yes
African buffalo	700–800	50–100	Woodland/grassland	Defence	Follower	Yes

physiological adaptations needed for these two modes of locomotion are incompatible (Taylor & Lyman, 1973; Slobodkin, 1974). One solution is to rely on manoeuvrability. When chased by predators that are faster than them, ungulates often perform sharp turns, forcing predators to slow down or risk overshooting and consequently to lose distance. Alternatively, prey that are unable to outrun predators in a direct chase may move into areas where the predator is at a disadvantage. Moose sometimes run into water when chased by wolves (Fuller & Keith, 1980) and dall sheep *Ovis canadensis* escape wolf attacks by moving onto steep rocky mountain slopes where their agility puts them at an advantage over less agile wolves (Murie, 1944).

With defence by either fight or flight, grouping may increase the effectiveness of the escape strategy. Large groups of ungulates are more effective at defending themselves than single animals because they can use a defensive formation in which their vulnerable rumps are protected by other group members (e.g. muskoxen *Ovibos moschatus* defending themselves against wolves). In addition, grouping may provide early warning of approaching danger and might confuse predators trying to select and attack a particular group member.

Table 5.10 Mean distance (± standard error) at which the first gazelle from a group detected an approaching cheetah, depending on the size of the group, and the type of group. From FitzGibbon (1990c)

Thomson's gazelle group size	Mean distance (m)	Sample size (*n*)
Alone	143 ± 12	44
2–5	170 ± 14	57
6–20	179 ± 12	60
21–50	215 ± 19	47
>50	225 ± 18	47
Mixed species Thomson's gazelles/ Grant's gazelles groups	214 ± 18	65
Single-species Thomson's gazelles groups	173 ± 7	17

THE EARLY-WARNING EFFECT

Stalking predators rely on surprise and are rarely successful if the prey is alerted to their presence before the final attack. Individuals can increase their probability of detecting predators by spending more time scanning the environment, but vigilance behaviour is costly because it interferes with foraging and other activities. By joining a group, an individual may be able to increase its probability of detecting predators and increase the distance at which detection occurs, without increasing its own level of vigilance, because in a group there are more individuals on the look out for predators at any one time (Powell, 1974; Lazarus, 1979). For example, larger groups of Thomson's gazelles detected approaching cheetahs at greater distances (Table 5.10).

Although important, the early-warning advantages of grouping may not be as great as they appear, because individuals in groups tend not to maintain the same high vigilance levels they adopt when alone. A decline in individual vigilance levels with increasing group size has been reported in 11 ungulate species (Elgar, 1989), and individuals may actually lower their vigilance to the point where groups do not experience improved predator detection (Barnard, 1980).

Where two species differ in their abilities to detect predators, one may gain an extra advantage from joining the other. As a result of their greater height and vigilance levels, Grant's gazelles *Gazella granti* detect approaching predators at greater distances than Thomson's gazelles and provide an early warning of danger for Thomson's gazelles that associate with them (Table 5.10) (FitzGibbon, 1990c).

THE CONFUSION EFFECT

When a large number of animals flee simultaneously in different directions, a predator might be sufficiently confused as to reduce the prob-

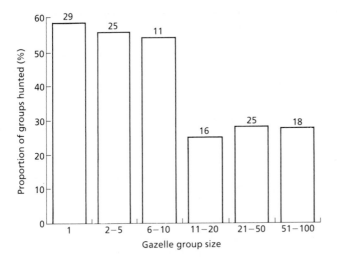

Fig. 5.5 The effect of prey group size on the percentage of Thomson's gazelle groups encountered by hunting cheetahs that the cheetahs chose to hunt (in 53 hunts). Numbers above the histograms denote the total number of groups in each size category encountered. From FitzGibbon (1990c).

ability of a successful attack. Predators may not only find it difficult to make a decision as to which animal to chase when faced with such a choice (the embarrassment of riches effect), but may also have difficulty in following the path of a chosen animal as it flees among many others (Pitcher, 1986). Grouped ungulates can exaggerate this confusion by scattering in all directions on being attacked. On the other side of the coin, however, is the risk that prey in large groups may themselves get confused and find their escape routes impeded by other group members. Crisler (1956) suggested that this was the reason that wolves were particularly successful at hunting large groups of caribou.

Under experimental conditions, grouping reduces the success rate of predatory attacks (Neill & Cullen, 1974; Kenward, 1978) but under natural conditions predators have the choice of prey groups and are likely to concentrate on the most vulnerable. If predators are less successful at hunting large groups, this may be reflected in a reduction in large groups hunted, as seen in cheetahs hunting Thomson's gazelles (Fig. 5.5), rather than a decrease in hunting success with increasing group size (FitzGibbon, 1988).

Reducing the probability of being selected in an attack

Grouping of prey frequently may also serve to reduce an individual's chance of being chosen as the focus of an attack (Hamilton, 1971). If a predator only attacks one individual from each group, the probability of being selected decreases as the reciprocal of group size (the dilution effect). To benefit from the dilution effect, it is necessary that attack rate does not increase in direct proportion to group size (after all, it is

---

Box 5.1 How prey 'talk' to predators

Ungulates show a battery of different anti-predator behaviour patterns in the face of danger. Springboks *Antidorcas marsupialis* bound along in a series of high leaps (pronking), impalas *Aepyceros melampus* make exaggerated jumps, white-tailed deer flash the white undersides of their tails (tail-flagging), and Thomson's gazelles stot by leaping off the ground while holding their legs straight. There has been much armchair speculation on the exact functions of these behaviours: that they signal the presence of danger to kin, that they induce prey to bunch together in flight, or even that they confuse the predator (for review, see Caro, 1986a), but they have received the attention of fieldworkers only recently (Bildstein, 1983).

Thomson's gazelles show a graded series of responses to different predators. They are most likely to stot to wild dogs, then to spotted hyaenas, and then to jackals and least likely to stot to cheetahs. They are more likely to stot to hunting than to non-hunting predators, but they do not stot in very dangerous situations when the predator is about to make contact. Based on observations which showed that wild dogs concentrated their chases on those gazelles that stotted at lower rates, and that gazelles stotted less in the dry season when they were probably in poor condition, stotting may serve to inform the dogs of the gazelle's ability to escape (FitzGibbon & Franshawe, 1988). In response to cheetahs, however, usually only one member of a group stotted, suggesting that group members were not competing to demonstrate their health. Cheetahs rely so much on a concealed approach that they usually abandoned hunts when gazelles stotted, suggesting gazelles were informing cheetahs that they had detected them (Caro, 1986b). These results demonstrate that prey may communicate different sorts of information to different predators because the type of danger they face strongly depends on the predator's hunting technique.

In addition, young gazelles stot at particularly high rates when they are disturbed from hiding and far from their mothers. As female antelopes have straight, thin horns specialized for stabbing, they can sometimes defend their fawns against predators (Packer, 1983), so fawns may be signalling to their mothers that they require assistance rather than 'talking' to predators.

Teasing apart the various hypotheses for stotting, leaping and rump-patch signalling is difficult because the predators' behaviour is often the clue to the function of the prey's behaviour; thus staged encounters with 'human predators' fill in only part of the story. So far, little attempt has been made to relate the distribution of anti-predator behaviour and morphology, such as white rump patches, in different species to the habitat, grouping patterns and principal predator species faced by ungulates.

---

possible that larger groups might experience higher attack rates than smaller ones, because they are more profitable to the predator; see Chapter 4). On balance, however, it seems unlikely that a group of 100 ungulates will experience 100 times more attacks than a solitary animal, especially if predators' success against small groups is higher (see above).

The benefit of the dilution effect can also be gained through membership of mixed-species groups, but the benefit derived will depend on the predators' preference for the constituent species. For example, lions in Nairobi National Park, Kenya, preferred to hunt wildebeests but not hartebeests *Alcelaphus buselaphus* (as measured by their representation in kills compared to their representation in the population), so it might pay a hartebeest to join a wildebeest group, but not vice versa (Rudnai, 1974; Gosling, 1980).

Within a group, individuals differ in their vulnerability to predation, and predators are frequently noted to select those individuals on the periphery of the group, e.g. cheetahs (FitzGibbon, 1990b). This may be simply because peripheral animals are closer to the predator when it encounters a group, or because selecting a peripheral animal minimizes the confusing effects of grouped prey. As a result of this selection of peripheral animals, prey individuals may compete for the relatively safe central positions (Hamilton, 1971). Individual group members may also be able to reduce their probability of being selected by spending more time vigilant. On 10 out of 12 occasions, for example, cheetahs chased those Thomson's gazelles that were less vigilant than their nearest neighbours (FitzGibbon, 1989).

The anti-predator behaviour of immature ungulates and their mothers

Ungulates are most vulnerable to predation during the first few weeks of life (Lent, 1974). They cannot run as fast as adults, and they are vulnerable to a greater range of predators as a result of their smaller size. It is common for more than 50% of infants from wild ungulate populations to be killed by predators before reaching adulthood, e.g. white-tailed deer (Cook *et al.*, 1971). It is not surprising, therefore, to find that infants and their mothers exhibit a number of anti-predator defences that are not shown by other members of the population.

FOLLOWING AND HIDING

Infants of ungulate species can be broadly categorized as *followers* (where the young accompany their mother soon after birth), or *hiders* (where the young remain hidden at some distance from the mother for the first few days of life but are visited by her; Lent, 1974; Leuthold, 1977). Following is generally found in migratory species or those that occupy open environments (Table 5.9). Following infants are precocial, and can outrun predators only a few hours after birth, e.g. wildebeest (Estes & Estes, 1979). They are often defended by their mothers or social group members, and tend to exhibit reproductive synchrony (Rutberg, 1987), which may reduce predation of infants because the sheer numbers of young leads to predator satiation (Estes, 1976; Ims, 1990).

Hiding is seen as an effective strategy for species that occupy closed environments of dense shrubs, and for small species that can lie concealed in herbs even in open habitats (Table 5.9). In hider species, the mother and infant are in contact only for brief periods to allow the fawn to suckle. In most species the habit of hiding has begun to decline by the third week after birth, but there is much variation within and between species (Ralls *et al.*, 1987). Fawns of hider species are extremely vulnerable when found out of hiding, and only 34% of Thomson's gazelle fawns aged 1–4 weeks escaped once chased by cheetahs. Offspring are cryptically coloured, select areas of tall vegetation in which to hide, and are only active for short lengths of time. Scent is often minimized through inactivity of the scent glands, and the ingestion of the young's faeces and urine by the mother. In Thomson's gazelles, hiding has been shown to reduce a fawn's probability of being located and killed by cheetahs (FitzGibbon, 1990a).

The effectiveness of hiding as an anti-predator strategy depends on the mother's ability to reduce the amount of information she transmits to predators as to the position of her infant: mothers do not remain close to infants (Byers & Byers, 1983) or consistently stare in their direction (FitzGibbon, 1988). Byers and Byers found that pronghorn antelope *Antilocapra americana* mothers gave away so little information about their fawn's location that it was not worthwhile for coyotes, their main predators, to search for fawns; instead they hunted for smaller and less nutritionally rewarding ground squirrels *Spermophilus columbianus*, which were more easily located.

If discovered out of hiding, infants may drop out of sight, adopting a prone position to avoid capture (Lent, 1974). While this tactic is common amongst hiders, it is also found in follower species, such as caribou. Infants that drop down when further away from predators are more difficult for the predators to find (Fig. 5.6). Mothers tend to exhibit increased vigilance levels compared to females without young (red deer; Clutton-Brock & Guinness, 1975), and this helps them to detect predators further away and give infants time to drop down. In some species, mother may defend their infants, particularly against secondary predators which pose little threat to adults. Thus, Thomson's gazelles defend their infants from jackals but not from wild dogs (Kruuk, 1972).

CONCLUSIONS

Understanding the population biology of large carnivores and their prey is advancing rapidly in three main areas. Direct observation and new techniques have shown that large carnivores can have a far greater impact on prey populations than was previously imagined. In many ecosystems they account for a higher fraction of all deaths than any

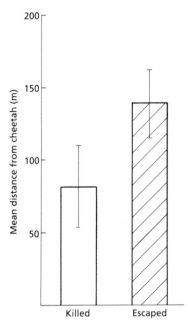

Fig. 5.6 Mean distance (± SE) from cheetahs that Thomson's gazelle fawns which were killed, and fawns which escaped from cheetahs, dropped down to hide. Killed, *n* = 7; escapees, *n* = 5. From FitzGibbon (1988).

other factor, and keep prey populations at low densities where food is not limiting. Large carnivores have been shown to select prey on the basis of its age, health, sex and oddity, and their preference for young animals suggests that they select on the basis of vulnerability. Age, sex and reproductive status of predators also has an important bearing on their feeding preferences.

New theoretical and empirical advances are being made concerning the extent of cooperative hunting in large carnivores and its role in the evolution of group living. Contrary to earlier dogma, carnivores hunt in groups because they live in groups, and not because they necessarily hunt cooperatively or experience greater per capita foraging returns than solitary individuals. Evidence now suggests that the adaptive significance of group living in many carnivores will be found in diverse reproductive behaviours, unrelated to food acquisition.

Apart from vigilance, grouping is the most widespread anti-predator defence amongst the principal prey of large carnivores. It both reduces the probability of being selected in an attack, and the success of attacks, and it is likely to provide a degree of protection from a wide array of predator species. For individuals, it is a low cost strategy that enables them to reduce the amount of time spent in costly vigilance behaviour and so leave more time available for feeding. The increased vulnerability of young ungulates has resulted in special anti-predator defences of mothers and infants.

While much is now known about the specific behavioural strategies used by ungulates to reduce predation, the extent to which risk of predation influences other aspects of behaviour, through its role in decision making, has yet to be investigated in detail (Lima & Dill,

1990). Predation risk is likely to influence such decisions as when and where to feed, what to eat, as well as the timing, form and extent of reproductive behaviour. The consequences of predation for prey populations extend far beyond the relatively few individuals killed, to influence almost all aspects of their behaviour and ecology.

ACKNOWLEDGEMENTS

We thank the Government of Tanzania, Serengeti Wildlife Research Institute and Tanzania National Parks for permission to carry out research. C.D.F. is grateful to St Catharine's College, Cambridge for funds to allow her to write this chapter collaboratively in the USA. We thank Tom Bergerud, Paule Gros, Warren Holmes, Hans Kruuk, Craig Packer, Herbert Prins, Bob Taylor and Richard Wrangham for helpful comments on the manuscript, and Tom Bergerud and Bill Gasaway for useful discussions.

6: Birds of Prey

IAN NEWTON

The subjects of this chapter include the 292 species that comprise the order Falconiformes. This order contains five families of varying degrees of affinity, namely the Cathartidae (New World vultures and condors), the Pandionidae (osprey), the Accipitridae (hawks, kites, buzzards, eagles and Old World vultures), the Sagittariidae (secretary bird) and the Falconidae (caracaras, falcons and falconets). All these groups have the same specializations for finding food, and for holding and tearing apart the bodies of other animals: acute vision, strong legs and feet equipped with sharp, curved claws, and a hooked beak. Yet they differ so much in other anatomical detail that they are almost certainly derived from more than one ancestral stock. In some respects, the convergence also applies to the owls, the nocturnal equivalents of the raptors, which I shall not discuss. Throughout, I shall use the term 'raptors' only for members of the Falconiformes.

Representatives of this order are found on all continents, except Antarctica, and in all habitats from dense jungle to open desert. Some species are noted for extreme stability in population, with breeding numbers over wide areas fluctuating by no more than 15% of the mean level over long periods of years. Pairs are often found at the same nesting sites year after year, and some of the cliffs occupied by peregrine falcons *Falco peregrinus* in Europe are mentioned in the falconry literature of earlier centuries. Other species, in contrast, show extreme nomadism, concentrating to breed in areas with temporarily abundant food supplies, and then moving on again. In natural conditions, raptors are clearly limited by food or other resources, but in parts of the developed world their numbers have been much reduced by human activities. These include not only the destruction of habitats (and food supplies), but also deliberate persecution (in the interests of game hunting) and pesticide use.

SEXUAL DIMORPHISM

In most raptors the female is the larger sex. This 'reversed size dimorphism' is probably connected with the raptorial life style, because it occurs among owls and skuas too, providing another example of convergence. Within the raptors, the degree of size difference between the sexes varies in relation to diet (Newton, 1979). In the world as a whole, there are about 75 raptor species that each live almost exclusively on a single kind of prey — on fish, reptiles, birds, and so on. If these specialists are arranged according to diet, grouping all species of like food together, dimorphism is seen to increase with the speed and agility of the prey (Fig. 6.1). At one extreme, those vultures and condors that feed entirely from immobile carcasses show no consistent size difference between the sexes or, if they do, the male is slightly bigger than the female, as in non-raptorial birds. Next come the snail-feeders, which show reversed dimorphism, but the female is only slightly larger than the male. The insect-feeders and the reptile-feeders show somewhat greater dimorphism, followed by the mammal-feeders and the fish-feeders and then the bird-feeders, which are the most dimorphic of all. The extreme is shown by the small bird-eating accipiters in which the female weighs nearly twice as much as the male, such as the European sparrowhawk *Accipiter nisus* and the North American sharp-shinned hawk *A. striatus*.

The link with feeding habits is apparent even among species in the same genus, if they take different foods. Among the falcons, for example, the insectivorous lesser kestrel *Falco naumanni* shows less size dimorphism than the rodent-eating common kestrel *F. tinnunculus*, and

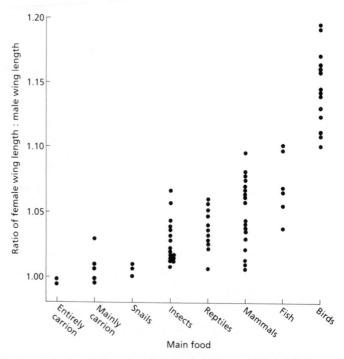

Fig. 6.1 Sexual size dimorphism in raptors in relation to diet. Each point represents one species, and only species which feed almost exclusively on one prey type are included. The faster the prey, the greater the degree of dimorphism in the raptor. In species with mixed diets, the degree of dimorphism is commensurate with the animals eaten. After Newton (1979).

this species in turn shows less than the bird-eating merlin *F. colum-barius* or peregrine *F. peregrinus*.

The relationship between diet and dimorphism is shown in Fig. 6.1 only for the 25% of raptor species that have restricted diets, but it also holds for the majority of species that eat two or more kinds of animal. In such catholic species the degree of dimorphism is commensurate with the kind of animals taken. Thus species which eat both mammals and birds are intermediate in dimorphism between the strict mammal-eaters and the strict bird-eaters.

Associated with the main trend is another, namely for the more dimorphic species to take larger prey relative to their body size than the less dimorphic ones (excluding the carrion-feeders). The snail- or insect-eating raptors capture prey weighing less than one-tenth of their own weight, but at the other extreme some of the bird-eaters take prey as heavy, or heavier, than themselves. Other species are intermediate. In other words, the more dimorphic the raptor, the faster and the larger the prey it takes relative to itself.

No one could argue over the relationship between dimorphism and diet, but there is no consensus over the evolutionary reasons behind it, nor why the female is the larger sex. There is a vast literature on the subject, mostly highly speculative. One problem is that almost every aspect of the biology of raptors is linked with diet, so many papers simply point to another correlation, claiming its responsibility in evolution for the pattern of dimorphism shown (for reviews see Snyder & Wiley, 1976; Newton 1979; Mueller & Meyer, 1985; Mueller, 1990). One consequence of reversed dimorphism is that females are normally dominant over males in raptors. Another is that, in the extreme species, notably the bird-eaters, the female takes partly larger prey than the male.

MATING SYSTEMS

Despite the size difference between male and female, sex ratios are usually equal or slightly in favour of females, at least among nestlings of the few species that have been studied (Table 6.1). Overall sex ratios are difficult to determine among adult birds, because the sexes often occupy partly different habitats or, outside the breeding season, partly different areas.

During the breeding season, the division of parental duties is more marked in raptors than in many other birds. The male provides the food, while the female stays at the nest, incubates the eggs and tends the young. This dichotomy persists until the young are about half-grown, after which the female helps with the hunting.

The majority of species that have been studied are monogamous. In some, such as large falcons, most individuals tend to keep the same mate year after year, with only occasional 'divorces', whereas in other species, such as sparrowhawks, mate changes are frequent (Newton, 1979). A few species, such as the black-shouldered kite *Elanus caeruleus*,

Table 6.1 Sex ratios of nestling raptors

Species	Locality	No. of broods	No. of males	No. of females	Ratio females:males	References
Hen harrier (Circus cyaneus)	Scotland, Orkney	?	1067	1180	1.11*	Balfour & Cadbury (1979), Picozzi (1984)
	Scotland, SW	14	17	25	1.47	Watson (1977)
Brown goshawk (Accipiter fasciatus)	Australia	?	41	56	1.37*	Aumann (1986)
Goshawk (Accipiter gentilis)†	Finland	378	568	526	0.93	Wikman (1976)
European sparrowhawk (Accipiter nisus)	Britain	651	1102	1061	0.96	Newton & Marquiss (1979)
	Denmark	?	147	121	0.82	Schelde (1960)
Cooper's hawk (Accipiter cooperi)	East USA	?	35	36	1.03	Meng (1951)
Harris' hawk (Parabuteo unicinctus)	Arizona	?	56	51	0.91	Mader (1976)
Bald eagle (Haliaetus leucocephalus)	Saskatchewan	?	53	50	0.94	Bortolotti (1986)
European kestrel (Falco tinnunculus)	Holland	684	1352	1268	0.94	Dijkstra (1988)
American kestrel (Falco sparverius)	East USA	19	37	34	0.92	Porter & Wiemeyer (1972)
	Utah	22	19	31	1.63	Smith et al. (1972)
	California	?	61	66	1.08	Balgooyen (1976)
Hobby (Falco subbuteo)	Germany	?	29	43	1.48*	Fiuczynski (1978)
Prairie falcon (Falco mexicanus)	Colorado	?	45	46	1.02	Enderson (1964)
	Idaho	?	188	199	1.06	Ogden (1975)
Peregrine (Falco peregrinus)	East USA	?	23	39	1.70*	Hickey (1942)
	Britain	159	176	187	1.06	Newton et al. (unpublished)
	Australia	?	495	545	1.10	Olsen & Cockburn (1991)
Merlin (Falco columbarius)	Britain	26	35	48	1.37	Newton et al. (1986b)

* Significantly different from equal ($P < 0.05$).
† Excludes some young from broods of one, in which sexing was thought to be uncertain.

may raise two broods in a year, and for the second brood the female may move to another male, leaving her first mate to raise the young (Mendelsohn, 1983). In the snail kite *Rostrhamus sociabilis*, either sex may abandon the brood to the first mate in order to nest again with a fresh mate (Beissinger & Snyder, 1987).

Polygyny, in which the male has more than one female at a time, has so far been recorded in about a dozen species. It is especially frequent in harriers, with some males having two or three mates. In the hen harrier *Circus cyaneus* occasional males have been found with up to seven females at once (Kraan & Strien, 1969; Balfour & Cadbury, 1979). The different females in a harem are not closely related to one another, and use different nests.

Polygyny tends to occur in good food conditions, when the male can feed more than one female at once, at least in the early stages of breeding (Newton, 1979). Thus in rodent-feeders it occurs most commonly in years of peak prey numbers (Korpimaki, 1988). Only after the primary female has laid does the male give large amounts of food to a second female (and so on), but after all his females have laid he tends to revert to the primary female, who is therefore more likely to raise young than subsequent ones. Secondary females may accept this arrangement because it is the only option they have to non-breeding, or because they are unaware at the start that the male is already mated (Korpimaki, 1988).

Polyandry, in which the female has more than one male at a time, has been recorded as a rare event in a few species, but is usual in the Galapagos hawk *Buteo galapagoensis* (de Vries, 1975; Faaborg, 1986). In this species up to four males may share the same territory and nest, copulate with the single female, and help with nest duties. The ecological conditions which favour this system are not clear, but, as polygyny is associated with food abundance, polyandry may arise in response to scarcity, so that more than one male is needed to provision a female. In the Galapagos hawk there is extreme competition for territories, and a surplus of males, but the sex ratio may be the result of polyandry rather than the cause.

One of the most unusual social systems occurs in Harris's hawk *Parabuteo unicinctus*, in the deserts of Central America, which may breed in groups of up to four individuals, with the primary pair and additional (mostly male) helpers, including young from a previous year (Mader, 1979; Bednarz, 1988). Once they achieve adult plumage (in the third calendar year) the extra males may copulate with the female and help to feed the young. This system is associated with group hunting, as the Harris's hawk is the only raptor known to hunt cooperatively like some mammals, such as wolves. In all these species the frequency of non-monogamous matings varies between areas, and in all monogamy also occurs. Non-monogamous relationships are easily overlooked, and may be more widespread than at present known. However, monogamy seems to be the norm in most raptor species.

In their dispersion patterns, raptors show the same broad relationships with food supply as do other birds, nesting solitarily when food is evenly distributed and predictable, and colonially when food is clumped and unpredictable. Given widespread nest sites, the three main patterns are as follows.

1 *Pairs spaced out in individual home ranges (or territories).* This seems to be usual in about 75% of the 81 raptor genera, including some of the largest, such as *Accipiter, Buteo, Aguila* and *Falco*. Each pair defends the vicinity of the nest and a variable amount of surrounding terrain, so that home ranges may be exclusive or overlapping. Through suitable habitat, the nests of different pairs tend to be spaced fairly regularly, at distances which vary with the species and area. Most species that show this dispersion system feed on live vertebrate prey, and show considerable stability in numbers and distribution from year to year. Individuals usually hunt and roost solitarily. Outside the nesting season, the pairs may remain together on their home ranges, or may split, each individual occupying a different range.

2 *Pairs unevenly distributed, occasionally in loose colonies, but hunting solitarily.* This system is sometimes shown, among others, by the kites *Milvus migrans* and *M. milvus, Elanus scriptus* and *E. leucurus*, and by the harriers *Circus aeruginosus, C. cyaneus* and *C. pygargus*. Groups of pairs may nest close together in 'neighbourhoods', and range out to forage in the surrounding area. The different pairs may hunt in different directions from one another, and several may hunt the same area independently, perhaps from time to time shifting from one area to another. The breeding groups usually contain less than 10 pairs, with nests spaced 70–200 m apart. In harriers, the tendency to clump is sometimes accentuated by polygyny, and in harriers and kites by the frequent need to concentrate in patches of restricted nesting habitat. Such species often exploit sporadic food sources, such as local grasshopper outbreaks or rodent plagues. They are nomadic to some extent, concentrating to breed wherever food is temporarily plentiful, so that local populations can fluctuate substantially from year to year. This does not occur throughout the range, however, and not all pairs of such species nest in groups.

Outside the breeding season, kites and harriers tend to base themselves in communal roosts, from which they spread out to hunting areas during the day. The roosts usually contain up to 20 individuals, occasionally more, and the occupants may range out more than 30 km to feed.

3 *Pairs nest in dense colonies and forage gregariously.* This system is shown by the small snail-eating Everglade kite *Rostrhamus sociabilis*; by the insect-eating kites of the genera *Elanoides, Gampsonix, Chelictinia* and *Ictinia*; by the insect-eating falcons, *F. naumanni, F. vespertinus, F. amurensis* and *F. eleonorae*; and by the large griffon vultures (*Gyps*). In these species, the pairs typically nest closer together (often less than 10 m apart) and in larger aggregations than those mentioned above. They also feed communally in scattered flocks or, in the case of vultures, spread out in the air, but crowded around carcasses. Feeding flocks are not stable, but change continually in size and composition, as birds join or leave. Colonies usually contain up to 20 or 30 pairs, but those of some *Gyps* vultures can be very much larger. The food sources of these various species are even more sporadic and fast-changing than those of the previous group. Food may be plentiful at one place on one day and at another place on the next. Such species roost communally at all times and, when not breeding, sometimes in enormous numbers. The insect-eating falcons in their African winter quarters may use the same roosts year after year, often containing thousands of individuals of several species. Such birds exploit local flushes of food, such as termites and locusts, and move around over long distances in response to changes in prey availability.

The above division of dispersion patterns into three categories is arbitrary, and in practice gradations exist between, on the one hand, exclusive highly defended home ranges spread evenly through the habitat and, on the other, dense breeding colonies and feeding flocks occupying only a fraction of the habitat at any one time. The former is associated with fairly uniform, stable and predictable food supplies, and the latter with unpredictable and sporadic superabundances. Rodent plagues, insect swarms, or large carcasses are all food sources which are irregular and continually changing in distribution, but on which, once located, many birds can feed together. Particular dispersion patterns are not necessarily characteristic of species, and the same species of eagle, buzzard, harrier, kite or falcon may adopt a different system according to how its food is distributed. Even the large steppe eagle *Aquila rapax*, which holds large territories on its breeding areas, occurs in flocks when eating termites in its African winter quarters, and the bald eagle *Haliaetus leucocephalus* in North America assembles in large numbers whenever concentrations of salmon are available. Nonetheless, many species show only one dispersion pattern over most of their range, or for most of the time.

DENSITIES

In the limitation of raptor breeding densities, within the habitats that are occupied, two resources seem of overriding importance, namely food supply and nest sites (Newton, 1979). The evidence for a link between density and food supply in areas where nest sites are not limiting is circumstantial, and based on the following findings.

1 An overall trend for large raptor species, which depend on large, sparse prey species, to breed at lower density than small raptor species, which depend on small, numerous prey species.

2 Area differences in breeding density within species that are associ-

ated with area differences in food.

3 Annual fluctuations in density within species that are associated with cyclical fluctuations in food.

4 Sudden and long-term changes in breeding density that are associated with sudden and long-term changes in food.

Body size and breeding density

Where nest sites are not limiting, the overall trend is for large raptor species to breed at lower density, in larger home ranges, than small species (Fig. 6.2). The huge African martial eagle *Polemaetus bellicosus* breeds at what must be some of the lowest natural densities for any bird: one pair $125-300\,\mathrm{km}^{-2}$, depending on area. The species lives mainly on game birds and mammals of up to several kilograms in weight. Various other eagles usually breed at densities of one pair $30-$

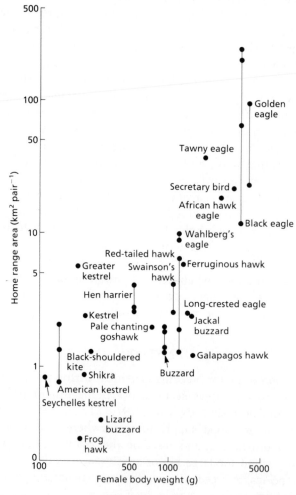

Fig. 6.2 Mean home range size in relation to body weight in raptors. The overall trend is for large raptors to breed at lower density, in more extensive home ranges, than small raptors. From Newton (1979).

$190 \, \text{km}^{-2}$, various buteonine hawks at densities of one pair $1-8 \, \text{km}^{-2}$, and small falcons and kites at one pair $1-3 \, \text{km}^{-2}$. The overall trend presumably holds because larger raptors generally eat larger prey species, and large prey species live and breed less abundantly that small prey species. So in each case, range size may be adjusted to food supply. The evidence is not decisive, of course, for it is possible that some unknown factor varies in parallel with food and causes the trend. A similar broad relationship between body size and range size holds among birds in general, and also among mammals (McNab, 1963; Schoener, 1968).

Regional variations in breeding density within species

Where nest sites are not limiting, regional variations in breeding density within species often correlate with regional variations in food supplies. In the sparrowhawk *A. nisus*, breeding densities in the woods of 14 different areas varied in relation to the local densities of prey birds (Newton *et al.*, 1986a). The hawks nested closest together, at higher density, in areas where their prey were most numerous (Fig. 6.3). The woods in all these areas were of roughly similar structure, and the differences in prey densities were associated with variation in elevation and soil type. Similarly, the densities of kestrels in different areas and in different years have been closely correlated with the densities of voles, which formed the major prey.

For other species the information is less quantitative, but is still consistent with the idea of a link between density and food supply. Thus a favourite prey of the buzzard in Britain is the rabbit *Oryctolagus cuniculus*, and buzzard breeding densities are generally higher in areas that have rabbits than in areas that lack rabbits (Newton, 1979). Likewise, in British peregrines, density is broadly related to land productivity

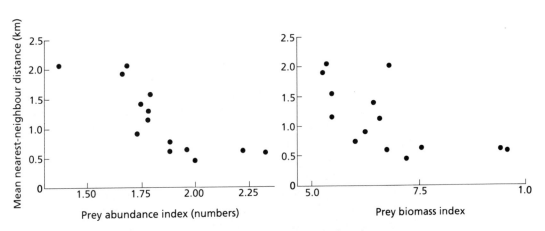

Fig. 6.3 Spacing of sparrowhawk nesting places in the woodland of 14 districts, shown in relation to food supply. Mean nearest-neighbour distances decrease (so densities increase) with rise in prey numbers and biomass. Relationship between spacing and prey numbers: $r = -0.77$, $P < 0.01$; between spacing and prey biomass, $r = -0.61$, $P < 0.05$. After Newton *et al.* (1986a).

and food supply, with the highest breeding densities along coasts rich in seabird prey, lower densities in hill areas of relatively high fertility or near to fertile valleys, and even lower densities in hill areas of low fertility or far from fertile valleys (Ratcliffe, 1972).

In addition, unusually high densities of raptors are invariably associated with an unusual abundance of food. The city of Delhi, in India, covers $150 \, km^2$, and in 1967 held 2900 raptor pairs, a density of more than 19 pairs km^{-2} (Galushin, 1971). These were mainly scavenging species, such as black kite *Milvus migrans* (16 pairs km^{-2}) and white-backed vultures *Gyps bengalensis* (3 pairs km^{-2}), but also included some other species. This high density was associated primarily with a huge amount of food within the city (mainly garbage and animal carcasses), but also with an abundance of nesting sites and an unusual tolerance by the human population.

In solitary nesting raptors the link between density and food supply is apparently brought about by spacing behaviour, as birds adjust their spacing (territory sizes) to correspond with the food situation. In permanent residents, such as golden eagles in Scotland, density is related to food supply in winter, when food is scarcest (Watson & Langslow, 1989); but in summer migrants, such as kestrels in Scotland, density is adjusted to food supply in spring, at the same time of settling (Village, 1989).

Annual variations in breeding density

Whereas for some species the evidence for density limitation in relation to food supply rests largely on long-term stability of breeding population, but at different levels in different regions, for other species it rests on annual fluctuations in density, which parallel changes in food. Most such species in the regions concerned have restricted diets based on cyclic prey. Two main cycles are recognized: (i) an approximate 4-year cycle of rodents on northern tundras and temperate grasslands; and (ii) an approximate 10-year cycle of snowshoe hares *Lepus americanus* in the boreal zone of North America (Elton, 1942; Lack, 1954; Keith, 1963).

Some grouse-like birds are also cyclic, but whereas in Europe they follow the 4-year rodent cycle, with peaks in the same years, in North America they follow the 10-year hare cycle. The populations of these various animals do not reach a peak simultaneously over their whole range, but the peak may be synchronized over tens or many thousands of square kilometres.

The main raptor species involved in these fluctuations are listed in Table 6.2, and all have been found to nest at greater density in years when their food is most plentiful. Sometimes the increase in numbers from one year to the next was so great that it must have been due mainly to immigration, while declines could be due to a combination of poor breeding, starvation and emigration (Sulkava, 1964). Moreover,

Table 6.2 Annual variation in breeding populations of raptors that exploit greatly fluctuating food sources

Species that eat rodents (approximately 4-year cycles)

Rough-legged hawk *Buteo lagopus*
- 0–9 pairs during 9 years, North Norway (Hagen, 1969)*
- 26–90 pairs during 34 years, Colville River, Alaska (Mindell et al., 1987)
- 10–82 pairs during 5 years, Seward Peninsula, Alaska (Swartz et al., 1974)

Hen harrier *Circus cyaneus*
- 10–24 females in 33 km² during 22 years, Orkney, Scotland (Balfour, in Hamerstrom, 1969)
- 13–25 females in 160 km² during 5 years, Wisconsin (Hamerstrom, 1969)*[†]
- 0–9 pairs in 6 years between 1938 and 1946 (Hagen, 1969)*

Kestrel *Falco tinnunculus*
- 35–109 clutches in 4 years; 109 clutches at vole density index 24, 97 at index 13, 35 at index 9, 50 at index 4, Netherlands (Cavé, 1968)*
- Approximately 20-fold fluctuation in index of number of broods ringed in Britain over 42 years, with peaks every 4–5 years (Snow, 1968)*
- 1–14 pairs during 5 years, North Norway (Hagen, 1969)*
- 1–16 nests during 12 years, Swabian Alps (Rockenbauch, 1968)*
- 28–38 pairs during 4 years, south Scotland, 9–28 pairs and 12–22 pairs in two areas during 6 years, southern England (Village, 1990)*
- 3–37 pairs during 8 years, south Finland (Korpimaki, 1985)*

Black-shouldered kite *Elanus caeruleus*
- Increase from 1 to 8 nests in 1 year, associated with rodent plague, South Africa (Malherbe, 1963)*
- 19–35 individuals, during 3 years Transvaal, South Africa (Mendelsohn, 1983)*

Species that eat gallinaceous birds and rabbits (4-year or 10-year cycles)

Ferruginous hawk *Buteo regalis*
- 5–16 pairs during 8 years in one area, 1–8 pairs during 3 years in another area, Utah (Woffinden & Murphy, 1977)*

Goshawk *Accipiter gentilis*
- 0–4 nests in 100 km² during 13 years; 2–9 nests in 200 km² during 7 years, in two areas of Sweden (Höglund, 1964)
- 1–9 nests in 372 km² during 4 years, Alaska (McGowan, 1975)

Gyr falcon *Falco rusticolus*
- 13–49 pairs during 5 years, Seward Peninsula, Alaska (Swartz et al., 1974)*
- 19–31 occupied cliffs and 12–29 successful nests during 4 years, Alaska (Platt, 1977)*
- 8–19 pairs during 27 years, Colville River, Alaska (Mindell et al., 1987)
- 19–31 pairs during 4 years, Brooks Range, Alaska (Platt, 1977)

* Prey population also assessed and related to raptor numbers.
[†] Excluding one year when population dropped from DDT poisoning.

some species, such as the goshawk *Accipiter gentilis*, fluctuate in density where they feed on cyclic prey, but remain stable in density where their food is more stable (often through being more varied). Such regional variation in the extent of annual fluctuations within species provides further circumstantial evidence for a link between breeding density and food supply.

Three steps in the response of raptors to annual variations in prey numbers can be recognized. Those birds which are subject to the most marked prey cycles show big local fluctuations in densities and breeding rates (e.g. grouse-eating goshawks in boreal regions); those subject to less marked prey cycles show fairly stable densities but big fluctuations in breeding rates (e.g. vole-eating buzzards in temperate regions); while those with fairly stable prey populations show stable densities and fairly stable breeding rates (e.g. bird-eating peregrines in temperate regions). Much depends on how broad the diet is, and whether alternative prey are available when favoured prey are scarce. The more varied the diet, the less the chance of all food species being scarce at the same time.

Long-term or sudden changes in food supplies

Long-term or sudden changes in raptor densities are often associated with human activities which change food supplies. The development of wild areas for agriculture almost always leads to a drop in the numbers of prey and in turn in the numbers of raptors. This is evident from the case histories of particular areas, and also from a comparison of natural areas with cultivated ones, a difference readily apparent but seldom documented. Conversely, the artificial inflation of food supplies can sometimes lead to high raptor breeding densities, as in the city scavengers mentioned above.

A specific example is provided by the buzzard, whose breeding densities in Britain fell after the viral disease myxomatosis reduced the numbers of rabbits *O. cuniculus*, which formed the main prey. In one area pair numbers dropped from 21 to 14 between one year and the next, as the disease swept through (Dare, 1961). Such instances add to the circumstantial evidence for a link between raptor breeding density and prey supply.

Shortage of nest sites

In some landscapes, raptor breeding densities are held by shortage of nest sites below the level that food supply would support. The evidence is of two types: (i) breeding raptors are scarce or absent in areas where nesting sites are scarce or absent, but which seem otherwise suitable (non-breeders may live in such areas); and (ii) the provision of artificial nest sites is sometimes followed by an increase in breeding densities.

Thus the kestrel *F. tinnunculus* increased in one year from less than 20 to more than 100 pairs when nesting boxes were provided in a

Dutch area with few natural sites (Cavé, 1968). Similar results were obtained with other populations of European kestrels, and also with American kestrels *F. sparverius*, ospreys *Pandion haliaetus* and prairie falcons *F. mexicanus* (Reese, 1970; Rhodes, 1972; Hamerstom *et al.* 1973; Village, 1983; R. Fife, unpublished). Provision of nest sites has allowed some species greatly to extend their breeding distribution. An example is the Mississippi kite *Ictinia mississippiensis*, which now breeds on the great plains of America 'in hundreds, if not thousands', in places where trees were planted by man (Parker, 1974). Likewise, nesting on buildings and quarries has allowed peregrines and lanners *Falco biarmicus* to occupy areas othewise closed to them through lack of cliffs.

In conclusion, the carrying capacity of any habitat for raptors is set by two main resources, food and nest sites, and whichever is most restricted can limit breeding numbers. On this basis much of the natural variation in breeding densities can probably be explained. Most of the evidence for this view is circumstantial, based on correlations. But the importance of nest sites has been confirmed by experiment, leaving the manipulation of food supply as an obvious future research need.

Much of the relevant research has been done on resident populations, and some migrant populations may be limited on winter quarters, and so unable to occupy their breeding habitat to the full. No cases are known, but the honey buzzard *Pernis apivorus* is a possible candidate. In addition, in some regions raptors are persecuted by man or poisoned incidentally by pesticides, so that their numbers are held by human action below the level that the habitat would support. The hen harrier *C. cyaneus* in grouse-hunting areas of Britain is an obvious example.

LIFE HISTORY STRATEGIES

Raptors range in weight from less than 100 g to more than 1400 g. This is an enormous spread compared to that of most other bird orders. Moreover, within the raptors the four main trends of life history are related to body size. The larger the species: (i) the longer the maximum life span, (ii) the later the age at which breeding begins, (iii) the longer a breeding cycle takes, and (iv) the fewer the young produced at each attempt. Near one extreme, small falcons live up to about 10 years, and begin breeding in their first year. They lay 5−6 eggs at 2-day intervals between each egg, and have incubation, nestling and post-fledging periods lasting about 28, 26 and 14 days respectively, bringing the total breeding period from the first egg to about 80 days. The maximum increase in population possible each year is thus 3.5−4.0 times the breeding population, assuming the parents survive. If more than one brood is reared each year, the potential rate of increase is even greater. At the other extreme, the large condors can probably live more than 40 years, and do not begin breeding until they are more than 5 years old. They lay only one egg at a time, and have incubation, nestling and post-fledging periods lasting about 55, 220 and 210 days. This brings the total breeding period from egg-laying to about 485 days, so that successful annual breeding is impossible. Ignoring the non-breeding immatures, the maximum possible increase in a population of condors is 50% in 2 years, or 25% in 1 year. (In practice, the spread of egg dates is such that some pairs could fit in two successful attempts in 3 years.) These two contrasting species lie at opposite ends of a spectrum, between which other species fit, showing a continuum of variation in life history strategy, evident within genera as well as within raptors as a whole.

The particular life history traits shown by different species greatly affect their population dynamics, the growth potential of their populations, and their ability to withstand human onslaught. In big long-lived species, population turnover is generally slow, with more overlap between generations and a more stable age structure, all of which tend to dampen short-term fluctuations in numbers. There also tends to be a relatively large non-breeding sector, consisting mainly of immatures. Less than half the total population may breed in any one year, producing only a small number of young. In small short-lived species, by contrast, population turnover is rapid, with less overlap between generations, a less stable age structure, and a high production of young, all of which facilitate short-term fluctuations in numbers. Most individuals that survive a winter will be capable of breeding if they can establish a territory, so that the non-breeding sector is usually small. Also, because of their fast breeding rates, small species can recover from a population low more quickly than can larger species.

In their population dynamics, the small raptors resemble songbirds, and the large raptors resemble certain seabirds. This last analogy extends especially to the small clutches, long breeding cycles and deferred maturity. In both groups, single egg clutches are frequent and, when two eggs are laid, often only one young is raised. Many large raptors show 'caenism', in which the first hatched young attacks and often kills its smaller sibling (Gargett, 1970; Meyburg, 1974; Newton, 1979). Moreover, the only birds other than certain large raptors whose breeding periods are known to last more than 1 year include some large seabirds, such as the king penguin *Aptenodytes patagonica* and great frigatebird *Fregata minor* (Stonehouse, 1960; Nelson, 1976).

Within each raptor species, food supply seems to have a dominant influence on breeding rate, regional and annual variations in productivity being correlated with regional and annual variations in prey supply, sometimes modified by weather (Newton, 1979). Moreover, the experimental provision of extra food in some species has led to an increase in breeding rate (Newton & Marquiss, 1981; Dijkstra, 1988).

From the many studies of breeding in birds of prey, summarized in Newton (1979), certain generalizations can be drawn. In good, as opposed to poor, food situations: (i) breeding densities are usually higher; (ii) more of the available nesting territories are occupied; (iii) more immature-plumaged birds breed; (iv) more of the territorial pairs lay eggs; (v) more of the pairs that lay eggs succeed in hatching and rearing their young; (vi) mean laying dates are earlier; (vii) clutches and broods are larger; (viii) nestling growth rates and fledging weights are greater; (ix) parental care is better, including defence against predators; and (x) repeat laying after failure is more frequent. Not all studied species showed all of these trends, but in some species not all aspects were examined. Also, the extent of annual fluctuations in breeding rate depends largely on the degree of dependence on a particular cyclic prey, with less extreme variation among populations that have alternative prey.

The effects of food on breeding rate may be modified by weather, as when mild springs advance laying in northern regions and cold springs delay it. Rainfall has been shown to influence the breeding of several species, partly because it reduces the availability of prey, and partly because it reduces the time that the birds have for hunting. On the other hand, the influence of predators and other factors on the breeding of raptors is mostly trivial, apart from human predation (Newton, 1979).

FOOD CONSUMPTION

Because all raptors eat similar food (animal carcasses), comparisons between species are fairly straightforward, especially in captive birds fed the same kind of meat. As expected, daily food intake varies with body size, the smaller species eating less than the large ones, but more in relation to their body weight. Small species, such as sparrowhawk, eat about 40 g of food day^{-1}, which is about 20−25% of body weight, whereas large species, such as golden eagle, eat about 200 g day^{-1}, which is some 5−6% of body weight. Moreover, the small species must feed more often: small sparrowhawks in good condition can survive only 3−5 days without food, whereas large eagles can fast for more than 2 weeks if need be.

The relationship between daily energy needs and body weight is shown for various northern hemisphere species in Fig. 6.4, based on studies of captive birds (Kirkwood, 1981). The amount by which these figures should be increased for free-living birds varies with the species, depending mainly on hunting techniques. In sit-and-wait hunters the daily energy needs in the wild are little more than in captivity, but in active hunters, such as hoverers, needs can be nearly three times as great (Kirkwood, 1981). There is also greater wastage in wild raptors, as parts of prey may be discarded or stolen by other species.

EFFECTS ON PREY POPULATIONS

Although raptors are often persecuted as predators of game or livestock, few detailed studies of their impact on prey populations have been

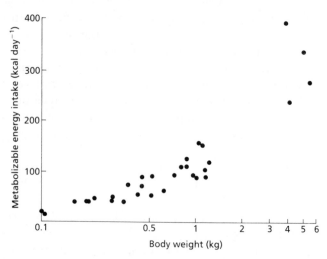

Fig. 6.4 Relationship between daily metabolizable energy intake and body weight in raptors (log-scale). Data are from captive birds kept under maintenance or near-maintenance conditions. The equation relating daily metabolizable energy need (ME_m) to body weight (W) in kilograms using log transformed data is: ME_m (kcal day^{-1}) = 110 kg$^{0.679}$ ($r^2 = 0.89$). Data from Kirkwood (1981).

made. Some of the most thorough concern the sparrowhawk, which is not an important predator of game, as it eats mainly songbirds. Sparrowhawks do not take their various prey species in proportion to the numbers of such species in the environment (Tinbergen, 1946). The extent of predation on any one species depends partly on the hunting behaviour of the hawk (attack modes and flight speeds) and partly on the behaviour and conspicuousness of the prey. Sparrowhawks find it easy to catch species which feed far from cover (e.g. house sparrow *Passer domesticus*) or are conspicuous (e.g. great tit *Parus major*), and take them in greater proportion than expected from their numbers in the area, whereas they find it harder to catch scrub-dwelling (e.g. wren *Troglodytes troglodytes*) or fast-flying species (e.g. swallow *Hirundo rusticola*). In consequence, sparrowhawks probably account for more than half the total mortality in certain prey species, and only a negligible fraction in others (Tinbergen, 1946).

Throughout the spring and summer, while they are breeding, sparrowhawks prey heavily on young songbirds, continually switching emphasis from one prey species to another, as each in turn produces young (Newton, 1986). Fledglings, which have just left the nest but are unable to fly properly, are especially vulnerable. Sparrowhawks near Oxford took up to one-third of all young great tits within the tit fledging period alone, and many others during the rest of the year, but still had no obvious effect on the numbers of pairs settling to breed each year (Perrins & Geer, 1980; McCleery & Perrins, 1991). However, during the early stages of breeding, some very local depression of tit breeding density became apparent within 60 m of active hawk nests, compared with densities elsewhere (Geer, 1978).

The main effects of the heavy predation on fledgling songbirds are to alter the seasonal distribution of mortality (many birds dying earlier than otherwise), to reduce the size of the post-breeding population peak, and to reduce the numbers dying from other mortality agents, such as starvation (Newton, 1986). When sparrowhawks were largely absent from south-east England due to pesticide poisoning in the years around 1960, no obvious increase occurred in songbird breeding numbers, and no obvious decline in such numbers occurred when sparrowhawks returned. This implied that, despite a heavy toll, sparrowhawks did not determine the level of the breeding populations of their most commonly taken prey.

On the coast of south-east Scotland, the impact of raptor (mostly sparrowhawk) predation on overwintering waders was estimated from frequent counts of kill remains and of the waders themselves (Whitfield, 1985). As in songbirds, species varied in vulnerability. The proportion of the total taken was estimated over two winters for redshank *Tringa totanus* at 20% and 16%, for turnstone *Arenaria interpres* at 4% and 5%, and for knot *Calidris canutus* at 1.5% and 0.5%. Over one winter the same proportion for purple sandpiper *Calidris maritima* was 1.5%, for dunlin *Calidris alpina* 4%, for grey plover *Charadrius squatarola* 2% and for ringed plover *Charadrius hiaticula* 19%. In at least four species juveniles were taken more than expected from their numbers, probably because they fed more near the edges of flocks and were slow to fly when a hawk approached. No mortality from hawk predation was found in three larger species. Thus sparrowhawks caused the bulk of winter mortality in the small–medium-sized waders studied, but this mortality was substantial in only two species. It could not be checked whether winter mortality was sufficient to reduce the breeding populations of any of the species examined.

Other work has concerned goshawks, which eat medium-sized mammals and birds, including game species favoured by human hunters. By use of radio-tagged falconry birds, Kenward (1978) found that goshawks tended to select the weaker individuals from woodpigeon *Columba palumbus* flocks. This was most evident when the pigeon was caught after a chase, as weak individuals were at a disadvantage compared to the rest of the flock; it was not apparent among pigeons caught 'by surprise' before they could fly far. By taking weak individuals, whose survival chances were low, the impact of goshawk predation on the pigeon population was reduced, because 28% of victims were already close to starvation, and beyond 'the point of no return' (Kenward, 1978).

In areas where pheasants *Phasianus colchicus* are released for hunting, a high proportion may be taken in winter by goshawks, which show both the numerical response to increased prey density (concentrating in areas of abundant prey) and the functional response (increasing the proportion of pheasants in their diet). In one Swedish study, goshawks killed 4–5% of released pheasants per month

(Kenward, 1977), and in another they killed 56% of wild hen pheasants during a winter (Kenward *et al.*, 1981). These losses were not mitigated by selection of birds in poor condition, probably because the pheasants were killed without a prolonged chase. Instead there was a preference for hen pheasants over the larger cocks.

By combining data from five Swedish and two German areas, it emerged that pheasants were a preferred prey for goshawks, which showed no tendency to switch to other prey when pheasant density was low (Kenward, 1985). Further work in good pheasant habitat on Gotland (Sweden) showed that, if the local density of goshawks was increased by high numbers of rabbits *O. cuniculus*, the hawks could reduce the number of pheasants below the level that could be replaced by breeding and thus cause a sustained population reduction. Temporary population reduction of ruffed grouse *Bonasa umbellis* numbers has been recorded during periodic goshawk irruptions in North America, in which the ratio of hawks to grouse is suddenly increased by hawk immigration (Keith & Rusch, 1988). However, these prey population reductions depended partly on prior mammalian predation which had reduced the breeding success of the prey, and mammalian predation alone can sometimes reduce game-bird populations, as shown by a predator removal experiment in Sweden (Marcström *et al.*, 1988).

Predation by goshawks and other predators may contribute to the cyclic fluctuations in hares and game-birds in the boreal zone of North America. As the numbers of hares rise, so do the numbers of goshawks, great-horned owls *Bubo virginianus* and other predators which eat them. Then, when hare numbers crash, the abundant predators switch to game-birds, reducing their numbers, and eventually declining themselves through starvation and emigration (Lack, 1954; Keith & Rusch 1988). In these circumstances, game-bird numbers can be said to be limited by predators, at least in the decline phase of the hare cycle. The same may happen in northern Europe, as generalist predators shift from their main prey, voles, to alternative prey including grouse, when voles get scarce (Angelstam *et al.*, 1984). In Sweden 'delayed density-dependent' mortality has been demonstrated for goshawk predation on hazel grouse *Bonasia bonasia* (Linden & Wikman, 1983). Further south in Europe, where raptors and other predators have a wider range of prey species, cycles in the numbers of individual prey animals are less marked or non-existent (Angelstam *et al.*, 1984).

The impact of any predator on its prey depends partly on the ratio of predator to prey numbers, as well as on the vulnerability of different sectors of the prey population. The idea that raptors and other predators remove the 'doomed surplus' was supported by work on red grouse *Lagopus lagopus scoticus*, in which winter predation was concentrated on non-territorial birds, most of which were destined to die anyway (Jenkins *et al.*, 1964). Moreover, when a territorial bird was removed, it was quickly replaced from the non-territorial contingent. In this study, however, breeding success was good, and both predation and shooting pressure were too light to remove all the doomed surplus, or to bring predators and hunters into conflict.

In various parts of Scotland, Redpath (1991) examined the impact of predation by hen harriers on the production of young by red grouse. Grouse chicks formed an important component of harrier diet, and in areas where harriers were present the ratio of young to adult grouse in July was 17% lower than in areas where harriers were absent. The difference could be attributed almost entirely to harrier predation. It was not clear, however, whether this predation reduced the numbers that were later shot by hunters, or the density of the next year's breeding population.

From such few studies, it is difficult to draw any generalizations about the impact of raptorial birds on their prey populations, especially as game-birds may be unusual in their vulnerability to predation. However, it is clearly no longer acceptable to maintain that predators take mainly the injured, diseased or dispersing prey (Errington, 1946), and do not depress prey populations. Although many bird populations are limited by food supplies (Lack, 1954; Newton, 1980), game-bird populations can be depressed locally or temporarily by predators, especially when the game-birds are secondary prey. Moreover, even when predators do not depress game-bird breeding populations, they may still be in competition with human hunters for the post-breeding population. It is not contradictory that predators might substantially lower game numbers at the time of the seasonal peak, but have no measurable effect on breeding numbers at the seasonal low.

The only satisfactory way to assess the effects of raptors and other predators on prey populations is by experiment, removing predators from certain areas, and leaving other areas as controls. Such experiments have been done several times, but to my knowledge only one showed that predators were depressing the breeding populations of their game-bird prey, and then it seemed that mammalian predators were more important than avian ones (Marcström *et al.*, 1988). In general, however, it is hard to separate the effects of raptors from those of other predators, and raptor predation on full-grown birds may be important chiefly when predation by crows and mammals has already greatly reduced the numbers of young produced.

EVOLUTIONARY ASPECTS

Raptors have another longer term influence on their prey species, besides affecting the pattern of mortality. By continually removing the more vulnerable individuals from the prey population, raptors (like other predators) act as an important agent of natural selection, gradually moulding the appearance and behaviour of the prey, and thus influencing the course of evolution. Considerable evidence has now accumulated that raptors can, in some circumstances, select sick, weak or otherwise odd individuals from the prey population (Dice, 1947; Eutermoser,

1961; Pielowski, 1961; Mueller, 1974; Kenward, 1978). Some of the odd prey included unusually coloured individuals giving firm evidence for the effect of predators in maintaining normal colour patterns in their prey.

The alarm calls, which many small birds give in response to a predator, and the fast zig-zag flights of waders over the seashore in the presence of raptors, have been interpreted as anti-predator, as has flocking itself. There is evidence for this latter view, in that single birds, or birds in small groups, are significantly more likely to be caught than are birds in large flocks, as shown by Kenward (1978) when flying his captive goshawk at woodpigeons. Moreover, shorebirds often change their dispersion patterns in accordance with predation risk: when under frequent attack by sparrowhawks, redshank abandon their territories and join flocks, while turnstones flock more tightly and range more widely (Whitfield, 1988). It is indeed not unreasonable to suppose that raptor predation has played a major role in the evolution of prey species, and has influenced much of the behaviour and colour pattern which we see in the prey species today.

7: Insectivorous Mammals

ILKKA HANSKI

It is a ravening beast, feigning itself gentle and tame, but being touched it biteth deep, and poisoneth deadly. It beareth a cruel mind, desiring to hurt anything, neither is there any creature it loveth.

[Topsell, 1607]

INTRODUCTION

Topsell's description of the greed of the shrew may fail as a piece of modern scientific writing, but this is of no concern to small, ground-dwelling insects. What matters to bugs and beetles is that large numbers of voracious shrews roam beneath the vegetation and grab anything not too big or too thick to kill. In the absence of more substantial prey, the European common shrew *Sorex araneus* will feed on small staphylinid beetles (Churchfield, 1982). One shrew needs to eat some 2000 of these in 24 hours, or roughly one beetle every 10 seconds of active foraging, just to stay alive (Hanski, 1984).

Of the 21 orders, 129 families and more than 4000 species of extant mammals, more than one-quarter (1200 species in 34 families and in nine orders) are primarily insectivorous (Corbet & Hill, 1980). Furthermore, many primates and rodents and other more omnivorous mammals include insects in their diet. Table 7.1 lists the families of predominantly insectivorous mammals that contain at least 20 species. These families fall into only three orders: Marsupialia (marsupials), Insectivora (shrews and moles) and Chiroptera (bats). Four other families in as many orders have specialized on ants and termites (Corbet & Hill, 1980). In addition, more than 200 mostly tropical species in 43 families have overcome the defences of ants and termites (Redford, 1988) to a sufficient degree to take advantage of their often enormous numbers.

Eisenberg (1981) has presented a classification of mammalian 'macroniches'. Of the eight kinds of insectivores, excluding the ant-eating species, four to five types occur in temperate latitudes and five to seven in subtropical and tropical regions. Curiously, northern South America is by far the richest region in foliage-gleaning and ant-eating insectivores, with eight and nine genera respectively, compared with none to three in the other zoogeographical regions. Non-flying insectivores are most diverse in south-east Asia (22 genera) and southern Africa (21), and least diverse in Europe (4). South America has no

Table 7.1 Families of insectivorous mammals with at least 20 extant species, and the four families which have entirely specialized on ants and termites. From Corbet & Hill (1980)

Family	Order	Geographical range	Species
>20 extant species			
Dasyuridae	Marsupialia	Australia, New Guinea	49*
Petauridae	Marsupialia	Australia, New Guinea	22
Soricidae	Insectivora	Eurasia, Africa, North America	246
Emballonuridae	Chiroptera	Tropics and subtropics	50
Rhinolophidae	Chiroptera	Old World, mostly tropics	69
Hipposeridae	Chiroptera	Old World, mostly tropics	61
Phyllostomatidae	Chiroptera	New World tropics, subtropics	140
Vespertilionidae	Chiroptera	Cosmopolitan	319
Molossidae	Chiroptera	Cosmopolitan, mostly tropics	91
Ant and termite specialists			
Tachyglossidae	Monotremata	Australia, New Guinea	2
Myrmecobiidae	Marsupialia	Australia	1
Myrmecophagidae	Edentata	South and Central America	4
Manidae	Pholidota	Africa, South-east Asia	7

* The number of described species has increased rapidly in recent years and is now more than 70 (C. Dickman, personal communication).

shrews, but their place is taken by small insectivorous marsupials and specialist cricetid rodents (Eisenberg, 1981). The smallest insectivorous mammals, bats and shrews, are most conspicuous in tropical and temperate regions, respectively. Small bats find it difficult to survive the long winters at high latitudes, whereas shrews find a relatively benign environment beneath a thick cover of snow (Aitchison, 1987).

Apart from some ant-eating species, insectivorous mammals are generally small and include the smallest living mammals: the shrews *Sorex minutissimus* (northern Eurasia) and *Suncus etruscus* (a widespread Old World species), the dasyurid marsupial *Ningaui timealeyi* (Australia) and the bat *Craseonycteris thonglongyai* (south-east Asia), all weigh around 2 g. The largest predatory beetles are larger than the smallest shrews! Insectivorous mammals have successfully occupied their niche for a long time (Eisenberg, 1981), and the pygmy shrew *Sorex minutus* is one of the oldest eutherian mammals (at least 4–5 million years; Sulimski, 1959). On the other hand, the *Sorex araneus* species group is notorious for its variety of 'chromosomal races' and sibling species, which are taken to be signs of recent and rapid chromosomal evolution (Hausser *et al.*, 1985).

Shrews and marsupial insectivores form guilds that exhibit strikingly regular differences in body size. The next section begins with a list of hypotheses about the ecological significance of size differences between congeneric and syntopic competitors. Three guilds of insectivorous mammals are then selected for a closer examination: *Sorex* shrews living in northern coniferous forests, dasyurid marsupials in Australia, and insectivorous bats more generally. The rest of the chapter is limited to a discussion of northern temperate shrews, partly out of personal interest and partly because of the amount and kind of information available for them. Special consideration is given to the role of insectivorous mammals in regulating forest insect populations.

THREE GUILDS OF INSECTIVOROUS MAMMALS

Even the most superficial study of shrews and dasyurid marsupials reveals striking differences in body size between coexisting species. There are five principal hypotheses about the significance of size differences within a guild of potentially competing species. (Kotler and Brown (1988) describe some other possible mechanisms not likely to be important in insectivorous mammals.)

1 The *classical view* of body size differences in competitors is based on resource partitioning: small predators are assumed to consume mostly small-sized prey, large predators are assumed to consume large-sized prey, and this partitioning of prey by size is assumed to facilitate coexistence (MacArthur, 1972). One complication is that because small predators may be unable to handle large prey, the large predators may have exclusive access to larger prey, giving them a competitive advantage (Wilson, 1975).

2 The *dominance hypothesis* proposes that large predators are dominant over small ones (Murray, 1971). Interference competition of this kind should decrease rather than increase the number of coexisting species. A classical example is the dominance of the stoat *Mustela erminea* over the weasel *Mustela nivalis* (Erlinge & Sandell, 1988).

3 The *productivity hypothesis* recognizes that individuals of small species are able to survive and breed in environments where food availability is too low to support individuals of large species (Thiel, 1975).

4 The *habitat selection hypothesis* assumes that habitat or microhabitat selection depends on the size of the species. There are many possible reasons why this might be so (Kotler & Brown, 1988), the productivity hypothesis being one special case.

5 The *predation hypothesis* suggests that predation is size-dependent (Kotler *et al.*, 1988), and if predation happens to be heavier on competitively superior species, then predation may facilitate coexistence.

These five hypotheses are not mutually exclusive. For example, hypotheses 2 and 3 in combination may explain regional coexistence of several species: the competitively superior (larger) species might exclude the smaller ones from high-quality habitats or microhabitats, but the inferior (smaller) species may find a refuge from competition in poor-quality habitats. To examine this combination of hypotheses in more detail, let us consider the following model (modified from Lande, 1987).

Assume that the environment is divided into discrete patches of the size of individual territories and that the patches can be of three qualities. The worst patches (a fraction u_1 of all patches) are unsuitable for both species, the best patches (a fraction h) are acceptable for both species, while the remaining patches (a fraction u_2) are suitable for species 2 (the smaller species) but unsuitable for species 1 (the larger, dominant species). Competition is asymmetric, so that species 2 can only establish itself in a patch in the absence of species 1, but species 1 can always replace an individual of species 2. Dispersing juveniles can search for m patches before perishing, unless they find an unoccupied, suitable patch and establish a territory. If p_1 is the fraction of patches occupied by the dominant species 1, Lande (1987) shows that, at equilibrium,

$$[1 - (u_1 + u_2 + p_1 h)m]R_0' = 1 \tag{7.1}$$

and

$$p_1 = 1 - \frac{1 - k_1}{h} \tag{7.2}$$

where R_0' is the net lifetime production of female offspring per female, conditional on the mother's finding a suitable territory (for details see Lande, 1987), and $k_1 = (1 - 1/R_0')^{1/m}$. The equilibrium fraction of sites occupied by species 2 is

$$p_2 = 1 - \frac{1 - k_2}{u_2 + (1 - p_1)h} \tag{7.3}$$

Figure 7.1 shows that, depending on the structure of the environment as reflected in the values of u_1, u_2 and h, there is a range of possible outcomes: (i) neither species is able to maintain a population in the environment; (ii) species 1 or 2 will occur alone; or (iii) the two species coexist in stable equilibrium. Extrapolating the model to guilds of many species, species diversity is expected to increase with increasing productivity at first, but eventually to decrease. Abramsky and Rosenzweig (1984) found just such a humped species richness curve for desert rodents. The prey-size partitioning hypothesis (hypothesis **1**) predicts that coexistence depends on the distribution of prey sizes. It is an empirical question as to how the prey size distribution changes with productivity, but in most cases the range of prey sizes is expected to become broader with increasing productivity, so that more predator species are expected to coexist in more productive environments.

Sorex shrews in northern temperate forests

Little ecological information is available for the many tropical shrews of the subfamily Crocidurinae, but many are thought to be uncommon and to have restricted geographical distributions (Corbet & Hill, 1980). In contrast, many temperate shrews in the subfamily Soricinae have transcontinental distributions and reach high densities.

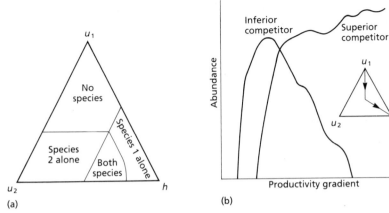

Fig. 7.1 (a) Coexistence of competitors under the combination of hypotheses 2 and 3 (equations 7.1–7.3) No species, species 1 (superior competitor) or species 2 (inferior competitor) alone, or both species may occur in the environment depending on the relative frequencies of high-quality (h), low-quality (u_1) and intermediate habitat patches (u_2). (b) A hypothetical example of changing abundances of an inferior and a superior competitor along a 'productivity gradient' (changing proportions of low-quality, intermediate and high-quality patches along the line in the insert). The vertical axis gives the fraction of suitable patches occupied by each species.

Sorex shrews have an exceptionally simple life cycle. Young shrews disperse from their natal area in search of a feeding territory, which they occupy and maintain during the winter (Croin Michielsen, 1966; Hawes, 1977). Territory ownership is essential for successful over-wintering (Hawes, 1977). All males and nearly all females delay maturation until the following spring (Pucek, 1959), when females maintain territoriality but males become more mobile and greatly expand their home ranges (Hawes, 1977). The breeding system is one of promiscuity. Shrews have postpartum oestrus, and females can produce two or even three litters in rapid succession. Most shrews die before the end of their second summer.

The *Sorex* shrews typically coexist with *Clethrionomys* and *Microtus* rodents, and the three genera comprise a continuum from insectivorous shrews to omnivorous *Clethrionomys* to herbivorous *Microtus*. In this comparison (Table 7.2), the food resources of the shrews are of the highest quality but also the most likely to become limiting. Several authors have reported a positive correlation between the density of shrews and the abundance of their prey (Holling, 1959a; Judin, 1962; Butterfield *et al.*, 1981). The food of shrews is not uniformly distributed, but tends to be concentrated in patches (e.g. under piles of decomposing vegetation), which may make territoriality profitable. Territoriality has the additional advantage of allowing the shrew undisputed access to a thoroughly familiar area in winter, when long-distance movements would be dangerous (Hawes, 1977). In summary, the strong territoriality exhibited by immature shrews contributes to a lower density than

Table 7.2 Comparison between three common genera of small mammals in northern temperate regions

	Sorex	*Clethrionomys*	*Microtus*
Food habits	Insectivorous	Omnivorous	Herbivorous
Food distribution	Patchy	Intermediate	Uniform
Territoriality			
Immature individuals	Yes	No	No
Mature individuals	Females	Females	Males
Social inhibition of			
maturation	Yes	Yes	No
Maximum densities (no. ha^{-1})	50	100	200

Note. There is variation in the breeding system of *Clethrionomys* and especially *Microtus* species (e.g. Viitala & Hoffmeyer, 1985). The information in the table refers to the common European species *C. glareolus* and *M. agrestis*.

tends to be found in *Clethrionomys* or *Microtus* and gives rise to relatively stable populations.

Although winter is the period when food availability is lowest, strict territoriality means that winter survival is likely to be density-independent. This prediction is borne out by 15 years of data from Finnish Lapland (Fig. 7.2). In contrast, 'breeding success', the proportional change in population size from early summer to early autumn, is strongly density-dependent (Fig. 7.2; similar results have been reported by Pernetta, 1977). Breeding success combines natality, which is lower

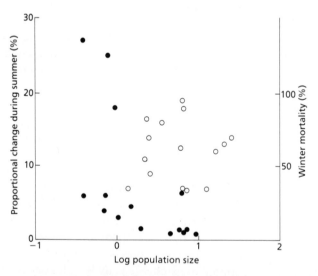

Fig. 7.2 Winter survival and proportional change in population size from early summer to early autumn in the common shrew *Sorex araneus* in Finnish Lapland during 15 years. Population size on the horizontal axis is either spring density (for change during summer, ●) or autumn density (for winter mortality, ○). From Kaikusalo & Hanski, (1985). Winter mortality is density-independent but summer change is density-dependent.

at high density (Sheftel, 1989), and mortality of young soon after leaving their natal area (Pernetta, 1977).

The relative stability of shrew populations is disrupted in areas where the microtine rodents have cyclic dynamics, as for example in northern Fennoscandia (Hansson & Henttonen, 1985; Hanski *et al.*, 1991); here the shrew populations tend to fluctuate in partial synchrony with the rodents (Hansson, 1984; Kaikusalo & Hanski, 1985). The differences between shrews and rodents in diet, parasites and life cycles strongly suggest that the cause of the observed synchrony is shared predators (Hanski, 1987a). Only a few species of predators feed on shrews when other prey are available, but when rodent populations crash, most predators take at least some shrews (Erlinge, 1975; Korpimaki, 1988). Incidentally, the most likely competitors of shrews in temperate coniferous forests are the wood ants, *Formica* species which the shrews (other than the smallest species, *Sorex minutissimus*) are apparently unable to feed upon.

The holarctic coniferous forests support a guild of shrews with some five species typically coexisting at one place; the record number of coexisting species is nine in central Siberia (Sheftel 1983, 1989). Figure 7.3 compares the morphology of the guild of *Sorex* shrews from northern Europe (Finland) and western Canada (Alberta), where there are five and four species, respectively (excluding the large water shrews

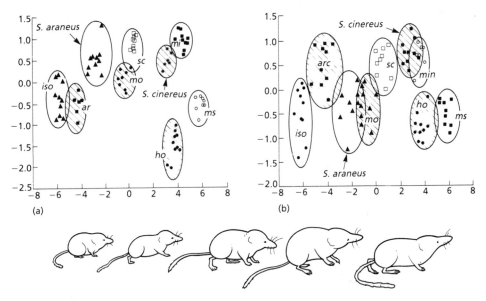

Fig. 7.3 Principal component analysis of metrical measurements of 17 skull and 15 post-cranial skeletal traits for five Eurasian and four North American (shaded) *Sorex* species (ho, *hoyi*; ms, *minutissimus*; mi, *minutus*; sc, *caecutiens*; mo, *monticolus*; ar, *articus*; is, *isodon*; 10 young individuals of each species measured). The horizontal axis is the first principal component (a measure of general size). Below the panels, the five European species are depicted in the order of increasing size. The numerically dominant species on the two continents are *S. araneus* and *S. cinereus*. Data from I. Hanski (unpublished).

Neomys fodieus and *Sorex palustris*. In both guilds, the species are well separated in size. Only the two smallest species, *S. minutissimus* in Eurasia and *S. hoyi* in North America, are clearly distinct in shape, and although these species are not related to each other, they deviate in the same direction from the rest of the species (Fig. 7.3).

Habitat selection in shrews supports restricted hypotheses **2** and **3**: larger species tend to be restricted to the most productive habitats, while smaller species are relatively or even absolutely more numerous in less productive habitats (Fig. 7.4). The model described by equations 7.1–7.3 assumes interspecific territoriality, which has been reported for shrews in North America (Hawes, 1977) and Siberia (Moraleva, 1987). The smaller species are not entirely excluded from the best habitats in Fig. 7.4, but this figure summarizes results for a large number of studies, and there is unknown variation in microhabitat quality within one 'habitat'. More detailed observations from uniform patches of high-quality habitat have revealed complete dominance by one large species. For example, the largest species in Fig. 7.3, *S. isodon*, may occur alone in high density in restricted areas of high-quality habitat (U. Skarén, personal communication).

At the continental scale, it is puzzling that the numerically dominant shrew from Europe to western Siberia is *Sorex araneus*, a relatively large species (8–12 g, Fig. 7.3), whereas the numerically dominant species in North America is a small one, *S. cinereus* (3–5 g, Fig. 7.3).

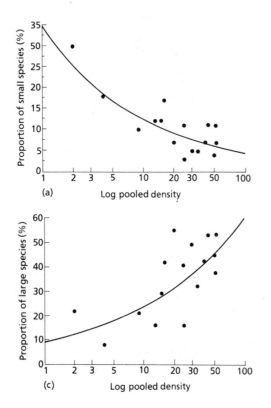

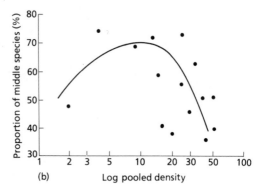

Fig. 7.4 Relative abundances of (a) small (adult weight <6 g, four species), (b) medium (7–9 g, three species), and (c) large species (adult weight >10 g, four species) of *Sorex* shrews against the logarithm of their pooled density, a measure of habitat suitability for shrews. The data points are average values for 16 habitat types, calculated from results for 93 localities in northern Europe and Siberia. From Sheftel & Hanski (unpublished data).

Clearly the reason for this difference is not that there are no large species in North America or small species in Eurasia (Fig. 7.3). A possible explanation, however, is a major difference in the availability and kind of food for shrews between the two continents. Apart from the area of North America to the west of the Rocky mountains, and the eastern seaboard, where rainfall is high and forests are productive, most coniferous forests of the interior of northern North America appear to be less fertile and probably have less food for shrews than their European and west Siberian counterparts. Earthworms (Lumbricidae), which are an important food source for *S. araneus* and other large species (Judin, 1962; Rudge, 1968), are believed to have became extinct in northern North America during the Pleistocene glaciations, and now occur only in limited areas where they have been re-introduced by man (either from Europe or south of the Appalachians in North America; Reynolds, 1977). This continent-wide difference in the shrew assemblages is in accordance with the productivity hypothesis (hypothesis **3**): relatively low food availability may limit the success of large species of shrew over vast areas of coniferous forests in North America.

Marsupial insectivores: Dasyuridae

There are more than 70 living species of dasyurid marsupials in Australia and New Guinea (C. Dickman, personal communication), varying by more than three orders of magnitude in size, from *Ningaui timealeyi* (2 g) and small *Planigale* species 3–4 g) to *Sarcophilus harrisii* (8 kg; Lee & Cockburn, 1985). The smaller species are entirely or primarily insectivorous, filling the same niche as shrews, while the larger species are equivalent to mustelids and small viverrids. All the species are opportunistic predators (Fox, 1982a; Lee & Cockburn, 1985). The insectivorous species display a range of foraging behaviours similar to that observed in shrews: the relatively large *Antechinus swainsonii* forages amongst forest litter, sometimes digging to obtain prey, while the smaller *A. stuartii* hunts at the soil surface and in trees (Lee & Cockburn, 1985). Such microhabitat partitioning may facilitate coexistence (hypothesis **4**; Dickman, 1986a, 1988a).

The dasyurid marsupials demonstrate a wide range of life history patterns: from monoestrous to polyoestrous species; from semelparous to iteroparous species; and from species with a very restricted breeding season to species that breed continuously (Lee *et al.*, 1982; Lee & Cockburn, 1985). The dramatic and complete post-mating mortality of males in some species of *Antechinus* has received considerable attention recently (Lee *et al.*, 1977; Braithwaite & Lee, 1979). Lee and Cockburn (1985) suggest that during the breeding season males commit a maximum amount of time to mating. As a result, they feed less, and produce glucose from tissue protein instead. A general stress response leads to gastrointestinal ulcerations, to suppression of the immune and inflammatory responses, and rapidly to the inevitable death of sexually active males. Females are not affected in the same way, and depending on the species, a fraction of females survives until the following breeding season. This contrast in the life histories of males and females, combined with the regular natal dispersal of juvenile males but strong philopatry of females, leads to high levels of intrasexual competition between mothers and daughters. Local resource competition amongst females (Cockburn et al., 1985) as well as interspecific competition (Dickman, 1988b) have been invoked as explanations of the remarkably wide variation in sex ratio observed in Antechinus.

Dasyurid populations are believed to be regulated primarily by intraspecific competition (Lee et al., 1977; Fox, 1982a; Godsell, 1982), with predation playing only a relatively minor role (Guiler, 1970; Braithwaite, 1973). Interspecific competition also occurs. Dickman (1986a,b; Dickman et al., 1983) has studied competition between two species, Antechinus stuartii (15−30 g) and the larger A. swainsonii (40−50 g). In one experiment, Dickman (1986a) removed one of the species from several relatively isolated 1.7-ha plots. Figure 7.5 shows that after the breeding season following the removal of the larger species, A. stuartii attained densities 60% higher than on the control plot, but there was no such increase in the density of A. swainsonii in a reciprocal experiment. The increase in the density of A. stuartii was correlated with enhanced survival of the weaned young, increased radius of movements and home ranges, and shifts in habitat use and diet. Other but less conclusive results on interspecific competition have been reported by Van Dyck (1982) for A. stuartii and A. flavipes and by Fox (1982b) for A. stuartii and Sminthopsis murina. These results support the hypothesis of asymmetrical interference competition by the larger species (hypothesis 2). Hall (1980) has shown that, contrary to hypothesis 1, the two species A. swainsonii and A. stuartii do not coexist by partitioning food on the basis of size, although as might be expected, the larger species uses more of the largest prey items.

The densities of small dasyurids are generally lower than the densities of shrews, at around 5 ha^{-1} or less (Fox, 1982a; cf. Table 7.2), though some much higher values have been reported, e.g. 45 ha^{-1} in A. stuartii (Wood, 1970) and 80 ha^{-1} in A. minimus (Wainer, 1976). As with shrews, densities may be directly related to food supply (Guiler, 1970). Dickman (1989) has demonstrated that supplementary food improves the survival, increases the body weight and decreases the movements of A. stuartii. Generally, smaller species occur in the less productive habitats or microhabitats (Fox, 1982a,b; Dickman, 1986a), supporting hypothesis 3, and since competition to occur mostly by interference (hypothesis 2), regional and local coexistence appear to be enhanced by the combination of hypotheses 2 and 3 rather than by food-size partitioning (hypothesis 1).

In the relatively well-studied temperate communities in Australia, the number of co-occurring dasyurids in one habitat is typically small,

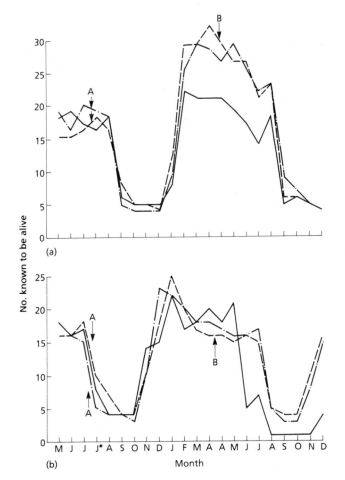

Fig. 7.5 Population size of *Antechinus stuartii* (a) and *A. swainsonii* (b) on semi-isolated plots from which the other species was removed at a point indicated by A (in both cases two experimental plots and the same control plot). The density of the smaller species (*A. stuartii*) increased after the breeding season following the removal of the larger competitor, but there was no reciprocal effect. Breeding occurs in winter and spring (August–November), and juveniles appear from December onwards. Control, ——; experimental removal plots, – – – and –·– (two replicates). From Dickman (1986a).

between one and three, although regional species diversity is high (Fox, 1982a, 1987). As we have seen, the number of co-occurring shrew species in northern coniferous forests is higher, typically from four to six. One possible explanation is the smaller average size of shrews than dasyurids, making small-scale microhabitat variation, in food availability, for example, more significant to shrews than to dasyurids. Baynes (1975; cited in Fox, 1982a) described 11 dasyurids from four habitat types in south-western Australia. Each habitat had one to three species, no species occurred in two or more habitats, and the species within habitats were well separated in size (Fox, 1982a shows that the minimum weight ratio was 1.5). To test whether the observed size separation was better than expected if the habitat assemblages had been drawn randomly from the regional species pool, I counted for each co-occurring species pair how many other species in the regional species pool were of intermediate size. The observed two smallest values were one and two species (sum three), while the probability of obtaining an equally or more extreme result was 0.058. Baynes' results thus suggest that co-occurring dasyurids do form non-random assemblages with respect to size. In view of the observational and experimental results on competition, and the lack of any information on other plausible processes, it seems reasonable to conclude that interspecific competition affects the species composition of temperate dasyurid assemblages (see also Fox, 1982a). The situation may be different in the arid regions of Australia, where densities of insectivorous marsupials are low, $<1 \, ha^{-1}$ (Dickman, 1989), but where local diversity may reach four to five species (C. Dickman, personal communication). It is possible that lack of significant interference competition at these very low densities contributes to the relatively high number of co-occurring species.

Insectivorous bats

Insectivorous bats are different from shrews and dasyurids in some obvious, and in some less obvious ways. Bats are able to fly, and can exploit the often large populations of flying insects, from which they select primarily moths and beetles. Like shrews and dasyurids, insectivorous bats are generalists (Fenton, 1982), though they may show short-term and seasonal specialization on an abundant prey species (Vaughan, 1977). Bats display more variation in their mode of foraging than shrews and dasyurids. At one extreme are the high, fast and direct-flying Molossidae (Freeman 1981a), which are the nocturnal counterparts of swifts. At the other extreme are the gleaners, which pick their food from the ground or from foliage (Fenton, 1982). Some bats employ the 'flycatcher strategy' (Vaughan, 1977).

The use of echolocation by insectivorous bats to detect their prey makes the interaction between the predator and they prey more intimate than the interaction between non-flying insectivores and their prey. (Incidentally, short-range echolocation, probably used for navigation in complex habitats, has been demonstrated also in shrews; Buchler, 1976). In principle, a predator capable of Doppler-shift compensation could use wing-beat information to identify prey species (Fenton, 1982), but this ability is not likely to be of great adaptive value in nature, where prey populations are continuously varying in time and space. More importantly, the use of echolocation means that the prey has the potential to detect an approaching predator, and the scene is set for adaptations and counteradaptations in the prey and the predator (see, for example, Fenton 1982). Ears sensitive to ultrasonic sound are widespread in insects such as moths, possibly indicating the extent of the selection pressure imposed by bats. In the vicinity of large colonies of bats the pressure is not difficult to imagine: Kunz (1974) estimated that the 50 000 *Myotis velifer* in Kansas consumed 15 tons of insects during the warm season.

Bats are almost exclusively nocturnal but, within this constraint, the insectivorous species have particularly plastic circadian rhythms, possibly related to variability in the availability of their prey species (Erkert, 1982). Because of their nocturnal activity, bats have few important predators. There is little likelihood of competition between insectivorous bats and nocturnal birds, nor between bats and diurnal birds, since most insect species are either nocturnal or diurnal in their flight periods.

An important consequence of nocturnal foraging in diurnal fasting, which poses as great a problem to small bats as it does to small shrews. Small insectivorous bats have solved this problem by entering daily torpor in order to save energy. Temperate bats also face seasonal fluctuations in food availability, but here the responses have diverged (see, for example, McNab, 1982): some bats migrate south for winter, some enter long, seasonal torpor, while others maintain year-round activity (the latter option is not open to bats in northern temperate regions).

A basic difference to non-flying insectivores is the generally much greater longevity of insectivorous bats than other mammals of similar size. Even the small species may live for 15 years or longer (Tuttle & Stevenson, 1982). McNab (1982) concludes that the generally low rate of metabolism and the tendency of small insectivorous bats to enter daily and (in many temperate species) seasonal torpor are related to their exceptional longevity. Great longevity is associated with low fecundity, and most bats produce only 1 young litter^{-1} (Tuttle & Stevenson, 1982). However, the diversity of social systems in insectivorous bats is impressively large, ranging from year-round harem systems to entirely solitary species in which the two sexes come together only for copulation (Eisenberg, 1981).

Field studies on the ecology of insectivorous bats present substantial technical difficulties, which may explain why many authors have resorted to morphometrical studies in their attempts to understand the ecology of bat assemblages. Wing morphology is obviously related to flight (Findley & Wilson, 1982), and hence to the mode of foraging. Jaw and tooth morphology are related to diet (Freeman, 1979, 1981b), and morphological similarity in mouthparts is assumed to be related to similarity in food (Findley & Black, 1979). Nonetheless, assemblages of insectivorous bats typically consist of many similar species, with only a few distinctive, apparently more specialized, species (Findley, 1976; Fenton, 1982). Thus, in striking contrast to their greater diversity in behaviour, life history and social systems, guilds of coexisting bats typically consist of species of rather uniform body size. At first sight, this is all the more surprising, because the number of co-occurring bat species may be an order of magnitude greater than the number of coexisting shrews, and more again than the number of dasyurid species.

One explanation may be simply that there are more species of bats in total, hence also more coexisting species of similar size. Another explanation is related to the fact that bats require a particular site for their diurnal roost. Many species of insectivorous bats congregate in vast numbers in the roost sites — in southern USA for example, a single cave may house up to 20 million individuals of *Tadarida brasiliensis* (Davis *et al.*, 1962). Populations of many bat species may be limited by the availability of roost sites (e.g. caves, crevices, or tree cavities) and not by food, although some species use external roost sites, like tree foliage, which are presumably less limiting. Roost site limitation is also suggested by the readiness of many bat species to adopt a variety of human structures as roosts (Kunz, 1982). Different species tend to use different roost sites (Graham, 1988), and this will amplify the importance of intraspecific competition relative to inter-specific competition, further facilitating coexistence (Ives, 1988). Even if populations were limited by food rather than by roost sites, intra-specific aggregation at the same roost sites would still mean that intraspecific competition for food was more important than interspecific competition, since individuals from the same roost are more likely to forage with one another than in the company of other bat species. Kunz's (1974) observations on colony size and dispersal distances indicate the intensity of intraspecific competition within large colonies, and show how competition may lead to temporary or permanent split-up of colonies.

FORAGING ECOLOGY OF SHREWS

Like dasyurid marsupials and insectivorous bats, shrews are opportun-istic predators. Studies of their diets have typically documented dozens of prey types, even if the taxonomic resolution of these studies rarely reaches even the generic level (Rudge, 1968; Pernetta, 1976; Butterfield *et al.*, 1981). There is a certain amount of differentiation in prey selection due to differences in foraging behaviour and microhabitats: larger species of shrew spend more time in underground tunnels search-ing for earthworms, for example, while the smaller species prey on insects between the litter and the soil surface (Okhotina, 1974; Pernetta, 1976).

There seems to be little difference in the 'profitability' of different prey types to different species of shrew (Dickman, 1988a, with the exception that the largest species do poorly on the very smallest prey (Fig. 7.6), and the smallest shrews cannot use the largest prey. For example, the pygmy shrew *Sorex minutus* has been observed to spend 10 minutes in trying to kill *Geotrupes stercorosus*, a large and sturdy beetle, without doing more harm than breaking one of its legs (I. Hanski, personal observations). A notable feature of the results in

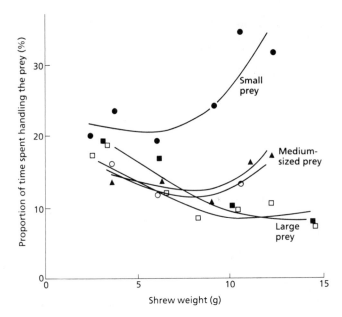

Fig. 7.6 Profitabilities of different-sized beetles to six species of shrew varying in size (*Sorex minutissimus, S. minutus, S. caecutiens, S. araneus, S. isodon* and *Neomys fodiens*). The beetles are, in the order of increasing size, *Cercyon* sp. (●, 1.3 mg), *Aphodius prodromus* (▲, 4.5 mg), *A. fimetarius* (○, 9.8 mg) *Sphaeridium* sp. (□, 14.3 mg) and *A. fossor* (■, 35.2 mg). Profitability is measured, on the vertical axis, by the percentage of 24 hours needed for killing, handling and eating the prey type to satisfy the size-dependent energy demands of the shrew. Data from I. Hanski (unpublished).

Fig. 7.6 is that all species did best on the largest prey types. This result is consistent with the observations that shrews generally prefer large prey items in experiments (Barnard & Brown, 1981) and in the field (Platt & Blackley, 1973; Dickman, 1988a; Parmenter & MacMahon, 1988).

Positive correlation between the numbers of shrews and the abundance of their prey suggests that while shrews may be food-limited, their prey is not usually predator-controlled, not at least by shrews. In support of the latter view, Platt and Blackley (1973) found that shrews (*S. cinereus*) caused no significant reduction of the biomass of prey insects, although shrew predation decreased the frequency of the dominant, large-sized prey. Shrews may take proportionately more large prey items simply because they are easier to detect than small ones. One large prey item may be equivalent to 10 small ones in energy content, and concentrating on large prey when they are readily available may reduce total search time. Large prey items, such as insect pupae, can be easily stored, and such food caches may be critically important in helping the shrews to survive short term fluctuations in food availability. Preference for insect pupae has important implications for the potential of shrews to regulate forest insect populations.

Small versus even smaller shrews

Shrews are neither nocturnal nor diurnal, but are active throughout the 24 hours. Their activity cycle is short, consisting of about 1 hour of foraging, followed by 1 hour of sleep (Saarikko & Hanski, 1990). The length of the sleep remains remarkably constant, but the length of the foraging bout is variable, and is especially long when food availability in the environment is changing (Table 7.3). This may indicate an increased level of environmental sampling by the shrews.

The smallest shrews weigh less than 2 g and consume more than twice their body weight in 24 hours (Hanski, 1984). The small species starve after a few hours of food deprivation. Larger species weigh 10–20 g, and although they are still small by mammalian standards, they can survive on their body reserves for 20 hours or more. Hanski (1985a) modelled the expected responses of small and large shrews to food shortages. Assume that the shrew is faced with the decision between continuing to forage, or resting until the food shortage is over. Hanski (1985a) concluded that the shrew should opt to rest rather than forage if

$$1 - (1 - p)^T > \frac{p}{p + (1 - p)q} \tag{7.4}$$

where T is the starvation time in hours, p is the probability that food availability returns to a high level in the next hour, and q is the probability that a foraging shrew dies in the next hour when food availability is low (e.g. because it is eaten by a predator). This simple model makes two testable predictions: (i) because T increases with body size, larger species of shrew are predicted to be more likely than small ones to decrease their activity during periods of low food availability; and (ii) the shorter the average length of food shortage (larger p), the more likely any shrew is to respond by decreasing activity. In a comparison of small and large species under the same experimental conditions, it was found that, as predicted, small shrews increased, while large species decreased, their activity in response to short-term food shortages (Hanski, 1985a). The duration of food shortages also had the predicted effect on the activity of the common shrew *S. araneus*: the length of the foraging bout was increased when food availability was constantly rather than only temporarily low (Table 7.3). It is worth reiterating here that the temperate shrews (subfamily Soricinae) have no capacity to enter torpor, unlike their tropical relatives (subfamily Crocidurinae; Vogel, 1980).

Shrews and other insectivorous mammals often find their food in places where it would be risky to stay for long periods of time, as, for instance, on the open forest floor. In these circumstances, a food item is picked up and carried to a spot where the shrew is safe from predators and competitors. McLeod (1966) demonstrated that while the spatial distribution of sawfly cocoons in the forest floor is initially

Table 7.3 Comparisons of the lengths of foraging bout and sleep in the common shrew *Sorex araneus* under different levels of food availability. In A, the same low food availability (reward probability 0.18 per visit to a feeder) is either constant for several days, or temporary, interrupted by higher level of food availability. Note that the foraging bout is longer in the former. In B, four situations are compared: in B1 food availability is low (0.10) during the entire foraging bout; in B2 food availability is high (0.50) during the entire foraging bout; in B3 food availability has changed from low to high during the foraging bout; and in B4 food availability has changed from high to low during the foraging bout. Note that a change in food availability increases the length of the foraging bout, probably because of increased sampling of the environment by the shrew. From Hanski & Saarikko (in preparation)

Environment	Reward (probability)	Length of foraging bout (minutes)	Length of sleep (minutes)
A1 Constantly low	0.18	60.9	64.3
A2 Temporarily low	0.18	49.5	62.0
ANOVA:			
Treatment (P)		0.079	0.159
Individuals (P)		0.000	0.000
B1 Low availability	0.10	43.5	64.0
B2 High availability	0.50	36.7	60.0
B3 Change: low to high	0.10 > 0.50	70.8	62.8
B4 Change: high to low	0.50 > 0.10	55.4	60.7
ANOVA:			
Treatment (P)		0.001	0.811
Individuals (P)		0.015	0.001

more or less random, the distribution of cocoons detected and consumed by shrews (*S. cinereus*) is highly clumped. The cocoons that were eaten were found in heaps of various sizes under logs, in old tree stumps, and suchlike places.

A simple extension of this behaviour is short-term food caching. If a group of desirable prey items is detected, it may be better to collect them all as quickly as possible rather than to eat them on the spot. Caching of insect pupae is common in all shrews (Buckner, 1959, 1969; McLeod, 1966; Hanski & Parviainen, 1985), but it is especially characteristic of small species of shrew, at least under laboratory conditions (I. Hanski, personal observations). This suggests that one function of caching might be to limit the access of stronger competitors to food. Barnard *et al.*'s (1983) experimental results with *S. araneus* demonstrated a dramatic increase in caching in the presence of a competitor. However, there is also another explanation for caching by small shrews; short-term food caches may serve the same function as body reserves in larger species, reducing the risk of starvation. Hanski (unpublished data) compared the responses of two species, *S. araneus* and *S. caecutiens*, to temporal variation in food availability. Relative to the larger *S. araneus* (8–12 g), two individuals of the smaller

S. caecutiens (3–6 g) increased their foraging activity with declining food availability (as expected for a small species), but in a striking contrast, two other individuals *decreased* their activity as food availability declined. These two small individuals, however, had accumulated food caches when food availability was high, and when food became scarce they started to deplete their caches.

Functional and numerical responses

Holling's (1959) classical experiments on three species of small mammals preying on the cocoons of the European pine sawfly *Neodiprion sertifer* demonstrated sigmoid (Type 3) functional responses in the field (Fig. 7.7), consistent with laboratory observations. Sigmoidal functional responses may be caused in a variety of ways, including learning (search image) and changes in the movement patterns of predators with varying prey density. Even the fine details of the spatial distribution of prey can have a profound influence on predation, as Buckner (1959) demonstrated in showing that the rate of predation on sawfly cocoons decreases significantly with distance from small-mammal tunnels. The size of the insectivorous mammal is likely to affect its responses to the spatial distribution of prey, as small species may survive on widely scattered food items, whereas large species require either larger prey items or larger patches of prey to survive and

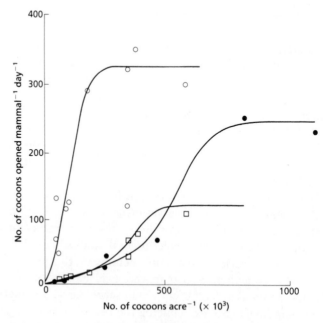

Fig. 7.7 Functional responses of *Sorex cinereus* (□), *Blarina brevicauda* (o) and *Peromyscus maniculatus* (●) to varying density of sawfly cocoons in pine plantations. From Holling (1959a).

breed. Hanski and Parviainen (1985) planted sawfly cocoons singly, in groups of 10 cocoons and in larger groups of 100 cocoons in the forest floor. The small *S. caecutiens* (3−6 g) consumed mostly single cocoons and cocoons in small groups, while the much larger bank vole *Clethrionomys glareolus* (15−35 g) took cocoons chiefly from the larger groups. It was not clear why the small species used so few cocoons from the large groups, but interference by the larger bank vole may have been important. Similar size-related differences in foraging behaviour have been reported for seed-eating desert rodents (Price, 1978).

The abundance of outbreak forest insects, many of which are potentially important prey items for shrews, may vary by two or three orders of magnitude. However, many of the outbreak forest insects are accessible to insectivorous mammals for only a short period of time, which means that a strong numerical response based on variation in predator birth and death rates is unlikely. For instance, no numerical responses of insectivorous mammals were reported following increased numbers of spruce budworm in Canada (Morris *et al.*, 1958). Holling's (1959a) results on *Sorex* and *Peromyscus* suggested a numerical response to a seasonal prey, *Neodiprion sertifer* cocoons, but he worked in pine plantations where there was little alternative food and where, during a *N. sertifer* outbreak, many sawfly cocoons may have exceptionally stayed in prolonged diapause and therefore have been available to shrews for most of the year (Hanski, 1988).

INSECTIVOROUS MAMMALS AND FOREST INSECT OUTBREAKS

Certain observations on forest insect outbreaks are consistent with the hypothesis that some insect populations are regulated by insectivorous mammals. Other data suggest that insectivorous mammals make at least an important contribution to the limitation of the pest populations.

1 Outbreaks tend to start in, or are restricted to, low-productivity habitats, e.g. pine forests on bogs and sandy soils (Hanski, 1987b).

2 Outbreaks are often preceded by a series of dry summers (Marninat, 1987). Although often attributed to changes in plant quality, both these observations are equally well explained by predation. Insectivorous mammals are typically scarce in low-productivity habitats (Bess *et al.*, 1947; Hanski & Parviainen, 1985; Hanski, 1990), and shrew populations generally decrease during exceptionally dry summers, due to a general decrease in food availability (Myllymäki, 1969; Pankakoski, 1984).

3 Many forest outbreak species have a short pupal period, while congeneric non-outbreak species have long pupal periods (Hanski, 1987b). A short pupal period increases the probability of an outbreak due to a short vulnerable period to predation (Buckner, 1955; Hanski, 1987b).

4 Many outbreak species have gregarious larvae, which may increase the probability of the species escaping from the control of local predators by increasing variability in the growth rate of the prey population (Hanski, 1987b).

Rate of predation by insectivorous mammals

Insectivorous and omnivorous mammals often inflict a high rate of mortality on insect pupae in the forest floor, while insectivorous birds prey primarily on larvae in the foliage. Depending on the biology of the prey species, either insectivorous birds or mammals could have the greatest potential to regulate the insect population. Small mammal predation on sawfly cocoons is typically 50% or more, occasionally up to 95% (Table 7.4). Lack of significant numerical (demographic) responses by small mammal populations, however, means that predation is likely to be relatively unimportant during outbreaks as a result of predator satiation. The important question is: what role do small mammals play in preventing insect populations from increasing to outbreak levels?

The sigmoid functional response which characterizes at least some interactions between small mammals and their insect prey (Fig. 7.7) have the potential to stabilize the prey population at low density. Apart from Holling's (1959a) classic work, however, few studies have examined density-dependence in the predation inflicted by insectivorous mammals. Olofsson (1987) used life-table analysis to investigate the different mortality factors acting upon the pine sawfly *N. sertifer* populations in Sweden. Only predation by small mammals, probably

Table 7.4 Rate of predation by shrews and omnivorous small mammals on the pupae of forest insects. The period over which predation rate has been estimated varies from one to several months, but generally most predation occurs in 1–2 months in late summer and early autumn. 'Exp' indicates that the data are from an experiment (usually a set of tagged cocoons placed in the forest floor for the pupal period)

Predator	Prey	Cocoon density $(1000\,ha^{-1})$	Percentage predation	Reference
Small mammals	Sawfly cocoons	<1 *Exp	99	Hanski & Otronen (1985)
Small mammals	Sawfly cocoons	<1 *Exp	72	Buckner (1958)
Small mammals	Sawfly cocoons	1 *Exp	68	Hanski & Parviainen (1985)
Small mammals	Gypsy moth pupae	Sparse	70	Campbell & Sloan (1977)
S. araneus	Winter moth pupae	? Exp	45	Buckner (1969)
Shrews	Sawfly cocoons	30	95	Olofsson (1987)
Shrews	Sawfly cocoons	43	95	Buckner (1964)
Shrews	Sawfly cocoons	150	55	McLeod (1966)
Small mammals	Sawfly cocoons	206	71	Kolomiets *et al.* (1979)
Shrews	Sawfly cocoons	1400	45	Buckner (1964)
Apodemus	Sawfly cocoons	c. 1500	50	Obrtel *et al.* (1978)
Small mammals	Sawfly cocoons	1700 Exp	45	Schoenfelder *et al.* (1978)
Small mammals	Sawfly cocoons	15 000	20	Griffiths (1959)

* The density estimate does not include the naturally occurring cocoons; the true density is likely to be between 1000 and 10000 cocoons ha^{-1}.

largely by shrews, was density-dependent. However, plotting the results of Holling (1959a) and Olofsson (1987) in the same figure with a sample of other results from various studies on pine sawflies strongly suggests that the positive density-dependence found by Holling and Olofsson (at low prey density) may be the exception rather than the rule: the general pattern appears to be inverse density-dependence (Fig. 7.8). Hanski (1990) has discussed in detail the reasons for the discrepancy between Holling's (1959a) results and the general pattern in Fig. 7.8.

Local versus regional regulation

Although local regulation of a prey population at low density requires the operation of one of more density-dependent factors, it would be wrong to dismiss predation by shrews and other insectivorous mammals as unimportant if no density-dependence was detected, or even if there was inverse density-dependence as suggested by the results in Fig. 7.8 and found by Elkington (personal communication) for small mammal predation on the gypsy moth *Lymantria dispar* in North America (but see Smith, 1985). Inverse density-dependent predation is unlikely to lead to a locally stable, low equilibrium, but it may lead to regionally low density of the prey. Predator–prey models have examined the hide-and-seek game which some specialist predators and parasites may play with their prey (reviewed by Taylor, 1988). However, models of mobile specialist predators are inappropriate here (see Chapter 3). The

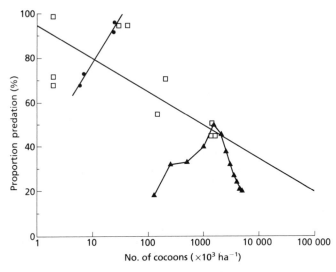

Fig. 7.8 Rate of predation by small mammals on sawfly cocoons (mostly *Neodriprion sertifer*) as a function of cocoon density in three sets of data: Holling's (1959) study (▲); Olofsson's (1987) study (●); and a compilation of several studies (□, from Table 7.4). The inverse density-dependence in the latter set of data is significant ($t = -3.52$, $P = 0.007$).

potential regulating power of insectivorous mammals rests on their relatively stable local populations and on the potentially high rate of predation on the preferred prey. The prey dynamics of this kind of interaction are encapsulated in a simple metapopulation model, as shown in Box 7.1.

The model assumes that the environment is divided into three kinds of patches: empty sites, sites with few prey, and sites with many prey (local outbreak). Empty sites may turn, in the model, to small local populations even in the absence of dispersal from large populations. The motivation for this assumption is the possibility of prey survival at very low density in some refuges, at such low densities that the population is likely to go undetected. For instance, Campbell and Sloan (1977a) report that small gypsy moth populations survive at sites with an abundance of protected resting and pupation locations above the forest floor.

For a range of parameter values, the model has two alternative stable states (Box 7.1), corresponding to regional rarity and to a large-scale outbreak of the prey species, respectively. Local prey populations

---

Box 7.1 A metapopulation model for pest outbreaks

Assume that the forest is divided up into patches, and that E, S and L are the fractions of habitat patches which are (apparently) empty, have a small population and a large (outbreak) population of the prey, respectively (thus $E + S + L = 1$). Now assume that

$$\frac{dE}{dt} = dS - (mL - i)E$$

$$\frac{dS}{dt} = (mL + i)E + cL - (b + d)S - fLS$$

$$\frac{dL}{dt} = bS + fLS - cL$$

where the rate parameters have the following interpretations. Small populations may either go extinct (parameter d) or become large ones, the latter either due to their own growth (b) or due to immigration from existing large populations (f). Large populations may become small ones (c), while empty sites may turn to small populations, either due to dispersal from large populations (m) or due to the prey having survived at very low density in an apparently empty site (the latter assumption introduces some local density-dependence and prevents metapopulation extinction). Predators are assumed to be continuously present at all sites, making the likelihood of a small prey population going extinct large. For a set of parameter values this model has two alternative, non-trivial stable equilibria (Fig. B7.1). In the low (endemic) equilibrium the metapopulation mostly consists of

continued on p. 185

Box 7.1 *contd*

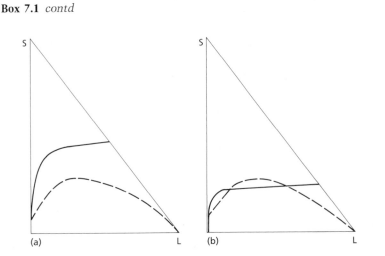

Fig. B7.1 Isoclines of the model. The two isoclines give the zero growth curves of the fractions of small (S) and large (L) populations. In (a) there is one positive equilibrium corresponding to low regional density of the prey (parameter values: $d = 0.1$, $m = 0.2$, $i = 0.005$, $c = 0.5$, $b = 0.03$ and $f = 1.0$). In (b) there are two stable positive equilibria, corresponding to low and high (outbreak) density of the prey, separated by an unstable equilibrium. The difference to (a) is in the higher rate of dispersal from large to small populations and a consequent increase in the probability of small populations becoming large (parameter values: $f = 2.0$, other parameters as in (a)).

a small number of small local populations, while in the high (epidemic) equilibrium many or most local populations are large. A regional outbreak may develop if a certain number of local populations escapes from the control of their local natural enemies. Dispersal from these populations may help neighbouring populations escape, rapidly leading to a regional epidemic (Hanski, 1985b). In the real world, the outbreak may be ended by deteriorating food resources, not taken into account in this model.

may escape from the control of their predators, and dispersal from such outbreak foci may help other local populations escape from the control of their local predators, resulting in a regional outbreak. A possible example is the pine sawfly *N. sertifer* in northern Europe. In most years, the species appears to be entirely absent from large areas (M. Varama, personal communication). It is of course impossible to be entirely sure about negative observations, but because the larvae of this species are gregarious, they are relatively easy to detect and hence unlikely to be overlooked. At shorter or longer intervals, the sawfly has spectacular regional outbreaks, when the density may reach several million cocoons per hectare (Juutinen, 1967).

Experiments

Experiments on forest insects and their mammalian predators can attempt either to change the suitability of the environment for the predators, or to manipulate the numbers of predators directly. Hamilton and Cook (in Buckner, 1966) report on experimental habitat alterations in larch plantations in New York State in North America. Nest boxes were provided for the deer mouse *Peromyscus maniculatus*, and branches were piled on the forest floor to encourage the red-backed vole *Clethrionomys gapperi* and the masked shrew *Sorex cinereus* to breed. The control measures were successful in increasing the numbers of these species of small mammal, and in reducing the numbers of larch sawflies, in comparison with plantations where nothing was done. Unfortunately, no quantitative data are presented.

Predators may be removed experimentally to study their effects on prey populations. Campbell and Sloan (1977b) report that the density of the gypsy moth egg-masses was 6.6 times higher in experimental study plots from which small mammals had been removed than in control plots. In other plots, both mammals and insectivorous birds were controlled, and the density of egg-masses was 8.8 times higher than in the control plots.

There were annual outbreaks of the larch sawfly *Pristiphora erichsonii* in Newfoundland in the 1940s and 1950s (Carroll, 1964; Warren, 1970). In 1942, Balch (in Carroll, 1964) realized that there were no shrews in Newfoundland (located 25 km from the coast of Labrador), and as the importance of shrews as sawfly predators became recognized, the idea was conceived of introducing the widespread *S. cinereus* to Newfoundland. The first attempt in 1957 failed, but in September 1958, 12 females and 10 males were successfully introduced to a bog in Newfoundland, confined by two roads and a river. At least six females and five males survived the winter, and by autumn 1959 130 shrews were recorded at the initial point of introduction (Warren, 1970). Subsequently the shrew population increased to a peak density of nearly 50 ha^{-1} in 1962 and started to spread at the rate of 20 km year^{-1} (Buckner, 1966). Two other populations were established by moving shrews from the original site of introduction. Carroll (1964) reported predation rates on sawfly cocoons averaging 44%, a typical figure for other studies (Table 7.4). It is most unfortunate that no quantitative studies have been reported on forest insect outbreaks in Newfoundland during the past decades. However, it is believed that the outbreak frequency and severity have decreased since the introduction of *S. cinereus* (Warren, 1970; Hudak, personal communication).

CONCLUSION

Insectivorous mammals are opportunistic generalists, with the notable exception of some tropical taxa that specialize on ants and termites, an abundant and predictable food resource. Both the ant-eating and generalist insectivorous mammals have successfully retained their macroniches for millions of years. Insectivorous mammals include species representing a range of life histories, from some shrews and dasyurid marsupials which live for only 1 year to some equally small insectivorous bats which may live for 15 years or more.

Guilds of coexisting shrews and insectivorous marsupials typically consist of a few species with conspicuous differences in their body sizes. Large species are competitively superior to small ones (interference), while small species may survive and breed in habitats and microhabitats where food availability is too low for larger species. Regional coexistence of many species of shrews and dasyurid marsupials varying in size is more likely to be due to these mechanisms than to partitioning of prey species by size.

In contrast to shrews and dasyurids, guilds of insectivorous bats consist of many species of more or less the same body size. Size differences may be less important in bats because of increased importance of intraspecific competition. Bats may compete for roost sites, which are often species-specific, and, since they forage from these roosts in largely conspecific groups, they are exposed to intraspecific competition for food more often than to interspecific competition. Patterns of high species diversity support the notion that coexistence of many species is no problem in bats: assemblages of insectivorous bats may contain an order of magnitude more species than do guilds of shrews or marsupial insectivores.

Insectivorous mammals show strong functional and spatial responses to changing numbers of particular prey species, but they do not generally demonstrate pronounced numerical responses, because most of their prey types are seasonal. Large prey types are often strongly preferred, even if they do not represent higher profitability than other kind of prey. The pupae of many outbreak forest insects belong to this category, and several types of observational and experimental data indicate that shrews and other insectivorous small mammals may contribute significantly towards the regulation of these insects at low population densities.

8: Marine Mammals

SIMON NORTHRIDGE & JOHN BEDDINGTON

INTRODUCTION

The marine mammals are a diverse group of animals, which includes 77 species of the order Cetacea (whales, dolphins and porpoises), 35 species of pinnipeds (seals, fur seals, sea lions and walrus) and four species of the order Sirenia (dugongs and manatees). In this chapter we consider the predatory marine mammals, the cetaceans and the pinnipeds.

These mammals have colonized almost all marine areas of the globe, from Arctic to Antarctic ice edges, from coasts and some rivers and lakes, to deep oceanic waters. They feed on an enormous range of fish, molluscan, crustacean and other food species, from surface waters to depths of 2000 m or more in the case of the sperm whale *Physeter macrocephalus*. Some, such as the killer whale *Orcinus orca* will occasionally prey upon other marine mammals, and even birds and turtles. Clearly with such a wealth of possible predatory interactions, it would be impossible to discuss all of them adequately in the space available here. We have therefore taken a few examples to consider the ways in which prey species may influence marine mammal behaviour and population dynamics and, more generally, the ways in which predation by marine mammals might affect their prey.

THE INFLUENCE OF PREY POPULATIONS UPON MARINE MAMMALS

In the course of evolution, food species may affect many aspects of a predator's biology, including its behaviour, distribution, physiology and morphology. The two groups of mammals we consider here, the cetaceans and the pinnipeds, although taxonomically disparate, exhibit some interesting similarities that are associated, in one way or another, with their feeding. Most obviously, perhaps, they exhibit similar morphological adaptations to locomotion, and hence to feeding, in an aquatic environment.

One less immediately obvious similarity between the two groups is their large size; this is also closely linked to their feeding habits, and it is this theme which we explore here. Even the smallest cetacean, the vaquita *Phocoena sinus*, which inhabits the Gulf of California, grows to about 1.5 m and 50 kg. The smallest pinnipeds are even heavier, and an adult Baikal seal *Phoca sibirica*, for example, may weigh 80–90 kg. Although small marine mammals, these are very large animals in comparison with other 'small' predators (see Chapters 9 and 10). At the other extreme, the largest of the pinnipeds is the southern elephant seal *Mirounga leonina*, males of which may exceed 3.5 tonnes, while the blue whale *Balaenoptera musculus* may grow to exceed 150 tonnes. In comparison the largest terrestrial herbivore, the African elephant *Loxodonta africanus* may reach 7.5 tonnes (Nowack & Paradiso, 1983), and the largest terrestrial carnivore, the Siberian tiger *Panthera tigris altaica* a mere 306 kg (Mazak, 1981).

Another characteristic of marine mammals is their relatively thick subcutaneous layer of blubber. Bowhead whales (*Balaena mysticetus*), for example, may have a blubber layer up to 50 cm thick, and even small cetaceans like the harbour porpoise (*Phocoena phocoena*) may have a blubber layer 2 or 3 cm thick. The blubber layer has long been recognized to act as a thermal insulator in marine mammals. It has also been proposed that the large size of marine mammals is a thermo-regulatory adaptation to the aquatic environment in which they live, because a large body volume has the effect of reducing the animal's surface to volume ratio and may therefore conserve heat (Bryden, 1972).

For some time it was also thought that the problem of heat con-servation for a homeotherm living in a marine environment was re-sponsible for a higher basal metabolic rate in marine mammals than in other mammals (Scholander, 1940; Kanwisher & Sundnes, 1966). Clearly such large, fat, predatory mammals might be expected to consume large amounts of food, especially given their relatively elevated energetic requirements. Because of this, there has been considerable interest in the ways in which marine mammal predation might affect the popu-lation dynamics of their prey, particularly where the main prey are fish. More recently, however, explanations of marine mammal size, proportion of fat and energy requirements have been modified somewhat.

The importance of the insulating properties of blubber, for example, may well have been over-emphasized. Kanwisher and Sundnes (1966), Tomilin (1967), Gaskin (1972) and Brodie (1975) have all suggested that cetacean blubber may be more than is required to insulate these animals from the thermal rigour of their environment. Indeed, in the tropics, or when exerting themselves, cetaceans may have to dissipate heat through the blubber layer. Blood supply to the dermis, through the hypodermis or blubber layer, enables the animal to keep cool.

Kanwisher and Sundnes (1966) proposed that the maintenance of hydrostatic buoyancy might be the primary purpose of the blubber layer of cetaceans, while Brodie (1975) suggested that its function as a food store to enable balaenopterids to feed on temporarily abundant food supplies might be of primary importance. It has subsequently become increasingly clear that fat deposits are extremely important as energy stores for many other marine mammals as well, especially where seasonal migrations occur.

The function of large size, if not for thermoregulatory reasons, is less clear. Clearly a large size helps to conserve heat by reducing the surface to volume ratio, but this does not explain why a blue whale grows to exceed 150 tonnes. Body size of this scale is clearly more than adequate for thermoregultory processes. Indeed, the fact that some of the largest and some of the smallest marine mammals coexist in polar waters suggests that large size is not simply an adaptation to survival in cold water.

Thus, although the minimum size for a marine mammal may be constrained by thermal regulation, and blubber has an important role in insulating all marine mammals, it has become clear that neither of these two aspects of marine mammal physiology is primarily thermo-regulatory, at least in adults.

The supposition that marine mammals might have disproportion-ately high metabolic rates (Sergeant, 1969) has also been revised recently. Lavigne *et al.* (1986) have suggested previous estimates of high basal metabolic rate (BMR) for marine mammals were due to methodological problems in measurement, and that marine mammals in fact have metabolic rates which are in line with terrestrial mammals, according to Kleiber's rule (Kleiber, 1975).

These three aspects of marine mammal physiology and morphology, their size, energetic requirements and fat content, are all of crucial importance in interactions with their prey. Recent studies have demon-strated how marine mammal populations may be limited by these three factors and the availability of their food.

Lockyer (1987), for example, has demonstrated that the blubber thickness of fin whales *Balaenoptera physalus* is correlated with their reproductive success. Anderson and Fedak (1985) have shown that body size in male grey seals *Halichoerus grypus* is also correlated with reproductive success, and Boyd (1984) has shown that the time of implantation in female grey seals is dependent upon the rate of weight gain during the spring. These studies show how the reproductive success of marine mammals may be linked directly to feeding success.

It is important to know how the availability of prey influences feeding success in marine mammals. For example, Kenney *et al.* (1986) demonstrated the energetic problems faced by right whales *Eubalaena glacialis* in attempting to locate copepod swarms of adequate density to support their food requirements. As an introduction to this question, we consider how the distribution of food may influence the success of marine mammals of different sizes. To do this we use two examples: the first involves some of the largest marine mammals; and the second, one of the smallest species.

Baleen whales and krill in the Southern Ocean

One of the most interesting and most frequently studied assemblages of marine mammals is found in the ocean waters surrounding the Antarctic continent. These marine mammals have been extensively exploited by humans in the last hundred years or so, with the result that considerable changes in numbers of some species have occurred.

The Southern Ocean ecosystem is dominated by one food species, *Euphausia superba* (Krill), a pelagic crustacean growing to about 6 cm in length and living for several years (Rosenberg *et al.*, 1986). Krill are enormously abundant and are the principal food of a large number of bird, fish, squid and mammal populations (Marr, 1962; Laws, 1977). Amongst the mammals, several species of baleen whale appear to be largely dependent upon this one species for their survival. Southern right whales *Eubalaena australis*, humpback whales *Megaptera novae-angliae*, fin whales *Balaenoptera physalus*, blue whales *Balaenoptera musculus* and minke whales *Balaenoptera acutorostrata* all feed extensively on this prey species (Nemeto, 1959), and it is the baleen whales which we consider in more detail here.

The fact that so many large predators feed on a single prey species suggests that there may be competition for food, and we might expect to observe a number of factors that limit the extent of direct competition between predator species. For example, Nemoto (1959) has described different modes of feeding in baleen whales. Right and sei whales feed in a different manner from the other species, 'skimming' rather than 'gulping' their prey. This 'skimming' habit is said to enable whales to feed on prey which are relatively dispersed, whereas 'gulping' whales feed on dense swarms of prey. In this way competition between 'skimming' and 'gulping' whales could be limited, but this does not explain how competition is limited among the four species of whale described as 'gulpers' (minke, fin, humpback and blue whales).

Besides exhibiting different feeding behaviours, there is some evidence that whales of different species consume krill of different size classes. Laws (1977) states that blue whales feed on first-year krill of 20–30 mm in size, whereas minke whales feed on smaller krill 10–20 mm in size, and that the two species avoid competition in this way. However, Basson (1989) has shown that by the start of the austral summer in October, krill will usually have reached 10 mm in length, and that growth is so rapid that by the end of the summer they will measure 30 mm, spanning the range of the proposed whale preferences within 6 months. Thus the apparent distinction in size classes of krill consumed by these two species of whale may be of little significance, because both are evidently consuming 1-year-old krill. Laws (1977) also points out that in some areas different whale species evidently do feed on the same size classes of krill.

Another difference among the baleen whales concerns the timing of their migrations from warm tropical or subtropical waters to Antarctic waters in the summer. Within a species, for example, larger whales tend to reach higher latitudes than smaller whales, and pregnant females tend to arrive earlier than lactating females (Laws, 1960; Mackintosh, 1965; Dawbin, 1966). Laws (1960) also provides evidence that there is some segregation by size, with larger whales displacing smaller ones from the centre of feeding grounds. Between species too, the larger species tend to reach higher latitudes sooner (but some minke whales may overwinter in Antarctic waters; Ohsumi *et al.*, 1970). Although such spatial segregation may be one way that competition is limited, it is clear that there is still considerable overlap in distributions between species. Minke and blue whales, for example, are found in the highest latitudes at the same time. Futhermore, krill are distributed over wide areas of the Southern Ocean, so that even whales that feed at widely separated localities may still be feeding on the same krill population, and could therefore still compete with one another.

None of these explanations resolves adequately the problem of how the krill resource is partitioned among potentially competing whale species. It may be that a seasonal superabundance of krill is sufficient to limit competition between whales, or that environmental fluctuations continually alter equilibrium conditions for the competitors (Hutchinson, 1961). Neither of these explanations seems likely, however. Given the relatively long life of individual whales, and their capacity to live for months with little feeding (Bryden, 1972), any seasonal superabundance of food would be expected to be exploited by increased numbers of whales. Indeed, Laws' (1960) observations of displacement of smaller whales by larger whales from the centre of feeding grounds would support the idea that food for whales in the Antarctic is a limiting resource, and that food is unlikely to be superabundant. Furthermore, the relatively low reproductive rates of these whales, and of marine mammals in general, their large body size, delayed reproduction, and long lives, all characterize them as *K*-selected species (Pianka, 1970; Estes, 1979). This suggests that these mammals' environment is relatively stable, that population sizes will tend to grow to near the carrying capacity, and that intraspecific and interspecific competition are likely to be keen (Pianka, 1970).

This paradox of how a relatively stable environment with a single food source can support several predator populations may be resolved when one considers the implications of whale body size. Brodie (1975) has pointed out that size is important as a means of buffering whales from seasonal shortages of food, but size also plays an important role in the short term. It might appear that smaller whales would be at a reproductive advantage in competition with larger whales, because their absolute energy requirements are smaller. This argument, however, ignores the rate at which food is acquired by whales, and when this is examined, a different picture emerges.

Lockyer (1976, 1981) has presented some illuminating information on baleen whale energetics. She examined the filter volume (the volume of water filtered per mouthful) of blue, fin and minke whales of different sizes. She also provided estimates of the annual and daily energy requirements of whales of the same sizes, estimates of costs of locomotion, and of the calorific content of krill. From these considerations, she suggested that whales in the Antarctic required about 30–40 (35) g kg^{-1} body weight day^{-1} of krill, for a nominal 120-day 'season' spent in Antarctic water. She also suggested that on average, a 'gulp' took approximately 30 seconds to complete, regardless of whale size.

Using these data, she estimated the time taken to fulfil the average daily requirement of whales of different sizes, assuming a constant feeding rate, for different krill densities. In Fig. 8.1 these rates have been plotted for a typical blue whale and a typical minke whale of 28 and 5 m in length respectively.

It is immediately clear that blue whales are more 'efficient' than minke whales, in that their larger size and proportionally greater filter volume enable them to satisfy their projected daily energy requirement more quickly for any given level of krill density. Minke whales, on the other hand, require a relatively longer period of time to acquire necessary amounts of food. Lockyer assumed that the quantity of food required is a constant proportion of body weight for whales of all sizes. In fact Kleiber's rule might suggest that this proportion should decrease with increasing body size, so that it is likely that the discrepancy depicted in Fig. 8.1 should be even greater. Lockyer also suggested that, on the basis of some anecdotal accounts, a typical krill swarm density might be around 2 kg m^{-3}. Assuming such a concentration to be constantly available, both minke and blue whales should be able to feed themselves adequately, although minke whales would require a feeding bout of around 3 hours to do so.

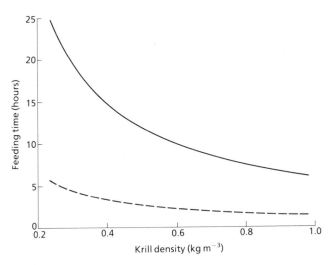

Fig. 8.1 Feeding activity required for adequate energy intake based on estimated average daily energy intake. Minke, ———; blue, – – –.

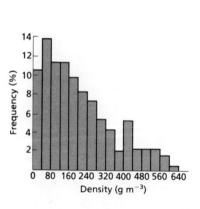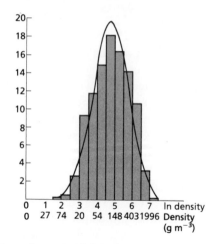

Fig. 8.2 Krill swarm densities. From Kalinowski & Witek (1981).

Of course, as Lockyer points out, this conclusion takes no account of the fact that krill swarms are unevenly distributed. Until recently, the nature of this distribution was unknown, but Polish research in Antarctic waters has addressed this point. Kalinowski and Witek (1981) suggest a log normal distribution of krill swarm densities, with an estimated mean krill density of only around 143 g m^{-3} (Fig. 8.2). Clearly, Antarctic krill swarms of this mean density would present energetic problems for whales, especially for those of the size of minke whales.

Figures 8.1 and 8.2 suggest only a relatively small proportion of krill swarms are sufficiently dense for whales to consume and meet their energetic requirements. Furthermore, because of their differences in body size, the proportion of profitable swarms is much smaller for minke whales than for blue whales. This analysis, however, assumes that whales are able to locate themselves permanently in areas of high krill density. In reality, of course, swarms are limited in size, and will be depleted by predation. Other important characteristics of krill distribution are therefore the swarm size and the distance between swarms. Again, Kalinowski and Witek (1981) suggest that swarm sizes are lognormally distributed, with a mean mass of approximately 1.7 tonnes (Fig. 8.3). While such a swarm of average size would be adequate for a minke whale's average daily food requirement, Lockyer (1981) estimates that a blue whale of 28 m needs an average daily intake of more than 5 tonnes of food, or at least five or six such average-sized swarms.

Blue whales, minke whales, and by analogy whales of intermediate sizes too, might therefore be expected to have rather different predatory responses to krill swarms of different sizes and densities. Blue whales might be expected to prefer swarms which are larger in size, but not necessarily denser, than those preferred by minke whales. Minke whales on the other hand, are likely to avoid even large krill swarms if the swarm density is too low. In order to illustrate the potential division

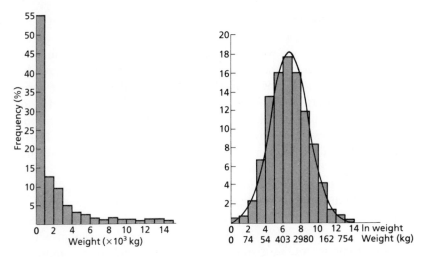

Fig. 8.3 Krill swarm sizes. From Kalinowski & Witek (1981).

of such resource, Fig. 8.4 shows the types of krill swarm, in terms of density and size, which might be preferred by minke and blue whales. This is based on the assumption of Lockyer's average daily energy requirements, and the assumptions that whales will only choose swarms which are both massive enough to provide a whole day's energy requirement and dense enough so that no more than an arbitrary 5 hours' feeding is required daily. Such assumptions are highly simplistic, because they make no allowance for the ways in which preference for krill swarms might be influenced by the availability of alternative krill swarms, and these would clearly be expected to influence a predator's choice (Chapter 4).

These observations suggest a possible mechanism whereby competition between the baleen whale species is limited. Assume that krill swarms are distributed such that those swarms which were large

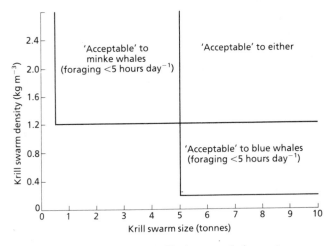

Fig. 8.4 Krill swarm characteristics suitable for two whale species.

enough to be economical for blue whales are not dense enough for minke whales, and those which are dense enough for minke whales are too small for blue whales. In this way, the resource could be partitioned amongst different whale size classes more or less discretely, depending upon the relationship between swarm density and swarm size. Unfortunately, Kalinowski and Witek provide no indication of the relationship between krill swarm size and density, nor on the spacing of krill swarms. However, it would seem that the size of krill swarms is more critical to larger whales, whereas the density of swarms is more critical for smaller whales.

Many of these assumptions have not examined critically. For example, we discuss the feeding of whales as if the only constraint is to achieve the intake of a fixed amount of food per day. A more realistic approach would entail examining feeding strategies in which the rate of energy intake is maximized under a range of assumptions about the distribution of krill swarms. Mangel (1988) and Butterworth (1988) have approached some aspects of this problem through their simulation models of krill trawling operations by Japanese and Soviet vessels in the Southern Ocean. These models have yet to be applied to whales, whose behaviour is substantially more difficult to monitor than that of fishing vessels.

The harbour porpoise

We have stressed the importance of body size in determining the feeding strategy of whales, and a consideration of the largest of the marine mammals revealed that large size has clear advantages where prey are patchily distributed. We turn now to one of the smallest marine mammals, the harbour porpoise *Phocoena phocoena* whose adults rarely exceed 2 m or 90 kg.

The harbour porpoise is distributed primarily in coastal waters in the temperate and subarctic waters of the northern hemisphere. In the Atlantic, the limits to its range are generally thought to be Senegal and North Carolina in the south, and Novaya Zemlya and Baffin Island in the north. In past centuries this was evidently a common mammal in European waters, but its range and abundance appear to have declined markedly (Andersen, 1982; Kayes, 1985; Evans *et al.*, 1986) although few quantitative data are available to substantiate this impression.

Few studies have been made of harbour porpoise feeding habits, but anecdotal accounts indicate that in the waters around Britain at least, herring *Clupea harengus* was an important part of the diet (Lydekker, 1895; Martin, 1981). Studies by Rae (1965, 1973) in Scotland indicate that clupeids including sprats *Sprattus sprattus* and herring, and other small shoaling fish such as whiting *Merlangius merlangius* and sandeels *Ammodytes* spp. predominated in the stomachs he examined. In the North Sea in particular, herring may well have been an important part of the diet of harbour porpoises. This fish not only has a high calorific value (>1500 kcal kg^{-1} for example), but used also to be the most abundant fish species in the North Sea, with a biomass in excess of 3.6×10^6 tonnes in the early part of the twentieth century (ICES, 1975). Furthermore, herring is a schooling fish, forming very dense aggregations, often near the surface, with schools of several thousand tonnes having been reported (Muus & Dahlstrom, 1974). On the basis of the considerations in the previous section, herring seems likely to be a very useful prey item for marine mammals in general.

The biomass of herring in the North Sea declined dramatically during the 1960s and 1970s, to reach a level of less than 250 000 tonnes by 1977, as a result of increased fishing pressure (Saville & Bailey, 1980). Such a substantial decline in the abundance of one of its most important prey items might be expected to have affected the harbour porpoises in the North Sea.

The effects of a decline in the abundance of a favoured food item will depend upon the way in which the decline is manifest. Figure 8.5 illustrates two extreme ways in which an increase in fishing might affect a stock of fish. In Fig. 8.5a,b the stock is reduced to one-half by removing one-half of the schools, whereas in Fig. 8.5c,d the stock is reduced to one-half by halving the size of every school. For an individual predator these two extreme options have rather different energetic consequences.

The energetics and behaviour of harbour porpoises have been studied in the Bay of Fundy in Canada by Gaskin and his colleagues. Yasui and Gaskin (1986) published an energy budget for harbour porpoises in which they estimate the costs of BMR, reproduction and two types of activity, which they call low activity and moderate-to-high activity. Low activity was assumed to indicate a resting state (0.6 m s^{-1} average speed), whereas moderate to high activity (1.5 m s^{-1} average speed) was assumed to include foraging. Two porpoises radio-tagged by Read and

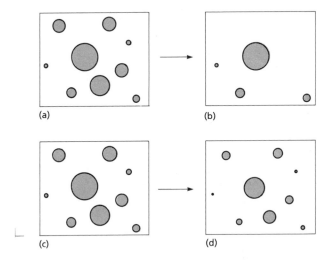

(a) (b)

(c) (d)

Fig. 8.5 Two possible extreme effects of exploiting fish schools.

Gaskin (1985), for example, spent 35% and 36% of the time between 0600 and 2400 hours in the latter state. The cost of low activity was estimated to be 12 kcal hour^{-1}, and the cost of higher activity states was estimated to be 73 kcal hour^{-1}. In addition to these costs, which exclude growth and reproduction, the BMR requirement was estimated at around 57 kcal hour^{-1}.

Read and Gaskin (1985) suggest harbour porpoises forage only during the hours of daylight and dusk, because after herring school structure breaks down, which makes efficient predation more difficult. This might suggest that each day porpoises have to relocate a herring school. Clearly the distance between schools will affect both the average time taken for an individual porpoise to locate a school, and its energy expenditure. Line A in Fig. 8.6 illustrates the way in which the amount of time spent foraging affects the amount of energy required above the resting requirement (1656 kcal day^{-1}: BMR plus low activity) for a non-reproducing adult. An increase in average foraging time from 8.4 hours (35%) to 16.8 hours (70%) of the day would require an extra 512 kcal day^{-1}, or an increase 23% energy required daily.

A decrease in the overall amount of food, if it is achieved by a decrease in the number of schools, may then affect porpoises in two ways.

1 As the total food biomass is decreased, so porpoises might be expected to compete for food with one another, leading to decreased individual and population growth rates, and increased mortality rates.
2 Unless there is an alternative prey source, individuals may be expected to generate higher energetic costs in feeding themselves, thus exacerbating the first effect. Only if the decrease in the overall amount of food is achieved without any reduction in the number of schools could the energetic costs of foraging be maintained.

Yasui and Gaskin (1986) also investigated the cost of reproduction for a female harbour porpoise. They estimated that the energetic costs of pregnancy and lactation were approximately 8.8×10^4 and $29 \times$

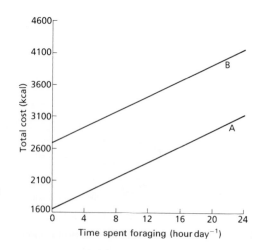

Fig. 8.6 Energetic cost of activity by *Phocoena*: the effect of changing foraging time. A, non-reproductive adult; B, reproductive female.

10^4 kcal respectively, partly spread over 2 years. This compares with an annual average requirement of a non-reproducing animal of the same size of around 90×10^4 kcal. On an annual basis, a female which reproduces every year may require about 40% more energy than a non-reproducing female of the same size. This increased cost for a reproducing animal suggests an increased rate of energetic intake. Watson and Gaskin (1983) found that female harbour porpoises with young spent a much larger part of the day foraging, and up to 5 hours per day more than animals without young. Clearly this increased rate of foraging will also incur additional energetic costs. Average daily energetic costs for a reproducing female foraging for various amounts of time are plotted in Fig. 8.6 (line B). Even spending just a few hours a day in medium to high activity, a reproducing female requires as much energy as a non-reproducing animal which is foraging continuously.

The combined costs of reproduction and increased foraging due to a reduced food supply could be of great significance to an animal the size of a harbour porpoise. Marine mammals of this size may be considered to be close to the lower size limit of what is energetically feasible given the constraints of food availability. For such a small marine mammal, the energetic costs of reproduction, and of locomotion are relatively high. Lockyer (1981), for example, estimated the costs of fetal growth and lactation for a 64.5-tonne female fin whale to be around 63×10^6 kcal spread over 2 years, while the costs of growth and other metabolic processes for the same animal might exceed 500×10^6 kcal over the same period. The additional costs of reproduction for a fin whale are only 12–13% more than non-reproductive metabolic costs (Table 8.1).

The small size of the harbour porpoise suggests that these animals are not as well buffered against reduced prey abundance as larger marine mammals might be. In order to reproduce successfully a female porpoise may have to increase her net energetic intake by around 40% compared with a non-reproducing animal, and this will require a relatively large increase in the time spent feeding. This in turn will generate additional foraging costs. In an environment where food is distributed patchily in time and space, any increase in the time spent locating food patches, or any decrease in the profitability of individual patches, is expected to have a greater than proportional effect on harbour porpoise reproductive success, as a result of the animal's small size. The larger marine mammals tend to have lower marginal costs for successful reproduction and for increased foraging, and this effectively buffers them from spatial and temporal fluctuations in food supply. The thick blubber layer, characteristic of marine mammals, will also enable the rate of energy expenditure to exceed the rate of energy replacement over an extended time period. This suggests that large size and thick blubber layers are adaptations to feeding on prey which are patchily distributed in space as well as time, rather than simply adaptations for thermoregulation (see above).

Table 8.1 Comparative energetic costs of a large and a small cetacean

	Harbour porpoise	Fin whale
Length (m)	1.3	22
Weight (kg)	25	65 000
BMR		
kcal hour^{-1}	57	2.17×10^4
kcal year^{-1}	48.7×10^4	190×10^6
Typical slow locomotion cost		
kcal hour^{-1}	12	1750
kcal hour^{-1} kg^{-1}	0.5	0.026
Typical fast locomotion cost		
kcal hour^{-1}	7.3	10.8×10^4
kcal hour^{-1} kg^{-1}	2.9	1.66
Typical cost of gestation (kcal)	8.8×10^4	3×10^6
Typical cost of lactation (kcal)	29×10^4	63×10^6
Total reproductive cost (kcal)	37.8×10^4	66×10^6
Total reproductive cost as percentage of BMR	77	35

EFFECTS OF MARINE MAMMAL PREDATION ON PREY SPECIES

Although the effects of marine mammal predation on the population dynamics and abundance of their prey are of considerable interest, they are not well understood. The effects of marine mammal predation are not easy to measure or to demonstrate.

1   Most marine mammals have varied diets, so that characterizing the mortality suffered by any one prey population is difficult.

2   Most marine fish have a wide variety of predators throughout their lifetimes, so that marine mammal predation is difficult to distinguish from other potential sources of mortality.

3   Fish have highly plastic life history parameters; growth and mortality rates, for example, may vary enormously with fish population size.

4   The types of experiment which it would be useful to perform in order to determine the effects of marine mammal predation (e.g. exclusion experiments) are not usually practicable. Even where such experimental situations may arise, as when a marine mammal population is destroyed, the interpretation of any changes in fish population dynamics will be difficult (because of the first three factors mentioned above).

Theoretical studies of the effects of marine mammal predation on fish stocks have concentration on attempting to assess the extent to which marine mammal predation might reduce yields to a fishery (Beddington *et al.*, 1985). Typically, in such studies, yield to a fishery might be expressed in terms of the growth and mortality rates, and the numbers of fish recruiting to the population. For example, yield from an exploited fish stock might be written as:

$$Y = FRW_\infty \left[\frac{1}{Z} - \frac{3 \exp(-K\,tr)}{Z+K} + \frac{3 \exp(-2K\,tr)}{Z+2K} - \frac{\exp(-3K\,tr)}{Z+3K} \right]$$

where Y is yield in weight of fish, F is the instantaneous fishing mortality, R the number of recruits to the fish population, W_∞ the expected weight of a fish at age infinity, K is a stock-specific growth constant (from von Bertalanffy's (1938) growth equation), tr is the age at recruitment and Z is the total instantaneous mortality rate ($Z = F + M$, where M is the natural mortality rate).

The contribution which marine mammal predation makes to such a fish population model is confined to its effect on the natural mortality rate, M. The natural mortality rate can be divided into two components, M^1 and M^*, for example, where M^* is the component attributable to marine mammals and M^1 is the remaining or residual natural mortality rate (Beverton, 1985). Although several models have examined the effects of various assumptions about the relationship between M^* and predicted yield (ICES, 1981; Allen, 1985; Beverton, 1985), it has proved difficult to estimate M^* under field conditions. This is because it is difficult to estimate mortality rates of fish with any accuracy, and because of a lack of accurate information on the feeding behaviour of marine mammals.

In a review of marine mammal–fishery interactions, Beverton (1985) reported only one study in which predation by marine mammals was demonstrated to have had an effect on a fish population. Predation by common seals *Phoca vitulina* on fish in the freshwater Lower Seal Lake, Quebec, meant that the fish fauna of that lake was depauperate compared with nearby lakes from which seals were absent. The most vulnerable prey species, lake trout *Salvelinus namaycush* exhibited all the signs of sustained severe depletion, including increased growth rate, reduced age at maturity, the absence of older age classes, and high specific fecundity. Their habit of spawning in large groups was considered to have made them more vulnerable to seal predation than brook trout *S. fontinalis*, which spawn in scattered locations in brooks. Brook trout were therefore less susceptible to predation by seals, and appeared in fish samples in much higher proportions than in other nearby seal-free lakes (Power & Gregoire, 1978).

More recently, Basson (1989) has demonstrated a possible decrease in both the mortality and growth rates of Antarctic krill *Euphausia superba* during the period 1928–35 following the initial depletion of baleen whales in the Southern Ocean. However, no clear continued trend in these parameters could be detected, perhaps because of increased exploitation by other predators that became more numerous, following the demise of the baleen whales.

Another approach has been to try to estimate the scale of the impact of marine mammal predation on an entire community rather than on its constituent prey species. Estimates of total marine mammal predation have then been compared with the 'predation' by commercial fisheries on the same fish stocks as a crude indicator of marine mammal impact on an ecosystem, and as a means of assessing the likely level of competition between fisheries and marine mammals. This approach has been adopted in the Bering Sea, which has one of the highest marine mammal densities in the world. McAlister and Perez (1976) estimated that pinnipeds alone consume around 2.8×10^6 tonnes of finfish annually, an amount which exceeds the commercial finfish landings from that area (Lowry, 1984). Laevastu and Favorite (1977) considered predation by marine mammals to be sufficiently important to be one of the primary inputs for a dynamic ecosystem model of the Bering Sea. Although marine mammals are clearly an important component of this ecosystem, Lowry and Frost (1985) still concluded that 'even in the simplest cases it has proven extremely difficult to measure the magnitude of competition or to assess the likely effects on fishery harvests'.

Within the marine mammal community of the Bering Sea, however, there are at least two species which are more amenable to study than many of the others. Both the walrus *Odobenus rosmarus* and the grey whale *Eschrichtius robustus* have relatively restricted diets and consume benthic, sessile or weakly motile organisms. The effects of their predation on these benthic communities are therefore relatively easier to study, than with wide-ranging animals that feed on more dispersed and mobile prey.

Oliver and Slattery (1985) studied the effects of grey whale browsing on the sea bed. Grey whales feed by sucking sediment into their mouths and filtering out food items, thereby excavating pits in the sea floor. These pits are highly characteristic and may cover more than 30% of the sea bed in some areas. They found that prey consisted primarily of ampeliscid and other amphipod crustaceans which form dense tube mats on the sea floor. Destruction of these mats by grey whales led to a colonization by other scavenging invertebrates, and later by tube dwelling ampeliscid amphipods again. Oliver and Slattery concluded that grey whale feeding has a dramatic effect on the structure of benthic communities, and may enhance the population size of several secondary prey.

Walruses also feed on benthic organisms, especially bivalve molluscs such as clams. Fay *et al.* (1977) found that areas of high walrus density in Alaskan waters coincided with areas of relatively lower bivalve mollusc density, and concluded that walruses had a significant impact on the mollusc fauna. In a later study, walruses were found to be taking a wider range of food items, including fish and other non-molluscan foods in their diet (Fay & Stoker, 1982). In Bristol Bay, Alaska, Fay and Lowry (1981) suggest that walruses impose a mortality rate on surf clams *Spisula polynyma* which is substantially greater than that required to produce the maximum sustainable yield, and that the walruses are therefore over-exploiting their clam prey.

In other areas of the world, and for marine mammals which have more plastic feeding habits, marine mammal predation may be a less significant influence on community dynamics. In the North Sea, for example, studies of the grey seal *Halichoerus grypus*, which is one of the most numerous of the few marine mammal species inhabiting the area, indicate that predation on most fish stocks is only a small proportion of the commercial fishery catch (Hammond & Harwood, 1985). Detailed studies of seal faeces over a number of years and a number of collection sites have indicated that the most commonly eaten food species are sandeels *Ammodytes* spp. These accounted for more than 60% of the food of grey seals at four sites in the North Sea in the early 1980s, with a total estimated grey seal consumption of sandeels of 35 000 tonnes. This compared with commercial fishery landings of more than 620 000 tonnes of sandeels from the North Sea in 1982. Sandeels are also consumed by a large number of seabirds and other predators (Furness, 1984). Other fish species consumed by grey seals in the North Sea were found to be eaten in much smaller quantities.

Harwood and Croxall (1988) examined the likely extent of competition between grey seals and commercial fisheries in the North Sea by comparing estimated consumption of different fish species with commercial catches of those species. The ratio of these two values was taken to be equal to the ratio $M^* : F$. Although this ratio exceeded 100% for flounders *Platichthys flesus* around Orkney, in only a few other cases did it exceed 10%, and in these cases the fish stocks were not of major economic importance. In the stocks that dominate the North Sea ecosystem numerically and economically, the ratios were considerably lower (cod 4.9%; whiting 1.9%; haddock 1.2%, plaice 0.1%; herring 0%; mackerel 0.3%). Harwood and Croxall state that 'in general the mortality caused by seals is one or two orders of magnitude less than that caused by the fishery'. In such a situation it seems evident that, at an ecosystem level, grey seals in the North Sea are unlikely to be a major factor in influencing community dynamics.

On the basis of such a few studies it is impossible to generalize about the effects of marine mammal predators on their prey. The wide range of feeding habits exhibited by marine mammals means that different species will influence their prey in different ways. For some prey populations, like surf clams, marine mammal predation may be the most important source of adult mortality. For other prey populations, mortality due to marine mammal predation is swamped by other sources such as fishing mortality. For the majority of marine mammal populations, there is no clear indication of how severely they might affect their prey populations.

The impact of marine mammals on the evolution of their prey remains even more obscure. It might be interesting to speculate upon the implications of marine mammals seeking out dense concentrations of prey, as baleen whales appear to be. Basson's (1989) study suggests that, historically at least, baleen whales contributed a significant part of overall krill mortality, and this might therefore be expected to exert some selection pressure for less dense swarms. A shift in the krill swarm density distribution (Fig. 8.3) might, in turn, be expected to have a dramatic effect on the reproductive success of baleen whales, particularly the smaller ones. One can only speculate that during their coevolution in the Southern Ocean, predation by baleen whales on dense swarms of krill must have been counter-balanced by other factors which make more diffuse swarming less successful for the krill, but such factors remain obscure at present.

CONCLUSION

In our two examples we have considered the relationship between the size of marine mammals and the availability of their prey. Larger marine mammals seem to be better adapted to feeding on more diffuse prey types than smaller marine mammals. This may be why the larger marine mammals are primarily oceanic, where prey patches are widely spaced, while the smaller ones are more coastally distributed, where prey tend to be more densely aggregated. These considerations suggest that population dynamics and evolution of marine mammals are considerably influenced by prey distribution, rather than simply by the amount of prey available, or its productivity. In particular, their generally large size and high fat content seem to enable marine mammals to exploit seasonal and patchy food resources that smaller or less fat mammals would be unable to do.

The extent to which marine mammals' predation influences the population dynamics of their prey is more difficult to determine. There have been very few documented cases in which marine mammal predation has been shown to have a significant effect on growth or mortality of their prey, although in some cases the scale of predation would make such effects seem likely.

Although we can suggest that the spatial and temporal distribution of prey is important to most marine mammals, it will be impossible to determine the extent to which, at an evolutionary level, such animals are influenced by their marine mammal predators until we know much more about the factors which govern the schooling behaviour, school sizes and densities of these prey species.

ACKNOWLEDGEMENTS

We would like to thank Marinelle Basson, Andrew Rosenberg and Joanna Seckerwalker at the Renewable Resources Assessment Group, and Tony Martin at the Sea Mammal Research Unit for helpful comments on an easier draft of this chapter. Financial assistance from the World Wide Fund for Nature (UK) is also gratefully acknowledged.

9: Marine Invertebrates

A.J. UNDERWOOD & P.G. FAIRWEATHER

INTRODUCTION

There is ample evidence that invertebrate carnivores are numerically and functionally important members of marine assemblages (Paine, 1966; Connell, 1972). For example, filter-feeding plankton make up a large proportion of the animals in the open ocean, while benthic filter-feeders effectively couple pelagic and benthic habitats, i.e. their feeding causes the transfer of energy from plankton to benthos (Crisp, 1964). On longer time scales, predation is credited as the cause of many evolutionary changes, such as the modification of behaviour (Curio, 1976) or morphology (Vermeij, 1978) of prey as a result of predator–prey 'arms races'. Stanley (1973) even speculated that predation in marine environments was the main force driving the earliest radiation of eukaryotic life on Earth.

Here, we shall consider some of the well-documented cases that indicate the styles and consequences of predation by marine invertebrates. Following Paine (1980), we define predation as the consumption of living animals (prey) by carnivorous animals (predators). Our own research has concentrated on benthic predators on hard substrata, but we believe these are exemplary of many general features of the biology of marine invertebrate predators. We hope that our biases will spur others more familiar with pelagic predators and predators in soft substrata to make the appropriate comparisons and to fill in the idiosyncrasies of those systems.

TAXONOMIC TYPES OF PREDATORS

Marine invertebrates include nearly all phyla of animals. At least some representatives of each phylum are carnivorous, either as filter-feeders, straining prey from the water column, or as predatory hunters (see Table 9.1). Not all, however, have been studied. For example, field studies of benthic predators have concentrated on starfish and gastropods, while crabs and octopuses are less well understood. Some smaller predators (e.g. many of the wormlike phyla) have rarely been studied experimentally. Amongst the filter-feeders, copepods and bivalves have received most attention. Many types of pelagic, filter-feeding or hunting predators await field investigation. Field studies investigating the full

Table 9.1 Phyletic distribution of marine invertebrate predators. The presence and absence of different types of predators (after the schema of Hughes, 1980a) in various marine phyla. Each type represented is indicated as +; − indicates a type not found in that phylum. The phyla Pogonophora, Pentastomida and Tardigrada have few or no predators

Phylum	Ambushers	Hunters		
		Searchers	Pursuers	Grazers
Protozoa	+	+	−	−
Porifera	+	−	−	−
Cnidaria	+	−	−	−
Ctenophora	+	−	−	−
Platyhelminthes	−	+	−	−
Nemertea	−	+	−	−
Rotifera	−	+	−	−
Aschelminthes	+	+	+	+
Annelida	+	+	−	+
Mollusca	+	+	+	+
Crustacea	+	+	−	+
Chelicerata	−	+	−	−
Uniramia	+	+	−	−
Entoprocta	+	−	−	−
Ectoprocta	+	−	−	−
Brachiopoda	+	−	−	−
Phoronida	+	−	−	−
Priapulida	−	−	+	−
Sipunculida	−	+	−	−
Echiura	−	+	−	−
Echinodermata	+	−	+	+
Chaetognatha	−	+	−	−
Hemichordata	+	−	−	−
Urochordata	+	−	−	−

taxonomic range of invertebrate predators at one site, even in commonly studied, rocky intertidal habitats, are relatively few (see, for example, Edwards *et al.*, 1982; Menge, 1982).

TYPES OF PREDATION

Hughes (1980a) introduced a useful schema for categorizing different types of predators. He used the mobility of predators and the time they spend handling (i.e. subduing and ingesting) prey to distinguish four types: grazers, ambushers, searchers and pursuers. *Grazers* live upon their prey as a substratum and only feed on small amounts (e.g. nudibranch seaslugs and cowries on sponges). Endoparasites represent the most extreme form of the group described as grazers. *Ambushers* are sessile or sedentary and therefore must wait for prey to come to them (e.g. anemones), or must actively filter seawater. *Searchers* actively patrol for prey and spend relatively little time ingesting prey (e.g. some crabs). *Pursuers* (such as starfish and whelks) also actively hunt, but spend much of their time subduing their prey. The major taxonomic groups of marine invertebrate carnivores are assigned to these types in Tables 9.1 and 9.2.

Other, more intricate attempts at categorizing the ways in which predation can occur have involved more variables, including other aspects of the behaviour of the predators and their prey. These include the type of habitat, the relative sizes and mobilities of predator and prey, whether the predator is resident or transitory in the habitat, the types of refugia for prey and feeding rates of the predators (some of these are illustrated in Table 9.2). For example, it is well known that muricid whelks such as *Nucella lapillus* tend to specialize on barnacles and mussels as prey. They are relatively small and so take a restricted range of sizes of prey, usually foraging during high tide close to crevices or other shelter. They attack one item at a time, taking several to many hours to attack and gain access to the tissues through the shell of the prey. Their limited mobility tends to restrict them to areas close to shelter, leading them to have patchy effects on populations of prey. Factors like slow rates of feeding mean that individual whelks cannot kill large numbers of prey during short periods of time (Paine, 1971).

In contrast, starfish such as *Pisaster ochraceus* are generalists, but often east competitively dominant species of prey (Paine, 1974). This has led to them being described as 'keystone predators' — species of predators that have potentially greater influence on the structure of an assemblage than simply causing a reduction in the abundance of their prey (see Chapter 17 and below). Starfish are large compared to most of their prey, and so can take a wide range of sizes of prey (Paine, 1976). Starfish forage low on the shore during high tide, but, using their many tube-feet, they can collect several prey at once, and can consume them later during low tide. This, combined with their great mobility, allows starfish to forage selectively, clearing many prey and leaving obvious foraging swathes corresponding to the paths taken by individual starfish while feeding. Thus, starfish also promote spatial patchiness, but on a different scale from that produced by whelks.

Whatever schemes of classification are proposed, there will always be problems in trying to encapsulate the complexities and subtleties of individual predators into categories based on a few simple taxonomic, behavioural or ecological criteria. For example, although many crabs can properly be identified as searchers or pursuers (as in Table 9.1), there are, in fact, two major ways in which crabs attack their prey. One sort of crab crushes the prey after seizing it in well-muscled and thickened claws. The other type of crab is a 'peeler' that uses fine claws to chip away the edge of shelled snails and thereby gradually expose the tissues of the prey (see review by Vermeij, 1978). The two types of predation have very different consequences for the evolution of body form in the prey (see below), yet would appear identical under the classification in Tables 9.1 and 9.2.

Table 9.2 Predatory characteristics of different taxa. Only adults are considered. Many marine animals have predatory, planktivorous larval stages. Type classification from Hughes (1980a)

Phylum	Taxon	Examples of genera	Type	Main habitat	Size relative to prey	Main prey types	Comments
Cnidaria	Anemone	Actinia Anthopleura	Ambush	Benthic	Same or larger	Plankton Mussels	Sessile
Cnidaria	Coral	Porites	Ambush	Benthic	Larger	Plankton	Sessile
Cnidaria	Jellyfish	Cyanea Physalia	Ambush	Pelagic	Larger	Plankton Fish	Swims
Platyhelminthes	Flatworm	Notoplana	Search	Benthic Infaunal	Larger or same	Barnacles Worms	Active
Nemertea	Nemertean	Nemertes Paranemertes	Search or graze	Benthic Infaunal	Larger	Barnacles	Active
Mollusca	Mesogastropod whelk	Polinices Charonia Cabestana	Pursue or graze	Infaunal Benthic	Larger or same	Bivalves Echinoderms Ascidians	Bores into shell of prey
Mollusca	Neogastropod whelk	Thais Nucella Morula	Pursue or graze	Benthic	Larger or same	Barnacles Mussels Snails	Bores into shell of prey
Mollusca	Coneshell	Conus	Search	Benthic	Larger	Worms Fish	Uses toxic stings
Mollusca	Nudibranch	Archidoris	Graze	Benthic	Smaller	Sponges Cnidarians	Live on their prey
Mollusca	Mussel	Mytilus	Ambush	Benthic	Larger	Plankton	Filter-feeders
Mollusca	Octopus	Octopus	Search	Benthic	Larger	Molluscs Crabs	Gathers more than one prey item at a time
Crustacea	Copepod	Calanus	Ambush or search	Pelagic	Larger	Plankton	Filter-feeders
Crustacea	Barnacle	Balanus	Ambush	Benthic	Larger	Plankton	Filters water column
Crustacea	Crab	Cancer Ozius	Search	Benthic	Larger	Mussels Snails	Crushes or peels prey
Crustacea	Lobster	Homarus	Search	Benthic	Larger	Bivalves	Crushes prey
Uniramia	Waterstrider	Halobates	Search	Surface of sea	Larger	Plankton Larval fish	Feeds on prey trapped in surface film
Echinodermata	Starfish	Asterias Pisaster	Pursue	Benthic	Larger	Mussels Snails	Gathers more than one prey item at a time
Chaetognatha	Chaetognath	Sagitta	Search	Pelagic	Larger	Plankton	Voracious feeders
Urochordata	Salp	Pegea	Ambush	Pelagic	Larger	Plankton	Form massive chains

Width of diet

Predatory behaviour depends primarily on how many species of prey are eaten; a dietary generalist (Emlen, 1973) eats a wider range of prey than a dietary specialist. Some components of a diet may be obligatory, and if a predator must include certain items of prey in its diet (e.g. in order to obtain sufficient vitamins or minerals), then its behaviour is likely to include long-term and wide-ranging searching for that type of prey, regardless of the availability of other types.

The places where predators actually forage can greatly influence any predator—prey interaction, and can confuse assessment of the width of a diet relative to the types of prey available. An obvious reason why particular types or sizes of prey are never consumed is that their distribution falls outside the foraging ambit of the predator. Thus, the highest intertidal locations on a seashore are often a refuge from predation for animals like barnacles (Connell, 1961). This is because most predators of barnacles are marine invertebrates or fish whose activities are confined to areas permanently underwater. Their opportunities to forage in intertidal areas (i.e. during high tide) become progressively more limited, the higher up the shore they move. This is especially so where handling times are longer than the duration of the period underwater during high tide at any given height on the shore. For example, many muricid whelks exploit the limited mobility of their sessile or sedentary prey by drilling holes through their defensive armour. Drilling, however, takes a long time (see, for example, Connell, 1970; Fairweather & Underwood, 1983).

In many marine habitats, large ambushing predators, such as some sea-anemones that sit in crevices and wait for prey to tumble in (Dayton, 1973), are less likely to show clear-cut preferences for different types of prey. All opportunities for such 'sit-and-wait' predators to feed must, presumably, be taken (see Chapter 13). In contrast, ambushers such as filter-feeding barnacles, can show considerable selectivity among different types of prey (see, for example, Anderson, 1981a).

Preferences and choice of diet

Murdoch (1969) demonstrated that some intertidal whelks would switch between different types of prey in such a way that the diet usually over-represented the more abundant of the two types of prey available. Such switching tends, in theory, to preserve the width of diet in any area, because the attention of predators is focused away from less abundant prey, which are therefore less likely to be driven to local extinction before there has been new recruitment to their populations. This kind of stabilizing predation has received little attention in field experiments in marine habitats.

The choice of prey species by marine predators is determined, at least in part, by morphological constraints. Predators cannot usually eat prey conspicuously larger than themselves (but see Table 9.2). Also, prey often defend themselves chemically or by other means. Thus, a common feature of marine predatory behaviour is that predators attack only certain sizes of prey out of the range of prey available. This is often revealed by a comparison between the size-frequency distributions of prey eaten and prey available (where *available* prey are those that are present, findable and of appropriate sizes for the predator to be able to attack; see, for example, Fairweather, 1985). Optimal foraging theory suggests that dietary choices are made to maximize the rate of gain of energy (Chapter 4) but other measures of prey value may be appropriate for particular marine invertebrate predators (Hughes, 1980b).

The mobility of predators has another important influence on the type of predation, because mobile predators can show a numerical response to density of prey. Predators may move to, and congregate at, sites where their prey are aggregated (Fairweather, 1988a). This has obvious advantages in reducing the time individual predators must spend searching for prey. It may lead to increased rates of mortality of prey from predation when predators redistribute quickly, but conversely, may reduce the average rate of predation if predators are less mobile (see Chapters 5, 10 and 13).

Another important aspect of behaviour that influences the range of prey taken is whether the predators forage singly or in groups. Where predators forage singly, the *average* diet may be very wide because different individual predators (however narrow their individual diets) feed on different things or on different proportions of the available prey (perhaps because the prey are patchily distributed). Where predators forage in groups, they may tend to have narrower ranges of diet because each individual in the group eats whatever the group finds. Alternatively, the groups may encounter, and feed on, a wider variety of prey, because the group may be attracted to any item that attracts any individual member of the group. This has been proposed as the reason why some predatory whelks form large aggregations while foraging (Jenner, 1958; Crisp, 1969).

These contrasting effects of schooling behaviour on diet-breadth have not been explored experimentally for marine predators. McKillup (1983) investigated an intertidal whelk *Nassarius pauperatus* that showed single or group foraging in different habitats. On some shores, food (carrion) was distributed patchily at point sources. On other shores, the whelks fed on microalgae which were more evenly distributed over the substratum. McKillup was able to demonstrate genetic components in differences in behaviour of whelks that usually form groups to feed as opposed to those that usually feed singly.

CHOOSING PREY

Patterns of foraging by predators are often altered by their current level of hunger. Hughes and Dunkin (1984a) described laboratory experiments in which recently fed whelks *Nucella lapillus* moved around much less than those that were hungry. The hungry animals also moved faster along straighter paths, which would presumably result in hungry whelks leaving an area, which should increase the chances of finding something to eat (on the basis that departure from an area where a predator is hungry is not likely to make matters worse, but might result in finding a patch of food). The less directional and slower movements of sated animals presumably tend to keep them in areas where food is available, and may result in patchy patterns of mortality suffered by the prey. Whether these tendencies in searching for food are general is unclear, and extrapolation from the laboratory to the field may not be straightforward because movements from one place to another are probably coupled with increased risks of mortality, e.g. natural enemies, storms, etc. (Menge, 1978; Menge & Lubchenco, 1981).

Many marine predators show well-developed sensory mechanisms and can detect their prey at a distance. These include the tactile system used by cephalopods (Wells, 1978), detection of burrowing prey by sound (Kitching & Pearson, 1981) and a variety of chemical stimuli. For example, Pratt (1974) demonstrated in laboratory experiments that the whelk *Urosalpinx cinerea* moved towards water that contained chemicals from prey species on which the whelks had previously fed, but showed less response to water containing chemicals from other potential prey. Crisp (1969) described the sensory system of the mud-whelk *Nassarius obsoletus* that enables the animal to discriminate among water-borne chemicals derived from potential prey.

Several authors have demonstrated that individual whelks are often very selective in the place at which they attack their prey. For example, Black (1978) found that the distribution of holes drilled by the whelk *Thais orbita* into the shells of limpets was non-random. Whelks were more likely to drill through those parts of the shell that would give direct access to energy-rich tissues (such as the digestive gland and gonad) of the prey. This behaviour presumably increases the relative gain of food per unit time spent drilling. How the predators determine the best sites for drilling through the prey's shell is not yet known.

LEARNING

There are potentially great advantages for marine predators that are able to learn how to discriminate among different types or sizes of prey, because different prey offer different rewards of energy and require different amounts of expenditure to obtain the food. It is not surprising, therefore, that many predators show sophisticated capacities to learn how to discriminate among individual items of prey. The most detailed study of this has involved the whelk *Nucella lapillus*.

When previously fed on mussels *Mytilus edulis* for 60 days and then presented with a different prey (barnacles, *Semibalanus balanoides*), whelks mostly ate the larger prey and usually drilled holes in the shell to gain access to the flesh. In contrast, whelks that had previously been feeding on the barnacles usually ate intermediate sizes of barnacles and usually prised the prey open using the foot and proboscis, rather than the more time-consuming method of drilling them (Dunkin & Hughes, 1984). There is no doubt that the difference was due to previous experience (i.e. learning) by those whelks that had previously been fed (had been 'ingestively conditioned') on the barnacles.

The immediate advantages of the more rapid method of handling the prey include decreased risk of death while exposed to inclement weather and to predators on the whelks themselves. Also, the more quickly each individual item of prey can be handled, the more time there is for the predator to find another meal. Another advantage of rapid entry to the prey was observed by Dunkin and Hughes (1984). After they had spent time drilling through the shell and gaining access to the flesh of the prey, some 12% of whelks were displaced from the food by other whelks. The extra time taken in drilling rather than prising open a shell was sufficient to increase the likelihood of competitive displacement of the drilling predators.

The mechanisms by which predators learn how to handle unfamiliar prey are not clear. Individual whelks took 25−30 days to drill about five or six mussels. At the end of this period, the whelks were more efficient at opening their prey, and had learned to drill through the thinner parts of the shell or to drill over the prey's digestive gland. These skills result in an increase in 'profitability' of individual prey of about 17% (Hughes & Dunkin, 1984a). In contrast, individual crabs *Carcinus maenas* took about 1−1.5 hours to break open mussels *Mytilus edulis* and consume them. After a period of time, the crabs decreased their handling time by about 60−70%, i.e. they learned to be more efficient (Cunningham & Hughes, 1984). The intriguing thing about these results is that, in each case, despite very great differences in the time they took to handle their prey, the two types of predators required some five or six trials with unfamiliar prey before they learned how to deal with them efficiently.

INDIVIDUAL DIETS

Diets of individual predators or groups of predators vary from time to time and place to place, making the interpretation and prediction of diets and foraging strategies extremely difficult. West (1988) monitored the movements and feeding of individually tagged whelks *Thais melones* on a tropical shore. She found that individuals varied considerably in the ranges and types of prey taken. Some were relatively generalist, eating a variety of different species of prey. Other whelks, however, were specialist, and tended to consume only one or a few types of prey. To add to the complexity, different individuals tended to concentrate on quite different subsets of the prey available. This sort of study raises considerable questions about methodologies used in describing and contrasting diets in different places or at different times, or by animals of different ages. The average diet is not likely to mean a great deal for individual predators; it reflects neither the actual diets of any particular specialist predators, nor the diet of those individuals that are more catholic in the range of prey they eat. Procedures for dealing with this kind of problem will have to be developed before comparisons of food-webs become realistic.

West (1988) suggested that the differences between individuals were maintained by the fact that each individual increased its efficiency by repeatedly handling the same type of prey, because this reduced the time taken to identify, attack and consume the prey. This is in line with the observations of Hughes and Dunkin (1984b) in laboratory trials that *Nucella lapillus* would continue to feed on the same prey as that previously encountered and consumed, regardless of the prior conditioning to different prey. This could explain how individuals eventually acquire their particular feeding preferences, and could also explain how and why predators start to switch to relatively more abundant prey (Murdoch, 1969).

The other area of marine predation that has received attention is the effect of different diets, or mixtures of diets, on the predators themselves. Moran *et al.* (1984) found considerable differences in the rates of growth and survival of marked whelks feeding on different types of prey on different shores. Experimental manipulations of available prey revealed that the whelks were influenced by their prey and not simply by being in different environments. Growth was fastest when the whelks were feeding on barnacles and was slower when the whelks were eating tubeworms. Mortality of whelks feeding on tubeworms was much greater than for those predators eating oysters, a much-preferred prey. Thus, the eventual sizes reached and frequency distributions of sizes of the predators were very much influenced by the types of prey available to them, and by the interaction of this with their preferences among different types of prey.

CONSEQUENCES OF PREDATION

Populations of prey

It is important to establish the degree to which predatory marine invertebrates reduce the densities of their prey populations. In addition, some predators (sometimes called 'hyperpredators') feed on other predatory animals. Thus, Feare (1970) reported intense predation by shore birds on the whelk *Nucella lapillus*. In consequence, where the birds were active, the survival of barnacles and mussels (the prey of whelks) might be expected to be greater. Heller (1975) suggested that predation explained why different coloured shells of *Littorina saxatilis* were found on shores with different background colorations. The predators were supposedly crabs, which search for prey using visual cues, and the crabs were thought to be preferentially consuming the inappropriately coloured morphs on any shore. In both these examples, the arguments are appealing, but there is no experimental evidence to support the contentions.

Kitching *et al.* (1966) demonstrated that the thickness of shells of whelks *Nucella lapillus* was influenced by predation by crabs. In experimental transplants from exposed shores where whelks have thinner shells, to sheltered locations where crabs are active and most whelks have relatively thick shells, they demonstrated that crabs found the thin shells much easier to open. Thus, selection against thin-shelled whelks was presumably a result of predation. They also demonstrated a 'balancing' selection against thick-shelled whelks in exposed locations, because thicker shelled individuals had smaller feet and thus could not cling on to the wave-stressed surfaces as efficiently as whelks with thinner shells. Thus, there are several genetic influences of predation at work in marine habitats (see below).

Sizes of prey

Connell (1972, 1975) has attempted a general synthesis of the influences of predation on the sizes of prey. His conclusions were that predation is most important in those parts of the prey's environment that are physically benign. As a result, there is a general tendency for prey to be young and small (and not yet consumed by their predators) or to be large (too big to be eaten, having escaped predation during the vulnerable size classes). Large individuals of a species of prey are often able to escape their predators because they have reached sizes too large to be attacked or ingested in the time available for feeding by the predator (Paine, 1976). This would explain the frequently observed, non-continuous distribution of sizes in many populations of marine invertebrates, that have small and large individuals, but often lack animals of intermediate sizes (see cited examples in Connell, 1972). It is not clear that this interpretation is general, however (Underwood & Denley, 1984). For example, inconsistencies and vagaries of recruitment of species with widespread planktonic offspring can also lead to deficiencies in some year-classes, giving rise to the same pattern of distribution of sizes (see Underwood & Fairweather, 1989).

MORPHOLOGY OF PREY

Predation has often been cited as a major selective force in the evolution of body forms of prey (Chapters 1 and 2). Much of the observational evidence for this has been reviewed by Vermeij (1978), who suggested that the armature and sculpture of shells of gastropods were evolved in response to the intensity of predation by crabs and fish. He also analysed differences from one ocean to another in this predator–prey 'arms race'. Thus, where crushing crabs are prevalent, gastropods tend to have greatly thickened shells. Where 'peeling' crabs are numerous, snails tend to have thickened lips to their shells. Where fish are important predators, gastropods commonly have large outgrowths on the shells which increase the effective diameter of each individual, making it invulnerable to all but the largest individual predators.

It is impossible to obtain incontrovertible experimental evidence that predation has caused these patterns in morphology, but there are some instances of marine predation that are associated with very rapid changes in prey morphology. For example, Seeley (1986) described marked differences in the shape and thickness of shells of a gastropod in collections made between 1871 and 1984. These changes took place within only a few generations of the snails, and coincided with the appearance of a predatory crab in the geographic range of the snail. In areas where the crab has not yet appeared, the snails were of the earlier, thinner morphology and, in experiments, were much more vulnerable to predation by the crab.

The other sort of evidence on the influence of predation on prey morphology comes from experiments that reveal how the observed morphology reduces the rate of predation. Thus, Palmer (1979) removed the outgrowing spines from shells of gastropods and showed that they then fell prey to fish that could not attack them when their shells were intact. This sort of evidence cannot reveal why and how the spines originally evolved (and whether or not this was a direct response to the evolution of predators), but it does provide evidence for predation causing the continued maintenance of the existing morphology.

BEHAVIOUR

The behaviour of prey can also be influenced by the activities of predators. For instance, sensing their enemies at a distance is likely to make any defensive or avoidance behaviour more likely to be effective. Many marine invertebrates show defensive reactions in the presence of predators, often detecting the predator at a distance by chemosensory means. Some species of prey show marked agitation, increased speed of locomotion and other defensive patterns of behaviour that are likely to remove the prey from the predator or otherwise to prevent the predator from being able to attack, or both (see review by Bullock, 1953, for marine gastropods). Such behaviours are not always successful, of course, and increased rates of movement can sometimes increase the risk of predation, e.g. jumping by snails sometimes places them nearer the predator than they would otherwise have been (Kohn & Waters, 1966).

Although specific types of behaviour are often observed in the presence of predators, it is not always easy to demonstrate that these bring about any reduction in mortality due to predation. Dayton's (1973) graphic account of sea urchins 'stampeding' when confronted with a single starfish is a warning about this. As a result of their panicky behaviour, many more urchins fell into the mouths of ambush predators (large sea anemones) than would have been consumed by the hunting predator.

Nevertheless, Phillips (1975, 1976) was able to demonstrate that intertidal limpets *Acmaea* spp. showed escape responses in the presence of predatory starfish. In the field, when starfish were experimentally tethered so that they could not move away, nearby limpets moved upshore. This behaviour would have the effect of decreasing mortality due to predation because predators are generally less active at higher levels in intertidal habitats (see above).

One of the more complex predator-avoidance behaviours is exhibited by snails which climb up the stalks of plants in salt-marshes to positions well above the level of water during high tide to avoid being reached by predatory crabs (Hamilton, 1976; Warren, 1985). The snails have well-developed visual responses to the grass-stalks which enable them to do this (Hamilton, 1976).

EVOLUTION OF DIVERSITY OF PREDATORS

The other area of interest in marine assemblages is the observation that there are more types of predators in some habitats and at some latitudes than others. Connell (1970) suggested that the greater diversity of intertidal whelks on the west coast of the USA than on the west coast of northern Europe was possibly a response to the greater consistency of recruitment of species of prey in the former habitats. Thus, where prey arrive regularly and predictably from year to year at all levels on a shore, predators that specialize in consumption of prey at different heights can persist. Where the vagaries of prey recruitment are large, such specialized types of predators would not be able to survive for long periods. Paine (1966) also considered this issue with respect to the productivity of different marine environments, and suggested that the diversity of predators would be greater where there was greater productivity, supporting enlarged and more complex food-webs.

Structure of assemblages

Predation is a process responsible for a number of changes in the structure of marine assemblages of species. Predation (or exploitation; Addicott, 1974) may result in decreased abundances, altered size-frequency distributions (and associated reproductive output and productivity) of prey, increased availability of any resources used by the prey (food, space, nesting sites) and, possibly, local extinctions of prey with concomitant declines of richness of species in areas where predation is intense. Demonstration that predation is occurring does not, however, suffice to indicate that the more complex interactions are also occurring.

DISTURBANCE

Because predators are capable of removing elements of assemblages of species, thereby freeing resources, some authors have considered predation to be similar in its effects and its importance to physical disturbances (like storms or oil-spills) that kill some species. Predators act in patchy ways to reduce abundances of species that are occupying space or exploiting resources. Under the disturbance model, predators are not selective and do not focus their attention on particular types of prey; they simply remove a variety of animals from the system. As a result, they can create a mosaic of patchiness, favouring the persistence of early colonists. When predation is very intense, some species of prey are eliminated entirely from an assemblage. Thus, the potential for increased or decreased species-richness depends on both the range of prey available and on the intensity of predation.

Equating predation and physical disturbance is misleading, however, because it ignores the feedback that exists between populations of prey and their predators. Thus, describing predation as a disturbance implies that the predators are external and uncoupled from the dynamics of their prey. Yet it is widely recognized that the behaviour, growth, reproduction, morphology and biochemistry of the predators are intimately associated with responses to different types of prey (see above; Connell, 1972; Moran et al., 1984).

In many marine habitats, predators can have major indirect influences on species other than their prey, because they remove organisms that would otherwise occupy space or consume other resources. Many authors have demonstrated the influence that predation can have on competition for food and space. Predators of mussels can increase the survival of barnacles (which are often an alternative prey; Menge, 1976). Predation on barnacles can increase the abundance of large limpets which are not usually an alternative prey (Underwood et al., 1983).

KEYSTONE PREDATORS

The preceding remarks applied to generalist, relatively unselective predators. Where predators select and attack preferentially the dominant competitor for resources in marine assemblages, predation can have more far-reaching consequences. Predators that remove competitive dominants from assemblages have the potential to increase dramatically the richness of species present and to increase the complexity of the web of interactions involving those species. The first demonstration of this keystone predation was by Paine (1966, 1974) in experimental removals of the starfish *Pisaster ochraceus* from shores where it consumed mussels *Mytilus californianus*, thereby creating space for all the species that were normally outcompeted by the mussels. The effect of starfish removal was a dramatic rise in the number of species present in the experimental plots. This model of predatory interactions requires that the assemblage is at or near some competitively determined equilibrium in the absence of predators. Thus, the impact of the keystone species, which selectively removes the dominant competing species, is inevitable. There has been debate about whether there is convincing evidence that the starfish *Pisaster* really does prefer the mussels as prey (Fairweather & Underwood, 1983) and whether other processes, particularly differences in the rates of recruitment of the prey, might not be more important than the influences of the predator (Underwood & Denley, 1984). There has also been discussion about the methods used to measure the relative importance of different types of predators in marine assemblages (see the interchange between Edwards *et al.*, 1982 and Menge, 1982).

BEHAVIOUR OF PREDATORS AND THE STRUCTURE OF ASSEMBLAGES

Where the distribution of predators is influenced by the physical or other characteristics of their environment, predation can have localized and patchy effects on prey and can thus serve as a major element causing heterogeneity in natural assemblages. For example, some inter-tidal whelks are unable to forage at high levels on the shore during low tide, because of the risk of desiccation whilst out of the water. Where some physical refuge provides shelter against inclement weather, the predators are able to persist and thus to attack prey that would other-wise be safe. Dayton (1971) found that whelks were more likely to be able to attack barnacles at high levels on shores where patches of anemones provided cracks and crevices that were moist during low tide, because the whelks could shelter among the anemones. It is not clear to what extent the whelks might have contributed to the persist-ence of the anemones themselves, since it is possible that the whelks might have been removing barnacles that were potential competitors for space, thus preserving the populations of anemones.

Moran (1985) has unravelled much of the complexity of sheltering behaviour in the whelk *Morula marginalba*, which aggregate in cracks and crevices during periods of low tide when the weather is warm. Sheltering behaviour was influenced by the time of day of low tide, so that when tides were low during the middle of the day, and whelks would be exposed to greater risks of desiccation, they were much more likely to be in shelters. The effects on prey were dramatic, because prey individuals near shelters were consumed very rapidly, but prey at sufficient distance from a shelter were invulnerable because the predator could not find, attack and consume them before the weather changed and the predator retreated to shelter again. This created 'haloes' of free resources around shelters that could then be occupied by other non-prey species, or by the juvenile stages of prey before they are vulnerable to predation (Fairweather, 1988b). Garritty and Levings (1981) have described a more complex system where a grazing snail was the main prey of a tropical whelk. Around crevices, there were no or few grazers, so these areas supported greater abundances of algae. The whelks were therefore indirectly responsible for the creation of 'reverse haloes' where space was much *more* utilized near the crevices.

FOOD-WEBS

In recent years, there has been considerable interest in attempts to analyse the relationships between activities and diversity of predators and the structure of assemblages of species. This has been reviewed by Paine (1980, 1988), whose most recent analysis makes it clear that, at least for marine habitats, theory far outdistances empirical evidence. The construction of food-webs is fraught with difficulties, not only because of the variability of individual predators (West, 1988) and the immense problems of identifying prey when they are consumed quickly and hence under-represented in most forms of practical sampling (Fairweather & Underwood, 1983), but also because there are great difficulties in determining the strengths of relationships between different types of predators and prey (Paine, 1980). Again, there are important differences among individuals and among species in what they eat, when they eat, and where they eat it, and these are important components of any analysis. The practice of lumping species together into 'functional' or taxonomic groups for the purpose of simplifying analyses is not a useful method for understanding the behaviour of complex assemblages where many species play important, interactive roles (see Underwood *et al.*, 1983; Fairweather *et al.*, 1984; and Paine, 1988 in contrast to Paine, 1980).

One conclusion from the study of marine food-webs is that relationships based on experimentally verified functional links among members are probably the only useful sort available. Other representations, based on sampling and quantitative summaries, or on measures of flow of energy from one point in the web to another, are less likely to provide realistic pictures of the assemblage being investigated (Paine,

1980). There is also a major problem in determining how to deal with benthic filter-feeders, since filter-feeders usually provide almost unlimited connectance to the planktonic assemblages on which they feed. Discovering that all life in the sea is connected may make for realism, but does not facilitate theoretical development. On the other hand, denying connections among assemblages in different habitats will only result in incomplete, or irrelevant theory. Until these issues are resolved, analyses of the role of predation in affecting the structure and dynamics of marine invertebrate assemblages will continue to be vexing and incomplete.

PELAGIC PHASE OF LIFE OF MARINE PREDATORS

Some marine predators are holoplanktonic, i.e. their entire life cycle is spent in the plankton, and, in consequence, they have very widespread populations that move with their prey (for review, see Raymont, 1983). One of the features of many marine predatory animals, however, is that their offspring are dispersed during a planktonic stage. Many complete their development using energy provided in the egg, while others are planktotrophic (but they usually eat microalgae and are not predaceous as larvae). This life cycle imposes some difficulties on marine predators that are not shared by their counterparts in other habitats.

Consequences of dispersal

OPEN SYSTEMS

One immediate consequence of a pelagic period of development is that marine populations often maintain genetic continuity across enormous geographic regions (Scheltema, 1971). They also have open populations with considerable mixing and little opportunity for local adaptation to prevailing environmental conditions because the offspring will probably recruit to populations in quite different areas.

There are also considerable sources of mortality in the plankton, which, among other things, makes average life expectancy of the individuals very small until they have arrived back into the habitat occupied by adults. Vagaries of mortality are likely to lead to marked fluctuations in the numbers of larvae that survive to the end of larval development (see, for example, Underwood & Fairweather, 1989). The consequences of this variable recruitment for populations of marine species have been explored using theoretical models of open populations (Roughgarden et al., 1985); there are quite different outcomes for a population of predators when recruitment is small, large or variable.

As a result of variable mortality and the lack of tight coupling between population size and the numbers that recruit in the next generation, there are uncertainties in the replenishment of prey for many marine predatory species. This probably means that many marine predators must be able to withstand periods of lack of food if they are to persist in some habitats. Alternatively, predators must be able to capitalize on a variety of types of prey so that they can withstand periods of shortage of the more preferred, or nutritionally more desirable, types if these happen not to arrive in sufficient numbers.

Hagmeier (1930) observed a heavy settlement of the bivalve *Spisula subtruncata* in the Wadden Sea, coincident with heavy settlement of the predatory gastropod *Natica alderi*, which only eats bivalves, and the brittle star *Ophiura texturata*, which is omnivorous. After a while, the bivalves were all consumed and the *Natica* then starved to death. The brittle stars, however, persisted because they were able to eat other types of prey.

Finding a place to settle

CUES

Before prey or predators can recruit into their adult populations, they have to leave the larval, planktonic existence and make a successful change to a different mode of life. Hadfield (1978) reviewed the processes of settlement and metamorphosis using a predatory marine invertebrate as the major example. Among many other changes that are necessary before a pelagic larval form can function in the adults' habitat is the complete biochemical, physiological and anatomical modification of the digestive system to transform the animals from a micro-herbivorous to a predatory life style. The details of this transformation have not yet been fully explored for any marine predator.

A notable feature of metamorphosis and settlement is the use of cues provided by the type of food which the adult predators eat. Nudibranch gastropods are usually specialized predators feeding only on one or a few species of prey. Several examples exist where the larvae of these predators can only be induced to complete metamorphosis and to settle from the plankton when the prey consumed by adults are present. Thus, adults of the nudibranch *Adalaria proxima* eat a specific type of polyzoan *Electra pilosa*, and the larvae of *A. proxima* will only settle in laboratory containers when the polyzoan is provided (Thompson, 1958). Other cues have been reviewed by Hadfield (1978). The potential advantages of having responses to appropriate sources of food are obvious, especially if the food is distributed patchily in space.

Barker (1977) was able to rear starfish *Stichaster australis* in the laboratory and to have them complete settlement when an encrusting calcareous alga *Mesophyllum insigne* was provided. He also found newly settled juveniles in the field on patches of this alga, cues from which seem to cause behavioural responses by the settling larvae. The starfish emerge from the patches of algae and begin to feed on mussels after 7–8 months on the shore, eventually becoming fully carnivorous after about 15–18 months from settlement (Barker, 1979). This is a difficult pattern of settlement to understand because it implies that the larval starfish respond to cues from plants that do not carry with them any certainty of availability of suitable prey after metamorphosis is complete.

THREATS DURING SETTLEMENT

The other problem faced by settling marine invertebrates is that their new habitats are often dominated by predators, notably filter-feeders. Thorson (1950) discussed anecdotal evidence for massive mortality suffered by all small marine invertebrates due to the effective filtration of water by filter-feeders, but it has been difficult to demonstrate a decrease in the rate of settlement of larvae due to the presence of predatory benthic filter-feeders in experimental studies (Young & Gotelli, 1988).

There are other threats to settling larvae. Yamaguchi (1974) demonstrated that juvenile crown-of-thorns starfish *Acanthaster planci* fed on algae for several months before moving on to corals, on which the adult starfish feed. While the starfish are small, they are themselves vulnerable to attack by their prey, and suffer considerable damage and can actually be killed by the corals until they are large enough to bypass the prey's defences.

Recruitment interacting with predation

PREDATOR SATIATION

Variations in the intensity or timing of prey recruitment can influence the pattern or outcome of predatory interactions in several ways. When recruitment is sufficiently large, the prey arrive in numbers too great for the predators to be able to consume them, so large numbers survive. Such a case was described by Dayton (1971) when barnacles arrived on a shore in very large numbers. As a result, the area was subsequently dominated by adult barnacles. Unless the predators undergo a rapid numerical response, they will not damp out the increased recruitment, and large numbers of prey will survive long enough to become too big to be eaten.

For predators to show an appropriately large numerical response (other than by altering their behaviour; Fairweather, 1988a) they must

spread dispersal. Thus, numerical responses to increased availability of
prey are not likely unless the predators have direct development (in
benthic habitats) or the larval stages do not disperse over great distances.

DIFFERENTIAL SELECTION OF PREY

Fairweather (1985) demonstrated another important interaction between
recruitment and predation by a generalist predator. He showed that
barnacles *Tesseropora rosea* were more likely to survive their early
stages on the shore if there were limpets *Patelloida latistrigata* present.
The limpets were a highly preferred prey, and when limpets were
present in large numbers before the arrival of the barnacles, the survival
of the juvenile barnacles was markedly greater. Fairweather (1985)
showed that the differences in prey abundance caused by differential
predation persisted for long periods (at least 2 years after the predators
were no longer active in the area), and thus were part of the historic
explanation for some of the variations in abundance of marine organisms
from place to place.

CONCLUSIONS

The relationships between predators and their prey in marine habitats
are complex and diverse. Here, we have only explored a few of the
possibilities and have made no attempt to cover all of the marine
habitats in which predation by invertebrates occurs, nor all of the taxa
of marine invertebrate predators. Even so, it is obvious that the study
of predation is complex, and that the consequences of predation are
many and various.

Considerable progress has been made in understanding the behavioural and physiological processes that determine the choices made by
predators, yet there is still no complete model capable of predicting
the diet of a marine predator when confronted with alternative types
of prey. Nor has it yet been possible to integrate information about
differences among individuals within species (whether genetic, environmental, the result of learned behaviour, etc.) so that the relationships
between populations of predators and prey can be better understood.
For many marine populations, the variance in the timing and intensity
of recruitment is so great that this level of detailed understanding may
never be achieved for predators and prey under field conditions.

Examination of experimental studies of marine predators has increased our understanding of the predatory processes. Causes and cues
for learning to discriminate among various types of prey, and the
consequences of predation to the evolution of morphological and
behavioural traits are becoming clearer. Some marine invertebrates
provide excellent opportunities for studying rates of learning (e.g. they
are not excessively fast). They are also suitable for examining the

complexities involved in having different preferences for, and availabilities of, different types of prey. In many marine systems, both predator and prey densities can readily be manipulated in experimental tests of different predictions about predation.

Experimental manipulations in marine systems have increasingly demonstrated the need for consideration of alternative explanations and models for patterns in assemblages. They also demonstrate that experimental procedures are going to be amongst the most powerful methodologies capable of increasing our understanding of relationships among predators and their prey. This summary of predation in marine habitats serves to underscore many of the current areas of ignorance, but provides some indication of the kinds of problems that are tractable. Without considerably more comparative assessment from other marine habitats, the relative importance of predation, and its interactions with other ecological processes, will continue to be only partially understood. Progress to date, however, does indicate that this important class of interactions between species is worthy of considerably more empirical and theoretical study.

ACKNOWLEDGEMENTS

This contribution was supported by funds from the Australian Research Committee (individually to PGF and AJU), from the Institute of Marine Ecology and the Research Grant of the University of Sydney (to AJU) and from the Macquarie University Research Grant (to PGF). Part of the preparation of the manuscript was made possible by the award of a Distinguished Research Fellowship (to AJU) at the Bodega Marine Laboratory (University of California, Davis), for which the senior author is very grateful. We also thank Dr P.A. Underwood for encouragement and advice and Dr M.J. Crawley for his great patience during the long delays caused to the manuscript by the protracted illness of the senior author.

10: Predatory Arthropods

MAURICE W. SABELIS

INTRODUCTION

Many arthropods have a predatory life style in that they attack, kill and consume live organisms within a short span of time. As a rule they kill many different prey individuals in the course of their life time and utilize their food content for maintenance, growth or egg production. Predators, thus defined, include not only members of the third and higher trophic levels, but so too are members of the second trophic level; examples are the seed-eating ants and the phytoplankton-consuming water fleas.

An important correlate of prey capture success is the body size of a predator relative to that of its prey. Predatory arthropods tend to capture prey smaller than themselves, except when they can make a powerful poison, use a trap or hunt in groups. The length ratio predator : prey ranges roughly from 100 to 0.3 and may be as low as 0.1 for pack hunters such as the army ants. The ecological implications of these size ratios are profound. Suppose the correlation between body size and intrinsic rate of population increase is negative (see, for example, Southwood, 1976; Peters, 1983; but see Gaston, 1988), then numerical responses of predatory arthropods to prey density will tend to become more and more delayed with increasing predator : prey size ratio. In addition, the functional responses will increase over a larger range of prey densities since larger predators eat more prey. Clearly, body size may well be a key feature in understanding the dynamics of arthropod predator—prey systems. Moreover, it may well be decisive in determining the outcome of competitive interactions and thus in understanding the evolution of community structure. For this reason, body size relationships will be a recurrent issue in this chapter on predatory arthropods.

Among the arthropods there are two other types of representatives of the third trophic level : parasitoids and parasites. Parasitic arthropods differ from true predators in that they may not even kill their host and, if they do, it may take several generations of the parasite. Parasitoids, as a rule, kill their host in one generation. What makes them different from the true predators considered here is that they do not immediately consume their host (except for the case of host feeding in parasitoids), but oviposit while their larvae utilize the host as a food source throughout development. Consequently: (i) their body size is always smaller than that of their host (although the initial host stage attacked by a parasitoid may be smaller in some cases); (ii) only adult females search for hosts; and (iii) they may deposit one or more eggs per host. In contrast, many predatory arthropods have larger bodies than their prey, they may well have females, as well as males and juveniles searching for prey (though not necessarily), and, in most cases, they require more than one prey to produce one egg themselves. In a parasitoid's life, time is split in two distinct phases: a food conversion phase, when larva, and a search phase, when adult. In predators these phases often concur. This is important because food conversion takes much more time than attack and ingestion. Thus, while the food conversion rate cannot limit the host-killing rate of a parasitoid, it is the most important factor controlling the prey killing rate of a predatory arthropod at high prey densities. Also, parasitoids are obligate carnivores in the larval phase, whereas predators are not necessarily so. Predators may well survive or even develop and reproduce on other food substances, such as fungi, pollen, nectar and other plant products.

Thus, there are reasons to suppose that for any given prey size predatory arthropods differ markedly from parasitoids and parasitic arthropods in their response to a change in prey density.

1 Their numerical responses tend to be more delayed and less steep than those of parasitoids (fewer offspring per prey killed)

2 Their functional responses tend to level off due to satiation, causing the maximum predation rate to settle at much lower values than expected from the predator's time budget for handling and searching.

3 Consumption of alternative food may further decrease the functional response to prey density, but it will promote survival (especially when prey density is low).

Whether these generalizations survive scrutiny and how they affect the characteristics of predator–prey dynamics, remains to be further explored in this chapter.

PREDATION

Phylogenetic distribution

Predatory arthropods are found among the crustaceans (water fleas, copepods, barnacles, crayfish, shrimp, crabs and lobsters; Pearse *et al.*, 1987), many of the chelicerates (including the horseshoe crabs, the sea spiders, the scorpions and pseudoscorpions, the wandering and web-building spiders, and several families/species of mites; Foelix, 1982; Hoy *et al.*, 1983; Pearse *et al.*, 1987; Legg & Jones, 1988; Gerson & Smiley, 1990), the centipedes (Pearse *et al.*, 1987) and several families/species of insects (Clausen, 1972). In some higher taxa virtually all species typically have a predatory life style; examples are the centipedes (Chilopoda), the spiders (Araneida) and the scorpions (Scorpionida). Among the mites (Acari) and the insects (Insecta) the predatory life style prevails only at lower taxonomic levels, such as family, genus or species, for example among the heteropteran insects all nabiid bugs are predators, several reduviid bugs are predators while others are blood parasites, and the black-kneed capsid, *Blepharidopterus angulatus*, is quite exceptional in being a predator among the predominantly plant-feeding mirid bugs.

What most arthropod predators have in common is the presence of extremities provided with structures resembling a pair of pincers. Such graspers are found on the front legs of the higher crustaceans, on both the pedipalps and chelicera of the scorpions and pseudoscorpions and only on the chelicera of the other chelicerates, whereas the centipedes have two jaw legs and several predatory insects have the two mandibles together forming one pair of pincers (such as ground beetles, tiger beetles, rove beetles and several ants). In some cases predatory insects have other extremities modified to form pincers: examples are the claw-like palps on top of the elongated and hinged labiae of dragonfly larvae, the raptorial front legs of nepid water bugs (water scorpions, water stick insects), the tibial and femoral modifications of the front legs of praying mantids and mantispids, and the hind legs of bittacid scorpion flies who hang down by their front legs waiting to catch a passing fly with their hind legs. Several other predatory insects attack prey without pincer-like structures; they manage to approach their usually slow moving prey and start to feed after injecting elongated sucking mouthparts (e.g. heteropteran predators, asilid or robber flies, predatory thrips, scorpion flies).

If arthropod predators differed only in the number of graspers, their positions and the extent to which they reach out from the body, then one might infer that, relative to its predator, mobility of the prey will decrease going from higher crustaceans, scorpions, pseudoscorpions via spiders, mites and insects with mandibular graspers to insects without graspers but with elongated mouthparts for sucking prey juices. This also implies a tendency to attack young prey stages with a body size smaller than its predator, because the younger the prey, the smaller their body size and the lower their mobility relative to the predator. Thus, a predator to prey size ratio exceeding unity may be expected, but, however sound the logic may be, this trend is blurred by a number of confounding factors, such as the musculature and use of extremities with graspers, multifunctionality of the graspers (e.g. role in mating, defence and competition), the injection of venoms, the use of traps, the mode of hunting (individual or group). The tentative conclusion emerging from these morphological considerations is that there are intrinsic reasons why predatory arthropods eat smaller prey; since they attack usually not by spitting, hitting, cutting, impaling, crushing (exceptions are crabs with massive chelas and lobsters with one of the two chelas adapted for crushing) but more generally by grasping, their prey should be relatively small and less mobile and even more so when the graspers do not reach out very far or are lacking entirely.

Relative body size and prey capture success

The observation that predatory arthropods eat prey of about their own size or more usually of a smaller size than themselves is supported by studies on scorpions (Polis & McCormick, 1986), spiders (Turner, 1979; Nentwig & Wissel, 1986), aquatic invertebrates (Warren & Lawton, 1987) and chewing insects (Enders, 1975), such as mantids (Holling, 1965; Holling *et al.*, 1976). Additional evidence comes from data on the natural enemies of arthropods known to be plant pests. Considering all the predatory arthropods that feed on aphids (Minks & Harrewijn, 1988), it is clear that they are almost always larger than the adult aphids (ladybird beetles, lacewings, gall midges, hover flies, predatory bugs, ground beetles, earwigs, spiders). In the few cases where they are not (e.g. trombidiid and erythraeid mites, very young developmental stages of some of the predators listed above), then they attack the similar-sized or smaller nymphal stages of the aphids only. A similar pattern can be observed in the natural enemy complex of spider mites, a family of exclusively plant-feeding mites that are much smaller than aphids (Helle & Sabelis, 1985). Their predators are usually smaller than the aphid predators (small ladybird beetles, small rove beetles, small gall midges, small hover flies, small lacewings or their young nymphal stages, predatory thrips, nymphs of predatory bugs and earwigs, some small spiders), but they are all clearly larger than even the largest spider mites. Among the mites there are several groups of predators with similar or slightly larger size, the most important of which are the phytoseiid mites. In the few cases where predators are smaller than adult spider mites (stigmaeid mites and a tarsonemid mite in the genus *Acaronemus*), they feed only on the immature stages of similar or smaller size.

Thus, many predatory arthropods eat prey of similar or smaller size. Generally, predator size is positively correlated to *mean* prey size, but the slopes differ depending on the mode of foraging and on the prey type (Hespenheide, 1973). These relations have been shown for arthropods as different as copepods (Wilson, 1973), water fleas, alder and snake flies, stoneflies, mayflies, backswimmers (for references see Warren & Lawton, 1987), predatory bugs (Evans, 1976a), damselflies (Thompson, 1975), phytoseiid mites (Bakker & Sabelis, 1989; Sabelis, 1990), robber flies, digger wasps (Hespenheide, 1973), antlions (Wilson, 1978), scorpions (Polis & McCormick, 1986) and wandering spiders (Enders, 1975; Nentwig & Wissel, 1986). Where recorded, the range of prey sizes increases with predator size; the *minimum* prey size tends to increase very little, but the *maximum* prey size increases steeply (Fig. 10.1).

In some of these groups maximum prey size is a few times larger than predator size. Examples are scorpions, robber flies, digger wasps and wandering spiders. Hespenheide (1973) and Enders (1975) argue that this is due to the use of powerful venoms. Although the venom

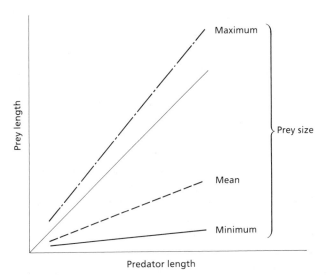

Fig. 10.1 Generalized relation of prey size to the size of its predator. Maximum prey size may be slightly larger than predator size and increases steeply with predator size. Mean prey size is always smaller than predator size and increases less steeply with predator size. Minimum prey size increases only very little with predator size.

cannot be injected until after contact, it appears possible to attack prey by surprise and subdue them by immediate paralysis. No experimental study is known which demonstrates by how much a venom increases the upper limit of the prey size taken. Offensive venoms probably evolved in situations where foragers had access to few alternative prey.

The use of traps is an alternative way to expand the range of prey sizes taken. Insects and spiders living near the water surface (back-swimmers, stick insects, pondskaters, pisaurid spiders) prey on arthropods that happen to drop and float on water. It is expected but not proven that they can feed on larger prey. Enders (1975) provides evidence that web-building spiders catch prey much larger than themselves, and larger than non-web spiders of similar size, such as wolf spiders and jumping spiders. In addition, they have overcome the problems of large and/or heavily defended prey by wrapping them in silk and/or by poisoning, thereby immobilizing their prey. Those web spiders which lack poison (Uloboridae) or wrapping behaviour (*Nephila* spp.) take prey smaller than their own size (Enders, 1975). Further consideration to the web and trap building arthropods is given elsewhere in this book (Chapter 13). Robinson and Valerio (1977) point out that silk may be used for the same purpose, but in an alternative way; some tropical jumping spiders attack their prey dorsally, followed by dropping on a dragline to dislodge their prey from the substrate. This 'drop and hold' strategy allows them to attack prey larger and heavier than they are themselves.

Sociality is another important way to increase the relative size of prey (Hespenheide, 1973; Enders, 1975; Griffiths, 1980c). Oster and Wilson (1978) distinguished five types of foraging among the social insects, each with important consequences for the range of prey sizes taken. Type 1 foragers are nestmates that hunt without trails. The term 'solitary hunter' would be misleading because their behaviour is unlikely to be independent of other nestmates and of the state of the colony as a whole (Traniello, 1989). Type 1 foragers are found, for example, among the anatomically primitive ant subfamilies Myrmeciinae and Ponerinae. They are specialized predators of arthropods and usually catch prey smaller than their own size. Type 2 foragers recruit other nestmates and cooperate whenever larger or well-defended prey is encountered. Examples are found among most kinds of ants and social spiders, such as *Agelena consociata* (Griffiths, 1980c). They select a broader array of dietary items than Type 1 foragers, but often smaller than Type 3 foragers which employ trunk trails as orientation devices from where they depart at intervals to forage in a Type 2 manner. Type 3 foragers have the broadest diets found among the social insects and are found among the anatomically more evolved groups of ants (e.g. myrmicine ants such as *Atta* and *Crematogaster*, formicine ants such as *Formica* and *Lasius*, and some dolichoderine ants). They are observed to drag prey loads up to 15 times heavier (Traniello, 1989). Type 4 foragers hunt in a Type 2 manner, but after recruiting nestmates they assault the prey *en masse*. They are predators specialized on exceptionally difficult prey. Examples include termite-raiding ponerine ants, ant-feeding cerapachyine ants and sowbug-hunting species of the ant genus *Leptogenys*. They have overcome the resistance of exceptionally large prey by an increase in the number of attackers rather than by gigantism. Type 5 foragers are group hunters. They proceed in bands or even armies of millions of individuals, seizing nearly all arthropods above some critical size and even sluggish reptiles and other vertebrates. As they deplete local prey areas, they frequently have to change the nest site. Examples include army ants in the higher doryline and ecitonine subfamilies. Group hunting probably leads to the most spectacular increases in relative prey size. Drastic decreases in relative prey size may be associated with increases in tempo of prey transport. Ants that lift rather than drag their prey may gain by speeding up transport at the expense of maximum prey size.

Oster and Wilson (1978) noted that size distributions among the worker ants tend to be monomorphic when they specialize on particular prey (i.e. Type 1 and 4 foragers), but polymorphic when prey of various sizes are included in the diet (i.e. Type 2, 3 and 5 foragers). They argue that continuous polymorphism serves at least in part to match a wider array of food items and that this adaptation may originate when competition is reduced. As an alternative to this ecological release hypothesis they consider the possibility where recruitment and assembly would otherwise be too expensive either in terms of energy and time invested or in terms of mortality from attack by natural enemies. A cost−benefit analysis of producing larger workers versus recruiting a larger number of small workers has not yet been carried out and may be complicated by the possibility that cooperative retrieval may serve to decrease interference competition through resource defence rather than to move prey with greater energetic efficiency (Traniello, 1989).

While the range of prey sizes taken tends to increase with body size and use of graspers, venoms, webs and group attack, this relation should not be taken too strictly. Each of these traits may have other functions as well and may therefore evolve under a variety of selection regimes. Body size may evolve not only through its effect on the range of prey sizes, but also depending on how it influences mating success, intraspecific and interspecific competition, the risk of hyperpredation, etc. Modifications of graspers, the use of venoms and nets, and social organization may also have various other functions than to increase maximum size of prey. For example, not all pincer-like structures serve to grasp prey; the abdomen of an earwig ends in a forceps that is thought to assist in folding wings and to function in defence when upon disturbance they are raised forward over the body in a scorpion-like fashion. Stinger venoms of scorpions probably not only serve to paralyse prey, but also to kill competitors and deter hyperpredators. Webs may also serve a suite of functions other than catching prey, such as a refuge to escape from predators, protection of offspring and a role in courtship behaviour (Foelix, 1982). Also, sociality may have an impact on interspecific competition and defence against hyperpredators. For example, in several ant species the largest individuals comprise a soldier caste that defends the minor workers and the food source during the mass retrieval of larger food items. The role of these soldiers is entirely supportive; they seldom if ever forage and retrieve food items on their own.

In conclusion, body size may well have an impact on the range of prey sizes taken, but its effect is modified by various factors, such as production of venoms, building of webs and formation of hunting groups. Moreover, body size has a number of other consequences; it may have an impact on dispersal capacity, competitive ability and the risk of hyperpredation. Hence, one should not expect to find general trends in predator-to-prey body size, and caution needs to be exercised in selecting the group of species to be compared.

Foraging costs and benefits

Foraging strategies principally differ in the amount of time and energy devoted to locomotion. At one extreme there are the sit-and-wait or ambush predators. They wait for their prey to come and therefore specialize on prey that actively move around; search costs are minimized at the expense of a lower rate of encounter with prey, especially the soft-and-sluggish prey types. By their immobility and crypsis they manage to catch or attack their prey by surprise, but to do so they have to spend much energy in quickly catching and subduing their prey. Examples are scorpions, crab spiders, web spiders, cheyletid mites, praying mantids, pit-trap building larvae of antlions and tiger beetles. Reactive distances are of the order of the length of their grasping extremities except for the web spiders which respond to prey-borne vibrations in their web. Although they sit and wait, several of these predators displace themselves at intervals to other areas where it may be more profitable to forage. For example, antlions may relocate their pits (Matsura & Takano, 1989) and spiders may replace and relocate their webs (Enders, 1976).

The next category consists of predators that alternate searching and sitting and may locate prey in either phase of activity. They spend much energy in pursuit. Most frequently they are predatory arthropods that locate prey by vision, such as some carabid beetles (*Notiophilus*, *Elaphrus* and adults of tiger (cicindelid) beetles) and the jumping (salticid) spiders. Reactive distances may be several times the body length of the predator. Other examples are pondskaters, which respond to water surface vibrations originating from arthropods struggling with water surface tension.

The central place foragers form a separate category. They usually spend much energy in roaming around in search for prey, but they are capable of finding their way back to their nest or preselected site for oviposition. Examples are ants and digger wasps. They may locate prey from quite a distance but pursuit and attack follow only when close to their prey. A further subdivision of foraging behaviour involving co-operation has been discussed above.

The largest category comprises predators that do not spend much energy in pursuit, but possibly more so in search. They may locate prey from considerable distances (often by smell), but do not attack until very close to their prey. Examples are wandering spiders, predatory mites, many ground beetles, ladybird beetles, lacewings, predatory bugs and earwigs.

Finally, there are the filter-feeders with many examples among the lower crustaceans. Some spend much energy in swimming and gain by sieving a larger volume of water (water fleas), whereas others actively use their extremities to stir the medium and thereby increase the volume of water passing the sieve (barnacles). For obvious reasons no energy is spent in pursuit and reactive distances are virtually zero.

If we classify prey according to their ability to resist predation (poisonous versus palatable food content; soft versus hard integument), their ability to escape predation (agile versus sluggish prey) and their body size relative to that of the predator, then one may expect the ambush predators and pursuers to take the more agile prey even when they are large and well defended, the central place foragers to take a broad range of prey sizes, the searchers to take the soft, sluggish and relatively smaller prey and the filter-feeders to concentrate on very small prey. Exceptions and difficulties in classification are bound to be found, but the general point of specialization on broad categories of prey will probably hold (see also Enders, 1975; Griffiths, 1980) and represents an improvement on the classification discussed earlier, which was based on grasper morphology alone. Clearly, in addition to the graspers one should also consider sensory apparatus, behaviour and energy spent in various activities as these are important determinants of capture success.

Although much information is available on the amount of time spent in searching, pursuing, capturing and eating, very little is known about the costs involved, let alone how these costs are related to prey size or type. The benefits of extracting food from various prey types are also poorly understood, but there is good evidence that the cumulative amount of biomass ingested follows an asymptotic exponential relation to feeding time. Yet, in theory, this information on costs and benefits is crucial in understanding the evolution of feeding time and ingestion efficiency. Following others (Cook & Cockrell, 1978; Sih, 1980), Lucas and Grafen (1985) applied Charnov's marginal value theorem to the case of feeding on single prey items captured at different time intervals and extended this model by including both time and costs involved in search, pursuit and consumption. Their model predicts that: (i) optimal searchers and pursuers should spend more time feeding per prey when costs of search and pursuit increase provided that these costs exceed the costs of consuming prey; alternatively, (ii) if these costs are lower than consumption costs, they should spend less time in feeding for a given increase in search and pursuit costs. Under the conditions of either of the previous two predictions; (iii) feeding time is expected to decrease with an increase in the ingestion rate. These largely untested predictions demonstrate the need for quantitative data on costs involved in various foraging phases, as well as the relevance of the above classification of foraging strategies. Lucas (1985b) further argues that some ambush predators like antlion larvae (but unlike the praying mantids studies by Holling, 1965, 1966) can attack new prey even while eating a prey captured earlier. In that case search costs are not relevant, but between prey arrivals there is a waiting time, which also entails a cost. Lucas (1985b) showed how this cost and others enter the optimality criteria for when to leave the current prey and attack a newcomer under various prey density regimes; the antlions appeared to discard their prey somewhat earlier than predicted, which Lucas (1985b) suggested to be adaptive in that it is more easy to catch prey while empty-handed than while eating.

Two predictions stand out as being of particular interest to what follows (Lucas & Grafen, 1985): for searchers and pursuers (with low eating costs relative to costs of search and pursuit) and for ambush predators: (i) feeding time per prey; and (ii) percentage food extracted should decrease with increasing prey density. Both predictions agree with the facts in that shorter feeding times and partial prey consumption are commonly observed in many different predatory arthropods, such as backswimmers (Cook & Cockrell, 1978; Giller, 1980; Sih, 1980), nepid water bugs (Bailey, 1986), coccinellid beetles (Cook & Cockrell, 1978), diving beetles (Formanowicz, 1984), ground beetles (Ernsting & Van der Werf, 1988), antlions (Lucas, 1985b), predatory mites (Sabelis, 1986, 1990) and water fleas (Johnson *et al.*, 1975). The agreement is qualitative, however. Quantitative tests often fail in one way or another. Further progress awaits a more detailed understanding of how food extraction rates change with the satiation level and how food quality changes in the course of the extraction process (Formanowicz, 1984).

Observations on predatory mites further challenge optimal ingestion models (Sabelis, 1986). These predators utilize most of their food for egg production and their rate of food absorption through the gut wall depends linearly on the amount of food present in the gut. Thus, to maximize egg production they should maximize the rate of food absorption from the gut by: (i) improving digestion capacity; and (ii) keeping their gut filled to capacity. If the predator reaches satiation and there is still food left in the prey, the predator has the option to continue feeding so as to compensate for the food cleared from the gut. The cumulative gain and its relation to the feeding time is given in Fig. 10.2a. At the start of feeding, the predator is unlikely to be in a satiated state. Hence, ingestion increases the food content of the gut, but with diminishing returns. When the predator reaches satiation, the cumulative gain curve shows a linear trajectory with positive slope; here, ingestion compensates digestion and egestion. When the ingestion rate drops below the rate of gut emptying, the linear trajectory becomes curved again. The optimal feeding time is given by Charnov's marginal value theorem. The ratio of total food gain and total time spent searching and feeding is maximized for long feeding times with a sudden shift to short feeding times as prey density increases and searching time decreases beyond a critical value set by the slope of the linear part of the gain curve. Direct observations of such sudden shifts in feeding time are not available, but it provides a possible explanation for the fact that predatory mites foraging at high prey densities tend to feed for much shorter periods per prey and leave much of the prey's food content unutilized (Sabelis, 1986, 1990). The properties of predatory mites that led to this explanation are shared by many predatory arthropods. They generally have digestion rates that depend on gut fullness and frequently face situations where the food deficit of the gut is much smaller than the prey's food content, especially at high prey densities. Hence, excessive waste of the prey's food content may therefore be a typical property of predatory arthropods, as optimal foragers.

What determines the shape of the functional response?

Ever since Holling's seminal papers (Holling, 1959b, 1961) the predator's time budget has been considered to determine the shape of the functional response curve. When the rate constant of prey capture a and

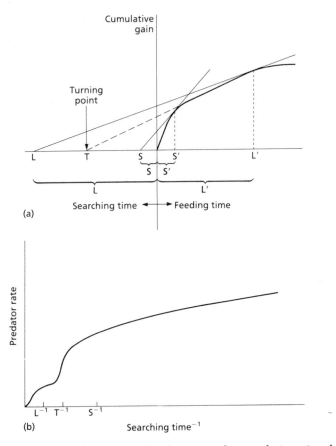

Fig. 10.2 (a) Application of marginal value theorem to the cumulative gain relation with feeding time when the searching time is either long (L) or short (S). The optimal feeding times for short and long searching times are indicated by S′ and L′. The cumulative gain curve is special in that it has a linear traject in between non-linear trajectories. If the tangent equals the (positive) slope of this linear traject then any feeding time associated with the linear traject of the gain curve is optimal. The searching time where there is a set of optimal feeding times rather than just one represents a turning point (T), where predicted optimal feeding times switch from very long to very short. Note that the cumulative gain curve changes in shape with the food deficit of the gut at the start of feeding, and that the gain rate decreases with the degree of gut fullness as ingestion is much faster than digestion. (b) Using Holling's disc equation with the searching time equal to the inverse of aD (instantaneous search rate times prey density) and the handling time as predicted from the marginal value theorem, the shape of the functional response is calculated. It results in a double plateau functional response because the upper asymptote depends either on very short feeding times or on very long feeding times. The L, T and S searching times given in (a) are indicated on the x axis.

the handling time T_h do not depend on prey density, predation rate increases with prey density at a decelerating rate to an upper asymptote. (Type II response). If these components do depend on prey density, then other types of functional responses are obtained, such as a sigmoid or Type III response when T_h decreases or a increases with increasing prey density (see Chapter 3). Let us first consider T_h-related causes of response shapes that deviate from Type II. According to the optimal ingestion model discussed above one would expect: (i) handling times to decrease with prey density giving rise to a sigmoid functional response; and (ii) handling times to be subject to a sudden shift to lower values giving rise to a double plateau in the response curve (Fig. 10.2a,b).

The first prediction does not meet the facts, and most functional responses published to date are close to a Type II curve (see, for example, Arditi, 1982). Hassell *et al.* (1977), however, presented data on lower crustaceans (*Cyclops, Daphnia*), coccinellid beetles and two hemipteran predators (*Notonecta, Plea*) whose functional response curves showed a propensity towards sigmoidy. Evidence for sigmoid responses of predatory mites is lacking or unconvincing because of insufficient data in the low density range and through the lack of properly standardized, experimental procedure (Sabelis, 1985, 1986, 1990). Hassell *et al.* (1977) and Arditi (1982) review various reasons why Type III responses may hitherto have been overlooked. Whether prey density-related changes in T_h play a role in causing sigmoid responses remains to be assessed, but we shall return to this point later.

The second prediction also has little support (see, for example, Arditi, 1982). However, some authors argued that the functional response has a double plateau (Sandness & McMurtry, 1970; see also Hassell *et al.*, 1976), as expected if the plateaus were set by the predator's time budget (Fig. 10.2b). Sandness and McMurtry (1970, 1972) observed that at low densities the predatory mites return to a dead prey several times to feed on it and spend a long time feeding on it on each occasion, but at high densities the predators are frequently disturbed by other prey bumping into them, and hence feed for less time on each occasion and return less often to dead prey. These observations are in support of a shift in feeding time with prey density, as predicted from optimal feeding time models.

The results of Sandness and McMurtry (1970) do not stand alone. For example, Kuchlein's data on the functional response of a predatory mite to the density of eggs of the two-spotted spider mite may be interpreted as showing a second rise at high prey densities (Metz *et al.*, 1988a). Data on damselfly larvae foraging for water fleas (Johnson *et al.*, 1975), on wolf spiders and crab spiders foraging for fruit flies (Haynes & Sisojevic, 1966; Nakamura, 1974; Hardman & Turnbull, 1980) also admit a similar interpretation.

Now, supposing that handling and feeding times change with prey density as predicted, can it explain why some functional response curves seem to be sigmoid, have a double plateau or show a second rise? The answer is negative because the total amount of time spent in prey handling often comprises a small proportion of the total time available, even at high prey densities. Thus, the plateaus of the functional response of an arthropod predator cannot be set by the predator's time budget. For example, predatory mites spend less than 10% of their time in feeding and handling of prey and they encounter many more prey individuals than they attack and feed on, even when the prey is virtually defenceless (e.g. eggs of the two-spotted spider mite); thus, although they would have plenty of opportunity, they spend little time in feeding. Hence, time budgets for searching and feeding activities appear irrelevant as an explanation for the existence of plateaus in functional response curves of predatory mites. Similar arguments can be given for a number of other predatory arthropods, such as praying mantids (Holling, 1965, 1966), backswimmers (Zalom, 1978), coccinellids (Dixon, 1970) and ground beetles (Mols, 1988).

Functional response curves of predatory arthropods are therefore more likely to be shaped by the search rate alone. Following Holling (1959b, 1961, 1965, 1966) the search rate can be split into various components, several of which may depend on the predator's gut fullness as a measure of its motivational state. Examples are walking speed, walking activity, prey detection distance, capture success rate and spatial coincidence of predator and prey. The effect of gut fullness s on these components is such that the rate constant of prey capture $g(s)$ decreases with s, as shown in a number of studies on searching behaviour (Holling, 1965; Curio, 1976; Sabelis, 1986, 1990). Figure 10.3 and Box 10.1 describe a predation model that takes into account the dynamics of gut fullness and the associated changes in the rate constant of prey capture. It proved to be quite successful in predicting the shape of the functional response of predatory mites and praying mantids (Curry & DeMichele, 1977; Metz & Van Batenburg, 1985a,b; Sabelis, 1986, 1990; Metz et al., 1988a). Though successful in its predictions and elegant as a description of rather complicated processes, the model is still too complex to provide direct analytical insight into the relative importance of underlying processes. The model is also too complex to serve as a function in simple models of predator–prey dynamics. However, this should not deter researchers from developing these models, since it may still be possible to obtain simple functions by deriving limiting equations or other approximations from the full model. This approach is illustrated by the work of Metz and Van Batenburg (1985b) and Metz et al. (1988a), who derived the following equation of the functional response F for the case where the time spent handling can be ignored and the food deficit of the gut is always smaller than the food content of the prey (i.e. when prey density is high or gut capacity is relatively small):

$$F(D) = d/[\ln(m/c) + \sqrt{bdc^{-1}D^{-1}}] \tag{10.1}$$

Here m represents gut capacity, c is the level of gut fullness above which the rate constant of prey capture $g(s) = 0$ (so-called capture threshold), D is prey density and d is the rate constant of digestion and egestion (or gut emptying), while $b = -0.5 \pi/g'(c)$ and $g'(c)$ is the differential of function $g(s)$ at $s = c$. For large D the functional response approaches an upper asymptote set by $d/\ln(m/c)$. The shape of the functional response is much like a Type II response, but the plateau is

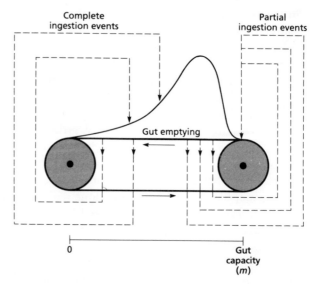

Fig. 10.3 A conveyor belt representation of the predation process for a predator that ingests part of the food content of its prey whenever it reaches gut capacity (m). The length of the belt represents the state space of gut filling. As indicated by the bar below the belt the state space goes from zero (0) to gut capacity (m). The conveyor belt is loaded with a heap of sand representing the distribution of predators over the state space. Sand is transported from right to left to mimic gut emptying. Upon arriving at the lower end of the belt the sand is assumed to stay on the belt (no deaths from starvation). Ingestion of a prey is represented by an instantaneous shift of 'sand' from left to right along the belt (dashed lines). The distance of the shift is equivalent to the food content of the prey. In cases where this distance is such that the sand would be dropped at the right side of the belt, it is deposited at the right end of the belt. These cases represent partial ingestion events.

Box 10.1 A Markov model of state-dependent predation

Suppose that the level of gut fullness is a good measure of the motivational state of a predator. It is denoted by s (satiation level), which ranges from 0 to m, the maximum gut capacity. The rate of gut emptying is represented by $f(s)$, where $f(s) = -ds$ and d is the rate constant of gut emptying. Ingestion of food from a prey occurs instantaneously and the size of a full upward jump is denoted by w. In case the ingestion jump exceeds the gut deficit $(m - s)$, the jump halts at m and an amount of food equivalent to the gut deficit is consumed rather than the full amount of food in the prey. The rate constant of prey capture is given by $g(s)$ and thus becomes a function of the satiation level. Usually, $g(s)$ is relatively high for s close to zero (empty gut) and then decreases for larger s. When $g(s)$ becomes zero for $s < m$, this critical value of s is called the capture threshold denoted by c. The rate of prey capture for a predator at s is equal to $g(s)D$, where D is prey density.

Imagine a large number of predators, each one foraging in a separate arena with the same constant density of prey (immediate

continued on p. 239

Box 10.1 *contd*

replacement of prey killed), then, although the predators all started foraging at the same level of gut filling s, their state of gut fullness is likely to vary from predator to predator due to the stochasticity of the search and capture process. Thus, there is a probability distribution of predators over the state space. The population state is denoted by p, where $p(t,s)ds$ denotes the fraction of the predator population with a satiation level between s and $s + ds$. Clearly, being a probability distribution, the sum of all $p(t,s)$ in the state space from 0 to m should equal unity and $p(t,s)$ is equal to zero for all values of s outside this state space.

With this notation the partial differential equation describing changes in the probability distribution over the state space now reads:

$$\frac{\delta p(t,s)}{\delta t} = -\frac{\delta f(s)p(t,s)}{\delta s} - Dg(s)p(t,s) + Dg(s-w)p(t,s-w)$$

The first term represents the net flow through gut emptying, the second term represents the flow to higher satiation levels through prey capture and ingestion, and the third term represents the flow of predators that arrive at exactly satiation level s through ingestion of w food units per prey. For $s < w$, the last term should be dropped from the equation, and, for $s = m$ we get the boundary condition:

$$-f(m)p(t,m) = \int_{m-w}^{c} Dg(s)p(t,s)ds$$

where the left-hand term represents the number of predators stacked to undergo digestion starting from state m, and the right-hand integral represents the sum of all predators reaching gut capacity through full or partial ingestion of their prey. Since prey density and the transition functions f and g do not change with time t, the probability distribution will converge to a steady state. The steady-state distribution $\hat{p}(s)$ can be found by setting the differential equations equal to zero and solving for $\hat{p}(s)$ (while taking care that $\hat{p}(s)$ should sum to unity). The functional response is now obtained from the steady-state distribution by:

$$F(D) = \int_{0}^{c} Dg(s)\hat{p}(s)ds$$

This completes the model description, as developed by Curry and DeMichele (1977), Taylor (1984), Sabelis (1981, 1985, 1986, 1990), Metz and Van Batenburg (1985a,b) and Metz *et al.* (1988a).

---

now set by the rate constant of gut emptying d, rather than by the predator's time budget. Perhaps more surprisingly, the plateau also depends on the existence of a capture threshold $s = c < m$. If $c = m$, then the shape of the functional response $F(D)$ simplifies to a square root function of D:

$$F(D) = \sqrt{b^{-1}\,cdD} \tag{10.2}$$

This result becomes more transparent by realizing that $g(s)$ exceeds zero whenever digestion and egestion bring s below gut capacity ($c = m$). Thus, the functional response curve cannot have a plateau; it keeps on increasing with prey density while the amount of food consumed per prey becomes less and less; ultimately, the predator becomes a parasite, as the victim does not die from the damage inflicted by the predator.

These simple models demonstrate the importance of the rate constant of gut emptying d and the capture threshold c in determining the shape of the functional response of a predatory arthropod. The biological function of the former is obvious, but what about the latter? Suppose that there are costs involved in detecting, subduing, consuming and digesting prey, then it pays only to invest when these costs are offset by energy returns from feeding. Since the amount of energy gained decreases with increasing gut fullness, there should indeed be a threshold level of gut fullness above which a predator refrains from attacking a prey. This capture threshold will become lower and lower, the larger the prey is and the more effective its defensive strategy. Conversely, for easy prey, the capture threshold will be close to or equal gut capacity. This tentative argument leads to the testable prediction that, going from large to small, armoured to soft bodied and poisonous to palatable prey, functional responses change from saturation curves to square root functions. This may explain why functional responses of adult predatory mites with respect to early developmental stages of the two-spotted spider mite (eggs, larvae, first nymphs) are of the square root type (Metz *et al.*, 1988a; Fig. 10.4a), but of the saturation type for adults of the prey (Sabelis, 1986, 1990; Fig. 10.4c). Most interestingly, while Sandness and McMurtry (1970) interpreted their data on predation of adult predatory mites on young nymphal spider mites as evidence for a double plateau functional response (Fig. 10.5a), there is an alternative explanation provided by the square root function (Fig. 10.5b) and there is no reason to assume that one is better than the other. It illustrates that interpretation of functional response shapes can be highly arbitrary and should be accompanied by quantitative models based on observations of the underlying behavioural and physiological mechanisms.

Additional evidence for square root type responses comes from Johnson *et al.* (1975) who reported wasteful killing of water fleas by damselfly larvae. More critical tests are needed. One approach is to assess capture thresholds directly by experiment and then measure the shape of the functional response. However, this is fraught with problems, such as how to decide whether $g(s)$ equals zero. Another approach is to modify predator strength relative to that of its prey. For example, if one could eliminate the defence of a weak and small prey, then the functional response may change into a square root type. Such a trend may be present in functional responses of predatory mites to the density of spider mite eggs in absence or presence of web. This web is produced by the prey and decreases the capture success rate of the

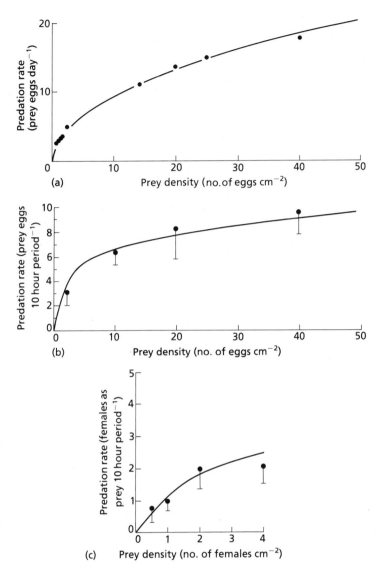

Fig. 10.4 Functional responses of predatory mites (Acari: Phytoseiidae) to the density of spider mites (Acari: Tetranychidae) for three different cases. (a) Females of *Typhlodromus occidentalis* Nesbitt consuming eggs of *Tetranychus urticae* Koch in absence of web. (b) Females of *Phytoseiulus persimilis* Athias-Henriot consuming eggs of *T. urticae* in presence of web. (c) Females of *P. persimilis* consuming females of *T. urticae* in presence of web. ●, mean; bar, standard deviation; line, model prediction. Data from Metz *et al.* (1988a) and Sabelis (1986).

predator (Sabelis, 1985). In the absence of web the functional response is clearly of the square root type (Metz *et al.*, 1988a; Fig. 10.4a), whereas in presence of web it seems more close to the saturation type (Sabelis, 1990; Fig. 10.4b). Another example of modifying predator-to-prey strength is by increasing temperature and thereby agility and strength of any poikilotherm arthropod, while using an immobile prey. A thought provoking example is recently provided by Eggleston (1990) who studied blue crabs eating juvenile oysters at low, intermediate

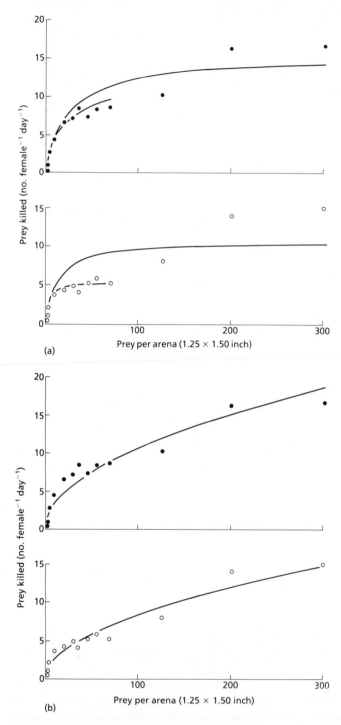

Fig. 10.5 Functional responses of two species of phytoseiid mites (adult females of *Amblyseius largoensis* and *Amblyseius concordis*: ● and ○ respectively) to the density of spider mites (protonymphs of *Tetranychus pacificus*). (a) Lines represent best fitting curves of Holling's disc equation. (b) Lines represent best fitting curves of the 'square root of prey density' equation discussed in the text. Data from Sandness & McMurtry (1970, 1972).

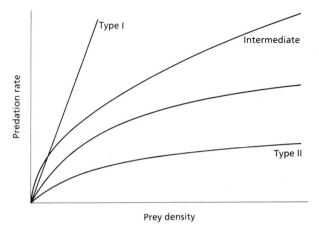

Fig. 10.6 Type I, Type II and the intermediate (square root type) functional response types. For explanation see text.

and high temperatures and found functional responses to change from saturation type to one without an upper asymptote.

Hence, there are reasons to propose a new type of functional response intermediate to the linear type (I) for filter-feeders and the saturation type (II) commonly found in predatory arthropods; it is typified by not having an upper asymptote and to approximate a square root function of prey density provided prey density is not too low (Fig. 10.6). Furthermore, one may hypothesize that, going from small to large predator to prey sizes (or other measures of relative strength), functional responses may change from a linear shape, via a square root shape to a saturation shape with lower and lower upper asymptotes. In none of these cases does proportional mortality increase with prey density; it either remains constant or declines. Hence, one may ask when to expect density-dependent predation.

Positive density-dependent trajectories of the functional response

Much of what has been discussed above applies to high prey densities only. For some predator–prey systems this is all that matters because prey distributions are patchy and prey density within a patch area has a rather constant value. For example, aphids and spider mites tend to form dense aggregations within which the numbers per unit area have a characteristic value for the species or species–substrate combination. If a predator happens to enter such an aggregation it may deplete prey supply in its direct environment, but as it moves on within the prey aggregation it experiences locally high prey densities again and again. Predation rates can then be approximated by the upper asymptote of the functional response curve. In other cases prey populations may settle at low densities over a large area. For example, several authors (Baars & Van Dijk, 1984; Lenski, 1984b; Pearson & Knisley, 1985; Sota, 1985) have emphasized that ground beetles predominantly experience conditions of low prey availability, and that these are marginal for development and/or reproduction.

Can there be positive density-dependent predation responses over this range of prey densities? One would expect so, but as yet the evidence is meagre. Hassell *et al.* (1977) found increasing proportional mortality in the low prey density range for a number of predatory arthropods, but the underlying mechanisms are still to be elucidated. Given the statistical problems in distinguishing between response shapes, insight in the causal mechanisms is indispensable. For predatory arthropods there are several possibilities. The first category consists of mechanisms that change the position of the upper asymptote of the functional response. One may think of changes in metabolic rate in response to food shortage. This would produce a double plateau in the functional response with a probably sigmoid trajectory in between the plateaus. Evidence for state-dependent metabolic rates comes from starvation experiments with spiders (Foelix, 1982). Absence or decrease in prey may also induce lower metabolic activity associated with a reproductive diapause. Danks (1987) reviewed the evidence for lacewings, coccinellids, carabids and predatory mites. Hence, there are good reasons to suppose that the digestion rate constants become a function of prey density. Increased food retention in the gut and oosorption are other, but essentially similar, mechanisms that may contribute to positive density-dependence in the low prey density range. Another category of mechanisms concerns changes in the rate constant of prey capture $g(s)$ and its relation to gut fullness s. However, such changes are likely to keep pace with changes in metabolic rate constants and therefore do not form a fully independent category unless the search rates change as a consequence of density-dependent changes in prey behaviour. When there are alternative prey, then the rate constants of prey capture may change due to prey selection behaviour. Methods to detect such changes are only beginning to be developed. Cock (1978) proposed to assess functional responses to the density of each prey type, estimate the two parameters required for a descriptive predation model and predict predation in prey type mixtures, by assuming that prey preference does not change as a result of the prey types being presented together. Deviations from the predictions imply either that search rate depends on total prey density or on prey type composition or on both. Dependency on total prey density is commonplace among predators. To eliminate this cause of deviating predictions Sabelis (1985, 1986, 1990) used the same approach, but employed the more complex model discussed in Box 10.1. Since this model takes into account the gut fullness and the way that gut fullness changes with total density, a more precise method is obtained to analyse when prey type composition alone might modify prey preference. Application of the method so far is limited to predatory mites and has shown that preference for various stages of two-spotted spider mites does not change as a result of the prey stages being presented together (Sabelis, 1986, 1990), but preference for either of two phytophagous mite species (fruit-tree red spider mites and apple rust mites) may well be affected by prey species composition (Dicke *et al.*, 1989). Whether prey type composition affects prey preference such that functional responses have trajectories exhibiting positive (prey) density-dependence, remains to be shown, but it seems likely to occur.

In conclusion, Hassell *et al.* (1977) were probably right in expecting functional responses to have trajectories with positive density-dependence. The mechanisms that can produce these effects seem to be widespread among predatory arthropods. However, a more precise assessment of the functional response over the full range of realistic prey densities, prey type compositions and environmental conditions remains to be done. It will probably show that functional responses can have much more complex shapes than the four basic types distinguished by Holling (see Box 3.2).

DEVELOPMENT AND REPRODUCTION

Predatory arthropods utilize the ingested food for maintenance, growth and reproduction. If growth stops and reproduction is absent, then the rate of predation decreases accordingly (as in males and old females). Beddington *et al.* (1976) proposed a simple model of energy allocation in which priority was given to energy needed for maintenance, while the surplus energy was used for growth or egg production. This results in the following linear relationships where I, A, E, G and R are rates of respectively ingestion, food absorption, egestion, growth and egg production, W_e is the weight of an egg at birth and the greek characters are constants:

$$G = \alpha(A - \beta)$$

and

$$R = \gamma(A - \delta)/W_e$$

with

$$A = I - E. \tag{10.3}$$

Here, growth and reproduction stop when A drops below respectively β and δ. Under steady-state conditions the ingestion rate is equal to the rate at which food is cleared from the gut by absorption through the gut wall ($A = d's$) and by egestion ($E = d''s$). Hence, the rate of absorption is given by:

$$A = (d - d'')\hat{s} = d'\hat{s} \tag{10.4}$$

The steady-state level of gut fullness is calculated using the model in Box 10.1. Alternatively, descriptive equations of the form $\hat{s} = mf(D)$ can be used (e.g. $m[1 - \exp(-\varepsilon D)]$). In this way, partial prey consumption can be taken into account when converting prey killed into amount of food ingested, absorbed and utilized. Figure 10.7 shows various types of relationships between the rate of oviposition and prey density. If it

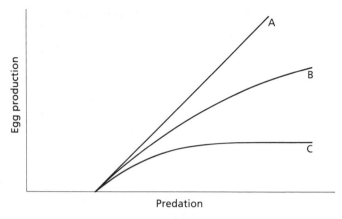

Fig. 10.7 Oviposition rate of predatory arthropods in relation to the rate of predation. Three types of relations are distinguished: A, linear (Beddington *et al.*, 1976); B, curvilinear without upper asymptote; and C, curvilinear with upper asymptote. All three types intersect with the x-axis at prey densities well above zero. Below the intersection point predatory arthropods allocate energy to maintenance processes only. Above this point the surplus energy is allocated to egg production. Note that it does matter whether the predation rate or the ingestion rate is given on the x axis; this is because only part of the prey is consumed and utilized for egg production. If the y axis represents development rather than egg production, then the relations are usually of the curvilinear type with an upper asymptote.

were possible to measure the actual amount of food extracted from prey, then a linear relation between oviposition and ingestion is expected. Ingestion and predation will also be linearly related when the amount of food consumed per prey is constant. However, curvilinearity will result when the amount of food ingested per prey decreases with increasing predation rates.

Development can be divided into feeding phases and non-feeding phases, e.g. egg stage and moulting stages. The total duration of the feeding phase depends on the egg-to-adult weight ratio (W_e/W_a). This ratio differs largely between predatory arthropods. For example, the scorpion *Paruroctonus mesaensis* increases 60−80 times in weight from 0.03 g (instar 2) to 2.0−2.5 g (non-gravid adults) (Polis, 1981), predatory mirids increase 15−20 times in length (Glen, 1973), whereas predatory mites in the family Phytoseiidae undergo a fivefold increase in weight (Sabelis, 1986), from 0.005 mg to 0.026 mg.

The egg:adult weight ratio may not always be a taxon-specific constant. As in bark beetles and drosophilid flies, size at maturity may well depend on the amount of food acquired during development, which in turn may affect fecundity of the adult female. Such effects are well known in parasitoids (e.g. host-size dependency of size and fecundity of the parasitoid), but surprisingly little in predatory arthropods. Possibly, predatory arthropods need not be flexible because the diets of successive stages usually overlap to some extent and because these diets are strongly determined by predator size. Hence, predators have little to gain by maturing at a smaller size. Indeed in hover flies, where larvae are predators of aphids and adults are pollen-feeders, pupal size and fecundity strongly depend on the amount of food acquired in the larval stage (see, for example, Cornelius & Barlow, 1980).

The rate of oviposition may not simply depend on age expressed in time units, but possibly also on age expressed in units of physiological effort. For example, in predatory mites of the family Phytoseiidae fecundity appears to be rather constant (and species specific) and the pattern of oviposition rates are little affected by intermittent starvation periods (provided not too long) suggesting that oviposition rates do not depend on age alone. There is strong evidence that the rate constant of food absorption d' bears a relation to the number of eggs produced in the past (Sabelis & Van der Meer, 1986). Thus, d' depends not on age, but evidently on some measure of physiological age, related to the wear and tear of the metabolic system. Whether physiological age plays a similar role in other predatory arthropods remains to be assessed.

Metabolic rates in predatory arthropods are generally set by the ambient temperature. Hence, predation, digestion, growth and reproduction are all strongly temperature-dependent, as shown in extensive studies of damselfly larvae (Thompson, 1978), predatory mites (see, for example, Sabelis & Van der Meer, 1986; Hayes & McArdle, 1987; Hayes, 1988), blue crabs (Eggleston, 1990). In some cases the effect of ambient temperature may be modified by thermoregulatory capacities through temperature preferences in combination with a certain degree of melanization of the integument. Thermoregulation occurs in tiger beetles (Dreisig, 1981) and ladybird beetles (Brakefield, 1985; Stewart & Dixon, 1989), and ants have developed a number of strategies to control nest temperature (Hölldobler & Wilson, 1990).

Several authors have demonstrated a negative relationship between intrinsic rate of increase and body size. However, Gaston (1988) argues that this relationship has been derived using data for species drawn from a number of phyla and covering several orders of magnitude in size. He analysed data from insects covering nine orders and found a negative relationship across orders, but no relationship within orders. This conclusion also applies to data on natural enemies of spider mites, a group of predators that are quite well studied with respect to their capacity of population increase (Helle & Sabelis, 1985). For example, phytoseiid mites are $c.$ 0.5 mm in length; their r_m values vary from 0.07 to 0.37 day (at 25°C) and the highest values come from some of the largest species (especially genus *Phytoseiulus*). Coccinellid beetles in the genus *Stethorus* are $1-1.5$ mm in length and have a maximum r_m of about 0.15 day (at 25°C).

Each of the components of the numerical response models discussed above may be related to body size, but little information is available from the literature. Gilwicz (1990) argued that food assimilation should increase more rapidly with body size, than does the respiration rate (see also Brooks & Dodson, 1965). Hence, the food density or ingestion threshold, where ingestion matches energy demands for maintenance and above which growth or reproduction occurs, may be lower for large-sized species than for small-sized species. Gilwicz (1990) found clear evidence for this trend among eight cladoceran species. A lower food threshold is an obvious competitive advantage for the large-bodied species, but this may be offset by the higher risk of predation by fish that is associated with being large.

SURVIVAL

Predator survival depends on food supply and escape from the predator's own enemies. Beddington *et al.* (1976) documented the effect of prey density on juvenile survival and development of a range of predatory arthropods. They concluded that there is a sigmoid relation between the proportion surviving to moult to the next developmental stage and prey density and no obvious, simple relationship between survival and the time spent between two moults. Similar conclusions can be drawn with respect to survival of the adults. The relation between adult mortality and age usually has a sigmoid shape too. Studies on predatory mites showed that the mean life span initially increases with decreasing prey density and then below a certain prey density threshold it decreases steeply to the value obtained in absence of prey (Sabelis & Van der Meer, 1986). An initial increase of life span with decreasing food supply was also observed in a carabid beetle (Ernsting & Isaacs, 1988). This indicates that some measure of the wear and tear of the physiological system could be useful in finding a simple description of mortality patterns. Sabelis and Van der Meer (1986) advocate usage of the cumulative number of eggs deposited up to the current age as a measure of physiological age in predatory mites. One may think of a more general measure, such as the amount of food that passed the gut wall or the weight of the individual arthropod plus the egg mass produced.

Escape from the predator's enemies is the other important component of predator survival. Several predatory arthropods have a large range of natural enemies, including a number of specialist predators. Examples can be found in Hölldobler and Wilson (1990) for ants, in Foelix (1982) for spiders, in Thiele (1977) for ground beetles, in Hodek (1973) for coccinellids, in Pearson (1988) for tiger beetles and in Kerfoot and Sih (1987) for aquatic arthropods. A rich variety of defensive strategies against hyperpredators can be found among the predatory arthropods. To illustrate this we follow a classification by Endler (1986b) who distinguished various phases in a successful predation event and the methods of defence by prey against predators at each stage.

Predatory arthropods may make themselves more difficult to detect by (apparent) rarity, immobility, crypsis and confusion. Examples are diapause in a freshwater copepod to avoid predation (Hairston, 1987); predator-induced vertical migration of water fleas (Ringelberg, 1991) and copepods (Neill, 1990) to escape from predatory fish during the daylight hours at the expense of a lower phytoplankton supply; immobility and crypsis of sit-and-wait predators such as mantids and crab spiders; swarm formation in whirligig water beetles (Gyrinidae) with likely confusing effects on predatory fish; and dense groups of water fleas, which may have a confusing effect on sticklebacks especially when the sticklebacks have to be cautious and watch for their predators (Milinski & Heller, 1978).

Predatory arthropods may also make it more difficult to be identified by their predators by masquerade, confusion, aposematism, Müllerian and Batesian mimicry (see Chapters 16 and 20). Examples are syrphid larvae, which are cryptic due to their colour patterns and shape (Rotheray, 1986); protective resemblance to bird dropping in the syrphid *Meligramma triangulifera* (Zetterstedt) (Rotheray, 1986); feigning death as in coccinellids (Hodek, 1973) and spiders (Foelix, 1982); the two-spot ladybird beetles, which are largely edible, polymorphic Batesian mimics of well-protected, monomorphic species such as the seven-spot ladybird beetle (Marples *et al.* 1989); and ant mimicry in salticid and clubionid spiders (Foelix, 1982, p. 250). Predatory arthropods may make it hard for their predators to come very close by sudden escape movements. Examples are hunters-by-eye, such as the jumping spiders and tiger beetles who jump or fly away when their enemies come too close.

Predatory arthropods may be hard to subjugate by their strength, their ability to struggle free, mechanical properties, noxiousness or even lethality. Examples are the armoured integuments of carabid beetles (Thiele, 1977), the anal secretions of lacewing larvae produced upon attack by ants (LaMunyon & Adams, 1987), the neck-teeth, elongated or pointed helmets that can be induced in water fleas by the presence of predators (Dodson, 1989b) and the abdominal forceps of earwigs. Once subjugated predatory arthropods may be emetic, poisonous or lethal. Examples are the emetic effects of alkaloids in ladybird beetles on insectivorous birds and the poison glands in the tail of scorpions (see also Blum, 1981 and Pasteels *et al.*, 1983).

Endler's classification should be supplemented with the case where prey warn each other of an approaching predator. Thiele (1977) and Spangler (1988) suggested that tympanic hearing and sound production in tiger beetles may function to alert conspecifics, as well as detect enemies, such as echolocating bats. Treherne and Foster (1980) showed that ocean skaters profit from being in a larger group as it increases the distance at which hyperpredators are detected. Investigations on ants have shown an elaborate communication system with alarm pheromones (Blum, 1981) to alert nestmates and the ability to specify which type of enemy is currently attacking and, therefore, which defensive strategy to use (Hölldobler & Wilson, 1990).

In general it is not an easy task to assess the adaptive significance of a trait supposed to be involved in defence. Endler (1986b) and

Pearson (1989) argue that a particular trait may have several functions (e.g. melanism may be involved in crypsis, mimicry and thermo-regulation) and defensive strategies may consist of various components that act in concert or each against a different type of enemy. For example, tiger beetles use body size, brightly coloured abdomens exposed in flight and defence chemicals against robber flies, whereas flight is the most important mechanism to escape from lizards and coloration is the major mechanism to deter birds. Some traits reduce the probability of attack by one type of predator, but simultaneously increase attack by another (Pearson, 1985). For example, flight helps to escape from lizards, but promotes attack by robber flies, which usually attack flying prey items (Pearson, 1989). The important lesson to be drawn is that predator survival may be a rather complex function of the abundance of each of its natural enemies. Whether prey supply and hyper-predator abundance interact in their effect on predator survival is not known, but it is likely that starved predators are more vulnerable to attack.

PREY DEFENCE AND ESCAPE

Following Endler's classification it would not be difficult to produce an even longer list of adaptive defences among the prey of predatory arthropods, but this list would be all too similar to the one produced above for predatory arthropods as prey for their predators. Instead, consider the more specific question about which defensive properties of the prey contribute to prey density-dependent trajectories of the functional response.

Negative dependence of the rate of predation on *local* prey density may arise when prey benefit from having lower per capita predation risks in a larger group. For example, aphids secrete siphuncular droplets upon contact with a predator. These droplets contain a volatile liquid, an alarm pheromone, that causes other aphids to move away often by falling or jumping. If aphid density were to increase, then the per capita risk of being eaten may decrease because the pheromone communication system is shared by a larger group of aphids. Kidd (1982) showed that grey pine aphids in large needle aggregations have a higher probability of escape from a syrphid larva than those in small aggregations; the alarm system of the grey pine aphid seems to be based only on direct neighbour-to-neighbour contact, rather than on pheromones. Lowered predation risks in larger groups may also be the result of cooperative defence, as is the case in sawfly larvae. Tostawaryk (1972) studied the functional response of a juvenile pentatomid predator to sawfly density and found some evidence for a dome-shaped (or Type IV) functional response. He attributed the decrease in predation rate at high prey densities to increased efficiency of group defence.

Prey defence or escape may also decrease in efficiency at higher prey densities. For example, increased competition among prey may cause them to stay longer in a developmental stage that is vulnerable to predation. Such effects may well be present in tadpoles as prey of dragonfly larvae. Travis *et al.* (1985) showed that odonate predation decreases with increasing tadpole size. They suggested that predation by odonates is an important selective force among anurans for rapid growth, so that the tadpoles can escape predation by growing too large for odonate larvae to handle.

AGGREGATION

Positive dependence on local prey density occurs when predators detect larger prey aggregations more easily than small ones or when predators stay longer in areas with high prey density than in areas with low prey density. Hassell *et al.* (1976) argue that after prey capture many predatory arthropods resume search by increased turning behaviour and a reduction of the speed of movement and that this behaviour decays to more straight walking paths if not reinforced by further location of prey. This behaviour is widespread among predatory arthropods (Kareiva & Odell, 1987) such as coccinellids (Nakamuta, 1985), anthocorids (Evans, 1976b; Lauenstein, 1980), chrysopids (Bond, 1980), syrphid larvae (Rotheray, 1983), carabids (Baars, 1979; Mols, 1987) and phytoseiid mites (Sabelis & Dicke, 1985). Responses to prey traces (honeydew, faeces, moulting skins, kairomones) seem to be quite similar (see, for example, Carter & Dixon, 1984; Sabelis *et al.*, 1984). Ernsting *et al.* (1985) studied the behaviour of a visually hunting carabid in response to cues left by springtails. They observed an increased frequency of stops, which provide better conditions for detecting prey movement.

Whether these behavioural mechanisms are sufficient to explain residence times of predatory arthropods in areas with different prey densities is not known. Sabelis (1981) found that computer simulations of tortuous walking paths of predatory mites predict much shorter residence times in leaf areas webbed and inhabited by spider mites than were actually measured. This difference may result from right-about turns observed at the edge of the web. Mols (1987) carried out similar simulations for carabid beetles in an environment with prey distributed in an aggregated way. He concluded that area-restricted search can promote the rate of prey capture. Kareiva and Odell (1987) modelled predator–prey dynamics as a diffusion process in a one-dimensional space. They showed that area-restricted search of adult ladybird beetles foraging for patchily distributed goldenrod aphids may well result in populations of predators flowing towards regions of high prey density. They feel that area-restricted search predation might *explain* the patchiness observed in prey distributions in an otherwise homogeneous vegetation domain. This assertion should challenge more critical experimental tests of predictions emerging from behaviour-based models.

Another level of explanation is obtained by applying optimal foraging theory to predict residence times within a prey patch. To date, however, very few studies of predatory arthropods (Bond, 1980) have been carried out to test experimentally whether patch residence times follow Charnov's marginal value theorem. There is no doubt that one reason for this lack of studies is the difficulty of measuring interpatch travel times and costs.

Predatory arthropods can be attracted to prey aggregations over distances much longer than their body length. There are several examples of predators using volatile kairomones emanating from their prey or their associates (Greany & Hagen, 1981; Sabelis & Dicke, 1985; Dicke *et al.*, 1991); clerid and other predatory beetles respond to the aggregation/sex pheromones emanating from bark beetles infesting pine trees; phytoseiid mites respond to kairomones emanating from host plants infested by spider mites; some coccinellid beetles respond to odour coming from cassava infested by mealybugs; chrysopid adults are attracted to a volatile breakdown product of a honeydew component, called tryptophan. The responses to kairomones may be of a quite complicated nature. Duelli (1984) showed that the adults of *Chrysoperla carnea* fly downwind, but upon sensing the kairomone they land first and then they move, in short bouts of flight, against the wind towards the kairomone source. The exact distances of attraction are not known, but it is of the order of several centimetres to several metres depending, among other things, on the amount of volatiles released. Hence, kairomones may also contribute to differential risks of prey aggregations being detected by predators.

Aggregatedness of prey distributions may reveal itself in different forms at various spatial scales. For example, two-spotted spider mites deposit their offspring within webs they produce on the underside of leaves. Once a leaf with web is found, it is very likely that others will be found nearby. Thus, there are two spatial levels where aggregation becomes manifest: webs on leaves and clusters of webbed leaves on plants. Predatory mites respond to the webs, the kairomones and the spider mites by performing behaviour leading to arrestment, but how do they promote arrestment within the cluster of webbed leaves? This is done in a remarkable way. When moving from leaf to leaf they pass plant area without prey cues except for the odour released from the spider-mite-infested plant sites. Phytoseiid mites foraging within the cluster are not likely to be hungry and in this state they do not disperse aerially, but move downwind in absence of kairomones or slightly upwind in presence of kairomones (Sabelis & Van der Weel, 1992). This response makes for a hitherto unknown mode of arrestment; when predatory mites happen to move out of the cluster area at the windward side of the cluster area, then the absence of kairomones induces downwind movement, which will bring them back into the prey cluster area. If the predators leave the leaf at the downwind side, then kairomones are in the air and elicit an upwind response. This behaviour is even more pronounced when the predatory mites become hungry, because in that state they always walk upwind. The consequence of this behaviour is anemotactic arrestment in a cluster of spider-mite colonies. It is likely that other such arrestment mechanisms operate at larger spatial scales, and that these have been overlooked, largely for want of a predator's eye view of prey distributions.

High numbers of predators in areas of high prey density need not be the result of attraction and arrestment alone; the numerical response to local prey density may further increase the size of an aggregation. For predators much larger than their prey one may expect the predator's intrinsic rate of population increase to be relatively unimportant. In that case local prey outbreaks may take place and their suppression relies entirely on the number of predators attracted and arrested. The predators may then prevent new prey outbreaks in the immediate vicinity of the initial outbreak (Kareiva & Odell, 1987). This type of pattern formation may be found in interactions between ladybird beetles and aphids. It may also occur in predators with broad diets, such as some carabid and staphylinid beetles that appear to aggregate at discrete patches of aphids (Bryan & Wratten, 1984). For predators that are close in size to their prey, capacities for population increase of the two interactants may be of the same order of magnitude. In that case it is the numerical response to local prey abundance rather than the attraction and arrestment of predators in search for prey aggregations that determines predator population size. As shown by a simple model in Box 10.2, suppression of the prey population now depends on the number of prey when the first predator arrives, the difference between their respective intrinsic rates of increase, the rate of predation and the arrestment of the predators and their offspring in the prey patch. Such

---

Box 10.2 A caricature of local predator–prey dynamics

Imagine a local population of prey with a high capacity of population increase. For the sake of simplicity assume that the prey population grows unlimited by food supply and that it can be suppressed only by their predators. The predators have a very low probability of finding the prey population, but once they invade the prey area they increase in numbers and their offspring do not disperse until after they have killed all prey. The rate of prey killing is constant, since the prey population spreads out while producing a constant number of offspring per unit area. Under these assumptions it is possible to obtain a simple answer to the following question: can a prey population be suppressed by a predator with a lower capacity of population increase?

Let α be the intrinsic rate of population increase of the prey and γ be that of the predator, while β is the rate of predation, then the following set of differential equations describe changes in the prey

continued on p. 254

Box 10.2 *contd*

population x and the predator population y (Diekmann *et al.*, 1988, 1989)

$$\frac{dx}{dt} = \alpha x = \beta y, \qquad \frac{dy}{dt} = \gamma y$$

Integration gives:

$$x(t) = x_0 \exp(\alpha t) - \frac{\beta}{\gamma - \alpha} [\exp(\gamma t) - \exp(\alpha t)],$$

$$y(t) = \exp(\gamma t)$$

where x_0 is the prey population size upon invasion by a predator ($y_0 = 1$). Now, it is easy to find an expression for T, the interaction time necessary to wipe out all prey ($x(T) = 0$):

$$T = \frac{1}{\gamma} \ln(y) = \frac{1}{\gamma - \alpha} \ln\left[1 + \frac{(\gamma - \alpha)}{\beta} x_0 \right]$$

The answer to the question posed above is obtained by deriving the condition for T such that, even though $\gamma < \alpha$, the time needed to wipe out the prey does not become infinitely long. This implies that the expression under the logarithmic operator satisfies the following condition:

$$0 < 1 + \frac{(\gamma - \alpha)}{\beta} x_0 < 1 \qquad \text{or} \qquad 1 < \frac{(\gamma - \alpha)}{\beta} x_0 < 0$$

$$\text{or} \qquad 0 < x_0 < \frac{\beta}{(\alpha - \gamma)}$$

Thus, given $\gamma < \alpha$, prey elimination occurs within a finite time span when x_0 is sufficiently small relative to β.

A tentative application to plant-inhabiting mites and their natural enemies (Helle & Sabelis, 1985; Sabelis, 1991) follows. Spider mites are phytophagous mites. Their most important predators are phytoseiid mites; they are of similar size and both have intrinsic rates of increase ranging between 0.1 and 0.35 day^{-1}. As a crude approximation of the maximum predation rate one may take $\beta = 100\gamma$. To eliminate prey with $\alpha = 0.35$ within a finite time span, the initial prey population x_0 should be lower than 3400, 600, 133 or 940 for predatory mites with $\gamma = 0.34$, 0.3, 0.2 or 0.1, respectively.

Other natural enemies of spider mites are coccinellid beetles in the genus *Stethorus*. They are two to three times larger than phytoseiid mites, have an intrinsic rate of increase of maximally 0.15 and a predation rate of maximally 80 spider mites day^{-1}. To eliminate prey with $\alpha = 0.35$ within a finite time span, the initial prey population x_0 should be lower than 400. Thus, the impact of coccinellid beetles on local populations of spider mites is always lower than phytoseiid mites with high γ. If coccinellid beetles were better able to find prey patches through their ability to fly, then they might compensate for their lower control potential by a higher tendency to aggregate in spidermite patches.

local population explosions of prey and then predators may also stop any incipient prey outbreak in the vicinity. This type of pattern is found in interactions between phytoseiid mites and spider mites on plants (Sabelis & Van der Meer, 1986). In a number of these acarine predator–prey systems the interaction ends with virtually complete extermination of the prey population, which then causes the predators to disperse. Why predatory mites should over-exploit local prey populations is not immediately obvious. On the one hand it may evolve for much the same reasons as virulence in host–pathogen systems (Levin & Pimentel, 1981). On the other hand, since predator dispersal relieves predation pressure on the prey population and consequently promotes growth of the prey population and thereby also of the predator population, a suite of dispersal strategies is possible where dispersal depends on prey density and eventually results in a much higher production of predators per prey patch (Sabelis & Van der Meer, 1986). Which, if any, of these strategies are evolutionarily stable is an open question for future research.

DISPERSAL

Dispersal of predatory arthropods may be active or passive. Passive dispersal occurs either by drifting in a flowing medium (air, water) or by hitching a ride on other organisms with an active mode of dispersal. Passive dispersal is widespread among predatory arthropods of minute size, such as the mites. For example, many phytoseiid mites disperse in air currents (Sabelis & Dicke, 1985), whereas macrochelid mites are predators of fly larvae and ride on dung beetles (Krantz, 1983). Spiders also drift on air currents, but their larger size requires the use of ballooning threads of silk to increase drag and remain floating in air (Richter, 1970; Van Wingerden & Vugts, 1974). Many of the lower crustaceans and young stages of higher crustaceans disperse passively in currents of water.

Active dispersal occurs by swimming, walking or flying. Ambulatory dispersal is always an option for the terrestrial crustaceans and chelicerates, whereas aquatic arthropods are often capable of swimming. The insects stand alone in being able to fly, albeit in the adult phase only. Partly, wings help insects to remain suspended in air currents, but at low wind speed wings allow directed flight, either away from an enemy or towards a positive stimulus. The capacity of dispersal by flight tends to increase with body size, but this trend is obscured in many cases. Wings and flight muscles entail a cost and this investment has to be weighed against alternative energy allocations. For example, brachyptery in carabid beetles is often found in stable habitats that are very isolated. (Thiele, 1977). If there were no cost, then selection would have favoured traits suppressing dispersal at the behavioural level. Several carabid species appear to be dimorphic or polymorphic with respect to wing size, suggesting either a temporary evolutionary state on the road to brachyptery or a balanced state with fitness returns for each dispersal strategy (Den Boer *et al.*, 1980; Harrison, 1980; Aukema, 1986, 1987; Desender, 1989).

An important point to note is that many population models of arthropod predator–prey interactions take aggregatedness into account, but do not include a separate dispersal phase for either the prey or the predators. During dispersal, the rates of encounter may be reduced to zero. Some predators, such as robber flies search and attack prey while flying, and similar risks are likely to confront predators that disperse by walking. Hence, there are good reasons to model a separate dispersal phase as representing a temporary withdrawal from the predator–prey interaction leading to redistribution in space (Sabelis *et al.*, 1991).

COMPETITION

Mutual interference?

When no clearly defined resource like food or space is in short supply, then the effects of predator density may be referred to as mutual interference. The best evidence for interference in predatory arthropods comes from experiments with plant-inhabiting predatory mites and backswimmers. Despite the provisioning of an ample supply of spider mites as prey, an increase in the density of phytoseiid females leads to a decrease of the rate of oviposition (Kuchlein, 1966; Eveleigh & Chant, 1982a,b), and to a lower proportion of daughters among the offspring (Sabelis & Nagelkerke, 1988). In contrast, if the predators are given the opportunity to move away, then the major effect is an increased rate of dispersal (Sabelis, 1981; Bernstein, 1984). Each of these effects may be viewed as the optimal solution to a combined sex ratio and oviposition game. Under conditions where offspring are likely to sib-mate, phytoseiid females produce more daughters than sons. Increasing the density of ovipositing females will decrease sib-mating and thus promote a tendency toward random mating. Under these conditions, females are expected to produce less female-biased sex ratios (i.e. closer to half sons and half daughters). It then also pays to lay fewer eggs at a given site both to avoid food competition among the offspring and to save eggs which can be laid elsewhere after dispersal. Sex ratio games can promote dispersal and decrease the clutch-size of eggs deposited at any site (C.J. Nagelkerke, in preparation). Thus, there may be resources in short supply (e.g. oviposition territories), but recognizing them depends on how one views the maximization of fitness. This is a serious objection against defining mutual interference in the way proposed at the beginning of this section.

The same point emerges from studies on age-dependent interference among backswimmers (Murdoch & Sih, 1978). The feeding rate of juvenile stages is greatly reduced in the presence of adults. Interference appears to result from the juveniles being less mobile and remaining longer at the edge of the habitats when adults are present. This behaviour may well be to avoid cannibalism by the adults and, if so, the resource of 'adult-free space' is in short supply. This makes the definition of mutual interference inappropriate for much the same reason as above.

The interference phenomena observed in predatory mites appear to have little impact on population dynamics. They will lead to a more even exploitation of the unevenly distributed prey within a cluster of spider-mite infested leaves, because interference is avoided through within-cluster dispersal. Predator density may be high enough for interference nearing the moment where the prey population is wiped out by the predators, but this condition lasts only for a short period and, as far as we know, local prey populations will be exterminated anyway, as the predator population by far exceeds the numbers needed for overkill (Sabelis & Van der Meer, 1986). Hence, interference in predatory mites is important for understanding the evolution of individual traits, but seems to be irrelevant for understanding local predator–prey dynamics. This is not so for the age-dependent inter-ference observed in backswimmers (Murdoch & Sih, 1978) and in fact leads to serious complications in formulating population models. Sih (1981) suggests that since the magnitude of interference decreases with increasing numbers of prey, interference may play a critical mediating role in determining functional and numerical responses of the predators.

Intraspecific predation

Intraspecific predation or cannibalism can be viewed as a way of killing competitors and thereby gaining access to a larger food source or as a way of obtaining scarce nutrients. In the first case cannibalism is an extreme form of interference competition, while in the second it relates to dietary demands. Most if not all cases of cannibalism include both killing and eating, so that the two functions are hard to separate. Among the predatory arthropods, cannibalism is widespread and asymmetric in that the larger sized (or less vulnerable) individuals eat the smaller sized (Fox, 1975; Polis, 1981). It is suggested to be a major mortality factor in natural populations of copepods, dragonfly nymphs, backswimmers, red wood ants, carabid larvae and scorpions (Mabelis, 1979, Heessen & Brunsting, 1981; Polis, 1981; Van Buskirk, 1989; Orr et al., 1990). Driessen et al. (1984) provide some evidence that the cannibalistic interactions during red-wood-ant wars serve to meet protein demands; red wood ants become less cannibalistic by the end of spring when they switch to forage for bibionid larvae and caterpillars, as soon as they are available as prey.

Cannibalism occurs in several other predatory arthropods, but the evidence comes from laboratory observations only and the relative importance under field conditions is not known. Examples are heteropteran bugs, phytoseiid mites, predatory soil mites, spiders and pseudoscorpions.

Predation is not the only form of interference between arthropods. Examples are web robbery in spiders (Foelix, 1982), contest competition for nets between larvae of caddis flies (Hildrew & Townsend, 1980) and pit relocation of antlion larvae as a consequence of dropping sand into their current pits (Matsura & Takano, 1989). As noted earlier for juvenile notonectids, the effect of interference may also be the avoidance of it. Legg and Jones (1988) suspect this to be the case for less aggressive and correspondingly less well-armoured species of pseudoscorpions, such as *Chtonius ischnocheles* (Hermann), which tends to avoid fights by quickly running backwards.

Intraguild predation

Intraguild predation is the killing and eating of species that use similar resources, and thus are potential competitors (Polis *et al.*, 1989). It differs from classical predation because the act reduces exploitation competition for limited food resources. Thus, its impact on population dynamics is more complex than either competition or predation alone. Polis and McCormick (1987) provide data on intraspecific and intraguild predation in scorpions. They emphasize that the success of cannibalistic attacks is determined by relative size. Within a guild of scorpions this appears to lead to rather complex interactions, such as age/size-mediated reversals, where young scorpions begin life as prey of another species, but later become the predator of that same species. These complex interactions may be important in explaining numerical dominance patterns in scorpions. Polis and McCormick found that *Paruroctonus mesaensis* is by far the most abundant of four species coexisting in a Californian desert. Yet as an adult, it is of intermediate size, only half the adult body mass of the largest species, *Hadrurus arizonensis*, which makes up less than 1% of all scorpions in the desert. The largest species takes 4–5 years to mature, while the more numerous, intermediate species reaches maturity in less than 2 years. This may contribute to differential intraguild predation, since *H. arizonensis* begins life as prey of *P. mesaensis* and later becomes the predator of *P. mesaensis*. Such age/size-mediated reversals and the time needed to increase in size may help explain the rarity of *H. arizonensis* (Polis & McCormick, 1987). Differential intraguild predation is also thought to explain competitive exclusion phenomena in laboratory experiments with populations of two phytoseiid species competing for one prey, the Pacific spider mite (Yao & Chant, 1989).

Interference by predation has a direct advantage in that the aggressor gains food. The disadvantage is that the aggressor runs the risk of being damaged or even falling prey itself. If there is a net gain to attacking a species of the same guild, then selection will favour interference by predation provided inclusion in its diet results in an overall net gain. If not, interference may evolve because of its effect on the availability of resources. Yet, this is not immediately obvious, since the resources freed by the aggressor are shared by all other members of the population. Only if the aggressor is able to monopolize the resources is this kind of interference competition likely to evolve. Note that this condition is more likely to hold when the prey distribution is patchy.

Another line of evidence for interference competition comes from the existence of territorial strategies in ants. Hölldobler and Lumsden (1980) describe various strategies ranging from absolute territories to spatiotemporal territories and discuss simple cost–benefit models to illustrate the economic defensibility of each type of territory. As these territories lead to monopolization of food within the territory, it is clear why such territories have evolved.

Exploitation competition

There is considerable evidence that predatory arthropod species exhibit reduced overlap in the range of prey types taken when they coexist with other predators in the same habitat (Wilson, 1978). Pearson (1980) found geographical variability in mandible lengths of tiger beetles, and at any one site, mandible lengths were clearly different from the most similar sized congener. Also, their ratio was different from a random sample of mandible lengths of species pairs. Juliano and Lawton (1990) studied various aspects of body form in assemblages of diving beetles and tested for deviations from appropriate null models. They found evidence for limiting similarity within groups of coexisting beetles. Lawton and Hassell (1984) review many other examples, but note that evidence of limiting similarity is lacking for phytophagous insects. The explanation may be that most phytophagous species are normally kept at low densities relative to their potential food supplies by a potent combination of natural enemies (Strong et al., 1984). Hence one may expect that interspecific competition would be of little importance for phytophagous arthropods, but it may well be important for the predatory arthropods that feed on them (but see Crawley & Pattrasudhi, 1988).

There are many observations that challenge the conventional view on the prevalence of limiting similarity. Studies on predatory arthropods are particularly weak because of insufficient knowledge about the diets and the preferences of the predators involved. For example, according to Thiele (1977, p. 130) the majority of investigations reveal that carabid species of similar body size show very similar preferences with respect to prey in any one habitat. However, even the cases which Thiele considered to be the most thoroughly investigated, lacked an adequate method for assessing prey preferences.

Studies on optimal diets may provide indirect evidence for exploitation competition, and exploitation competition provides a general mechanism for driving the evolution of prey preferences in multiprey/multipredator communities. It may cause prey preferences to differ between two or more predator species with overlapping ranges of prey sizes or types. Usually, optimal diets are defined by the food value of each prey type (e.g. energy gained per handling time) and by a measure of the abundance of each prey type. Prey species not in the optimal diet may be included depending on: (i) the associated energy gain per time spent handling; and (ii) the abundances of the prey species in the optimal diet (but not the abundances of prey species that are not part of the diet; see Chapter 4). Elner and Hughes (1978) studied how shore crabs selected between different size classes of mussels, and found some support for selection of the most profitable mussel size. Predictions from optimal diet theory often fail to match experimental results so nicely. For example, Dicke *et al.* (1988) carried out preference analyses in three different ways: (i) by choice tests in a Y-tube olfactometer (Sabelis & Van de Baan, 1983); (ii) by predation experiments in mono and mixed cultures of prey types (Sabelis, 1990); and (iii) by electro-phoretic diet analysis of field-sampled individuals. They studied three species of phytoseiid mites, *Amblyseius finlandicus* (Oudemans), *Amblyseius potentillae* (Garman) and *Typhlodromus pyri* (Scheuten), and offered two species of prey, rust mites and European red mites. The first predator species showed a slight but consistent tendency to prefer rust mites over European red mites in all three tests. The second showed a strong preference for European red mites, whereas the third showed a less pronounced preference for European red mites. Re-productive success on each of the prey types was tested under conditions of ample prey supply. It was found that rust mites were the more profitable prey in terms of reproductive success for all three predator species (Dicke *et al.*, 1990), and these results were in sharp contrast to the observed preferences. An explanation may be found by modelling the evolution of prey preferences in *dynamic* predator–prey systems consisting of two or three predator species and two prey species. In this type of model the evolution of prey preference faces a boomerang effect. Once a particular preference has become frequent in the popu-lation, it causes a decrease in the abundance of the preferred prey type, which may eventually become so scarce that a predator does better by preferring the alternative prey type. This type of interaction between prey preference, food value of prey and the abundance of the preferred prey provides plentiful opportunities for further research on the evol-ution of prey specialization in predator–prey communities.

Elucidation of prey preferences and their proximate and ultimate causes seems to be a major task for future research. Many of the taxa that include predatory arthropods contain some species that are obviously specialists. For example, among the carabid beetles there are genera specialized on hunting snails (larvae and adults of *Cychrus*) and springtails (*Notiophilus*) (Thiele, 1977); among the ants there are species of the genus *Leptogenys* specialized on sowbugs, *Amblyopone* specialized on centipedes, *Discothyrea* and *Proceratium* specialized on spider eggs; also there are several genera specialized on either termites or ants (Hölldobler & Wilson, 1990); among the plant-inhabiting predatory mites there are species, such as *P. persimilis* with a strong preference for spider-mites in the genus *Tetranychus* (Helle & Sabelis, 1985); among the spiders the genus *Mastophora* (a bolas spider) stands out as an extreme specialist in that they produce a mimic of the sex pheromone of moths (Spodoptera) and therefore lure and catch male moths only; the pirate spiders (Mimetidae) prey exclusively on other spiders which spiders belonging to the Zodaridae hunt only ants (Foelix, 1982); among the crabs some species, like *Cancer pagurus*, have massive chelas that are used to crush thick strong shells of snails and clams (Pearse *et al.*, 1987). Several other examples of specialization are available, but the vast majority of predatory arthropods have rather broad diets consisting of various types of prey (Slansky & Rodriguez, 1987). Some predatory arthropods even include fungi and plant foods, such as fruits, pollen, nectar and even leaf tissue. How predatory arthropods select their prey or other foods is poorly understood, let alone the role that exploitation competition might play in structuring the evolution of prey preferences.

MUTUALISM

Predatory arthropods may be engaged in many types of mutualistic interactions (Boucher, 1985), but the one of interest here is their engagement in protective mutualisms. Predators may protect other organisms from their natural enemies and may get something in return from the organism that gains the protection, such as transport, shelter, nutritious substances or perhaps nothing more than a signal betraying the presence of prey. Examples include the following.

1 Carrion beetles of the genus *Necrophorus* who carry and feed mesostigmatid mites of the genus *Poecilochirus*, which in turn feed on eggs of carrion flies thereby reducing competition with larvae of the carrion beetle (Wilson, 1980).

2 Plants with structures containing cavities (e.g. hollow thorns of *Acacia*) that are utilized by ants as nest sites (Beattie, 1985).

3 Plants which produce food bodies that are fed upon by ants (Beattie, 1985).

4 Plants with extrafloral nectaries or honeydew-producing Homoptera that provide food for a suite of natural enemies including ants, lacewings, coccinellids, etc. (Boethel & Eikenbarry, 1986).

5 Plants with very small cavities or hair tufts at vein junctions that are inhabited by mites that feed on pollen, fungi or other mites (Pemberton & Turner, 1989; O'Dowd & Willson, 1989).

6 Plants that are induced to produce volatiles after attack by phytophagous mites, thereby helping predatory mites to find their prey and perhaps even to discriminate between different prey species (Dicke & Sabelis, 1988, 1989, 1992).

7 Caterpillars of the butterfly family Riodinidae which literally call for the help of ants by producing substrate-borne vibrations using vibratory papillae (De Vries, 1990).

It is one thing to show that the interaction between two species conveys benefits to both, but it is another to prove that the traits involved in the interaction are still beneficial when all other effects are taken into consideration. Extrafloral production of nectar and herbivore-induced production of volatiles may lead to responses of several arthropods, not only the beneficial ones. These may include other herbivores, hyperpredators and hyperparasitoids; it remains to be shown whether there is a net benefit to the plant producing the extra-floral nectar and the volatile substances. Moreover, when protection is gained by providing nectar, shelter or signals and is shared by neighbouring plants, then it is to be expected that other plants will cheat by producing none or less so that plant populations and communities become polymorphic with respect to these traits (Sabelis & De Jong, 1988). This is an important message to field ecologists attempting to assess aggregative responses to prey density in the field. If natural selection were to maintain variability in plants with respect to their investment in attracting and arresting predatory arthropods as bodyguards, then aggregative responses of predators in the field may well be less pronounced or even obscured.

CONCLUSION

Predatory arthropods have been shown to exert a major impact on structuring aquatic and terrestrial communities (Huffaker & Rabb, 1984; Kerfoot & Sih, 1987). Spectacular examples are provided by the impact of army ants on local insect faunas (Sudd & Franks, 1987), the large-scale successes in biological control of cottonycushion scales by vedalia beetles in citrus (Caltagirone & Doutt, 1989) and the control of spider mites by phytoseiid mites in various crops (Helle & Sabelis, 1985). Quite commonly, predatory arthropods drive their prey locally extinct and then go locally extinct themselves. It is unfortunate that so much of the hard evidence comes from laboratory or greenhouse experiments rather than from the field. Examples are phytoseiid mites (Sabelis & Van der Meer, 1986; Nachman, 1987a,b), backswimmers (Murdoch et al., 1985) and water fleas (Murdoch & Scott, 1984). Field observations have provided indications of local extinctions of predators, even when the predator is thought to be quite polyphagous. Examples are carabid beetles (Den Boer, 1990), army ants (Sudd & Franks, 1987) and possibly also vedalia beetles (Murdoch et al., 1985). Less violent fluctuations are expected to occur in predatory arthropods that exhibit broad diets and territorial behaviour, such as many ants (Hölldobler & Lumsden, 1980; Hölldobler & Wilson, 1990) and web spiders (Provencher & Vickery, 1988). Exactly how the traits of these predators and their prey contribute to produce the observed dynamics is still poorly known.

Advances in knowledge have been gained through a combination of modelling and population experiments to test model predictions. Some of the better studied predator–prey systems are ladybird beetles and aphids as prey (Gutierrez et al., 1981; Gutierrez & Baumgaertner, 1984a,b; Gutierrez et al., 1984), phytoseiid mites and spider mites as prey (Sabelis et al., 1988a,b), water fleas and phytoplankton as prey (Kooyman & Metz, 1984; Metz et al., 1988b) and damselflies (Crowley et al., 1987). However, these modelling efforts were all rather system-specific and therefore did not provide hypotheses with any great generality. The way forward is likely to come from an approach which starts by asking what predatory arthropods have in common, and then asking questions about how specific traits of particular groups of predators modify the conclusions drawn for the more general case.

The following traits appear to be essential for understanding arthropod predator–prey interactions.

1 *Body size of predator and prey.* Relative body size influences prey capture success, the range of prey sizes taken, partial prey consumption, capture thresholds, prey preference, survival, investment in defence, dispersal capacity and competitive ability.

2 *Egg : adult weight ratio.* This ratio is fundamental to arthropod life styles. It determines: (i) the duration of pre-adult development; as well as (ii) the quantity of offspring to be produced from the amount of resources allocated to reproduction. As a measure to characterize life styles, however, it lacks generality because the weight of eggs and adults is phenotypically plastic, and can be affected by environmental conditions such as food scarcity.

3 *Rate of food digestion and egestion.* This trait is a major determinant of: (i) the upper asymptote of the functional response; and (ii) the rate of development and reproduction. It may depend on age, stage, state (diapausing, migratory, hungry), diet composition and, as in all poikilotherms, on temperature. Digestion appears to be slower than ingestion. The rates of these processes often differ by orders of magnitude, making time budget considerations superfluous. At the time scale of digestion, ingestion can be represented as a point event.

4 *Allocation decisions.* The way that the absorbed food mass is allocated to maintenance, growth, energy reserves (e.g. for diapause), reproduction and sex, may be described by a set of decision rules that characterize the life history strategy.

These four components may suffice to describe the strategy of a predatory arthropod and should serve as a basis for assessing evolutionarily stable prey selection strategies, population dynamics and ultimately also community structure.

The essence of a predator's performance is certainly not captured by considering patterns of energy flow only. There is also a spatial dimension to it. Predatory arthropods respond to differential prey abundance at various spatial scales. These responses may well be understood from an optimal foraging point of view. If prey aggregation were to lead to a higher risk of predation for prey in relatively dense aggregations, then selection would favour prey genotypes which avoided these aggregations and the selection process would go on until the risk of predation was equal for all prey individuals. Yet, prey aggregations are commonly observed to exist. So, either aggregation itself must bring advantages which compensate for the higher risks of predation, or the aggregation site confers advantages that are not available elsewhere (e.g. a food source of higher quality). Explaining distribution patterns of predator and prey starting from individual foraging tactics seems to be an important task for future research; the more so because of the impact of patchiness on the evolution of various traits (Wilson, 1980) and the population dynamics of predatory arthropods and their prey (Hassell, 1978).

11: The Population Biology of Insect Parasitoids

MICHAEL P. HASSELL & H. CHARLES
J. GODFRAY

INTRODUCTION

Parasitoids are some of the most abundant of all animals, probably comprising 10% or more of all metazoan animals. They occur in a number of different insect groups, and are recognized by their characteristic life cycle, which has aspects in common with that of both predators and parasites. The adult female lays her eggs on, in or close to the body of another arthropod (usually an insect), which is eventually killed by the feeding parasitoid larva. Like most true parasites, all the energy and nutrients necessary to complete larval development are acquired from a single host, but like predators they almost invariably cause the death of their host.

The parasitoid life style has evolved many times in the insects, but most species are found in the Diptera (two-winged flies) and Hymenoptera (sawflies, wasps, ants and bees). Within the Diptera, the majority of species occur in two families, the Bombylidae and the Tachinidae. The Hymenoptera is divided into the primarily herbivorous Symphyta (sawflies) and the Aculeata, which contains the Apocrita (ants, bees, solitary wasps, social wasps and their allies) and the Parasitica, an enormous assemblage of species, almost exclusively parasitoids, arranged in about 45 families and including the ichneumons, braconids, chalcids, pteromalids and fairy flies, as well as many other less familiar groups. Askew (1971) provides an excellent introduction to the biology and systematics of parasitoids, as do Gauld and Bolton (1988) for the hymenopterous species. Clausen's (1940) classic work contains a wealth of fascinating information on the natural history of parasitoids.

Mobile hosts are usually stung and paralysed by the female parasitoid prior to oviposition. In some cases the paralysis is temporary and the host recovers and continues feeding, even though it contains an immature parasitoid. The advantage to the parasitoid of such a life history (called *koinobiont* by Askew and Shaw (1985) is that the female is able to attack hosts that are too small to support a developing parasitoid. The parasitoid remains dormant in the host until the host has grown to a sufficient size to allow its development. *Idiobiont development* occurs when no host growth occurs after parasitism. Parasitoid life histories can be classified by the host stage attacked, e.g. *egg parasitoids, larval parasitoids*, etc. Some parasitoids with koinobiont development attack one stage but emerge from and kill another stage; these are referred to by terms such as *egg—larval parasitoids*. Although in most species it is the adult female which locates suitable hosts, in a few species it is the highly modified (*planidial*) larvae that are responsible for host location.

Many parasitoid females (*ectoparasitoids*) lay their eggs on the surface of the host and the larvae develop by feeding through small punctures in the host's cuticle. Alternatively, *endoparasitoids* inject their eggs into the host's body and their larvae feed internally. Endoparasitoids commonly show koinobiont development, in contrast to the idiobiont development of ectoparasitoids, and are highly adapted to survive within the host's body where they have to withstand attack by a cellular immune system. The complexity of the biochemical battle between host and parasitoid has only recently begun to be revealed (Strand, 1986). For example, a number of species are known to inject a symbiotic virus into the host, which destroys the ability of the host to encapsulate the parasitoid's eggs (Stoltz & Vinson, 1979; Edson *et al.*, 1981).

Either a single larva (*solitary* parasitoids) or more than one larva (*gregarious* parasitoids) may successfully develop in a single host, and broods of several thousand have been recorded from a few gregarious species. The largest clutches are known from species which lay one or a few eggs in a host, but whose eggs divide asexually to give rise to a large number of identical larvae (*polyembryony*). If a previously parasitized host is discovered by a second female of the same species, and if this female deposits a second egg (or clutch of eggs in the case of gregarious species), *superparasitism* is said to have occurred. If the second female is of a different species, the term *multiparasitism* is used.

Some parasitoids, known as *hyperparasitoids* or *secondary* parasitoids, have become specialized to develop on other parasitoids attacking a host. Of these, *facultative* hyperparasitoids attack a healthy host or another parasitoid with equal readiness, while *obligate* hyperparasitoids only attack other parasitoids. A very few *tertiary* parasitoids (those that attack hyperparasitoids) are also known. One family of parasitoid wasps (Aphelinidae) contains many species in which the female develop as normal endoparasitoids of scale insects, while the males develop as hyperparasitoids of females (such species are known as *heteronomous* hyperparasitoids).

In the first part of this chapter, we describe recent research into the behavioural ecology of parasitoids and their importance as experimental models for testing evolutionary theory. The second part is concerned with the population dynamics of parasitoids and how they may regulate the abundance of their hosts and affect the structure of multispecies systems.

Finding a host

Consider the problems facing a newly emerged female parasitoid about to begin searching for hosts — a small insect searching for another small insect in a highly complex environment. To add to her difficulties, the host may well have evolved a variety of means to disguise its whereabouts. How will natural selection mould the behaviour of the searching parasitoid to overcome these problems and to maximize the number of located hosts? Two major classes of adaptations can be predicted. First, the parasitoid should make use of any information in the environment concerning the direction or presence of suitable hosts. Second, the parasitoid should respond to the spatial distribution of hosts and to the presence of other searching parasitoids in a way that leads to the maximum encounter rate with hosts.

The study of the stimuli used by parasitoids to locate hosts has advanced considerably in recent years as applied entomologists have sought ways to attract parasitoids into crops for pest control (see the series of reviews in Nordlund *et al.*, 1981). By far the commonest means of host location are via chemical cues, but it is now known that some species of parasitoids employ sight, sound, heat or vibration as clues to a host's presence. The chemicals involved, often called *kairomones*, may be volatile substances that operate as attractants over long distances by imparting directional information, or short-range arrestants that are typically of low volatility and inform the parasitoid of the immediate presence of a host.

Chemicals from a variety of different sources can assist parasitoids in host location. In the absence of direct cues emanating from the host, the parasitoid may rely on cues from the host environment. Thus *Drosophila* parasitoids are attracted to odours from rotting fruit or vegetation where their hosts are likely to occur. Van Alphen and Vet (1986) have made a special study of these parasitoids; they discovered that many of them are habitat-specialists rather than host-specialists, and only react to odours coming from certain types of *Drosophila* habitat. They even discovered a new species by virtue of its host location behaviour, which was only subsequently confirmed morphologically (Vet *et al.*, 1984).

More directly associated with the host are chemicals found in host frass or linked with host oviposition. For example, moths frequently dislodge scales while ovipositing, and in several species these are known to contain an arrestant chemical which causes egg parasitoids in the genus *Trichogramma* to increase their searching effort in the immediate neighbourhood. It also reduces the probability of parasitoid dispersal (Lewis *et al.*, 1971).

The detection of chemical cues by parasitoids is combined with behavioural adaptations for host location. Where the cue is directional, the parasitoids obviously move towards its source. In the absence of directional information, arrestant stimuli may elicit area-restricted search, i.e. foraging that is concentrated in a limited area. Both such responses occur, for example, in the case of the ichneumonid, *Venturia* (= *Nemeritis*) *canescens*, which attacks stored-product insects living in grain and similar substrates. The long distance attraction is probably towards a combination of chemicals emitted by the grain and from the hosts' maxillary gland during feeding. When close to a patch of hosts, the parasitoids respond strongly to an arrestant chemical (contained within the maxillary gland secretions) by increased frequency of probing movements and by a tendency to 'turn back' at the boundary of the patch (*klinokinesis*) (Corbet, 1973; Waage, 1978). A behavioural model to explain such searching patterns of parasitoids has been developed by Waage (1979a). He suggested that the response of a parasitoid to the arrestant chemical wanes during search within a patch until a threshold is reached, at which point the insect leaves the patch. The discovery of a host acts to increase the responsiveness to the arrestant and thus leads to the insect staying longer on patches containing greater numbers of hosts.

It has been known since the 1930s that parasitoids can learn to associate new cues with the presence of hosts, although only recently has the frequency of learning in parasitoids been appreciated (van Alphen & Vet, 1986). In a series of classic experiments, Arthur (1966) showed how the parasitoid wasp *Itoplectis conquisitor*, which attacks a variety of caterpillars hidden in leaf rolls, learns to recognize the morphology of host retreats. Arthur constructed artificial host retreats of various shapes, sizes and colours, and showed that the discovery of a host in one type of retreat led subsequently to an increased tendency to search for structures of a similar design. In a related vein, a number of species have been shown to orientate towards particular volatile chemicals, having previously responded to the same substance in the vicinity of a host.

The hosts of parasitoids are frequently distributed across the environment in discrete patches. How long should a parasitoid remain in a patch before dispersing and searching for a new patch? This problem is very similar to the patch-use problem in classical foraging theory (see Chapter 4), yet the marginal value theorem has seldom been applied to parasitoids searching for hosts, principally because one of the main assumptions of patch-use-models — that the optimal behaviour of an individual can be calculated without reference to the activities of other individuals — will rarely be met in studies of parasitoids. This assumption fails because in most host–parasitoid systems, host patches are often discovered by more than one parasitoid, and thus the optimal residence time on the patch is likely to be affected by the behaviour of the other individuals. Foraging in such competitive situations is often studied using the ideal free distribution (Chapter 4). Again, at least in its original form, this does not apply directly to typical instances of parasitoid foraging since it assumes non-depletable patches. Workers attempting to predict how natural selection should mould parasitoid behaviour in patchy environments have thus been forced to develop their own theory.

Cook and Hubbard (1976) and Comins and Hassell (1979) studied simple analytical models in which groups of parasitoids forage in depletable patches. There were assumed to be no travel costs. These models predict that all parasitoids should first search in the best patch until it has been reduced to approximately the same level as the next best patch (the exact level depends on the detailed assumptions of the model). The parasitoid population then divides between the two patches until both are depleted to approximately the same level as the third patch, and so on until all patches are equally exploited. In effect, the parasitoids adopt a sequence of ideal free distributions which are readjusted as resource depletion proceeds. More recently, Bernstein *et al.* (1988) have constructed a series of similar, but more realistic, simulation models, in which they relax the assumption that each parasitoid has an exact knowledge of the quality of the environment and is able unerringly to select the optimal patch. This is achieved by incorporating a learning process into the model. They conclude that as long as the environment is not depleted very quickly, the pattern of sequential ideal free distributions predicted by the simpler models is relatively robust. An obvious next step is the incorporation of travel costs between patches into this type of analysis.

Host acceptance

On discovering a host, a parasitoid has to decide whether or not to oviposit. Hosts will vary in quality, and while it is obvious that a parasitoid should always attack the best quality hosts, and avoid hosts where the probability of larval survival is zero, it is less obvious how it should deal with hosts of intermediate quality. The problem of host acceptance is clearly similar to the optimal diet problem in classical foraging theory (Chapter 4). The question 'how many food items should be incorporated into the diet to maximize the rate of gain of energy?' is replaced by 'how many host types should be incorporated into the host range to maximize the rate of gain of fitness?' Here, 'fitness' might be defined as the number of offspring produced by a searching wasp, weighted by their probability of survival.

The optimal diet model applies exactly to parasitoids whose reproductive success is limited by the time available for search and egg-laying, and which search for and attack a large number of hosts in their lifetime (van Alphen & Vet, 1986; Charnov & Stephens, 1988). For such parasitoids, the major predictions of this model (ranking by profitability, the 'zero-one' rule and independence of inclusion and encounter rate; see Chapter 4) all apply. Charnov and Stephens (1988) have also shown how the optimal diet model can be extended to incorporate the dynamics of egg production where a high rate of production increases the risk of mortality. They also point out that other modifications of the optimal diet model, such as the incorporation of recognition time and non-random host discovery, are probably applicable to parasitoid foraging. Classical diet models assume that the optimal behaviour of the individual is independent of its energy reserves, egg supply and other measures of its condition. In general, this is likely to be untrue (i.e. some parasitoids will be egg-limited rather than time-limited). A new generation of foraging models based on dynamic programming techniques is currently being developed to examine these problems (Iwasa *et al.*, 1984; Mangel, 1987, 1989), but it is still too early to judge the importance of these techniques for studies of parasitoid foraging.

Tests of optimal host-range models are still in their infancy. Van Alphen and his colleagues (reviewed in van Alphen & Vet, 1986), working with *Drosophila* species and their parasitoids, have shown that the rank preference of hosts matches their rank profitability, and that the acceptance of an inferior host type depends on the abundance of superior hosts. One particularly interesting test is Janssen's (1989) field study of optimal diet in *Drosophila* parasitoids. Using a portable binocular microscope, he was able to follow these very small insects in the field and to measure their encounter rate with different host species. The profitability of the different host species were measured in the laboratory. His conclusion was that the encounter rates in the field were so low that even hosts with tiny profitabilities should be parasitized.

We now turn to one particular question of host acceptance that has engendered a considerable amount of research: when should a parasitoid superparasitize a host? As long ago as 1934, Salt discovered that the egg parasitoid *Trichogramma* was able to detect the presence of parasitoid eggs within a host and thereby to avoid superparasitism. Since then, the avoidance of superparasitism has been recognized in a large number of parasitoids (Salt, 1961; van Lenteren, 1981). Indeed, until recently, avoidance of superparasitism was considered to be the rule for parasitoids, and any observed instances of superparasitism were considered to be maladaptive mistakes.

Today, superparasitism is analysed within the same cost–benefit framework as other questions of optimal host choice (van Alphen & Visser, 1990). The benefits of superparasitism are normally considerably less than the benefits of parasitism of a fresh host. This is because the progeny of the superparasitizing individual are frequently at a competitive disadvantage to the elder, immature parasitoids already resident in the host. However, superparasitism will be favoured if these benefits outweigh the costs of superparasitism. These costs are normally *opportunity costs*; the wastage of eggs that could have been placed in unparasitized hosts, or the wastage of time that could have been used searching for fresh hosts. There may also be a direct cost if there is a possibility of *self-superparasitism*, which occurs when a female attacks a host in which she has oviposited previously. Experimental studies of superparasitism are reviewed by van Alphen and Visser (1990).

Clutch size and sex ratio

Once a parasitoid has decided to oviposit on a host, it must decide how many eggs to lay and, in the case of hymenopterous parasitoids, what sex of progeny to produce. This second decision arises since Hymenoptera have a haplodiploid genetic system, under which males develop from unfertilized, haploid eggs and females from fertilized, diploid eggs. A hymenopteran can thus control the sex of an offspring by choosing whether or not to fertilize an egg. Other parasitoids have a normal diploid genetic system where sex is determined by the random segregation of sex chromosomes.

In many groups of parasitoids, the distinction between solitary parasitoids where only a single individual develops per host, and gregarious parasitoids where more than one individual develops per host, is associated with a dichotomy in larval behaviour. The larvae of solitary species have well-developed mandibles and use them to destroy the larvae of other parasitoids (their own species as well as other species). In contrast, the larvae of gregarious species do not possess these 'fighting' mandibles. The distinction between the two types of larvae is important when examining the optimum clutch size of the parent. Consider a solitary parasitoid with fighting larvae ovipositing into a relatively small host. Now suppose that over evolutionary time the host increases in size, or the parasitoid switches to a larger host, so that the parent is selected to rear more than one parasitoid per host. The parent will only be able to achieve this gaol if selection acts simultaneously on the larvae to give up fighting. However, genetic modelling suggests that the conditions for the larvae to abandon fighting are very stringent. The reason for this is that any mutant allele for non-fighting will be at a considerable disadvantage because it will tend to be eliminated by the fighting allele. The solitary, fighting life history has thus the properties of an *absorbing state* — once a population has evolved this life history it is difficult to reverse the process (Godfray, 1987a). The observed clutch size of a parasitoid thus depends on its evolutionary history. A prediction of this argument is that hosts of the same size may be attacked by gregarious parasitoids or by solitary parasitoids that do not exhaust the host's resources. A comparative study of clutch size in the genus *Apanteles* by Le Masurier (1987) (Fig. 11.1) provides evidence in support of this prediction.

The optimal clutch size for a gregarious parasitoid can be studied using techniques first applied to birds (Box 11.1). As an example of a test of clutch size theory, we consider the bethylid wasp *Goniozus*

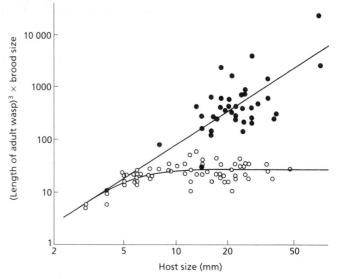

Fig. 11.1 The relationship between a measure of the total volume of parasitoids emerging from a host and the size of the host for 40 gregarious (●) and 64 solitary (○) species of the braconid wasp genus *Apanteles*. From Le Masurier (1987).

Box 11.1 Optimal clutch size in gregarious parasitoids

How many eggs should a parasitoid place on a host? Consider a parasitoid that attacks just a single host in its lifetime. It will be selected to maximize its lifetime reproductive success which, in this case, is synonymous with maximizing its fitness returns from a single clutch. A simple graphical model allows us to calculate the optimum clutch size (Fig. B11.1). The shaded bars represent the per capita fitness of the parasitoid larvae as a function of clutch size. The more parasitoids present in the host, the greater the competition and the lower the fitness of each individual. The parent's overall fitness can be calculated by multiplying the per capita fitness of the offspring by her clutch size (the total height of the bar). In the particular example illustrated here, the clutch size that maximizes fitness returns from the clutch is three. This result is known as the *Lack clutch size* after David Lack (1947) who argued that birds should be selected to maximize fitness returns per clutch.

Note that as each egg is added to the clutch to attain the Lack clutch size, the incremental gain in fitness to the parent is reduced. Thus, for parasitoids which produce more than one clutch, there may be circumstances when initiating a new clutch will be a more profitable strategy than laying the final eggs to achieve the Lack clutch size. For example, consider a parasitoid whose reproductive success is limited by her supply of eggs, and not by opportunities to lay clutches. As the wasp adds eggs to the clutch for decreasing increments in fitness, a time will come when greater fitness returns per egg can be gained by abandoning the present clutch and initiating

continued on p. 273

Box 11.1 *contd*

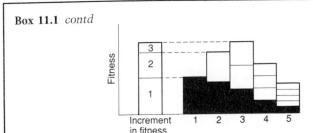

Fig. B11.1 Graphical model of optimum clutch size.

a new clutch. A similar argument can be made for parasitoids whose reproductive success is limited by time and which are selected to maximize their rate of gain of fitness (the argument in this case is essentially identical to the patch residence problem of classical foraging theory, see Chapter 4).

To conclude, parasitoids whose reproductive success is solely determined by their opportunities for producing clutches (i.e., their ability to locate hosts) will produce the Lack clutch size. When other considerations such as limited egg supply or limited time become important, clutch size below the Lack clutch size are expected (Charnov & Skinner, 1984; Parker & Courtney, 1984; Godfray, 1987b).

nephantidis (Hardy *et al.*, 1992). The wasp paralyses a microlepidopteran caterpillar and lays a number of eggs externally on the host. The wasp remains with the host until its offspring pupate, during which time it protects them from superparasitism and probably also from multiparasitism and hyperparasitism. This is a rather unusual behaviour for a parasitoid but suggests that the reproductive success of the insect is strongly limited by its opportunities for producing clutches. This argument indicates (see Box 11.1) that the wasp will produce a clutch of eggs that maximizes her fitness returns from that clutch, the so-called 'Lack clutch size'. Figure 11.2 shows that the wasp matches its clutch size to the size of its host, a qualitative prediction of the theory. Attempting to derive the Lack solution by manipulating clutch size and calculating the survival and fitness of members of artificial clutches showed that clutch size had no effect on larval survival but that large clutches produced small wasps. These wasps, in turn, survived for a shorter period of time and were less fecund than wasps developing in smaller clutches. The calculated Lack clutch size was, however, larger than than laid by the wasp. The chief problem, which has beset all experimental studies of clutch size, is that it has not yet proved possible to measure the fitness consequences of developing in a large clutch in the field: laboratory studies tend to underestimate the disadvantages of small body size.

Since Fisher (1930) showed that frequency-dependent selection should lead to an equal sex ratio, there has been great interest in

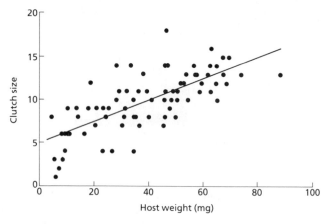

Fig. 11.2 The clutch size produced on hosts of different size by a gregarious bethylid wasp *Goniozus nephantidis*. From Hardy *et al.* (1992).

explaining the occurrence of biased sex ratios in plants and animals. Some of the best examples of biased sex ratios in nature come from parasitoid wasps, and the development and verification of theories to

---

Box 11.2 Parasitoid sex ratios 1: local mate competition

Fisher's (1930) argument for the prevalence of equal sex ratios assumed random mating. Hamilton (1967) first explored the sex ratio effects of frequent mating between siblings. Hamilton assumed that females settle in patches to reproduce in groups of n individuals (foundresses). All individuals produce equal numbers of progeny and these individuals develop and mate prior to dispersal. In the next generation, females form new groups of n foundresses at random.

Hamilton showed that this form of population structure led to a female-biased sex ratio. For a diploid genetic system, the optimal sex ratio is $(n - 1)/2n$ while for a haplodiploid genetic system the sex ratio is slightly more female biased (Taylor & Bulmer, 1980). When $n = 1$, i.e. matings are always between siblings, the expression $(n - 1)/2n$ predicts a sex ratio of zero. This is interpreted to mean that the parent wasp should only produce enough sons to mate all her daughters.

Hamilton called the process that gave rise to a female-biased sex ratio 'local mate competition'. It is now realized that whenever siblings of one sex compete among themselves to a greater extent than siblings of the other sex, the evolutionary stable sex ratio will be biased in favour of the sex that competes the less (Taylor, 1981; Bulmer, 1986). In the case of local mate competition, sons compete for mates whereas there is no competition among daughters. The sex ratio is thus biased towards females. An alternative, and equally valid, explanation of sex ratio biases with local mate competition in terms of hierarchical selection theory is provided by Frank (1985, 1986).

explain these biases has proved to be one of the most successful ventures in evolutionary ecology. Local mate competition and host quality are two of the most important factors that lead to sex ratio biases in parasitic wasps (Boxes 11.2 and 11.3).

A number of workers have carried out experimental tests of local mate competition theory in parasitoids. Werren (1983, see also 1980) studied the parasitoid wasp *Nasonia vitripennis*, which attacks house flies and related Diptera. He allowed different sized groups of parasitoids to attack patches of hosts in the laboratory, and recorded the sex ratio of their progeny. The results (Fig. 11.3) are in good agreement with theory. Werren (1983) also observed that nearly equal sex ratios were found in broods collected in the field from sites such as large carcasses and waste tips where many wasps were likely to be present. In contrast, broods collected from sites such as birds' nests, where foundress number was likely to be small, were female biased.

Box 11.3 Parasitoid sex ratios 2: sex ratio and host size

It has long been known that male parasitoids tend to emerge from small hosts and female parasitoids from large hosts. An evolutionary explanation of this observation was provided by Charnov (1979) who showed that it was a particular example of a wider class of phenomena in which the fitness of males and females were affected to differing degrees by the environment.

For many parasitoid species, longevity and fecundity increase with adult size. Thus both males and females benefit from developing in large hosts. However, there are strong reasons to argue that females gain more from developing in a large host than do males. This is because the major components of female lifetime reproductive success, in particular her fecundity, are more closely correlated to body size than the major components of male reproductive success, in particular mating ability. Charnov showed that under these assumptions female wasps will be selected to place daughters in larger hosts and sons in smaller hosts.

Charnov's host quality hypothesis makes two further predictions. First, there should be a sharp threshold host size below which males are laid and above which females are laid. In practice, however, individual variability between wasps will tend to result in a more fuzzy threshold. Second, the importance of host size is relative, not absolute. Thus, a medium-sized host might attract females when presented in combination with small hosts, but males when presented in combination with large hosts. The prediction that host size is of relative importance is a strong, testable prediction of the theory. However, a wasp species that invariably encounters the same host size distribution in nature might evolve to respond to absolute host size because in this case relative and absolute host size are always the same. A facultative response to relative host size is expected in species that encounter host size distributions that vary temporally or spatially.

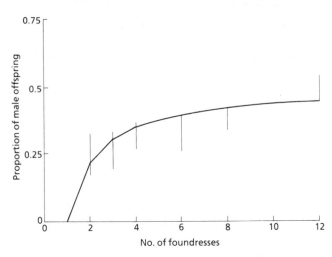

Fig. 11.3 The sex ratio produced by females of the pteromalid wasp *Nasonia vitripennis* in laboratory patches containing different numbers of foundresses. From Werren (1983). The curve is the expectation based on Taylor and Bulmer (1980).

The predictions of Charnov's host quality hypothesis were tested by Charnov *et al.* (1981) with two species of parasitoid that attacked stored product pests. The wasps were presented either with mixtures of large and medium hosts, or with medium and small hosts. Charnov's hypothesis predicts that the *relatively* larger hosts should always have a more female-biased sex ratio. Medium-sized hosts should thus have female-biased sex ratios when presented to the wasp in combination with small hosts, but male-biased sex ratios when presented to the wasp in combination with large hosts. The behaviour of one species confirmed the hypothesis, while the second species failed to produce females on *relatively* larger hosts but did produce females on *absolutely* larger hosts. The behaviour of the second parasitoid may still be adaptive if, in nature, it tends to encounter hosts with an unvarying size distribution (see Box 11.3).

THE DYNAMICS OF HOST–PARASITOID INTERACTIONS

Although there is still no clear consensus on the general importance of parasitoids for the regulation of their insect hosts, several lines of evidence point to their potential importance. In the first place, theoretical studies of host–parasitoid population dynamics clearly indicate that parasitoids have the capability of regulating their host populations. Second, some laboratory experiments have clearly shown host–parasitoid systems persisting under controlled conditions (Fig. 11.4). Third, the success of some classical biological control programmes indicates, at least in agroecosystems, the extent to which parasitoids can depress their host populations, with host and parasitoids then persisting at relatively low densities (Chapter 18). Finally, there is

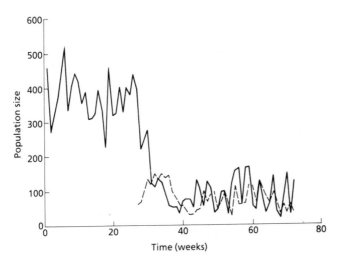

Fig. 11.4 Population dynamics of the bruchid beetle *Callosobruchus chinensis* (——) feeding on black-eyed beans, and its pteromalid parasitoid *Anisopteromalus calandrae* (– – –), in a laboratory system. The environment is patchy with 50 beans, each in an individual container with restricted access to both hosts and parasitoids. The parasitoids were introduced to the arena in week 26 after which the host population was maintained at a reduced equilibrium level. From V.A. Taylor & M.P. Hassell (unpublished data).

evidence from the injudicious use of insecticides that kill more natural enemies than pests; the resultant host resurgence suggests that they were previously maintained at low densities by the natural enemies (Wood, 1971; Conway, 1973).

More difficult to interpret are those field studies that examine the role of parasitism under natural conditions. Several surveys of average parasitism over a number of generations have failed to detect clear signs of parasitism being related to host density (Dempster, 1975, 1983). However, such literature surveys should be viewed with caution. There are, for example, difficulties in detecting density-dependent processes from time series data (Dempster & Pollard, 1986; Hassell, 1986, 1987; Mountford, 1988; Hassell *et al.*, 1989). There is thus a danger that the apparent absence of density-dependence from some life table studies may arise simply through inadequate analysis and/or insufficient data. In addition, there has probably been a bias towards choosing systems for long-term study where the host tends to be consistently abundant. It seems likely that abundant hosts will be resource limited rather than parasitoid regulated, since enemy-regulated species would be expected to persist well below their carrying capacities.

Examination of Fig. 11.4 prompts several questions.
1 What determines the degree to which a parasitoid population can depress average host population levels?
2 What factors are promoting the persistence of the interacting populations?

3 Can parasitoids under natural conditions play roles similar to those in Fig. 11.4 in reducing host abundance and regulating populations? With such questions in mind, we turn first to the predictions of some theoretical models with a view of establishing how, at least in principle, the different components of parasitism can influence host populations.

We begin with some contrasting approaches to modelling the population dynamics of host–parasitoid systems, all of which consider specialist parasitoids that attack only a single host species. The different approaches embrace dynamics at both *local* and *regional* scales. Taylor (1988) gives a succint summary of the differences between the two:

> 'Local populations ... are defined as the units within which occur reproduction, population regulation, and interactions such as predation, and within which the movements of most individuals are confined; there may well be further spatial structure [e.g. 'patches'] within these. Regional populations [or *metapopulations* as they are often called] ... are collections of local populations or cells linked by dispersal'.

There is no hard and fast boundary between the two scales, which may indeed overlap; but the distinction is a useful one in emphasizing one scale (local) over which a population forages for oviposition sites and in which various processes affect the stability of the local population, and one (regional) where there is dispersal between local populations, whose primary dynamical effect is regional persistence rather than local stability (Taylor, 1988, 1990).

Local population dynamics

We consider two approaches to investigating the dynamics of local populations of hosts and parasitoids. The first assumes that the populations have separate, synchronous generations (as in many temperate host–parasitoid systems) and consist of coupled difference equations in discrete time. These models are ultimately derived from the Nicholson and Bailey (1935) model. The second approach assumes that hosts and parasitoids reproduce continuously, and is derived from the familiar Lotka–Volterra predator–prey equations (Chapter 3), a coupled system of differential equations in continuous time.

DISCRETE TIME MODELS OF LOCAL POPULATIONS

A general framework for coupled host–parasitoid interactions with discrete generations is given in Box 11.4, together with a range of

---

Box 11.4 A general framework for coupled host–parasitoid interactions with discrete generations

Coupled host–parasitoid interactions with discrete generations are often represented by the following general framework:

continued on p. 279

Box 11.4 *contd*

$$N_{t+1} = \lambda g \, (fN_t)N_t f(N_t, P_t)$$
$$P_{t+1} = cN_t[1 - f(N_t, P_t)]$$

Here N and P are the host and parasitoid populations, respectively, within successive generations, t and $t + 1$; $\lambda g(fN_t)$ is the per capita net rate of increase of the host population (which is density-dependent when $g < 1$); c is the average number of adult female parasitoids emerging from each parasitized hosts (c depends therefore on the average number of eggs laid per host parasitized, the survival of these progeny, and their sex ratio); and the function $f(\cdot)$ defines the fraction of hosts that survive parasitism.

Some specific forms for the function $f(\cdot)$ are shown below.

$f(\cdot)$	Reference	Local stability properties $(g = c = 1)$
A $\exp(-aP_t)$: random host encounters, constant searching efficiency (a) where a is the proportion of hosts encountered per parasitoid per unit time	Nicholson & Bailey (1935)	Unstable
B $exp(-QP_t^{1-m})$: random host encounters, searching efficiency declines with parasitoid density $(a = QP_t^{-m})$ due to interference	Hassell & Varley (1969), Perry (1987)	Stable if $$1 > m > 1 - \left(\frac{\lambda-1}{\lambda \ln \lambda}\right)$$
C $\exp\left(-\dfrac{aP_t}{1 + aT_hN_t}\right)$: random host encounters, searching efficiency declines with host density due to effects of handling time (T_h)	Royama (1971) Rogers (1972)	Unstable
D $\left(1 + \dfrac{aP_t}{k}\right)^{-k}$: clumped distribution of attacks per host described by negative binomial distribution with clumping parameter k	May (1978)	Stable if $k < 1$
E $\displaystyle\sum_{i=1}^{n} [\alpha_i \exp(-a\beta_i P_t)]$: hosts and parasitoids interact over n patches with a fraction α_i hosts and β_i parasitoids in the ith patch. Random host encounters *within* patches	Hassell & May (1973, 1974)	Stable if $$\lambda \sum_{i=1}^{n} [\alpha_i(a\beta_i P^\star)\exp(-a\beta_i P^\star)]$$ $$< \frac{\lambda-1}{\lambda}$$

specific forms for the function, f, describing parasitism. The average population densities in these models are set by the balance between: (i) the *net* host rate increase; and (ii) the *overall* efficiency of the parasitoids. The net host rate of increase is the realized fecundity per adult reduced by the average of all mortalities other than that caused by the parasitoids. The overall parasitoid efficiency is the per capita searching efficiency (a in Box 11.4), reduced by any male bias in the sex ratio, and by any mortality affecting the female parasitoid progeny prior to their search for hosts in the next generation.

The original model of Nicholson and Bailey (A in Box 11.4) predicts an unstable interaction with expanding oscillations in host and parasitoid population densities. While this has been observed from a few simple laboratory experiments (see, for example, Burnett, 1958; Huffaker, 1958), it runs counter to the experience from other laboratory systems (e.g. Fig. 11.4), from the results of successful classical biological control programmes and from long-term field studies of natural interactions where hosts and parasitoids have often been observed to coexist over several generations without expanding population fluctuations. The search for mechanisms that might account for the persistence of predatory–prey interactions in general has prompted many of the recent developments of the basic Nicholson–Bailey framework. Several of these (e.g. B and C in Box 11.4) have retained the original Nicholson–Bailey assumption that parasitoids exploit their hosts at random (i.e. all host individuals are equally susceptible to parasitism), but have allowed for other general features of parasitism (e.g. parasitoid interference and handling time, respectively). Of these, interference is one mechanism which acts in a density-dependent way on the parasitoid population and could, at least in principle, strongly promote stability (Hassell, 1978).

Entries D and E in Box 11.4 differ in allowing non-random exploitation of hosts by the parasitoids. Hosts are thus assumed to vary in their susceptibility to parasitism, which may arise in several ways: for example, from heterogeneity in the spatial distributions of the interacting populations (see, for example, Hassell & May, 1973, 1974), from temporal asynchrony (Griffiths, 1969) or from phenotypic variability in the susceptibility of hosts to parasitism (Hassell & Anderson, 1984; Godfray & Hassell, 1990). The central conclusions from these models are straightforward. With random distributions of parasitism throughout the host population, the system collapses back to the unstable Nicholson–Bailey model that exhibits expanding oscillations around an unstable equilibrium. If the distribution of parasitism from patch to patch is sufficiently aggregated, however, stable populations are easily achieved. The essential ingredient is that some host individuals (by virtue of where, when or what they are) are more prone to parasitism than others. In the extreme case, some hosts will be completely protected (e.g. within a refuge). Such absolute refuges can strongly promote population regulation as long as the proportion of hosts so protected is not too small (too few hosts escape parasitism) or too large (too few hosts parasitized for regulation) (Hassell, 1978; Hassell & May, 1988).

Models involving heterogeneity in the distribution of parasitism between hosts have developed in quite different directions. Some have included specific details of distributions of host and parasitoids per patch (e.g. E in Box 11.4), while others have striven for simplicity, aiming to capture only the main dynamical properties of their more complex cousins. An example of such a simple model is that of May (1978) (D in Box 11.4), where the distribution of parasitoid attacks amongst the host population is assumed to follow a negative binomial distribution rather than a Poisson distribution. The precise relationship of this model to more explicit models of patchily distributed hosts and aggregating parasitoids has been discussed by Chesson and Murdoch (1986) and Hassell *et al.* (1991). Its value lies in providing a simple, albeit crude, means of representing the non-random distributions of parasitism that can arise in many different ways within a host population.

The theoretical interest in the dynamics of interactions in patchy environments has prompted many field workers to record the variation in levels of parasitism from patch to patch, in relation to host density per patch. Sufficient of these have now accumulated to fuel a number of recent reviews cataloguing the incidence of different patterns (Lessells, 1985; Stiling, 1987; Walde & Murdoch, 1988). These surveys emphasize that while there are many examples of spatial patterns of parasitism that are *directly* density-dependent, there are almost as many showing quite the opposite trend (*inversely* density-dependent), and even more showing no relationship at all (Fig. 11.5).

A popular interpretation of these data, guided by the earlier theoretical literature (see, for example, Hassell & May, 1973, 1974; Murdoch & Oaten, 1975), has been that only the direct density-dependent patterns promote the stability of the interacting populations. This, however, is not the true picture. Both inverse density-dependent patterns (Hassell, 1984; Walde & Murdoch, 1988), and variation in parasitism that is *independent* of host density (Chesson & Murdoch, 1986; Hassell & May, 1988; Pacala *et al.*, 1990; Hassell *et al.*, 1991), can be just as important as the direct density-dependent patterns to population regulation. This has obvious implications for the design of field studies on host–parasitoid systems, because no longer can the effects of such heterogeneity be inferred simply from the shape of the relationships between percentage parasitism and local host density.

We are left, therefore, with a picture where any non-random variability in levels of parasitism within a host population can in principle contribute to population regulation. This has recently been discussed in detail by Pacala *et al.* (1990), Hassell and Pacala (1990), Hassell *et al.* (1991) and Pacala and Hassell (1991). They have shown that the dynamical effects of *all* forms of heterogeneity illustrated in Fig. 11.5 can be assessed by the same criterion, which they call the '$CV^2 > 1$ rule'.

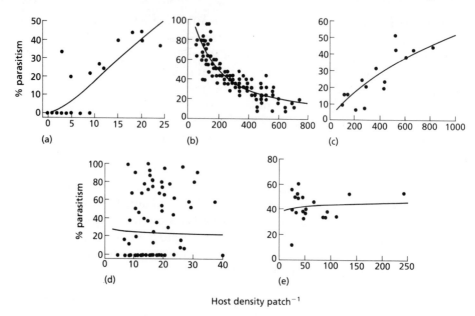

Fig. 11.5 Examples of field studies showing percentage parasitism in relation to host density per patch. (a) Direct density-dependent parasitism of cabbage rootfly (*Delia radicum*) by *Trybliographa rapae* (Jones & Hassell, 1988) ($CV^2 = 1.52$); (b) inverse density-dependent parasitism of gypsy moth (*Lymantria dispar*) eggs by *Ooencyrtus kuwanai* (Brown & Cameron, 1979) ($CV^2 = 0.37$); (c) direct density-dependent parasitism of the scale insect, *Fiorinia externa*, by *Aspidiotiphagus citrinus* (McClure, 1977) ($CV^2 = 0.34$); (d) density-independent parasitism of the gall midge, *Rhopalomyia california*, by *Tetrasticus* sp. (Ehler, 1986) ($CV^2 = 7.33$); (e) density-dependent parasitism of the olive scale (*Parlatoria oleae*) by *Coccophagoides utilis* (Murdoch et al., 1984) ($CV^2 = 0.05$). From Pacala & Hassell (1991) in which techniques for fitting curves are described.

This states that the host and parasitoid populations will be stable (in the sense that overall population densities remain roughly steady) if the coefficient of variation squared (CV^2 = variance/mean^2) of the density of searching parasitoids in the vicinity of each host exceeds approximately unity. Since the density of searching parasitoids in each patch can either be a random variable independent of local host density or a deterministic function of local host density, the CV^2 may be partitioned into two components: *host density-independent heterogeneity* (HDI) and *host density-dependent heterogeneity* (HDD), respectively. Comparable terms have been coined by Chesson and Murdoch (1986) who labelled models with randomly distributed parasitoids as *pure error models* and those with parasitoids responding to host density in a deterministic way as *pure regression models*. In an analysis of 65 data sets on the distribution of parasitism in the field, Hassell and Pacala (1990) found that in 18 $CV^2 > 1$ and in 14 of these HDI was the primary source of the stabilizing heterogeneity. Figure 11.5d provides a good example of this, showing density-independent variation in parasitism (HDI) to be much more important than density-dependent variation (HDD) and sufficient to cause a very high value for the overall heterogeneity of risk from parasitism ($CV^2 = 7.33$). Hence, although the data in Fig. 11.5d appear erratic, they actually contain more evidence of factors that could stabilize dynamics than do the examples in Fig. 11.5a–c where there is covariance between parasitism and local host density. Thus, contrary to the popular view, this analysis suggests that *density-independent* spatial patterns of parasitism may often contain more stabilizing heterogeneity than *density-dependent* patterns.

CONTINUOUS TIME MODELS OF LOCAL POPULATIONS

The models discussed so far are appropriate to specialist parasitoids with discrete generations that are synchronized with the generations of their host. A quite different framework is needed when hosts and parasitoids are no longer constrained in this way, and have several overlapping generations per year and unequal generation times. This is often found in the moist tropics where hosts and parasitoids may be active throughout the year, without the need for diapause during an unfavourable season.

The effect of natural enemies in continuous time is normally studied using variants of the Lotka–Volterra equations (Chapter 3). The original Lotka–Volterra model is, however, not very suitable for representing host–parasitoid interactions because of the importance of developmental time-lags. In the majority of hosts and parasitoids (and unlike many predator–prey systems), the length of the adult reproductive period is relatively small compared with the time taken to grow from egg to adult. Since these developmental lags can lead to qualitatively different dynamic behaviour to that found in models with no time-lags, it is particularly important that they are incorporated into the model structure (Murdoch *et al.*, 1987). Box 11.5 describes a simple model of a continuous time host–parasitoid interaction based on such a time-lagged Lotka–Volterra model.

One of the population processes found in such continuous time models that is not observed in discrete time models relates to the observation that generation cycles that may be caused and maintained by parasitoids and other natural enemies occur in a variety of tropical insects (Godfray & Hassell, 1989). Exploration of models of the type described in Box 11.5 shows that two conditions are required to obtain such cycles with periods of approximately one host generation (Fig. 11.6) (Godfray & Hassell, 1987, 1989; Godfray & Chan, 1990; Gordon *et al.*, 1991). First, the developmental lag in the parasitoid life history must be approximately 0.5 (or 1.5, 2.5, etc.) times the developmental lag in the host life history. Second, the reproductive lifespan of the adult host and parasitoid has to be relatively short compared with the developmental lag in order to prevent parasitoids reaching maturity over an extended period and thus tending to counteract the tendency for cycles (Godfray & Waage, 1991).

Box 11.5 A simple model of a host−parasitoid interaction in continuous time

Let us consider the following simple example (Godfray & Chan, 1990), and suppose that a population of P parasitoids attacks N hosts only in the adult stage. The instantaneous risk of parasitism is a function of both host and parasitoid density, $f(N, P)$. Hosts reproduce at a rate λ and suffer density-independent mortality at a rate μ_N; parasitoid adults suffer density-independent mortality at a rate μ_P. The time taken for a host to grow from oviposition to adult is τ_N and for a parasitoid τ_P. Finally, we assume no juvenile mortality in hosts or parasitoids.

The population dynamics of such an interaction are described by the equations

$$\frac{dN_t}{dt} = \lambda N_{t-\tau_N} - N_t f(N_t, P_t) - N_t \mu_N$$

$$\frac{dP_t}{dt} = N_{t-\tau_P} f(N_{t-\tau_P}, P_{t-\tau_P}) - P_t \mu_P$$

The three terms in the host equation are respectively, the recruitment to the adult host population which depends on births τ_N ago, the mortality of adults caused by parasitism and the instantaneous mortality caused by other, density-independent causes. The two terms in the parasitoid equation are first, the recruitment to adult parasitoids which depends on the hosts parasitized τ_P ago and, second, the instantaneous density-independent mortality affecting the adult parasitoids.

The only density-dependence in the system is introduced through the parasitism term $f(N,P)$. A simple result occurs if parasitism depends only on the density of parasitoids and $f(N, P) = aP^m$ where a is a measure of the searching efficiency and m is a measure of the strength of density-dependence acting on the parasitoids (see Box 11.4). When $m = 0$, the system is unstable, as for the equivalent discrete model. As m increases (i.e. as the overall parasitoid efficiency declines with increasing parasitoid numbers) the system stabilizes though the value at which this happens depends on τ_P and τ_N.

The dynamics of a more realistic model with parasitism affecting an immature stage is qualitatively the same as this simple model (Godfray & Hassell, 1989). Murdoch *et al.* (1987) analyse a related model based on the life cycle of a scale insect and its parasitoid.

Age-structured population models can reveal other interesting properties that are not apparent in simpler models. For instance, Godfray and Waage (1990), in modelling the interaction between the mango mealy bug *Rastrococcus invadens* (Hemiptera: Pseudococcidae) and two of its encyrtid parasitoids in West Africa, found that host equilibrium densities decreased as the severity of the density-dependence in parasitism increased. This is the opposite trend to that expected

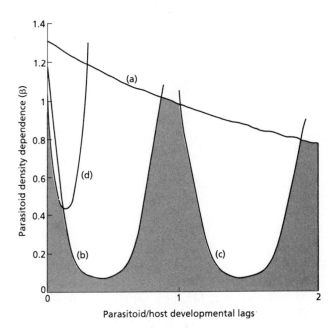

Fig. 11.6 Stability boundaries for the continuous time host–parasitoid model. The lines represent roots of the characteristic equation for different values of the strength of the parasitoid density-dependence (y axis) and the ratio of the parasitoid to host developmental lags (x axis). Stability is ensured when all roots have negative real parts (shaded area). Each root is associated with cycles of a characteristic length. (a) Lotka–Volterra cycles of period of about three or four host generations; (b) and (c) cycles of period one host generation; and (d) cycles with period of one-half of a host generation. Case (d) is probably not of biological significance.

from simple discrete-generation, host–parasitoid models. A further example is illustrated in Fig. 11.7 which shows the long-term erratic fluctuations observed by Utida (1950) in a laboratory population of a bruchid beetle (*Callosobruchus maculatus*) and its parasitoid (*Anisopteromalus calandrae*) cultured under controlled conditions. One interpretation for such irregular patterns of abundance is that they arise from environmental stochasticity (e.g. fluctuations in microclimate within the cultures). However, they could also be determined by the internal dynamics of the interaction, as illustrated recently by Bellows and Hassell (1988). They developed a detailed, age-structured host–parasitoid model based directly on a laboratory system using *C. maculatus* and a parasitic wasp (*Lariophagus distinguendus*) that is a close relative of *A. calandrae*. The predicted population dynamics (Fig. 11.7b) appear to be chaotic (see Chapter 19), which suggests that the irregular fluctuations observed by Utida may also result from deterministic chaos rather than from environmental stochasticity.

Regional population dynamics

The current interest in regional or metapopulations stems from the possibilities of persistence at this scale even if at the local scale the

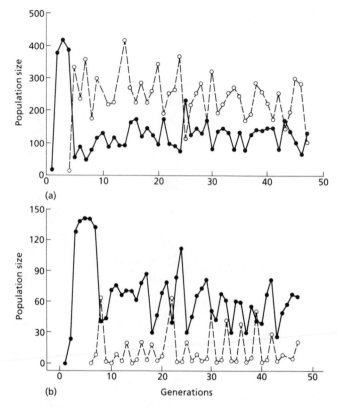

(a)

(b)

Generations

Fig. 11.7 (a) The observed population dynamics of *Callosobruchus chinensis* (—●—)
and its pteromalid parasitoid *Anisopteromalus calandrae* (—○—) from the
laboratory system of Utida (1950). (b) The apparently chaotic predicted dynamics for
the interaction between *C. chinensis* (—●—) and the pteromalid parasitoid
Lariophagus distinguendus (—○—) from the simulation model of Bellows and
Hassell (1988).

populations exhibit unstable fluctuations (Reeve, 1988, 1990). The
essential ingredient for this to be possible is some degree of dispersal
of individuals between local populations, such that the unstable population dynamics at the local scale are always out of phase and so
allow the persistence of the population at the regional scale.

Regional host–parasitoid models are broadly of two kinds (see
Taylor, 1988, 1990, for a review): (i) so-called 'cell occupancy' models
with no local scale dynamics; and (ii) models which explicitly describe
both local scale dynamics *and* movement between local patches. In
cell occupancy models local patches pass in sequence through a number
of states — from being empty, to being colonized by hosts, to being
found by parasitoids, to becoming empty again following the extinction
of the hosts by the parasitoids (see, for example, Maynard Smith, 1974;
Hastings, 1977; Zeigler, 1977; Crowley, 1981). Such systems are often
modelled in continuous time, and dispersal amongst patches can lead
to asynchronies in the state of the different patches. This, in turn, has
the potential to promote the persistence of the system as a whole.

In the second class of models, populations can rise and fall within
patches (or groups of patches, depending upon the units of local scale)
due to their internal dynamics, as well as by dispersal (see, for example,
Comins & Blatt, 1974; Zeigler, 1977; Crowley, 1981; Reeve, 1988,
1990). Once again, however, the general conclusion is that regional
persistence is promoted by a combination of interpatch asynchrony
and dispersal between the patches at the local scale.

Spatial models of host—parasitoid interactions have generally assumed that all hosts and all parasitoids in each generation redistribute
themselves amongst all available patches. This 'blanket' or global
redistribution is different from the linking of patches by some dispersing
or migrating individuals. These models, therefore, despite their spatial
structure, portray local population dynamics. Whilst global redistribution is true of many systems, others (e.g. scale insects and their
parasitoids) typically show less complete mixing, in which some of the
hosts and parasitoids tend to remain within the patch from which they
originated. Such linking together of patches by *incomplete* dispersal
introduces the possibility of different dynamics within local patches
compared to the regional population as a whole, and models of such
systems serve as a clear link between local and regional population
models. Reeve (1988, 1990) has demonstrated this by simulating host—
parasitoid interactions in a patchy environment, with specified fractions
of hosts and parasitoids leaving patches to enter a pool for subsequent
even redistribution amongst the patches (see also Hassell & May,
1988), with parasitism within patches defined by D in Box 11.4 and
with random parasitoid attack rates and host growth rates designed to
mimic the effects of environmental variability in the system. Reeve
found that locally unstable systems could sometimes persist indefinitely, provided migration rates were low and some environmental
variability present. Indeed, if the dispersal is by diffusion to nearest
neighbours, such systems can persist, albeit chaotically, even without
any variability other than in the initial distributions of hosts and
parasitoids (Hassell *et al.*, 1991).

Parasitoids in multispecies assemblages

Traditionally, the emphasis in population ecology has centred on the
dynamics of single-species and two-species systems, and little attention
has been paid to the more complex webs of multispecies interactions
of which hosts and their parasitoids are an integral part. One of the
major problems in contemporary population biology is to determine
the degree to which conclusions from simple population models can
be applied to these more complex systems. This is another facet of the
general question of scale in population biology.

Most hosts are attacked by a number of different parasitoid species
and most parasitoids have a variety of different hosts. In addition,
predators, competitors, food-plants and pathogens are all likely to

impinge on the dynamics of any given host–parasitoid system. There are several ways of studying the construction and behaviour of complex, parasitoid-dominated communities. One approach is to study the properties of naturally occurring parasitoid communities and try to deduce patterns and assembly rules from a comparative study of the subset of possible communities that are actually found. Another approach is to extend the studies of one- or two-species interactions to systems containing a larger number of species.

The majority of parasitoid community studies have concentrated on the parasitoids attacking a single host species: that is, they have abstracted one host and its parasitoids from the community within which it is embedded. Though interesting information can be obtained from this approach, e.g. the analysis of the host characteristics that lead to a large parasitoid complex (Hawkins & Lawton, 1987; Hawkins, 1988, 1990), very few studies have attempted to construct food-webs describing the interactions between guilds of hosts and parasitoids at a single locality (a notable exception is Askew, 1961). Again, though it is possible to obtain some insight into parasitoid community structure from existing data (see, for example, Askew & Shaw, 1985; Pimm *et al.*, 1991), our present knowledge of parasitoid community structure is too rudimentary to answer the questions posed at the beginning of this section.

In contrast, a growing number of studies have explored the effects of adding additional species into simple two-species host–parasitoid models. These studies include the following.

1 Two parasitoid species attacking the same host species (May & Hassell, 1981; Kakehashi *et al.*, 1984).

2 Host, parasitoids and hyperparasitoids (Beddington & Hammond, 1977; May & Hassell, 1981).

3 Competing host species sharing the same parasitoid species (Roughgarden & Feldman, 1975; Comins & Hassell, 1976).

4 Hosts attacked by both specialist and generalist natural enemies (Hassell & May, 1986).

5 Host, parasitoids and pathogens (Hochberg *et al.* 1989).

Interestingly, in several of these systems the dynamics are not merely the expected blend of the component two-species interactions. Rather, the additional non-linearities introduced by adding extra species may lead to a wide range of complex dynamical properties. Space precludes a detailed analysis of all this work but, as representative studies, we discuss two examples in some detail.

MIXING GENERALIST AND SPECIALIST NATURAL ENEMIES

Most insects are attacked by several species of parasitoids, including both specialists and generalists. Unlike the specialist dynamics described above, the dynamics of generalist parasitoids are characterized by their population size remaining relatively constant compared to the fluctuations in the abundance of a particular host species. A simple framework for the dynamics of a generalist parasitoid is outlined in Box 11.6. The generalists show a numerical response to host density, perhaps by 'switching' from feeding elsewhere or on other host species (Murdoch, 1969). Such numerical responses by generalist predators have been observed to take the simple form shown in Fig. 11.8a. Combining these with a typical Type II functional response yields relationships in which mortality is density-dependent at low host densities, but inversely so at high host densities.

In short, because they are uncoupled from their hosts, generalists can cause direct density-dependent mortality and this may be sufficient to regulate a host population (Fig. 11.9). There are none of the time delays inherent in coupled interactions, and hence no tendency to produce oscillations in population density. Such generalists are likely to be common in the real world. Although they feed on a range of host species, the abundance of a given host will sometimes have an effect on the parasitoid's reproductive success. Figure 11.8b shows an example where a clear indication of a time-lag introduced by prey density

---

Box 11.6 Models for a generalist and specialist attacking a common host (Hassell & May, 1986)

Let us assume a numerical response by a generalist in which the numbers searching at time t (G_t) is a simple monotonic function of host density (see, for example, Southwood & Comins, 1976): $G_t = h[1 - \exp(-N_t/b)]$, where h is the saturation number of parasitoids and b determines the typical prey density at which this maximum is approached. The numbers attacked by the G_t generalists is derived from entry D in Box 11.4 to give the population model: $N_{t+1} = \lambda N_t g(N_t)$, where $g = [1 + a_g G_t/k]^{-k}$. Locally stable equilibria can occur due to the density-dependence in this model (Fig. 11.9) and are made more likely by small handling times, high searching efficiency, a_g, a 'strong' numerical response (large h, small b) and low net host rates of increase, λ.

We now add to this picture by including a specialist, P, as well as the generalist. However, with more than one mortality acting, we need to be explicit on the relative timing of generalist and specialist in the host's life cycle. With the specialists acting before the generalists, we have:

$$N_{t+1} = \lambda N_t f(P_t)\, g(N_t f)$$
$$P_{t+1} = cN_t[1 - f(P_t)]$$

where the function $f(\cdot)$ is again given by entry D in Box 11.4. But with the generalists acting first, the model becomes:

$$N_{t+1} = \lambda N_t f(P_t) g(N_t)$$
$$P_{t+1} = cN_t g(N_t)\, [1 - f(P_t)].$$

The properties of these models are described in the text.

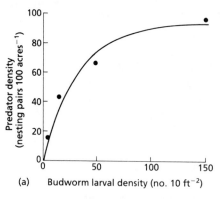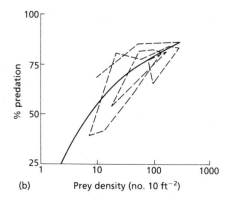

(a) Budworm larval density (no. 10 ft^{-2})

(b) Prey density (no. 10 ft^{-2})

Fig. 11.8 (a) The numerical response of the bay-breasted warbler (*Dendroica fusca*) in relation to density of third instar larvae of the spruce budworm (*Choristoneura fumiferana*) of foliage. (Data from Mook, 1963.) (b) Density-dependent predation of winter moth (*Operophtera brumata*) pupal density per generation. The dashed line serially joins the points for the 18 generations, and emphasizes the slight time-lag in the predators' response to prey density. From Hassell & Anderson (1989). Data and fitted curve from Varley & Gradwell (1968).

affecting natural enemy reproduction is superimposed on the overall density-dependent relationship. Such examples fall into an intermediate category between the generalists discussed here and the specialists of the previous sections.

A population model combining such generalists with the specialists of previous sections is summarized in Box 11.6. The following principal conclusions emerge.

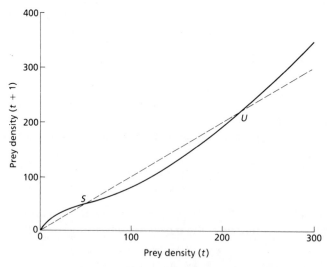

Prey density (t)

Fig. 11.9 The map of prey populations in successive generations, t and $t + 1$, obtained from the model in Box 11.6 with $b = 50$, $h = 105$, $a = 0.02$, $\lambda = 3$, $k = 100$ and $T_h = 0.2$. Stable and unstable equilibria are denoted by S and U, respectively.

1 A specialist can co-exist more easily in such a system if it acts before the generalist in the host's life cycle. This is simply due to the larger pool of hosts available. Conversely, if the host rate of increase is too low, or if the generalists are too efficient relative to the specialists, the specialist will be unable to invade and a persistent three-species interaction is impossible.

2 If the specialist acts before generalists, and if the generalists cause strong over-compensating density-dependence, there is the risk of higher host populations than would exist with the generalist acting on its own.

3 With the three species coexisting, there can either be only one equilibrium point (or stable cycle), or there may be more complex situations with a variety of possible alternative stable states. Thus, there may be two alternative persistent states, one with only the generalist and the other with all three species present, or two alternative three-species states in which the interaction may 'flip' between high and low levels if sufficiently perturbed.

4 A stable system with all three species can exist when one or both of the two-species interactions alone would be unstable.

In short, unlike the separate pairwise interactions each of which have rather straightforward dynamics, the combined three-species system presents a wider, and in some respects unexpected, range of properties.

HOST–PARASITOID–PATHOGEN DYNAMICS

Consider the case of a host insect with discrete generations attacked by a specific, synchronized parasitoid and also by a pathogen (virus, bacterium or protozoan). The pathogen is directly transmitted between hosts by a free-living infective stage that is released into the environment on the death of the infected host. A model of this system combines the continuous production of infective pathogen stages and the continuous mortality of infected hosts within a single host generation, with the discrete, generation-to-generation processes of host reproduction and attack by parasitoids (Hochberg *et al.*, 1989). A range of conditions define whether the pathogen and parasitoid can coexist on the same host population, or whether only one or the other can persist. Of particular interest is the wide range of dynamics than can occur on moving from two to three species; the addition of the third species (parasitoid or pathogen) may either stabilize the existing two-species interaction, or it may induce oscillatory, or even chaotic fluctuations, depending on the precise combination of parameter values (Fig. 11.10). The addition of an extra mortality factor may, or may not, further depress host abundance below the level attained in the absence of either enemy. This sounds a warning note for the use of such mixed natural enemy complexes in biological control programmes (see Chapter 18).

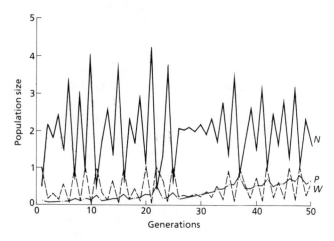

Fig. 11.10 Chaotic dynamics from an interaction of hosts (*N*), pathogens (*W*) and parasitoids (*P*) predicted by the multispecies model of Hochberg *et al.* (1989).

CONCLUSIONS

Recent advances in understanding how natural selection affects animal behaviour have given rise to the fields of sociobiology, evolutionary ecology and behavioural ecology. The analysis of parasitoid reproductive strategies has proved amenable to this behavioural/evolutionary approach because of a more direct link between behaviour and fitness than is typical of most applications of optimality theory. We expect new insights into parasitoid biology to come from this approach and for experimental studies using parasitoids to be increasingly used in the validation of general theory in behavioural and evolutionary ecology.

The role of parasitoids in the regulation of natural insect populations remains uncertain. Much of the evidence for the importance of parasitoids as regulatory factors comes from the analysis of theoretical models, together with some circumstantial evidence from biological control programmes and laboratory studies. Many existing long-term population studies in the field, instead of resolving the issue, have served mainly to highlight the shortcomings of our techniques of experimental manipulation and analysis, and to illustrate the inherent problems of working on natural communities. The testing of theoretical ideas using direct observations and experiments on field population systems is still in its infancy. It is, however, a crucially important step to take, and one on which the future understanding of population dynamics much depends.

12: Bloodsucking Arthropods

CHRISTOPHER DYE

INTRODUCTION

Blood is a rich cocktail of amino acids piped to within reach of the vertebrate skin, where a wide variety of parasitic arthropods make all or part of their living by chewing, sawing, lapping or sucking to obtain blood meals. Most importantly, the amino acids are the building blocks of egg proteins, though glycogen can be produced to supplement the fuel supply. The drawing off and refining of blood are hazardous processes, however. The cocktail may be spiked with antibodies, platelets, coagulants and other parasites, and it constitutes one of the tissues of a sophisticatedly protective host.

Red blood is the special resource in this chapter, but the arthropods that exploit it are a heterogeneous group (Mattingly, 1969; Price, 1980; Balashov, 1984). They have travelled along a variety of evolutionary highways (see, for example, Waage, 1979b on insects) and exhibit a broad spectrum of life histories. While tsetse flies (*Glossina* spp.) produce a single egg following a blood meal, hard ticks (Ixodidae) may produce up to 20 000. A female sandfly (*Phlebotomus* spp.) can expect to live a week or two, whereas soft ticks (Argasidae) can survive several years (15 years may be a laboratory record!). Female *Tunga penetrans* fleas spend their entire adult lives in an excavation under the surface of human skin, while blackflies (*Simulium* spp.) travel hundreds of kilometres on winds of the African Inter-Tropical Convergence Zone. Biting midges (*Culicoides* spp.) take around 0.2 mg blood at a sitting, but hard ticks can consume 10 000 times as much.

Some arthropods feed only on blood for their entire lives (e.g. the hemimetabolous ticks and triatomine bugs). Others feed on blood as adults, but spend their larval lives as predators or detritivores (e.g. fleas (Siphonaptera) and flies like mosquitoes, midges and horseflies).

Quite clearly, blood adds up to *reproductive value* (Fisher, 1930) in many ways. Over these diverse arthropod life histories there are no clear patterns of fecundity, longevity, dispersal or body size associated with the blood-feeding habit. Such difficulties in organizing animal life histories are commonplace, because there is more to population biology than can be embodied in simple classifications like the $r-K$ continuum (Begon *et al.*, 1986).

Thus, rather than being a unifying review, this chapter is structured around four topical questions. Which arthropod populations are regulated by their supply of blood? What effect do bloodsucking arthropods have on their host populations? What impact do vertebrate pathogens have on their arthropod vectors? How do vector-borne diseases influence host population dynamics? The final section of the chapter contains some examples which show how evolutionary inferences can be drawn from comparative ecology. The conclusion is that convincing patterns are only likely to emerge from comparisons made within narrowly circumscribed taxonomic limits. Whilst a single lens can be held revealingly to variation within species complexes or among geographic variants, there is no focal point for sensible comparisons of the blood-sucking arthropods as a whole.

REGULATION OF
BLOODSUCKING ARTHROPOD POPULATIONS

For ectoparasites which live entirely or mostly on their hosts (e.g. lice and the wingless Nycteribiid bat parasites) it is clear that an increase in the *number* of hosts would lead to an increase in the equilibrium population size of arthropods.

However, most blood-feeding arthropods spend a part of their lives away from the host. Immature *Aedes* mosquitoes are typically (though not exclusively) found in small bodies of water, such as treeholes and water storage jars. The have received special attention from research workers who have been drawn to their experimentally tractable habitats. Consequently, the regulatory processes operating during the pre-adult lives of, for example, *Ae. triseriatus* and *Ae. sierrensis* in North America, and *Ae. aegypti* in tropical Asia and Africa, are perhaps the best understood of all blood-feeding arthropods. Field studies have shown that both larval mortality and adult fecundity can change with larval density. Intraspecific competition for food is probably the main mechanism. Significant interspecific interactions, competitive and predatory, have been identified, but we do not yet know whether these are density-dependent (see the collection of papers in Lounibos *et al.*, 1985).

How commonly blood is a limiting factor is uncertain, even for those species which feed on blood alone. In order to test whether the amount of blood, or the number of resting sites in cracks in walls, was responsible for limiting the numbers of kissing bugs *Triatoma infestans* in a Brazilian house, Schofield and Marsden (1982) eliminated half the resting sites by plastering half the wall area to see if the number of bugs was reduced, and if so, by what fraction. The resident population of more than 2000 individuals was undiminished by the treatment, suggesting that some other limiting process was acting before space. There is now a good deal of circumstantial evidence to suggest that, for these triatomine bugs, the limiting factor will often be blood (Schofield, 1985).

Table 12.1 Feeding success of tsetse (*Glossina* spp.), and the irritability of an ox, at different fly densities. From Vale (1977)

	Bait	
	Ox only	Ox plus the 'odour' of six other oxen
ESTIMATED NO. OF FLIES ATTACKING		
G. morsitans		
males	13	31
females	33	95
G. pallidipes		
males	131	596
females	235	1134
ESTIMATED FRACTION FEEDING		
G. morsitans		
males	0.32	0.23
females	0.39	0.3
G. pallidipes		
males	0.51	0.39
females	0.53	0.31
OX TAIL FLICKS/REPLICATE	477	1361

More direct evidence that blood is limiting to populations of free flying insects has come from some exquisite field experiments with tsetse flies *Glossina morsitans* and horseflies. In Zimbabwe, Vale (1977) manipulated the number of tsetse approaching an ox by tethering it alone, or in the presence of 'ox odour' emerging from a tube which led to an underground chamber concealing six more oxen. While there was a 4.5-fold increase in the number of tsetse attacking in the presence of enhanced ox odour, this was accompanied by a 30% reduction in the feeding success of each individual tsetse (Table 12.1). A change in the frequency of tail flicking by the ox signalled the increasing irritation of the animal as it became more heavily besieged. A criticism of this experiment, and of many other experiments of this kind, is that the host animal is faced with unnaturally high attack rates. However, in another study Vale experimentally *reduced* the attack rate of tsetse by interrupting the arrival of insects by surrounding the animal with electric screens, in order to contrive attack rates that were less than or equal to those experienced naturally. As with enhanced fly densities, he found that the proportion of flies feeding was inversely related to the number of arrivals.

Waage and Davies (1986) (Fig. 12.1) also investigated the success of flies feeding on Camargue horses at natural densities. In what may be the only experimental study of interactions within a community of bloodsucking insect species, they compared the residence times of different kinds of tabanid flies attempting to feed in the presence and

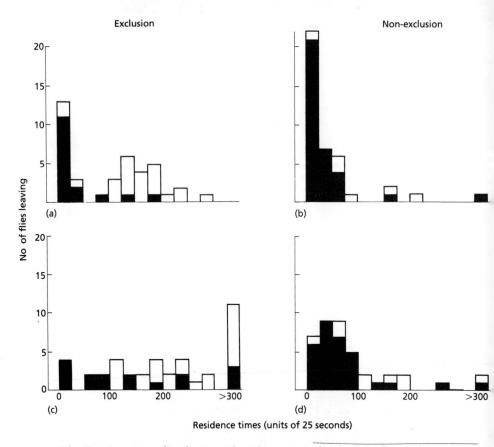

Fig. 12.1 Frequency distributions of residence times for the horseflies *Hybomitra expollicata* (a, b) and *Tabanus bromius* (c, d) in exclusion and non-exclusion treatments. Residence times are classified in 25-second intervals, and combined for times over 300 seconds. Residence times end when the fly leaves voluntarily (□) or as a result of grooming by the host (■). Each experiment used 40 flies. From Waage & Davies (1986).

absence of other species of horseflies. Longer residence times mean more blood consumed and more eggs laid. A fly which has fed for a longer period may also have to make fewer hazardous attacks during each ovarian cycle. When other flies were allowed unrestricted access to the horse, mean residence times for two species were 47 seconds and 83 seconds. But when other insects were prevented from landing by the experimenters wielding a fly-whisk, residence times rose to 101 seconds and 260 seconds. Apparently, when a fly is feeding with others, the horse is more likely to get rid of it by leg-kicking or head-slapping. Departures caused by direct interference by other flies were much rarer than departures resulting from grooming (e.g. 1% versus 62% for *Hybomitra expollicata*). Host irritability, therefore, was the main factor responsible for intraspecific and interspecific competition between these adult horseflies.

Suggestive though they are, these studies do not show definitely that competition for blood regulates population size. The work of

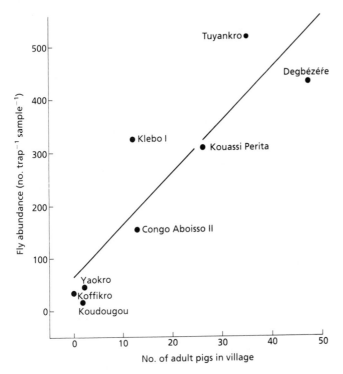

Fig. 12.2 Total number of tsetse flies *Glossina p. palpalis* per trap per sample as a function of the number of adult pigs in each village in Ivory Coast. From Rogers *et al.* (1984).

Rogers *et al.* (1984) in Ivory Coast has come closer to this goal. First, they observed a striking positive correlation between the number of pigs in a village and the number of tsetse caught there (Fig. 12.2). It seems likely that a decrease in the number of pigs in a village would affect tsetse population size by intensifying competition for blood meals. However, they also found a different relationship between ovarian age and wing-fray for populations of tsetse caught in the village and populations caught in the bush (Fig. 12.3). Tsetse resident in the village could easily take blood meals; in the bush hosts were relatively scarce. Experimental removal of the population of tsetse in one village that had a large pig population brought about a change in the ovarian-age–wing-fray relationship, which took on the characteristics of a local bush population (Fig. 12.3a): the regression coefficient increased from 0.274 to 0.323, even though the daily catch of flies hardly changed. Evidently, immigrant tsetse from the bush compensated for the removal of village flies. In contrast, experimental removal of the tsetse from a village that had a small population of pigs did not alter the relationship between ovarian-age and wing-fray, which was typical of that observed among local and relatively abundant bush flies (Fig. 12.3b). From a close analysis of the ovarian age data, Rogers *et al.* found that the mortality rate of bush flies was higher than that of village flies, 0.0289

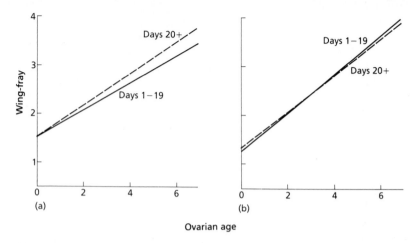

Fig. 12.3 Changes in wing-fray–ovarian-age regressions during removal trapping of tsetse flies at (a) Degbézéré and (b) Koudougou. Regression lines are significantly different in (a) but not in (b). From Rogers *et al.* (1984).

day^{-1} as against 0.0212 day^{-1}. The carrying capacity of a village for tsetse flies appears to be determined by the abundance of pigs; extra flies are relegated to the bush where their death rate is higher because blood meals are harder to find.

Ticks, and hard ticks in particular, remain on their hosts for much longer periods than flies, and there is scope for the host to add *acquired immunity* to its repertoire of defences. Much of the work on acquired immunity has been stimulated by the possibility of developing a tick vaccine, and has been directed at domestic animals, especially cattle. For example, Cheira *et al.* (1985) estimated that acquired resistance of Kenyan cattle to infestation with *Rhipicephalus appendiculatus* could reduce egg production by the ticks by as much as 98%.

Earlier work by Randolph (1979) on the interaction between *Ixodes trianguliceps* and mice is one of the few ecologically motivated studies of tick population regulation. In laboratory experiments, the percentage of ticks which engorged on mice was inversely related to the number of larvae which had previously attached to the host (Fig. 12.4a). But this result was not repeated with the normal wild host, the long-tailed field mouse *Apodemus sylvaticus* (Fig. 12.4b). Trager (1939) and Chabaud (1950) also reported acquired resistance on unnatural but not on natural hosts. As yet, we can only guess whether ticks which feed exclusively on wild animals are limited by processes that occur on or off the host.

BLOODSUCKING ARTHROPODS AND HOST POPULATION DENSITY

In general, the highest biting rates of bloodsucking insects are recorded at higher latitudes. Gillett (1971), for example, has calculated that

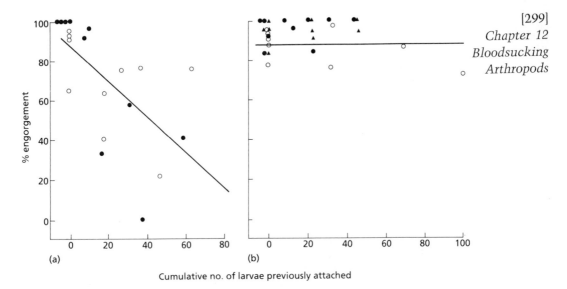

Fig. 12.4 Percentage of larval ticks *Ixodes trianguliceps* engorged as a function of the number previously attached. The hosts were (a) laboratory mice and (b) long-tailed field mice, *Apodemus sylvaticus*. Different symbols refer to different experiments. From Randolph (1979).

seasonally abundant north Canadian mosquitoes could inflict 9000 bites every minute on a naked man, consuming a quantity equal to about half his blood volume in less than 2 hours. In the face of this kind of onslaught, animals take various steps to avoid such grievous bodily harm. Among the *Curious Facts in the History of Insects* (Cowan, 1865, cited by Christophers, 1960) is Marcellinus' story about swarms of mosquitoes attacking the eyes of Mesopotamian lions with such ferocity that they were forced to take refuge in rivers and to drown, or go mad. A more reliable anecdote relates that porcupines take refuge from mosquitoes in trees (Marshall *et al.*, 1962). Musk oxen, reindeer and horses form tight groups, which reduces the biting rate suffered by each animal (Hamilton, 1971; Duncan & Vigne, 1979; Helle & Aspi, 1983). Camargue horses, when not feeding, gather on windy patches of bare ground where horseflies have difficulty landing on them (Duncan & Cowton, 1980; Hughes *et al.*, 1981).

In most examples of this kind, it is difficult to be certain how much the observed behaviour of host animals is due to pressure from biting flies, and it is impossible to translate such changes in behaviour into population dynamics. One spectacular exception was the destruction of the lion population of Ngorongoro Crater during the stable fly (*Stomoxys*) plague of 1962. In a year of exceptionally steady rainfall, wet mud at the edge of game watering holes provided durable breeding sites for *Stomoxys*. Park wardens observed lions climbing into trees and hiding in hyena holes in their attempts to evade the huge population of flies. The lions also changed their hunting habits; some were shot whilst raiding Masai camps while others left the area. One way or

Table 12.2 Nest usage in two colonies of cliff swallows. From Brown & Brown (1986)

Nest usage	Fumigated	Not fumigated
Colony 1		
Old nests present	89	155
Old nests unused	3	155
Old nests used	86	0
New nests built	4	0
Total active nests	90	0
Colony 2		
Old nests present	57	94
Old nests unused	18	91
Old nests used	39	3*
New nests built	0	0
Total active nests	39	3

*Two of these nests were adjacent to the line dividing fumigated and non-fumigated halves of the colony.

another, the stable fly epidemic reduced the lion population by an order of magnitude from 60 to six or so animals (Fosbrooke, 1963; Wildt *et al.*, 1987). As a result of inbreeding, the mean heterozygosity of the current Ngorongoro population (>100 animals) is just 37% of the larger population in the adjacent Serengeti Park. Thus, in addition to reducing the lion population, the stable fly plague had long-term effects on the lions by causing low genetic variability (Frankel & Soulé, 1981). For example, male lions may now suffer from elevated levels of abnormal sperm and diminished concentrations of circulating testosterone, a hormone which has a crucial role in spermatogenesis (Wildt *et al.*, 1987).

Bloodsucking insects can have chronic as well as acute effects on vertebrate populations. Brown and Brown (1986) found that colonially nesting cliff swallows (*Hirundo pyrrhonota*) avoid using nests that contain ectoparasites, particularly cimicid swallow bugs *Oeciacus vicarius*. Birds arriving at colonies in early summer occupied, almost exclusively, those nests which had been experimentally fumigated to kill their resident bugs (Table 12.2). Fumigated and non-fumigated nests were separated by less than 1 m in some cases. The result of attempting to use infested nests is shown in Fig. 12.5: nestling body weight was inversely related to the number of bugs per nestling. By comparing fumigated and non-fumigated nests in large swallow colonies, Brown and Brown found that the loss of body weight was associated with a reduction in survivorship of up to 50%. Bugs may play a role in regulating swallow colony size because bigger colonies had larger numbers of bugs per nest; ectoparasitism imposed a substantial, density-dependent cost to coloniality.

This conclusion does not, however, exclude the confounding effects

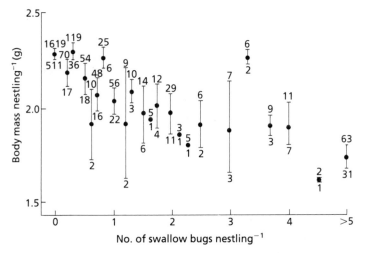

Fig. 12.5 Mean body mass (± standard error) of nestling cliff swallows (*Hirundo pyrrhonota*) aged 10 days plotted against the number of swallow bugs (*Oeciacus vicarius*) per nestling. Numbers of nestlings and nests sampled are shown above and below error bars respectively. Body mass declined significantly with increasing swallow bug infestation. From Brown & Brown (1986).

of a troublesome third party: vector-borne blood parasites. The next two sections concern the ways in which blood parasites might influence arthropod–vertebrate relationships.

THE IMPACT OF PATHOGENS ON THEIR ARTHROPOD VECTORS

For a long time, it was assumed that vertebrate diseases were not pathogenic to their arthropod vectors, flying vectors at least, because impaired vector performance would jeopardize transmission of the disease. In fact, it seems likely that those parasites which rely on the mobility of their vectors must often compromise infectivity to the vertebrate with pathogenicity to their invertebrate hosts, because both these parameters increase with parasite load. For example, *Culex quinquefasciatus* mosquitoes carrying a higher load of *Wuchereria bancrofti* are more likely to generate transmissible human filariasis (Southgate, 1983), but the insects are also more likely to be killed by the parasite before taking an infective bite on a human (Samarawickrema & Laurence, 1978).

A close look at Fig. 12.6 shows where the compromise is made in the transmission of bancroftian filariasis. Elevated mosquito mortality was detectable only at unusually high filarial loads, and was observable in the 10% of mosquitoes that contained more than about 20 first stage larvae. These 10% of the mosquitoes, however, harboured more than 50% of the parasite population.

Under controlled conditions, parasites have been shown to impair fecundity and flight ability, as well as longevity, and they can also

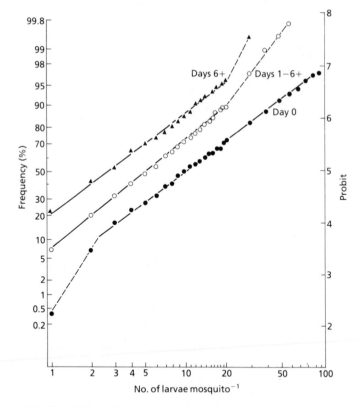

Fig. 12.6 Number of filaria larvae (*Wuchereria bancrofti*) in wild caught mosquitoes *Culex quinquefasciatus.* —●—, mosquitoes with microfilaria, —○—, short 'sausage' stage to short second stage larvae; —▲—, long second stage to infective stage larvae. Heavily infected mosquitoes are lost from the population. From Samarawickrema & Laurence (1978).

have a profound effect on feeding behaviour (Molyneux & Jefferies, 1986; Dobson, 1988a; Dye, 1990). In general, a heavy parasite load makes blood-feeding difficult, and the vector may be successful only after repeated attempts, if at all. Each attempt, however, represents an opportunity for transmission. For the parasite, therefore, multiple feeding during one ovarian cycle may well offset any concomitant increase in vector mortality. This remains to be tested directly by experiment.

The idea that parasites may manipulate their vectors takes its most extreme form in the work of Ribeiro *et al.* (1985) and Rossignol *et al.* (1985). They suggest that mosquito and disease might be mutually beneficial symbionts, rather than host and parasite. Their argument goes as follows. The disease malaria is a device to enhance parasite transmission to the vector. This can work first by making infected hosts more attractive to mosquitoes. Fevers, for example, raise body temperature, and it has been possible to demonstrate that mosquitoes do prefer to feed on infected hosts in some systems. Mahon and Gibbs (1982) found that mosquitoes preferred to bite hens with an arbovirus, rather than uninfected controls, and infected hens were more attractive to mosquitoes, even when the mosquitoes were prevented from biting. Likewise, Turell *et al.* (1985) discovered that mosquitoes chose lambs exhibiting symptoms of Rift Valley Fever in preference to uninfected lambs, and mosquito biting rate was related to the body temperature of infected animals (Turell *et al.* 1985).

In other cases, mosquitoes have shown no preference for infected individuals. In an area of Papua New Guinea where malaria and bancroftian filariasis are endemic, Burkot *et al.* (1989) could find no evidence that *Anopheles punctulatus* preferentially took blood from people with parasites. They attributed this to the fact that many of the infections were asymptomatic: there were no differences in body temperatures between aparasitaemic, lightly infected and heavily infected people. Much earlier, Young *et al.* (1948) had fed *An. quadrimaculatus* on people from North Carolina with recent histories of malaria. Asymptomatic persons were as infectious as, and five times more common than, those with clinical illness. Burkot and colleagues believe that a preference for malaria-infected persons, if it occurs at all, is more likely to be seen during epidemics, when most infections are symptomatic.

Besides making hosts more attractive, parasites can subjugate host defences when hosts are most infectious, and also change the haemostatic properties of blood. In laboratory experiments (Day & Edman, 1983; Rossignol *et al.*, 1985), and in theory (Daniel & Kingsolver, 1983), mosquitoes stand a better chance of taking a full blood meal from an infected individual, and of taking it more quickly. Ribeiro *et al.* (1985) have also studied the giving-up time, beyond which a hungry mosquito switches to another site, perhaps to an alternative host. They suggest that mosquitoes are more likely to give up before getting blood from an uninfected host, with the result that infected hosts are the source of a disproportionately large number of blood meals.

The advantage to a mosquito of feeding on more yielding, infected hosts is supposed to compensate for the costs of parasitism. These include impaired blood-feeding when parasites are transmitted back to the vertebrate. But the costs and benefits of parasitism have yet to be properly added up. The task of doing this will be formidable, but exceptional theories require exceptionally careful justification before they win acceptance.

The traditional view that viruses, bacteria, protozoa and helminths do *not* influence arthropod ecology and behaviour has been formally enshrined in numerous mathematical models of vector-borne disease that adopt this simplifying assumption. Having identified examples to the contrary, many authors have now pointed out that the early, simple mathematical models of transmission are wrong (see, for example, Molyneux & Jefferies, 1986). It is worth looking carefully at this assumption, and considering what difference it makes in terms of practical epidemiology.

Most quantitative studies of vector-borne disease epidemiology begin as variations on the theme of the classical Ross–Macdonald model of malaria (Macdonald, 1957; Anderson & May, 1991). At any one time, each person can be in one of three states of infection. The fractions of persons susceptible (x), and female mosquitoes latently infected but not yet infectious (y') and infectious (y) are assumed to change as follows:

$$\frac{dx}{dt} = \frac{(abpM)}{N} y(1 - x) - \rho x \tag{12.1}$$

$$\frac{dy'}{dt} = apx(1 - y - y') - ap\hat{x}(1 - \hat{y} - \hat{y}') - \mu y' \tag{12.2}$$

$$\frac{dy}{dt} = ap\hat{x}(1 - \hat{y} - \hat{y}')e^{-\mu n} - \mu y \tag{12.3}$$

where a is the daily biting rate of a female mosquito, b is the proportion of infected bites on man that produce an infection, p is the probability that a bloodmeal is taken on a person rather than on an animal, M is the size of the female mosquito population, N is the size of the human population, ρ is the per capita rate of human recovery (from infectiousness) and μ is the per capita mortality rate of female mosquitoes. The extrinsic incubation period n is the number of days required to produce infectious malaria sporozoites within a mosquito. The circumflex above variables is a device allowing them to have the values of n days ago, at the start of the extrinsic incubation period.

The components of transmission due to mosquitoes have been collected together as the Vectorial Capacity, C, which has become a guiding concept in malaria entomology (Garrett-Jones, 1964; Garrett-Jones & Grab, 1964; Garret-Jones & Shidrawi, 1969; Dye, 1992). Strictly, the Vectorial Capacity is the number of future inoculations which arise from the bites taken on an infectious person in one day. In the context of the present model, C is defined by

$$C = \frac{(M/N)a^2bpe^{-n\mu}}{\mu} \tag{12.4}$$

In turn, C is related to the *basic reproductive rate* of infection, R_0 (Anderson & May, 1991 and Chapters 14 and 15) by

$$R_0 = \frac{C}{\rho} \tag{12.5}$$

R_0 is the average number of secondary infections arising from each primary one in a host population that is assumed to consist almost entirely of susceptibles, and $1/\rho$ is the average number of days a case remains infectious. This simple model of transmission makes many assumptions, and some of these have already been questioned. As its shortcomings have been identified, more details have been incorporated into equations 12.1–12.5. These more complex models have been useful for exploring the dynamical consequences of, for example, non-random host-seeking by mosquitoes (Dye & Hasibeder, 1986; Hasibeder & Dye, 1988; Kingsolver, 1988b). However, with a growing number of parameters, it now seems *less likely* that embellishments of equations 12.4 and 12.5 will give satisfactory estimates of C and R_0 in practice. The reason is that some of the key parameters, like survival rate, are exceptionally difficult to measure in the field (see, for example, Najera, 1974; Garrett-Jones et al., 1980), and realistic variants of equations 12.4 and 12.5 include more of these difficult parameters.

In the history of vector-borne disease epidemiology, there have been few attempts to assign confidence limits to estimates of C and R_0. This poses the important question of what these entomological measurements really are useful for. Certainly, any estimates made from equations 12.4 or 12.5, (Milligan & Baker, 1988; Rogers, 1988), or from similar models need to be carefully verified. In many instances, it may well be better to estimate R_0 by adapting methods developed for *directly transmitted infections*. Some of these are based on the age-distribution of infection within the vertebrate host, which is often more easily and more reliably measured (Anderson & May, 1991).

VECTOR-BORNE DISEASES AND HOST POPULATION DYNAMICS

Much of the epidemiology of parasitic infections is founded on the fact that infection leads to disease. Theory, however, has been slow to acknowledge that disease may have consequences for the dynamics of host populations (Anderson & May, 1979). In the model of equations 12.1–12.3, birth and death rates are assumed to be equal and constant, irrespective of the parasite rate. Abandoning that assumption for directly transmitted infections has shown, for example, that cycles in populations of forest insects might be driven by viruses (Anderson & May, 1980, but see Fischlin & Baltensweiler, 1979), and that different patterns of dynamics are possible when there is a relatively invulnerable reservoir of pathogens. If it is generally true that vector-borne diseases have evolved to be more severe than directly transmitted ones, because transmission does not depend on host mobility (Ewald, 1983), then similar investigations of the population dynamics of vector-borne diseases are overdue.

In models of directly transmitted infections, the rate of transmission is usually assumed to be an increasing function of host population size N. Thus, host death rate could increase as transmission increases, and vice versa, and in principle the parasite could regulate the host population around some characteristic level of abundance. For vector-borne diseases, on the other hand, the first assumption has usually been that the rate of transmission depends upon the number of vectors per host, M/N. If parasites kill their vertebrate hosts, N declines, the rate of transmission *increases* and the host population declines further towards extinction, taking the parasite with it.

This scenario is unlikely and it encourages a closer look at the natural history of highly pathogenic vector-borne diseases. Linked host and vector populations are, in theory, one way of preventing extinction of the host by a vector-borne pathogen: the linkage makes the transmission term of a vector-borne disease model more like that of a directly transmitted infection. Some mutual dependency of host and vector is expected when an insect transmits a highly pathogenic agent and lives in close association with its host. This is the case with the domestic Indian sandfly, *Phlebotomus argentipes*, that is a vector of human visceral leishmaniasis, due to *Leishmania donovani*. Dye and Wolpert (1988) used the following system of difference equations to study the long-term dynamic behaviour of the disease in Assam:

$$S_{t+1} = S_t + aN_t - cI_t\left(\frac{S_t}{N_t}\right) - bS_t \tag{12.6}$$

$$I_{t+1} = I_t + cI_t\left(\frac{S_t}{N_t}\right) - (b + d)I_t \tag{12.7}$$

$$R_{t+1} = R_t + eN_t - bR_t \tag{12.8}$$

The total human population size (N_t) is the sum of people in three categories: susceptible (S_t), infectious (I_t) and resistant (R_t). Susceptibles and resistants are born at rates a and e. The disease was fatal, almost without exception, before the introduction of antimonial drugs around 1920 and, during the first epidemic of 1875–1900, the disease reputedly claimed up to 40% of the population of some districts of Assam. In the model, all infectious individuals die and the disease-induced death rate (d) is much higher than the natural death rate (b). The resistant category, R_t, includes several classes of people who, when infected with *L. donovani*, do not develop clinical symptoms and are never infectious to sandflies. These people act as a source of newborn susceptibles. Both sandfly and human populations are permitted to change in the model, but they are assumed to change in concert, so that the number of sandflies per person is invariable (included in c). The transmission term allows the long-term persistence of the disease, and the model has been used to suggest why Assam suffered a second large epidemic of kala-azar during the 1920s (Fig. 12.7). McCombie Young (1924) believed that the required number of susceptibles were generated by the renowned influenza pandemic of 1918–19. By contrast, a joint analysis of model and data suggests that the outbreak was the natural consequence of birth and immigration during the preceding two decades (Dye & Wolpert, 1988).

After reviewing the evidence, Holmes (1982) argued that vertebrate mortality due to pathogens generally plays an insignificant part in host population dynamics, because it is mainly compensatory, merely replacing other mortalities rather than adding to them. The exceptions he allows are pathogens in novel hosts. A celebrated example is the malaria invasion of the native Hawaiian birds (Warner, 1968), now

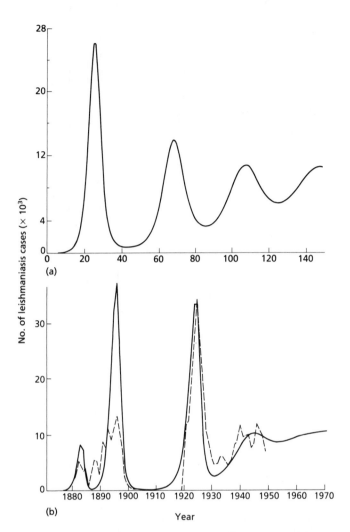

Fig. 12.7 The dynamic behaviour of a model of sandfly-transmitted human visceral leishmaniasis. (a) One infectious person introduced into a large community of susceptibles and resistants. (b) Best fit to the available data (dotted line), accounting for complexities such as immigration, and the introduction of drug treatment around 1920. See Dye & Wolpert (1988), for details.

well-supported by data (van Riper *et al.*, 1986; van Riper, 1991). *Plasmodium relictum capistranoae* is transmitted mainly by the mosquito *C. quinquefasciatus*, and is generally more pathogenic to native birds than to introduced species in the Hawaiian Islands. When challenged with the parasite, fewer of the native birds survive (Fig. 12.8a). Assembling the data, van Riper *et al.* (1986) propose that the distribution of bird malaria on the slopes of Mauna Loa is controlled by the processes depicted in Fig. 12.8b. Susceptible native birds like the apapane and iiwi are now restricted to drier, highland sites, though they make downward forays by day into areas that are occupied by mosquitoes at night. The Mauna Loa amakihi is a more sedentary bird whose present

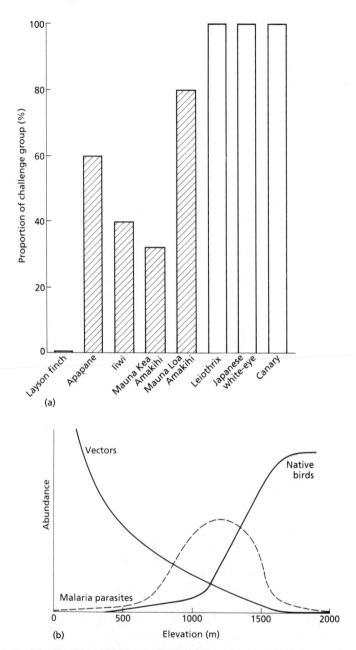

Fig. 12.8 (a) Percentage of Hawaiian birds of different species surviving experimental challenge with the malaria parasite *Plasmodium relictum capistranoae*. □, introduced birds; ▨, indigenous species. After van Riper *et al.* (1986). (b) A model of native bird abundance, malaria incidence and mosquito distribution on Mauna Loa, Hawaii. From van Riper *et al.* (1986).

day resistance (Fig. 12.8a) may be attributable to the greater selection pressure applied by continuous exposure. The distributions of the more malaria-susceptible native bird species are probably restricted by

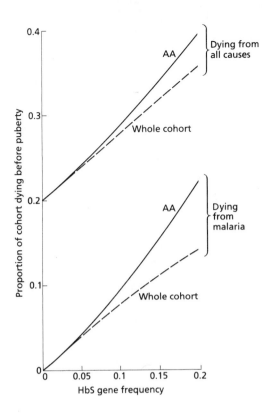

Fig. 12.9 Proportion of cohorts of children expected to die before puberty for given equilibrium frequencies of the sickle cell haemoglobin gene (HbS). The risk of dying from causes other than malaria was assumed to be 0.2 for AA and AS, and 1.0 for SS. From Molineaux (1985).

the reservoirs of parasites, and mosquitoes, which are maintained by the pool of less susceptible ones.

The polymorphisms maintained in human populations living in areas endemic for *P. falciparum* malaria signal other exceptions to Holmes' view. For example, persons homozygous for the sickle cell haemoglobin gene (SS) never survive to reproduce. Heterozygotes (AS) do survive; indeed, they constitute 20% or more of the population in some parts of Africa. Their abundance is explained by their substantial selective advantage over normal homozygotes (AA) in *P. falciparum* areas. Molineaux (1985) has estimated what that selective advantage must be, by calculating the malaria mortality rates of normal homozygotes necessary to maintain the sickle cell gene at observed frequencies (Fig. 12.9).

PATTERNS IN THE POPULATION BIOLOGY OF BLOODSUCKING ARTHROPODS

Those who must manage vector-borne diseases will be interested in how the answers to the four questions posed in this chapter influence control operations. Rogers and Randolph (1984a) have given an example of the application of studies on tsetse population regulation. Figure 12.10 shows the effect of releasing sterile males into a tsetse population that had a net reproductive rate of $R = 10$. The parameter b here

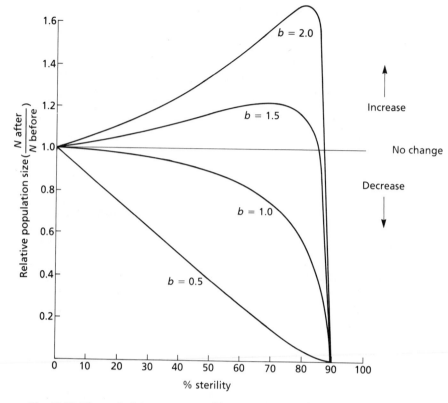

Fig. 12.10 Theoretical impact on equilibrium tsetse population size of varying degrees of induced sterility. The success of releasing sterile flies depends on the severity of density-dependence, an increasing function of b. From Rogers & Randolph (1984a).

measures the strength of density-dependent regulation acting over a whole generation, and includes all the processes which affect adult and pre-adult mortality and fecundity. It is likely to have a value close to 1.0 (exact compensation) for tsetse. The figure shows that introducing sufficient sterile males to reduce fertility to about 50% barely affects tsetse population size. Note that for overcompensating density-dependence ($b > 1$), releasing sterile males can actually increase the population density of wild flies.

Significantly, much of the information about the regulation of tsetse populations comes from data which were collected for other purposes (Rogers & Randolph, 1984b, 1985), so the cost of gaining the above insights amounts to little more than the cost of analysis. To date, bloodsucking arthropods have been subjected to rather few analyses of this kind. There has been, for instance, no serious study of regulatory processes in the *An. gambiae* species complex, which contains two of Africa's key malaria vectors.

The search for ways to organize the life histories of bloodsucking arthropods will have greater appeal to ecologists, who are not faced with the more immediate problems of managing vector-borne diseases. This chapter has uncovered few general patterns because, by picking examples from all the bloodsucking arthropods, the questions posed in the introduction have been applied too broadly. The following are two cases (among many) where patterns do emerge from restricted comparative work; the second example involves blood parasites, but the first does not.

Competition for blood, in the long term, is a force which is likely to shape the genetic components of arthropod life histories. A compelling example is provided by the evolution of *autogeny* in mosquitoes (the ability to produce eggs without a blood meal). O'Meara and Edman (1975) compared two populations of *Ae. taeniorhynchus* in south Florida, one at Big Pine Key, where there was a large population of deer, and a second at Flamingo, where deer and other large mammals were absent or rare. Collections of females in the field showed that 21% of the flies were engorged at Big Pine Key, but only 8% were engorged at Flamingo. Among the F1 progeny of individuals collected at the two sites, the frequency of autogenous females was greater from Flamingo (94% versus 74%), where the number of mosquitoes per host was much higher. When selection for autogeny was stronger, and autogeny was consequently more common in the population, each female mosquito was able to produce more eggs without taking blood (Fig. 12.11).

As we have seen, blood parasites can influence the feeding behaviour, survival and fecundity of vectors. But, in so doing, they will ordinarily impose small selection pressures. The reason is that arthropod infection rates are generally low, even when almost all vertebrates are infected. From the viewpoint of vectors, parasites are 'rare enemies' (Dawkins, 1982). This is so because the number of bites a mosquito can expect to

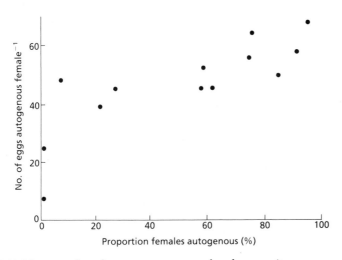

Fig. 12.11 Mean number of eggs per autogenous female mosquito (*Ae. taeniorhynchus*) as a function of percent autogeny found in different North American populations. Data from O'Meara & Edman (1975).

take in her life time (3—4) is much less than the average number required to pick up infection and become infectious (Nedelman, 1984). The relation between vector and parasite populations is therefore highly asymmetrical. Nevertheless, Grimstad *et al.* (1977, 1980) contend that La Crosse virus has been the selective agent of *refractoriness* (the inability to transmit virus) in some populations of *Ae. triseriatus* mosquitoes. Among 20 strains of *Ae. triseriatus*, the rate at which mosquitoes acquired infection when feeding through an experimental membrane varied from 40 to 93%. Additionally, the ability of different virus strains to generate fatal infections in mice varied between 20 and 90%. The variation was related to the distribution of endemic and non-endemic disease areas in North America, and both infection and transmission rates were lower in strains that had been collected from areas where the disease was endemic. This story is compelling, though the evidence is only correlative.

When we compare ticks and triatomines, or fleas and flies, the important variables change many at a time. To make progress in understanding and classifying the life histories of bloodsucking arthropods we must aim to make univariate comparisons, by using 'natural experiments'. The success of this kind of comparative work will hinge on making judicious taxonomic restrictions, on removing at the outset the confounding effects of allometry, developmental constraints and phylogenetic history.

ACKNOWLEDGEMENTS

This chapter is much the better for the comments and suggestions of Alan Clements, Clive Davies, Charles Godfray, Richard Lane, Colin Leake, Sarah Randolph and Chris Schofield.

13: Spiders as Representative 'Sit-and-wait' Predators

SUSAN E. RIECHERT

INTRODUCTION

Spiders exemplify a class of predators referred to as 'sit-and-wait' foragers. This foraging mode is defined as one in which no time or energy is expended in the search for food that is not simultaneously used in other activities (Schoener, 1971). Although no animal group fits this definition perfectly, many spiders come close. For them, the search for prey ends when suitable foraging sites are located, although to judge by their giving-up times, they continuously evaluate their chosen sites. There are various physiological and behavioural adaptations associated with this foraging mode. To understand predation by sit-and-wait foragers, it is necessary to have an understanding of these traits. This chapter, therefore, begins with an analysis of the various traits affecting the foraging behaviour of spiders and the influence that spider predation might have on associated prey population dynamics. The chapter ends with a consideration of the role of spiders in the invertebrate community with particular reference to their potential use as biological control agents.

FORAGING MODE

Early classifications of spiders suggested that web-builders fit the definitions of sit-and-wait foragers, whereas non-web-building spiders, commonly referred to as hunting spiders, are more active searchers (Bilsing, 1920; Turnbull, 1973). This view has largely been discredited with studies of the foraging behaviour of the hunting spiders. As with the web-builders, most hunting spiders ambush prey from a stationary foraging site, e.g. Lycosidae (wolf spiders) (Cragg, 1961; Edgar, 1969, 1970; Ford, 1977), Oxyopidae (lynx spiders) (Weems & Whitcomb, 1977), Salticidae (jumping spiders) (Givens, 1978; Hill, 1979). Possible exceptions are the sac spiders belonging to the families Anyphaenidae, Clubionidae and Gnaphosidae, which are reported to forage actively for larvae and eggs (Bushman et al., 1976; Richman et al., 1980).

Feast and famine existence

Insect population numbers and rates of growth vary greatly from place to place and from week to week, associated with seasonal and local weather conditions. Thus the spiders, which are often relatively long-lived compared with the insects they feed on, are subject to alternating periods of feast and famine. The physical environment also determines the activity levels of the insect prey, and hence their probabilities of encountering the sit-and-wait predator. This makes prey availability levels even more variable to this type of predator than to the more active searchers. Data available for a desert grassland population of the funnel-web spider, *Agelenopsis aperta* (Gertsch) (Agelenidae), exemplify the variance in encounter rates with prey associated with local weather conditions. Air temperature and percentage cloud cover were the major determinants of prey biomass encountered by this species (Riechert & Tracy, 1975). It is likely that spiders in more equable, non-desert environments encounter prey at less variable rates, but sit-and-wait predators will always be at the mercy of a wide range of environmental variables, and these will fluctuate at rates that are beyond their control.

Is the famine more frequent than the feasting for spiders? Data collected on prey availabilities to this desert grassland population of *A. aperta* and six others over time indicate that for five of the six populations, energy requirements are met by prey contacting the webs on less than 20% of the days they are foraging. No prey at all were encountered another 30% of the time (Fig. 13.1). Only one of the populations had consistently high encounter rates with prey. This was a population occupying a spring-fed woodland where the permanent water source provided a constant influx of insects into the system.

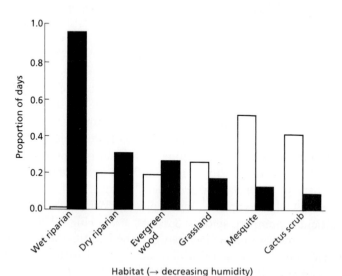

Fig. 13.1 Prey availability information for six populations of the aridlands spider *Agelenopsis aperta* (Agelenidae) based on sticky trap captures over two to nine field seasons for the different populations. Optimal energy needs refer to 22–23 mg dry weight of prey that this spider will consume when prey are present in unlimited supply. □ no prey encountered; ■ optimal energy needs met. From Riechert (1978).

Other examples show broadly similar patterns. At 75% of its web-sites, the rainforest spider, *Agelena consociata* (Agelenidae), did not encounter enough prey during the rainy season to enable it to build new web-traps to replace traps damaged by rains (Riechert, 1985a). This factor may underlie the group-living life style of this species, for less web investment is required per individual spider with increasing numbers of cooperating individuals, and there is a threshold web-trap size required to catch prey. That spiders rarely encounter an abundance of prey is further indicated by the fact that many workers have shown that the majority of spiders collected from the field are underfed (see, for example, Miyashita, 1968; Edgar, 1969; Anderson, 1974; Riechert, 1978). For instance, in comparing the body weights and dimensions of field-collected spiders of two species (the hunting spider *Lycosa lenta* (Lycosidae) and the cribellate web spider *Filistata hibernalis* (Filistatidae)) with individuals of known nutritional status in the laboratory, Anderson (1974) concluded that both species populations are near starvation in nature. Fujii (1972) reached a similar conclusion in comparing the growth rates of *Agelena opulenta* (Agelenidae) in the field and in the laboratory.

Adaptations to feast and famine

PHYSIOLOGICAL AND MORPHOLOGICAL ADAPTATIONS

Spiders exhibit several physiological and morphological adaptations that accommodate a feast and famine life style. Their food is broken down externally into a nutrient broth, for instance, which eliminates the bulky waste products that would take up valuable body space during a feast. Within the abdomen itself there is massive branching of the midgut so that large meals can be stored (Millot, 1949). Spiders also lack heavy sclerotization on the abdomen which, thus, can be distended during periods of food intake (Wilson, 1970). When combined, these features permit spiders to take very large meals. For instance, Anderson (1974) reports that spiders may increase their body weight by more than 50% with a single feeding. This represents a very high consumption rate for a predaceous arthropod, most of which exhibit relative consumption rates at the low end of the range recorded for arthropods $(0.002-6.9 \, \text{mg day}^{-1} \, \text{mg}^{-1})$(Slansky & Scriber, 1985).

Spiders, in general, have low levels of metabolism, an adaptation to existence under limited food. Under periods of food stress, Ito (1964) noted that metabolic rates can be lowered even further. Subsequent work indicated that metabolic rates can even be lowered without a corresponding decrease in activity (Miyashita, 1969; Anderson, 1974; Tanaka & Ito, 1982). The reductions noted in metabolic expenditures during fasting ranged between 17 and 63% for different species and sexes. Lowered energy expenditure in the assimilation of food contributes to this decrease (Nakamura, 1987), as does the utilization of fat during fasting as a metabolic fuel (Tanaka *et al.*, 1985). Work by Tanako and Ito (1982) indicates that recovery from a period of starvation is rapid: on average within 5 days of resumption of feeding, body weight and metabolic rates returned to pre-starvation levels in the wolf spider *Pardosa astrigera* (Lycosidae).

These changes in metabolic energy expenditures associated with starvation permit spiders to resist long periods of food deprivation. In fact, the survival rate of starved spiders is as high as that of non-starved spiders with the exception of individuals in early stages of development (Turnbull, 1962, 1973). Thus, spiders will survive with low food levels although both their growth and reproductive rates increase with the quantity of food consumed (Turnbull, 1962; Kajak, 1967; Riechert & Tracy, 1975; Wise, 1979; Vollrath, 1985; Hammerstein & Riechert, 1988; Miyashita, 1988).

Because sit-and-wait predators frequently suffer periods of food deprivation, one might expect them to exhibit adaptations to procuring a broad range of prey. An example of a physiological adaptation to taking a broad spectrum of prey is the use of venom in the capture process. The majority of spiders utilize venom in their capture of prey (the family Uloboridae is the exception). Venomization hinders the escape of the more mobile prey, and it is probable that the use of venom enables spiders to take larger and more difficult prey than would be otherwise possible. Most spiders can take prey that are 1.5 times their own body sizes, and some can take prey that are 2–3.3 times their body sizes (Nentwig & Wissel, 1986).

BEHAVIOURAL AND ECOLOGICAL ADAPTATIONS

Another adaptation for taking larger prey is the swathing of prey items in silk and glue by certain species. Many spiders exhibit some sort of wrapping behaviour (see Nentwig & Heimer, 1987 for a review). This behaviour is size dependent and it is utilized significantly more often in the capture of larger prey items (Robinson & Olazarri, 1971; Nentwig, 1985). It has the advantage that the prey items can be immobilized without direct contact (Robinson *et al.*, 1969; Robinson & Olazarri, 1971), which also permits the successful capture of prey types that have defensive chemicals stored in glands. Here, the noxious insects exhaust their chemicals as they struggle against the silk while the spider sits at a safe distance away. Nentwig (1987) reports that long legs serve the same purpose in some spiders (e.g. the Pholicidae, cellar spiders).

Diet choice

The adaptations mentioned above for taking large and even noxious prey are indicative of the degree to which sit-and-wait predators are generalist feeders. In his classic treatise on spiders, Savory (1928) states that spiders show 'no trace of discrimination'. This statement has held up well through the years as the diets of various spider species have been documented, but some web-spinners assiduously cut obnoxious prey out of their nests. In general, spiders include between 7 and 14 different orders of arthropods in their diets (Nentwig, 1987; Riechert & Harp, 1987), a much more diverse diet than would be needed to achieve dietary mixing for balanced nutrition. Dietary mixing to achieve an optimum amino acid balance has been reported for only one spider species belonging to the genus *Pardosa* (Lycosidae) (Greenstone, 1979). This wolf spider existed in an unusually simple ecosystem where it encountered only four potential prey species.

Those few spider taxa that do show specialized feeding attack prey types that are locally abundant. Most of the specialized foragers among the Araneae feed on ants, probably because ants are, once located, present in predictably high numbers (Glatz, 1967; Robinson & Robinson, 1971; Schneider, 1971; Edwards *et al.*, 1974; Harkness, 1977; Hölldobler, 1979; Mackay, 1982; Oliveira & Sazima, 1984). A few members of the large family Araneidae (orb weavers) specialize on adult lepidoptera. Two species build a 'ladder' web in which the sticky orb web is greatly extended vertically and flattened horizontally to permit capture of moths and butterflies that would normally merely lose scales and escape the web. The ladder web has evolved both in the New World and Old World tropics in araneid species (Eberhard, 1975; Stowe, 1978). Again, there are spiders that specialize in the capture of other spiders. For example, the pirate spiders (family Mimetidae) enter the webs of other spiders after mimicking prey contact or courtship to lure the web owners out into a vulnerable position (Czaika, 1963). Another spider specialist is the salticid *Portia fimbriata*, which feeds almost exclusively on other spiders using the same attack pattern noted for the mimetids (Jackson & Blest, 1982).

Note that in no case does the specialization go beyond the order level. Even within these relatively specialized spider taxa, the species accept other prey items in feeding experiments (see review in Nentwig, 1987). The spiders that come closest to specialization on a single prey species are the bolas spiders that have given up the orb web and merely use a thread with a sticky droplet at the end in prey capture. The droplet contains a pheromone that attracts male noctuid moths to it (Eberhard, 1977, 1980).

The diversity of prey types taken by spiders is exemplified in Fig. 13.2 using data from two populations of *A. aperta*. Note that the population inhabiting an area with abundant prey (the wet riparian habitat) does not show increased prey selectivity. Instead, it decreases its attack rate on the entire spectrum of prey types encountered. This is related to the fact that the two populations show genetic differences in their latencies to attack prey (Table 13.2). Individual spiders could readily obtain all of their energy needs from the small subset of these prey that could be handled most efficiently, or which pose less of a

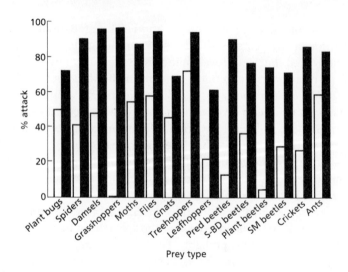

Fig. 13.2 Capture attempt rates over the spectrum of prey types available to two populations of the aridlands spider *Agelenopsis aperta* (Agelenidae). Based on observed natural encounters (minimum number = 10). S-BD Beetles, soft-bodies beetles; SM Beetles, beetles < 4 mm in length. Riparian population □ has high encounter rate, grassland population ■ a low encounter rate with prey.

threat of injury to the spider (Table 13.1). Why this behaviour is not more flexible, with individuals adjusting their attack rates in relation to temporal and spatial variation in prey abundances, is a mystery (see Chapter 4). It is plausible that possessing experience in handling a broad spectrum of prey types may be of benefit to the sit-and-wait predator during periods of prey shortage.

Foraging site selection

With a sit-and-wait predatory mode, the selection of a suitable foraging patch is all important to the success of an individual spider. The positioning of spiders within habitats is non-random, and is typically related to prey availability. Riechert (1976), for instance, found a significant positive correlation between the distribution of webs of *A. aperta* and the distribution of mammalian faecal material in a desert grassland habitat. Webs were also more abundant close to flowering herbs and shrubs. Insect prey densities were higher than average at all these sites (Riechert & Tracy, 1975). In studying the habitat-selection process, Riechert (1985b) found that this species cues not on the physical properties of cattle dung and flowers, but on vibratory and chemical cues emitted from the insects that are attracted to these objects. Other researchers have observed similar patterns. Turnbull (1966) in a study of a spider community in a pasture, for instance, noted mass movements of spiders from one section of a field to another with changes in local prey abundances and Morse and Fritz (1982) report that crab spiders of the family Thomisidae settle in vegetation that affords significantly higher levels of prey. The process of habitat selection by spiders is

Table 13.1 Profitability scores (seconds capture expenditure per joule of food obtained) for different prey types encountered by *Agelenopsis aperta* occupying a riparian habitat where prey levels are unusually high. Subset providing daily energy needs with minimum time and energy expenditures is more profitable; subset involving no risk of injury is less profitable; and subset involving risk of injury is least profitable

More profitable (scores > than median)		Least profitable (demonstrated injury risk)	
Prey type	Profitability score (seconds J^{-1})	Prey type	Profitability score (seconds J^{-1})
Damselflies	0.04		
Walking sticks	0.10		
Moths	0.12		
Treehoppers	0.17		
Lace wings	0.17		
Robber flies	0.21		
		Wasps	0.27
Flies	0.34		
Butterflies	0.41		
Larvae	0.56		
Grasshoppers	0.85		
Soft-bodied beetles	1.10		
Less profitable (scores < median)			
		Mantids	1.35
Predaceous bugs	1.64		
Crickets	1.69		
Plant bugs	2.32		
		Bees	2.42
Plant beetles	2.92		
Leafhoppers	4.39		
Small beetles	4.65		
		Ants	5.04
Spiders	6.02		
		Predaceous beetles	7.85
Parasitic wasps	19.21		

reviewed in detail by Riechert and Gillespie (1986). Briefly, the first decision in the habitat selection process concerns when to leave an existing foraging site in search of a new one. Some spider species may initiate ballooning activity (passive transport by air currents) while others move by ground when physical conditions or prey availabilities become unfavourable, e.g. increasing litter temperature and decreasing humidity (Braendegaard, 1937; Duffey, 1956; Eberhard, 1971; LeSar & Unzicker, 1978). It is difficult to determine the proximate factors that underlie such moves, but hunger has been directly demonstrated as underlying location changes in two members of the comb-footed spider family, Theridiidae (Turnbull, 1964; Hölldobler, 1979). Janetos (1982a)

Table 13.2 Comparisons of latencies to attack prey for two populations of the spider *Agelenopsis aperta* existing under different prey availability levels (means and standard errors)

Prey	Limited prey, grassland	Abundant prey, riparian	Significance
Field			
Crickets	2.6 ± 0.2	55.0 ± 0.3	*P* < 0.001
Ants	1.3 ± 0.2	25.3 ± 0.3	*P* < 0.001
Second generation laboratory reared			
Crickets	3.9 ± 1.0	34.5 ± 12.3	*P* < 0.005
Ants*	34.5 ± 1.8	183.0 ± 1.9	*P* < 0.05

*The spiders were reared on crickets in the laboratory, whereas they had had no experience with ants prior to the trials with them.

modelled the relationship between the frequency of foraging site moves and the amount of energy expenditure in the site (e.g. cost of building a web-trap of a given type) and prey abundance. He concluded that spiders occupying more favourable prey environments and with low investment in a foraging site should use hunger as a cue to moving, while those occupying more variable prey environments and/or with high investment should not. This hypothesis is in agreement with Caraco's idea of risk sensitivity in foraging (see Chapter 4). Gillespie and Caraco (1987) utilize risk sensitivity analysis in explaining population differences in the frequency of foraging site moves by the orb weaving spider *Tetragnatha elongata*. Individuals of a population occupying a more favourable foraging environment moved web-sites frequently compared to spiders occupying an area where prey numbers were low. Animals transplanted between habitats exhibited switches in behaviour that were appropriate to local prey availabilities. Stopping at a new foraging location once movement has been initiated involves selection of a microhabitat that provides those structural features requisite for attachment of the web-trap in the case of the web-builders and favourable thermal conditions and prey levels for all taxa. The process is initiated by a random search. *Agelenopsis aperta* circles the site of departure in increasingly larger arcs until either olfactory or vibratory cues or shade trigger an approach to a patch (Riechert, 1985b). Once in the patch, the spider moves along a temperature gradient, ensuring that the web is placed within a favourable thermal environment, but the generality of these behaviours is not known. Gillespie (1981) suggests that spiders with low energy investment in their webs choose sites by sampling the prey availability. Thus, for these spiders, foraging site selection involves the placement of a few strands of silk at temporary sites, presumably those chosen on the basis of a favourable temperatures and humidities. Her experimental work was with the cribellate spider *Amaurobius similis* (Amaurobiidae). Similar transient web-building activity has been observed in the comb-footed spider *Achaearanea tepidariorum* (Theridiidae; Turnbull, 1964). In contrast,

A. aperta has a permanent web-trap in which there is considerable energy investment. Once silk laying is initiated at a site, the site is seldom abandoned in search of a new one.

Having located a suitable site, the web-building spider can increase its foraging success at that site by orienting its web to make maximum use of air currents that will bring insects in contact with the trap (Eberhard, 1971) and to ensure that the spider has maximum time to engage in prey capture activities, given local thermal conditions (Krakauer, 1972; Riechert & Tracy, 1975; Carrel, 1978; Biere & Uetz, 1981). For example, at exposed desert sites *A. aperta* orients the funnel of its web in a north-westerly direction, an orientation that allows the sun to shine down the funnel for the shortest period during the day (late afternoon) (Riechert & Tracy, 1975). This helps to keep spider body temperature below a thermal limit during the day. The sun is also shining down the funnel of the web at a time when its heating will most benefit foraging, for at dusk the desert sky acts as a heat sink and temperatures drop rapidly. This directionality to the web is not obvious in more protected sites. In addition to web orientations, thermo-regulatory behaviour involves posturing and shuttling between patches of sun and shade to maintain body temperatures within a favourable temperature range for foraging (Humphreys, 1987).

Functional responses to prey densities

Spiders exhibit, at least when hungry, a Type 3 or sigmoid response curve (Chapter 3), in which an initial lag in capture rate is followed by an exponential rise in prey capture activity (Fig. 13.3). Holling (1959a) suggests the initial lag is associated with the experience a predator

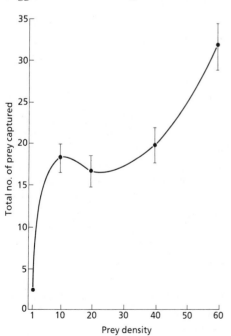

Fig. 13.3 Example of a sigmoid functional response curve exhibited by spiders: first day of prey capture by young orb weaving spiders (*Araneus diadematus*: Araneidae) following collection from the field. Prey density is number per 400 cm^2. From Smith & Wellington (1986).

needs to obtain with a new prey type before it can be efficiently utilize this prey. Increased capture efficiency and recognition of prey types as encounters with them increases has been noted for many spiders (Dabrowska-Prot & Luczak, 1968; Kiritani & Kakiya, 1975; Nakamura, 1977; Mansour *et al.*, 1980a; Smith, 1984b). The sigmoid response curve exhibited by some spider species, however, has been attributed to the influence of prey density on prey activity (Haynes & Sisojevic, 1966). At higher densities insects interfere with each other, exciting others to activity. Haynes and Sisojevic refer to this increased activity as 'effective density' and suggest that it results in increased encounter rates between spiders and their prey.

Another characteristic of the spider's functional response curve is a very high maximum attack rate (e.g. there is no hint of a plateau in Fig. 13.3) (see also Haynes & Sisojevic, 1966; Miyashita, 1968; Edwards *et al.*, 1974; Kiritani & Kakiya, 1975; Givens, 1978; Kajak, 1978; Smith & Wellington, 1986). As already discussed, morphological and physiological adaptations make spider ingestion of large meals possible. High plateaus in a functional response curve in spiders may also occur for at least two other reasons: (i) it may be more efficient to take less material from each prey item; and (ii) since digestion of the prey takes place externally, the spider may continue to kill prey before it becomes sated on a prey that it has captured earlier.

The sigmoid response curve exhibited by spiders when first brought in from the field appears to be lost with continued exposure to high prey densities; Smith (1984b) reports that the Type 2 response curve with a lower plateau is exhibited eventually. The type of response curve exhibited by spiders appears, therefore, to depend upon hunger level.

Numerical responses to prey densities

Considerable evidence exists that spider populations change in density in response to changes in prey availability, both by aggregation and, over longer periods, by enhanced reproduction. As described earlier, spiders tend to move out of habitat patches affording few prey to those offering higher prey densities (Turnbull, 1964; Riechert, 1976; Kronk & Riechert, 1979; Janetos, 1982a,b; Rypstra, 1983). Both Wise (1975, 1979) and Riechert (Riechert & Tracy, 1975; Riechert, 1981) have observed increased levels of food consumption with higher prey densities as a result of increased reproduction. In the case of *A. aperta*, there is a significant relationship between the level of prey consumption and the numbers and weights of offspring produced (Fig. 13.4) (Riechert, 1981; Hammerstein & Riechert, 1988).

Both aggregation and increased reproduction should lead to higher densities of predators in patches of habitat with higher densities of prey. The extent to which this correlation is realized, however, is limited by a variety of factors (see below).

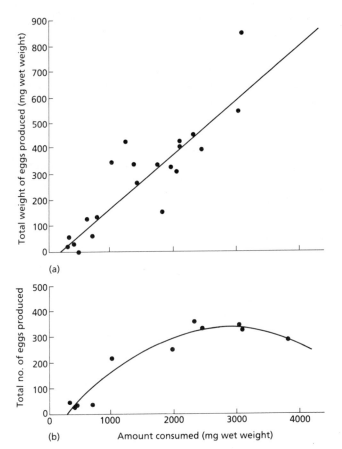

Fig. 13.4 Relationship between food consumption and fecundity for the funnel-web spider *Agelenopsis aperta* (Agelenidae). Prey consumption refers to period from penultimate instar through remainder of life. Regressions: (a) weight of eggs produced $y = 0.2x - 38.6$, $r = 0.877$, $P = < 0.01$; (b) number of eggs produced $y = -82.7 + 0.29x - 0.00005x^2$, $r = 0.969$, $P = < 0.01$. From Riechert & Tracy (1975).

Social structure

Spiders are not necessarily free to occupy the foraging sites of their choice, because intraspecific and interspecific competition may force them into less preferable foraging patches (Morse, 1980; Riechert, 1981, 1986). There is considerable evidence from spacing patterns of spiders (e.g. regularity in nearest-neighbour distances) to suggest the operation of either exploitation or interference competition (Riechert, 1982). Most of the competition appears to occur within species populations, and interspecific interactions are often predatory rather than competitive in nature (see, for example, Riechert & Cady, 1983; Wise, 1984).

The majority of web-spider species exhibit a territorial social structure involving the defence of an area larger than the web. The wandering spiders maintain a personal space or 'roving territory'. The size of the territory defended is dependent on prey availabilities (Riechert, 1978, 1982) and spiders compete for sites and associated territories that maximize their foraging success. Individuals that fail to obtain prime sites are forced into marginal habitats. Population differences in territory size have been maintained in individuals reared in the laboratory with unlimited food and with no experience with conspecifics (*A. aperta*: Riechert, 1986; Riechert & Maynard Smith, 1989; *Metepeira* sp. (Araneidae): Uetz *et al.*, 1982, 1986). Riechert (1981) suggests that territory sizes are adjusted over evolutionary time to match the lows in prey availabilities for different habitats, increasing the probability of survival to reproduction of territory holders during the worst possible times.

SPIDER PREDATION AND PREY DENSITY

The conditions under which predators may limit populations of prey have received considerable attention from ecologists. This is due, in part, to an interest in controlling pest outbreaks in agricultural systems. Both the theoretical (Hassell, 1978) and applied biological control literature (Caltagirone, 1981) emphasize that the ideal predator should show high specificity to a given pest and have the ability to track the density of prey (Fig. 13.5a) (Chapters 3 and 19). Both theoretical (see,

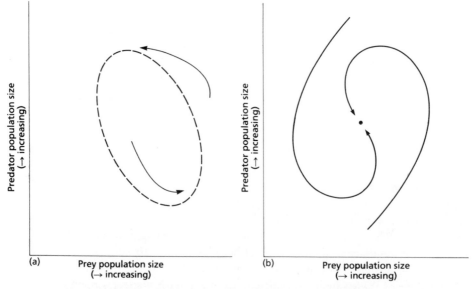

Fig. 13.5 Graphical representations of two kinds of stable predator–prey interactions. (a) Limit cycle in which prey and predator population numbers cycle out of phase with one another. (b) Stable equilibrium in which the corresponding population densities approach an equilibrium point. Arrows refer to respective population trajectories, approaching limit cycle (– – –) in (a) and equilibrium point (●) in (b). (See pp. 63–77.)

for example, Rosenzweig & MacArthur, 1963; DeAngelis, 1975; Tanner, 1975; Post & Travis, 1979) and empirical (see, for example, Kayashima, 1961; Kiritani & Kakiya, 1975; Edwards *et al.*, 1979; Mansour *et al.*, 1980b; Roach, 1980; Jones, 1981; Chiverton, 1986) evidence suggests, however, that assemblages of generalist predators can limit the growth of associated prey populations (Fig. 13.5b). In addition to its general significance to community structure and dynamics in natural eco-systems (Erlinge *et al.*, 1984), this form of control may be important in short-lived crops, where polyphagy may help maintain predator abundance when pest species' numbers decline (Murdoch, 1975; Ehler & Miller, 1978). Nevertheless, the importance of polyphagous predators in pest control is debated (see, for example, Riechert & Lockley, 1984; Spiller, 1986).

Most temperate spider taxa appear to meet the conditions considered necessary for them to deplete prey population numbers: they are present in diverse assemblages (Turnbull, 1973), and when hungry they will consume all prey types encountered that are within favourable size ranges (Riechert & Luczak, 1982). On the other hand, spiders are territorial and even cannibalistic towards members of their own species populations, a social structure that sets bounds on population growth that is independent of immediate food supplies (Riechert, 1982). Spiders also have long generation times relative to most of their insect prey. Both these traits mean that individual spider species tend not to exhibit density-dependent tracking of their prey (Riechert & Lockley, 1984). As might be predicted, therefore, successes in biocontrol using single spider species have been rare (Riechert & Lockley, 1984).

It is of considerable interest to understand how spiders influence the abundance of their prey in natural ecosystems. A significant effect of the generalist spider assemblage on associated prey populations was suggested as early as 1960s (Clarke & Grant, 1968), but experimental studies are few. Clarke and Grant removed spiders from an enclosed area and compared prey numbers over time in these areas with numbers in enclosures from which spiders had not been removed. They observed exponential rises in prey numbers in the removed area that were absent in the plots containing spiders.

A significant effect of spider predation upon pest populations in agroecosystems has only been suggested more recently (see, for example, Kiritani & Kakiya, 1975; Edwards *et al.*, 1979; Jones, 1981; Sunderland *et al.*, 1985; Chiverton, 1986). The greatest success in biocontrol using spiders has been associated with habitat manipulations. The Chinese, for instance, completed a province-wide experiment in rice paddies in which they 'encouraged' the spider fauna by providing teepee-like bamboo structures for shelter from unfavourable temperature extremes (Jones, 1981). When a pest outbreak was noted in a section of a paddy, the bamboo teepees were moved at midday, when the spiders would be sheltering in them, to the infested areas. By augmenting the spider assemblage, the Chinese were able to reduce their use of chemical pesticides by 60%. Sadly, this practice is being discontinued with the desire to make the Chinese agricultural system more westernized: the people no longer wish to expend the manpower required for the implementation of the control measure (Min Yao Liu, personal communication).

The effect of habitat manipulations on spider diversity and prey population density has recently been quantified by Riechert and Bishop (1990). They managed a garden in order to encourage a diverse assemblage of spiders to remain in the system following natural immigration. The system was chosen as an example of the worst kind of situation for spiders, namely a temporary habitat from which predatory arthropods are continuously disturbed by cultivation practices. If a significant effect was observed in this situation, then it might be generalized to other contexts. The aim was to create a favourable physical and feeding environment for the spider fauna as it immigrated into the garden from nearby habitats. Two types of habitat manipulations were tested: (i) mulch was laid between rows and around individual plants; and (ii) rows of vegetables were alternated with rows of flowering buckwheat. Both habitat manipulations were expected to reduce emigration of spiders from the treatment plots, but for different reasons: the mulch was expected to provide high humidities and protection from temperature extremes that spiders seek during unfavourable periods of the day and night (Riechert & Gillespie, 1986), while the presence of flowering plants was expected to attract pollinating insects that might serve as an alternative source of food for the spiders during periods of low pest densities. Plots were set up with these manipulations separately and in combination, and there were control plots with neither treatment. A final set of plots contained both flowering plants and mulch but the spiders were systematically removed as in the Clarke and Grant (1968) experiment already mentioned. The experiment was carried out in two contrasting habitats, hillside versus second river terrace.

The authors found that for all vegetable crops that showed some level of insect damage, spider numbers were significantly higher in plots with mulch added, and plant damage and insect numbers were significantly lower in these plots than in plots lacking mulch and the plots from which spiders had been removed (Fig. 13.6). These results are similar to those of Clarke and Grant's (1968) experiment which was completed in a natural ecosystem. They are also consistent with other findings on the effects of the spider assemblage on associated prey populations. Chiverton (1986), for instance, placed barriers in cereal fields to limit the entrance of crawling predaceous arthropods (spiders were one of three predominant types) into selected plots. This reduction of predation pressure brought about increases in cereal aphid density of two to six times that in unenclosed control plots.

Spiders and other generalist predators can have a direct effect on prey populations through consumption of prey, and an indirect effect by causing them to disperse from their food source (Nakasuji *et al.*,

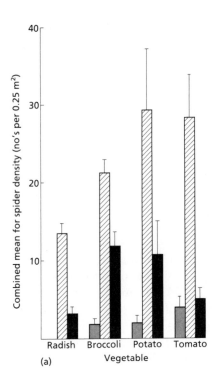

(a)

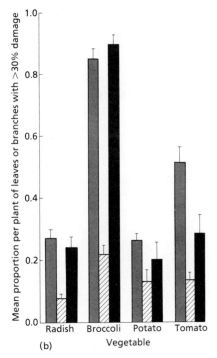

(b)

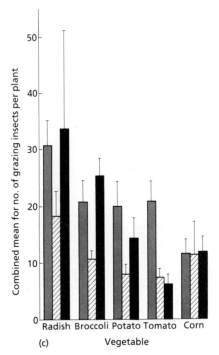

(c)

Fig. 13.6 Example of spider assemblage potential to control prey in a garden ecosystem with habitat manipulations completed to augment spider numbers and diversity. Bare ground (■), control plots; mulch (▨), treatment having a significant positive effect on spider numbers; mulch + flowers − spiders (■), treatment containing mulch from which spiders were removed. Data collected at maturity of the respective vegetable types in mixed vegetable garden systems in east Tennessee. T-bars refer to standard errors. (a) Spider densities; (b) plant damage (vegetable types showing some degree of damage in at least one plot shown); (c) pest insect numbers on plants.

1973; Mansour *et al.*, 1981; Dill, 1987). These disturbance effects are reported to cause indirect mortality since many of the dispersed individuals fail to relocate at a feeding site (Yamanaka *et al.*, 1972; Nakasuji *et al.*, 1973; Mansour *et al.*, 1981).

The consensus of these small-scale experiments is that spider assemblages can significantly reduce prey populations, even though spiders are sit-and-wait predators and do not exhibit density-dependent tracking of individual prey species. The depression of prey numbers achieved by spiders is small when compared to the successes achieved with some parasitoids (60–80% in the Riechert and Bishop (1990) study as compared to 95–99.7% in some parasitoids (Beddington *et al.*, 1978)). Nevertheless, the spider fauna in a system may serve as a buffer in limiting the potential for exponential growth in associated prey populations. Thus in the agroecosystem, the presence of a diverse assemblage of spider species at the time of pest invasion of the crop may prevent pest numbers from ever building up to damaging levels in the first place. The extent to which this role is realized for spiders in the agroecosystem depends on favourable habitat conditions as already discussed. The chemical applications common to agroecosystems are also critical limiting factors to the use of spiders in biocontrol. In an experiment recently completed in a garden system (Riechert, unpublished data), it was found that regardless of the timing of treatments with a broad-spectrum pesticide, all plots to which pesticides had been added had significantly lower numbers of spiders and species richness than did control plots. The future use of spiders as biocontrol agents requires the development of chemical pesticides that are targeted for specific pests, leaving the beneficial predatory assemblage intact to effect its control.

14: Macroparasites: Worms and Others

ANDREW P. DOBSON, PETER J. HUDSON &
ANNARIE M. LYLES

INTRODUCTION – IT'S A WORMY WORLD

The traditional ecological perception of parasitic helminths is of rather obscure group of 'wormy' creatures which make their living by feeding on excess food in the guts of their hosts. In this chapter, we attempt to redress this balance by showing that the parasitic helminths are a diverse assemblage of natural enemies, which cause a ubiquitous and constant drain on the energetic resources of most free-living organisms. Parasitic helminths may be considered as supremely adapted predators that constantly imposed a slow but steady drain on host resources, and thus significantly influence their host's energetics, behaviour, demography and evolution.

The diversity of parasitic helminths

Parasitism has evolved many times. It is the obligate life style in cestodes, trematodes and acanthocephalans and may be the commonest life style amongst the nematodes. The different taxonomic groups of parasitic helminths have distinct differences in their life history strategies, and in the numbers of different host individuals required to complete the parasite's life cycle. Many parasitic nematodes and all the monogenean trematodes have direct (or monoxenic) life cycles. In these species transmission between hosts proceeds either directly or via free-living larval infective stages. In contrast, the cestodes, digenean trematodes, acanthocephalans and some nematodes have indirect (or heteroxenic) life cycles. These species utilize one or more species of intermediate host, as well as one or more larval stages during transmission between successive definitive hosts. The different taxonomic groups of parasitic worms also exhibit characteristic differences in reproductive strategies in each host of their life cycle. The main features of these are illustrated in Table 14.1.

The average burden of parasites per host

Most vertebrates, and many invertebrates, accumulate parasites throughout the course of their lives. Both the diversity of parasite species and the burden of each species tend to increase with host age

Table 14.1 The reproductive strategies of different groups of parasitic helminths in each host of their life cycle

	Definitive	Intermediate
Monoxenic		
Monogenea	Monoecious (A : i,v)	—
Nematoda	Dioecious (A,T : i,v)	—
Heteroxenic		
Digenea	Monoecious and dioecious (A,T : v)	1st: asexual (m) 2nd: paratenic (i,v)
Cestoidea	Monoecious (A,T : v)	1st: paratenic (i) 2nd: asexual (v)
Nematoda	Dioecious (A,T : i,v)	1st: paratenic (i,m) 2nd: paratenic (v)
Acanthocephala	Dioecious (A,T : v)	1st: paratenic (i) 2nd: paratenic (i,v)

A, aquatic; T, terrestrial; i, invertebrate; v, vertebrate; m, mollusc. In all the heteroxenic hosts the presence of a second intermediate host is not obligatory, but occurs in many species. A *paratenic* host is one in which the parasite does not reproduce, but is used for transport to the definitive host.

(Fig. 14.1). In ecological terms, each host individual can be considered as a discrete patch of habitat supporting a subset of the total parasite assemblage present in the habitats used by the host. The size and diversity of these communities can be seen in a summary of the parasites of North American mammals (Table 14.2); this analysis was undertaken by collating data from 88 different parasite surveys. Although the specific burden of any host is dependent on a variety of different factors (e.g. the host's feeding strategy and social system), in general,

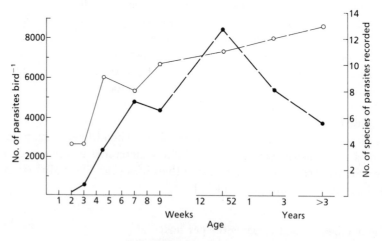

Fig. 14.1 The relationship between age and mean parasite burden for a population of brown pelicans (*Pelecanus occidentalis*). ○, data for mean numbers of helminth parasites per bird in each age class; ●, are mean numbers of parasite species per bird. Data from Humphrey *et al.*, (1978).

Table 14.2 The average parasite burdens of North American mammals. The data are from 76 populations of mammals comprising 10 species of lagomorphs, 22 rodents, 35 carnivores and 11 artiodactyles. Crude summary data are presented for both the total diversity of parasites present in the population sampled and an estimate of the average abundance in each host individual. In general, carnivores had the most diverse parasite faunas, lagomorphs the least diverse, while rodents and artiodactyles supported communities of intermediate diversity. Gregarious species tended to have significantly higher mean parasite burdens than 'asocial' species. Data from Lyles & Dobson (in preparation)

	Mean burden per individual host	No. of species per host population
Trematodes	108	1.8
Cestodes	140	2.8
Nematodes	117	5.3
Acanthocephalans	1	0.3
No. of parasite species	3	10

the average mammal harbours a multispecies community of several hundred parasitic worms.

The impact of helminth parasites

The mechanisms of parasite pathology differ widely between different systems. Many nematodes of veterinary importance produce significant reductions in the growth rate of individual hosts, the spiny heads of acanthocephalans produce extensive ulceration in the guts of infected individuals, while the asexual stages of the taenaid cestodes produce cysts in the viscera and even in the brains of their hosts. The majority of carefully controlled studies of the impact of parasites on their hosts suggest that host mortality rates increase with parasite burden (Fig. 14.2a,b) or cause significant reductions in the fecundity of infected hosts.

The energetics of parasitic helminths

Only a few studies have attempted to quantify the impact of parasitic helminths on their hosts in energetic terms. The data in Fig. 14.3 illustrate the rates of energy flow in a controlled laboratory system. Although the parasites only obtained 0.8% of the energy ingested by the host, they assimilated 88% of this energy and invested 45% in production (growth and fecundity). In contrast, the hosts assimilated 74% of the energy they ingested, but were able to invest only 5% of this in production. Complementary comparative studies of the body composition of parasitic and free-living platyhelminths (Calow & Jennings, 1974) suggest that parasitic helminths store significantly less

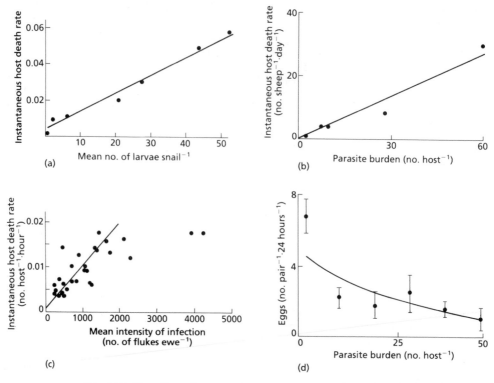

Fig. 14.2 The effect of parasite burden on host mortality: (a) the nematode *Elaphostrongylus rangiferi* in the snail *Arianta arbustorum* (after Skorping, 1985); (b) the cestode *Hymenolepis diminuta* in the beetle *Tribolium confusum* (after Keymer, 1980); (c) the trematode *Fasciola hepatica* in sheep (after Smith, 1984a). (d) The influence of parasite infection on host fecundity: the cestode *H. diminuta* in the beetle *T. confusum*. After Keymer (1980).

of their assimilated energy as lipid than their free-living relatives. The adoption of a parasitic life style has reduced the need to form lipid reserves and this allows parasitic species to produce a large number of eggs, more or less continuously.

LIFE HISTORY STRATEGIES OF PARASITIC HELMINTHS

One of the major energetic advantages of adopting a parasitic life style is that it permits a potentially enormous reproductive rate. Table 14.3 compares the fecundity of a number of free-living worms with the fecundity of similar-sized parasitic worms. These data suggest that parasites produce offspring at rates several orders of magnitude higher than free-living worms. However, parasite life histories involve a number of sequential 'birth' and 'survival' processes which allow the parasite to transmit itself between successive stages in the life cycle. Although the rates of egg production are high in parasites, the rates of transmission between successive hosts are usually low and transmission

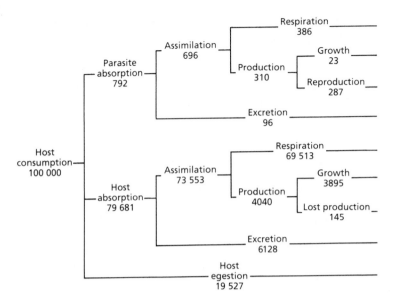

Fig. 14.3 The relative rates of energy flow in a laboratory host–parasite system: rats (albino, ASH/W strain) parasitized with the cestode *H. diminuta*. The units are arbitrary, relative to host consumption of 1×10^5. After Bailey (1975a).

Table 14.3 The comparative fecundity of free-living and parasitic flatworms. Fecundity data from Reynoldson (1983) and Crompton & Joyner (1980)

Species	Fecundity	Host
FREE-LIVING SPECIES	(young adult^{-1} breeding season^{-1})	
Turbellaria		
Polycelis nigra	2.5	
P. tenuis	1.0 (27)	
Dugesia polychroa	0.8–1.7 (36)	
Planaria torva	4.9–15.2 (67)	
Dendrocoelom lacteum	4.9–10.3 (33)	
Bdellocephala punctata	16.0 (75)	
PARASITIC SPECIES	(eggs worm^{-1} day^{-1})	
Digenea		
Schistosoma mansoni	100*	Hamster
Cestoda		
Echinococcus granulosus	600*	Dog
Hymenolepis diminuta	200 000	Rat
Taenia saginata	720 000	Man
Acanthocephala		
Moniliformis moniliformis	5 000	Rat
Nematoda		
Ascaris lumbricoides	200 000	Pig
Enterobius vermicularis	11 000	Man
Wuchereria bancrofti	12 500	Man

*These species also have a period of asexual reproduction in their life cycles.

stages have to successfully establish as adult parasites in a suitable definitive host before they can reproduce themselves and be counted as contributing to their parent's fitness. Parasite transmission must therefore be considered as an additional component of fecundity when determining the net reproductive rates of parasites. Initially let us consider the different birth, death and transmission processes that operate in the life cycle of a common parasite of game birds.

Reproductive rates of monoxenic macroparasites

Trichostrongylus tenuis is a parasitic nematode that lives in the caecal sac of game birds. It is a particular problem in red grouse, *Lagopus lagopus scoticus* (Hudson, 1986), where it causes the disease strongy-

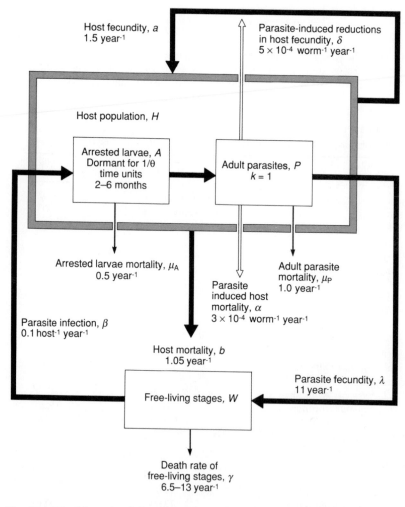

Fig. 14.4 The life cycle of the nematode *Trichostrongyle tenuis* in the red grouse (*Lagopus lagopus scoticus*). After Hudson *et al.,* (1989).

parasites living in the host's caecae produce eggs which are passed out
in the caecal faeces where they eventually hatch to produce several
moults of free-living larvae. The final larval stages migrate away from
the faecal pat and locate a heather plant, which they climb up before
encysting in the flower buds. These buds form the major component of
the diet of adult grouse, and transmission is completed when the birds
ingest an infected flower bud. Worm burdens in young birds build up
rapidly, and an adult bird may contain between 1000 and 20 000 worms in
its caecae. The frequency distribution of worms per bird is initially
random in immature birds, however, as the birds age, small differences
in susceptibility rapidly produce more aggregated distributions, a few
individuals harbouring heavy infections, while most birds harbour
light to moderate infections.

Effect on host survival and fecundity

Although each stage in the parasite's life cycle has a characteristic
mortality rate, an additional important form of mortality results from
interactions between the parasite and its host. In the case of *T. tenuis*
the parasite has been shown to reduce both host survival and fecundity
(Fig. 14.5, Table 14.4). These effects on the host's demographic rates
are of key importance in host–parasite relationships; they determine
both the impact of the parasite on the host population *and* the parasite's
own life expectancy and transmission success. This point may be
illustrated by deriving an expression for the basic reproductive rate of
the parasite, and by examining some models of its population dynamics.

Basic reproductive rate, R_0

The life time reproductive success of a parasite, R_0, can be formally
defined as the number of secondary infections produced by a single
infectious female worm when introduced into a wholly susceptible
host population (Anderson & May, 1979). Derivation of an expression
for R_0 for any parasite requires the construction of a set of equations
which describe the birth and death rates of the parasite at each stage of
its life cycle. Three coupled differential equations may be used to
describe the dynamics of the hosts, H, the adult parasites, P, and the
free-living population, W:

$$\frac{dH}{dt} = (a - b)H - (\alpha + \delta)P \tag{14.1}$$

$$\frac{dW}{dt} = \lambda P - \gamma W - \beta WH \tag{14.2}$$

$$\frac{dP}{dt} = \beta WH - (\mu + b + \alpha)P - \alpha\left(\frac{P^2}{H}\right)\frac{(k + 1)}{k} \tag{14.3}$$

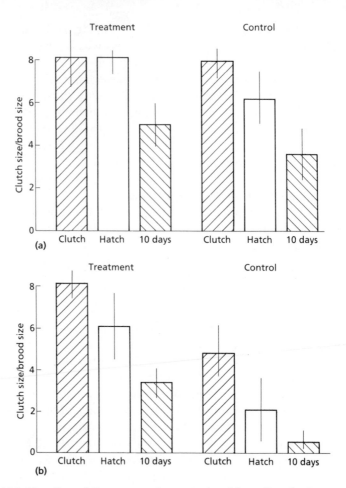

Fig. 14.5 The effect of *T. tenuis* on the survival and fecundity of red grouse. These results were obtained experimentally by orally dosing free-living birds with either an anthelmintic drug to remove the parasites or with water as a control. The experiment was repeated in two different years: (a) 1982; and (b) 1983. The data compare the fecundity of treated and control birds in terms of their clutch size, brood size and brood survival to age 10 days. From Hudson (1986).

The parameters of the model are defined in Fig. 14.4. The equations can be used to determine the growth rate of the parasite when first introduced into a host population of size H:

$$R_0 = \frac{\beta\lambda H}{(\mu_p + b + \alpha)(\gamma + \beta H)} = \frac{T_1}{M_1 M_2} \tag{14.4}$$

Here $T_1 = \beta\lambda H$, the rate of transmission of eggs from birth through to establishment in another grouse, and M_1 and M_2 are the mortality rates of each stage in the life cycle: for adult worms, $M_1 = (\mu_p + b + \alpha)$, for free-living larvae, $M_2 = (\gamma + \beta H)$. Essentially, the basic reproductive rate of the parasite is the product of the birth and transmission rates and the life expectancies of each stage in its life cycle. Increases in transmission efficiency and the life expectancy of each life-cycle stage tend to increase R_0, while increases in parasite-induced host mortality

Table 14.4 The effect of *T. tenuis* on the survival of red grouse. Data show the recapture rate of control and treated birds. From Hudson, 1986)

Year	Treated birds	Control birds	Significance
1982, number tagged	44	159	
2 months after treatment	4	9	
Nematode eggs g^{-1} faeces	5.0 ± 2.6	87.1 ± 1.4	$P < 0.01$
Number recaught (winter 1982)	12	30	
'Survival'	0.27	0.19	$P < 0.10$
1983, number tagged	37	115	
2 months after treatment	7	8	
Nematode eggs g^{-1} faeces	72.1 ± 1.4	223.4 ± 1.4	$P < 0.05$
Number recaught (winter 1982)	4	9	
'Survival'	0.11	0.08	NS

tend to reduce the basic reproductive rate. Estimates of R_0 for the grouse–*T. tenuis* system suggest a value of between 5 and 10 secondary infections (Dobson & Hudson, 1992). As an adult female worm produces around 250 eggs day^{-1} and lives on average for 12–18 months, her average lifetime reproductive success is several orders of magnitude smaller than even her *daily* rate of egg production.

Threshold for establishment, H_T

Because the expression for R_0 always contains a term for host population size, it is possible to obtain an expression for the threshold number of hosts required to just sustain an infection of the parasite. This is obtained by reorganizing equation 14.4 with $R_0 = 1$;

$$H_T = \frac{\gamma M_1}{\beta \,(\lambda - M_1)} \tag{14.5}$$

This expression suggests that increases in transmission efficiency and rates of egg production tend to reduce the threshold for establishment, while short adult life expectancy and parasite-induced host mortality tend to increase the size of the population required to sustain an infection. Further reductions in the threshold for establishment may be achieved by the ability of the nematode larvae to enter a state of 'arrested development' or *hypobiosis* immediately after having established in their definitive hosts. Studies in the north and east of Scotland, where both grouse densities and the winter survival of free-living nematode larvae are relatively low, suggest that this strategy is relatively widespread (Shaw, 1988).

Aggregation of the parasite in the host population

As with most helminth parasites, the statistical distribution of the adult parasites in the host population is aggregated and best described by a negative binomial distribution (Hudson, 1986). A few hosts harbour large parasite burdens, while the majority of hosts contain low or intermediate burdens. Small differences in susceptibility to infection are also important in producing these aggregated distributions (Anderson & Gordon, 1982). These differences may be genetic, behavioural or due to heterogeneity in the spatial distribution of the parasite's free-living infective stages. They have important dynamic consequences for the stability of the host−parasite relationship in that regulation of the parasite population is concentrated on the section of the population in heavily infected hosts. Parasite-induced host mortality of heavily infected host individuals thus operates as an important source of density-dependent parasite mortality, which acts in addition to other sources of regulation such as reductions in parasite fecundity or survival (Keymer, 1982).

Equilibrium host and parasite densities

Equations 14.1 to 14.3 may be used to derive expressions for the equilibrium numbers of hosts and parasites (Anderson & May, 1978):

$$\frac{P^\star}{H^\star} = M_P^\star = \frac{(a - b)}{(\alpha + \delta)} \tag{14.6}$$

$$W^\star = \frac{d'(a - b)}{\beta\,(\alpha + \delta)} \tag{14.7}$$

$$H^\star = \frac{\gamma d'}{\beta\,(\lambda - d')} \tag{14.8}$$

Here $M_P^\star$ is the mean parasite burden of adult grouse at equilibrium and $d' = \mu_p + \alpha + b + k'(a - b)\alpha/(\alpha + \delta)$, with $k' = (k + 1)/k$.

These expressions suggest that host population density varies directly with the degree of aggregation of the parasites $(k + 1)/k$, and inversely with both parasite fecundity, λ, and the rate at which infective larvae are ingested, β. Mean parasite burden $M^\star$ varies inversely with the parasite's ability to reduce host survival and fecundity. The high parasite burdens observed in this sytem suggest, therefore, that the per capita mortality rate of *T. tenuis* is relatively low.

Experiments with grouse that have been treated with antihelmintics to reduce their parasite burdens confirm that the parasites have only a low effect on host survival. However, these experiments also suggest that the parasites do have a significant effect on the fecundity of infected hosts (Fig. 14.5).

Parasites and cycles in grouse abundance

The long-term dynamics of the grouse nematode system are dominated by the relative effect of the parasite on host survival and fecundity. In any helminth system where the parasite has a more significant effect on host fecundity than on survival, there will be a propensity to oscillate (May & Anderson, 1978). In the grouse system, the long-lived free-living stages and territorial behaviour of the grouse interact synergistically with this oscillatory tendency to give stable limit cycles of grouse and nematode abundance. Models of the system produce cycles of abundance which are simiar in frequency and magnitude to those observed in long-term studies of individual grouse populations (Hudson *et al.*, 1992). Further evidence that the parasite is a cause of long-term cycles in grouse is provided by current experiments in which the treatment of grouse with antihelmintic drugs has stopped the decline phase of the grouse cycle in several populations (Hudson *et al.*, in preparation).

COMPLEX LIFE CYCLE (HETEROXENIC) MACROPARASITES

Although many nematodes and the monogenetic trematodes have direct life cycles, the commonest life cycles in the parasitic helminths are those where the parasite utilizes two or more species of host in sequence. This type of life cycle is obligate in the acanthocephalans, cestodes, digenetic trematodes and in many nematodes. The advantages of this type of life cycle may be considered by examining the life history of *Polymorphus minutus*, an acanthocephalan parasite that utilizes the freshwater shrimp *Gammarus pulex* as its obligate intermediate host, and a wide range of waterfowl and mammals as its definitive hosts (Crompton & Harrison, 1965). This lack of specificity for a definitive host is characteristic of many parasitic helminths with heteroxenic life cycles. Interpreting the adaptive nature of this type of complex life cycle (Fig. 14.6, Table 14.4) is best undertaken by deriving expressions for R_0, the basic reproductive rate of the parasite.

Basic reproductive rate of heteroxenic macroparasites

The derivation of an expression for R_0 for heteroxenic parasites is similar to that illustrated above for the monoxenic parasite *T. tenuis*, and results in an expression of the general form

$$R_0 = \frac{T_1 T_2}{M_1 M_2 M_3} \tag{14.9}$$

Essentially, the basic reproductive rate is the product of the birth and transmission rates (T_1 and T_2), and the net mortality rates (M_1, M_2 and M_3) of each stage in the parasite life cycle (Dobson & Keymer, 1985). In order for life cycles of this type to evolve, the transmission rate of the parasite using the intermediate host must produce a value of R_0 that exceeds that for the life cycle in the absence of the intermediate host. Increased mortality rates, incurred by the use of additional life history stages, must be compensated by increases in transmission

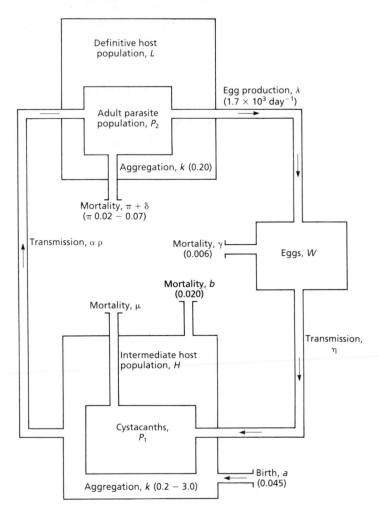

Fig. 14.6 The life cycle of the acanthocephalan *Polymorphus minutus*. The figures in brackets give approximate estimates of the birth, death and transmission rates per day at each stage of the interaction. No brackets indicates parameter estimates unknown. After Dobson & Keymer (1985).

efficiency or rates of fecundity (Dobson & Keymer, 1985). In the acantho-cephalans the vertebrate host, where the parasite reproduces sexually, was secondarily included into the life cycle (Crompton, 1985). In this case, evolution from a motile, free-living adult stage, to a relatively sedentary parasitic one, allowed a significant increase in fecundity and in the opportunities for individual parasites to find mates.

A further major advantage of complex life cycles is that they usually reduce the threshold at which the parasite is able to establish (Dobson, 1988a; Anderson, 1992). This allows the parasite to colonize smaller populations of hosts than might otherwise be possible, and to exploit ephemeral or migratory hosts that may only visit the parasite's habitat for short periods of time. In the case of *P. minutus* many of the waterfowl species it utilizes as definitive hosts are migratory species that move between different water-bodies at different times of the year. In order to increase its ability to colonize waterfowl hosts, the parasite has evolved the ability to modify the behaviour and chromatophore response of infected *G. pulex*, by altering their phototactic response and hence reducing their ability to camouflage themselves against the substrate. This increases the shrimps' susceptibility to predation by definitive hosts (Hindsbo, 1972; Moore, 1984). Digestion of infected intermediate hosts in the definitive host's gut leads to the release of the spiny-headed parasitic adult stages, which develop rapidly after attaching to the host's gut wall. The fertilized females produce large numbers of eggs which pass out of the gut with the host's faeces into the water where they are ingested by shrimps. Rapid amplification of the parasite can cause serious mortality in flocks of waterfowl (Itamies *et al.*, 1980).

Heteroxenic parasites with two reproductive phases

The digenean trematodes exhibit a further level of complexity in life cycle structure by possessing an asexually reproducing stage in their intermediate host (Fig. 14.7). All digeneans utilize molluscs as intermediate hosts, but molluscs were part of the life cycle before the secondary inclusion of vertebrate hosts (Gibson, 1981; Brooks *et al.*, 1985). The available phylogenetic evidence suggests that *asexual* reproduction in these species is also a secondary adaptation in response to the parasitic existence in the mollusc host (Shoop, 1988). The schistosomes are representative of this type of life cycle.

SCHISTOSOMIASIS IN SNAILS AND MAN

Schistomiasis is one of the most important human diseases in the tropics. Current estimates suggest that around 100 million people are infected with these worms. The sexually reproducing adult schistosomes live in permanent copula in the blood of infected individuals, and their eggs are passed in the urine of infected individuals. Pathology occurs as a consequence of the host mounting an immune response to the parasite, when an accumulation of eggs blocks the spleen and excretory organs of infected hosts. This leads to debilitation and ultimately to the death of infected hosts.

An important comparison can be made at this point between the digeneans and the acanthocephalans. Parasites of both these groups of parasites tend to be highly specific towards their intermediate hosts but fairly general in their choice of definitive host. In both cases, parasite burdens in the definitive host can rise to levels where they cause either significant morbidity or even mortality. In part this may reflect the shorter duration of evolutionary association between the parasite and its definitive host, but it may also reflect the lack of opportunity to coevolve with any one of its definitive host species.

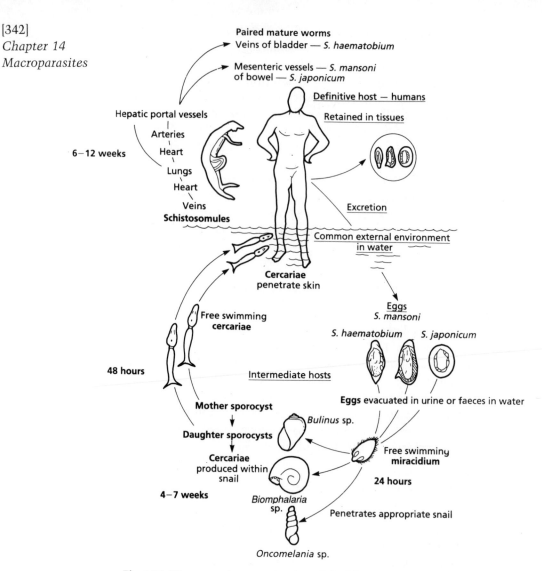

Fig. 14.7 Diagrammatic representation of the life cycle of human schistosomes. From Jordan & Webbe (1969).

However, the observed high levels of virulence may result from relatively weak selection on the parasite to develop a less damaging relationship with any one species of definitive host. Instead, selection may emphasize maximum utilization of these hosts as resources which can be turned into further parasite transmission stages. Understanding the relative importance of these different selection pressures requires further experimental and theoretical work.

ASEXUAL REPRODUCTION AND ITS INFLUENCE ON R_0 AND H_T

Expressions can again be derived for R_0 and H_T for life cycles containing two free-living stages and an asexually reproducing larval stage. As before, this expression consists of the product of the two transmission processes and the life expectancies of the four different stages in the life cycle (Dobson, 1988a; Anderson, 1992). More detailed analysis of this type of life cycle suggests that the evolution of asexual reproduction involves an important tradeoff between rates of production of free-living infective cercaria and the resultant pathology to infected molluscs (Pan, 1965; Wright, 1966; Minchella, 1985).

THE VERTEBRATE HOST'S IMMUNE RESPONSE TO HELMINTH INFECTION

An important feature of the relationship between the schistosomes and their vertebrate hosts is the host's immune response. The presence of the schistosomes in the host's body leads to the production of defence cells (antibodies), which attack and attempt to destroy the schistosomes. The immune system can be considered to act like the predator–prey relationship with the parasites acting as prey to the predatory antibodies. The population dynamics consequences of the human immune response to infection by helminths have been examined by Crombie and Anderson (1985) and Anderson and May (1985c). Because of the relatively short duration of the immune system's memory for helminth parasites, the dynamics of these systems are characterized by a tendency to oscillate when worm transmission rates are low, and a tendency to more stable dynamics when transmission is high (Fig. 14.8a). This analysis of the dynamics of the interaction between the parasite and the host's ability to mount an immune response provides an important explanation for the patterns of variation in parasite burdens with age observed in data collected from different human populations infected with schistosomiasis (Fig. 14.8b).

COEVOLUTION OF HOST–PARASITE RELATIONSHIPS

Unravelling the evolutionary interactions between the parasitic helminths and their hosts is a daunting task, not least because of the different degrees of specialism and generalism exhibited by various helminth taxa for their definitive and intermediate hosts. Although the techniques of phylogenetic systematics are beginning to produce some interesting examples of joint radiation in well-studied groups of hosts (e.g. pinworms and primates; Brooks & Glen, 1982; Fig. 14.9), it is likely that the majority of helminth species have yet to be described by taxonomists. This produces serious gaps and potential biases in studies which attempt to construct coevolutionary phylogenies for less completely studied groups of hosts (Gibson, 1987). Nevertheless, a number of detailed experimental studies of specific systems are beginning to produce valuable evidence about the genetics and coevolution of host–helminth relationships.

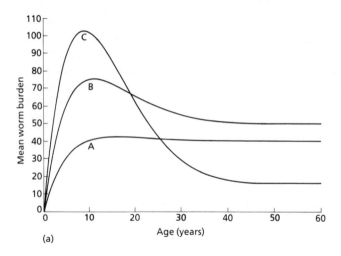

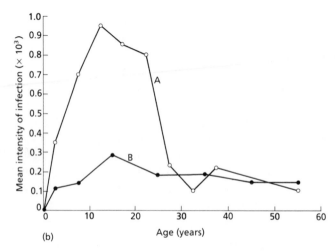

Fig. 14.8 (a) The expected relationship between age and parasite burden for the acquired immunity model described by Anderson & May (1985c) and Anderson & Crombie (1985). A corresponds to low transmission rates and short-lived but strong immunity; B to higher transmission rates and longer lasting immunity, while C corresponds to high transmission and weak but persistent immunity. (b) The observed relationship between host age and intensity of infection (eggs g^{-1} of faeces) for populations from areas with differing rates of *S. mansoni* transmission: A, high transmission rates; B, low transmission rates.

Coevolution between snails and schistosomes

The genetics underlying the relationship of the schistosomes and their hosts have been studied in considerable detail (Richards, 1975; Rollinson & Southgate, 1985; Woodruff, 1986). These studies suggest that both the establishment of the worm and its subsequent development are dependent upon the genotype of both the parasite and the host. The presence of parasite-resistant strains of hosts has led to the suggestion

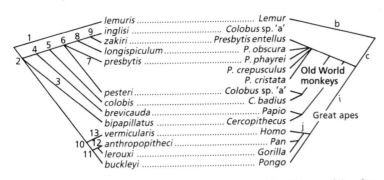

Fig. 14.9 The phylogenies of the pinworm genus *Enterobius* (Nematoda) and its primate hosts according to Mitter & Brooks (1983).

that schistosomiasis might be controlled by artificially increasing the proportion of resistant snail strains (Richards, 1970; Woodruff, 1986). A detailed study of the interaction between *Schistosoma mansoni* and its snail host *Biomphalaria glabrata* by Minchella and LoVerde (1981) demonstrated that snail strains that are 'insusceptible' to schistosome infection (i.e. unsuitable for parasite development) have *higher* reproductive success than susceptible strains when reared alone (Table 14.5). However, the presence of the parasite *reverses* the fitness advantage of 'insusceptible' snails in mixed host populations. Thus, the relative fitness of the different snail strains depends upon the presence of the parasite. In parasite-free environments, snail populations may be dominated by the 'insusceptible' snails, while the presence of schistosomes gives the 'susceptible' strains a frequency-dependent advantage which allows their numbers to increase in populations where the parasite is present.

Single locus evidence for susceptibility/resistance

Several studies have produced evidence to suggest that single locus traits may determine key aspects of susceptibility and resistance. Wassom

Table 14.5 Estimates of net reproductive rate (R_0) and intrinsic rate of increase (r) for a 30-week period for insusceptible and susceptible stocks of *Biomphalaria glabrata* with and without *Schistosoma mansoni*. After Minchella & LoVerde (1981)

Treatment	Single stock		Mixed stock	
	R_0	r	R_0	r
Unexposed insusceptible	13.95	0.74	7.97	0.49
Unexposed susceptible	7.13	0.75	5.20	0.39
Exposed insusceptible	9.77	0.64	1.76	0.24
Exposed susceptible	4.42	0.47	4.90	0.35

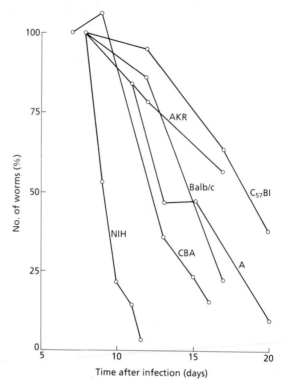

Fig. 14.10 The normalized course of experimental infections with the nematode *Trichinella spiralis* in different inbred strains of laboratory mouse. After Wakelin (1978).

et al. (1973) showed that susceptibility to the cestode, *Hymenolepis citelli*, in the deer mouse, *Peromyscus maniculatus*, was inherited in a simple Mendelian fashion, and Wakelin (1978, 1985) discusses other traits which may be inherited in this way. Laboratory studies with inbred mouse strains also provide clear evidence that the establishment and subsequent survival of parasitic worms is dependent upon the host's genetic background (Fig. 14.10). Similarly, evidence of predisposition to helminth infection in human and mouse populations has been obtained by comparing parasite burdens in individuals before and after reinfection following removal of parasites (Schad & Anderson, 1986; Scott, 1988).

Multiple locus evidence

Much of the evidence for multilocus control of genetic resistance to parasitic helminths comes from studies of parasites in hosts of agricultural and veterinary importance. The barbers pole worm, *Haemonchus contortus*, is a nematode that can cause significant reductions in the growth rate of sheep and lambs. Studies in Australia suggest that resistance of sheep to infection is controlled by a variety of genes

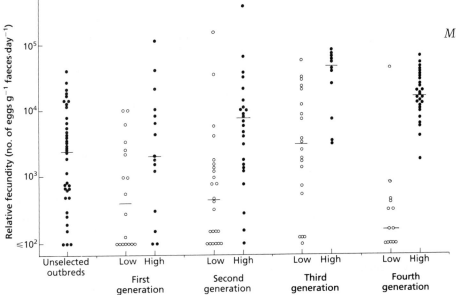

Fig. 14.11 The response of guinea pigs to selection for susceptibility and resistance to infection with *Trichostrongylus colubriformis*. The figure shows the relative fecundity of the worm (eggs g^{-1} faeces day^{-1}) in individual animals of the two selection lines. (After Rothwell in Wakelin, 1978.)

(Albers & Gray, 1986), and selective breeding experiments using laboratory populations of guinea pigs suggest that it should be possible to select host populations that are more (or less) resistant to worm invasion (Fig. 14.11).

Genetic variation in parasites and their hosts

Genetic variation in susceptibility in the host population will be of major importance in determining the statistical distribution of parasites in the host population (Anderson & Gordon, 1982). The failure of worms to mature in hosts that develop an effective immunity, and the survival rates of established worms, have both been demonstrated to be under genetic control (Wakelin, 1984). Variation in these traits between different individuals in free-living populations is likely to contribute significantly to observed degrees of aggregation of the parasite.

Relatively few studies have examined levels of genetic variability and heterozygosity in parasitic helminths. The studies of Rollinson and Wright (1984) and Mulvey and Vrijenhoek (1982) suggest that although populations of schistosomes from snails in isolated ponds tend to show low levels of variability, there are distinct differences in the gene frequencies of samples from each location. In contrast, Rollinson and Southgate (1985) found relatively high levels of genetic variability in infra-populations of adult schistosomes collected from lechwe (*Kobus lechwe kafuensis*).

These results are intriguing when considered in terms of the adaptations of parasites that permit them to establish in low density host populations. In fragmented habitats such as lakes, ponds or islands, where host populations may be either ephemeral or below the parasite's threshold for establishment, parasites will be strongly selected for their colonization abilities. Where rates of colonization are low, then those parasite individuals that are able to colonize isolated host populations will probably be able to infect the majority of hosts in that habitat. Low levels of genetic variability within hosts from the same habitat 'fragment' suggest that genetic drift and founder effects are important in determining the genetic structure of some parasite populations. In contrast, populations of parasites from migratory species, such as the lechwe, are likely to be acquired from a variety of locations and may be more heterogeneous. Further studies of levels of genetic variability of parasite populations at the within-host, within-host population and between-host population levels will be valuable in unravelling the genetics of these relationships.

FINAL DISCUSSION AND CONCLUSIONS

The last 10 years have seen a considerable increase in our understanding of the population dynamics of parasitic helminths and their interactions with their hosts. Much of this work has been stimulated by the mathematical framework developed by Anderson and May (1978). The field of helminth ecology continues to develop and to provide interesting new insights into the dynamics of plant–herbivore relationships (Grenfell, 1988), community structure and even animal behaviour (Milinski, 1984b; Moore, 1984).

Nevertheless, many fascinating problems still remain to be addressed on the ecology and evolution of the parasitic helminths as natural enemies. In particular, we are only beginning to understand the role of the host's immune response and its genetic control in determining the dynamics of the host–parasite relationship. Although there is some evidence to suggest that immuno-efficiency is dependent upon host nutritional status, our understanding of the energetic costs of the host's immune response is only rudimentary. Again, helminth parasites have been under-exploited as potential agents of biological control. Their relative absence from pest populations of introduced feral vertebrates (rats, goats, cats, dogs) on oceanic islands suggests that carefully designed, controlled release programmes might provide a powerful and effective way of controlling these pests (Dobson, 1988b).

15: Microparasites: Viruses and Bacteria

D. JAMES NOKES

INTRODUCTION

Viral and bacterial infections have long been associated with distinctive patterns of behaviour in host populations. We are all familiar with the classic form of the epidemic and the regular cycles in incidence displayed by these microparasitic infections within human host communities. Such phenomena have a well-documented historical record, examples of which are given in Fig. 15.1. This chapter is intended to provide some detail on the properties of the causative agents and the mechanisms of interaction between host and parasite populations that generate these often dramatic and memorable events in our lives.

Many of the factors governing what we think of as typical population biology of viruses and bacteria are embodied in two observations: (i) The aetiological agents are small directly reproducing microparasites; and (ii) classically, records of major epidemics and long-term cyclic incidence come from high density human populations, often in well-developed and urbanized societies, such as the UK and USA (see the Registrar General's Statistical Reports, Office of Population Censuses and Surveys, London and US Public Health Reports, Washington). Consider these two points, in turn, in more detail. Rapid multiplication (giving rise to short generation times) within the host individual, characteristic of microparasites, typically results in an acute infection, i.e. a crisis for the host, which if left unchecked will cause host death. High pathogenicity is therefore a common microparasite feature, and its population consequences have caught the imagination of us all. In the pandemic of influenza of 1919, for instance, it is estimated that over 20 million people died (McNeil, 1976); the Black Death in the middle ages is reputed to have reduced the population of Europe by one-quarter (Brayley, 1722). In the early 1960s there was a worldwide epidemic of rubella (German measles), which in the USA alone is calculated to have caused over 30 000 stillbirths and 20 000 cases of congenital rubella syndrome (Krugman, 1965). Today, infections such as measles, tetanus and polio, the many diarrhoea-inducing viruses and bacteria, e.g. rotavirus, *Vibrio cholerae* (cholera) and *Salmonella typhi* (typhoid), and acute respiratory infections, e.g. *Bordetella pertussis* (whooping cough) and respiratory syncitial virus, remain the dominant causes of roughly 14 million deaths annually of infants under the age of 5 years in the developing world (Evans, 1989; Grant, 1989).

[349]

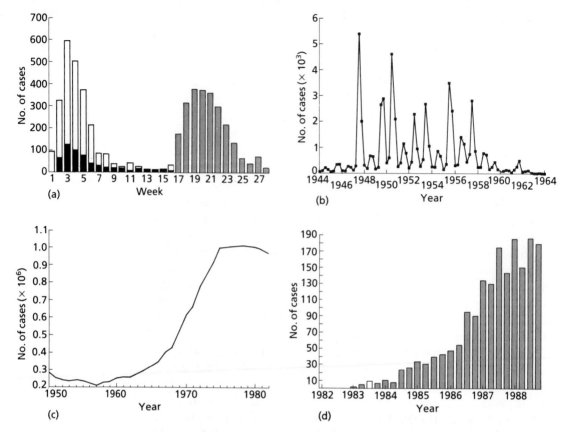

Fig. 15.1 Epidemic nature of microparasites. (a) Influenza pandemic of 1890–92. ■, 1890; ▨, 1891; □, 1892. After Stuart-Harris *et al.*, (1985). (b) Reported cases of poliomyelitis in England and Wales 1944–58 (Registrar General's Statistical Reviews, OPCS, London). (c) Cases of gonorrhoea in the USA 1950–82 (Centers of Disease Control, Atlanta, GA). (d) The AIDS epidemic in the UK (CDSC, London).

This pathogenicity has inevitably exerted an evolutionary pressure on the animal host to mount a cogent defence against the onslaught. In vertebrates, individuals respond to microparasite infection with a strong immunological reaction, and, if successful, recovery is usually accompanied by a long-lasting specific immunity to reinfection. For the parasite, the human host resource is in general only a short-term commodity, either because of host death or loss to infection through immunity. This in turn has accentuated the 'boom and bust' strategy of the microparasites adapted to a finite host resource (*r* strategists).

The necessity for close contact between susceptible and infectious individuals to effect transmission of many microparasites leads to a requirement of a threshold host density to allow propagation of an epidemic. Inevitably, the build-up of immunity to reinfection within a host population (herd immunity) following microparasitic invasion potentially leads to the demise of the infection. Persistence in the form of recurrent epidemic behaviour therefore requires the rapid replenishment of the susceptible pool (by births), which is of course more easily satisfied by a large population. Thus it would seem that many of the medically important childhood viral and bacterial infections such as measles, mumps, rubella and pertussis, which have short infectious periods and induce lasting specific immunity in the human host, are the product of modern urbanized societies, for in the past host densities would have been insufficient to enable persistence (Black *et al.*, 1974). Epidemic phenomena are rarely associated with the larger macroparasites which, in general, increase in numbers on an accumulative exposure basis as opposed to direct reproduction within the infected host. Thus worm infections are, in general, chronic and, in conjunction with the increased complexity of macroparasites over the viruses and bacteria, specific immunity is rarely effective.

Although the popular image of the behaviour of microparasites in host communities, referred to above, is consistent with the typical characteristics of these infectious agents, it does in fact overlook the rich variety that exists. Close inspection of viral and bacterial infections reveals a variety of patterns of behaviour: from transient invasions and outbreaks to long-term persistence; from chaotic incidence, to regular oscillatory, to more or less stable. The time span of an epidemic can be observed to vary from the short outbreak of *Salmonella* food poisoning, to a number of months for an epidemic of measles or influenza, to an extended period, perhaps for some years, as is currently exhibited by the human immunodeficiency virus (HIV). The exact nature of infection patterns in communities is dependent on numerous factors, the most important of which seem to be: (i) the typical course of events, following infection, within the host individual; (ii) the mode of transmission; (iii) the characteristics of the host populations in terms of their demography and social/behavioural structure.

The aim of this chapter is to draw the reader's attention to these major determining factors in an attempt to unravel some of the reasons for the similarities and differences in behaviour observed for different microparasites in a variety of host populations and reveal those characteristics which distinguish microparasitic population biology from that of other infectious disease agents and natural enemies. Much of the discussion will relate to the common childhood viral and bacterial infections of man, such as measles and rubella, that are transmitted directly between individuals that come into close contact, and induce lasting immunity to reinfection, for these provide the clearest picture of typical microparasite population biology. In other sections material is drawn from the literature on other microparasites exhibiting different life cycles and the range of host species widened. The potential for microparasites to influence host population growth is also considered, and, throughout, comments are made on the evolutionary pressures for change which are evident in host–microparasite interactions.

FACTORS UNDERLYING
MICROPARASITE POPULATION BIOLOGY

An understanding of the behaviour of microparasitic infections in host populations requires consideration of three main factors.

Natural history of the infection in the host individual

The biology of the interaction between the host and parasite following infection is the main factor determining the course of events, such as durations of latency and infectiousness and the development of immunity, observed for a specific parasite or class of parasites in a particular host species. This is illustrated schematically in Fig. 15.2 for a typical microparasite infection in a mammalian host. Factors such as the site of invasion and exit from host (via the skin, respiratory surface or gut lining), and the destination of the parasite within the host (e.g. the site of entry, organs such as lungs and liver, and of reproduction, and immunologically 'hidden' sites) will have a direct bearing upon not only the form but the temporal nature of the interaction. For example, with infections of the body surface (e.g. common cold), multiplication at the site of entry leads to rapid development of disease symptoms and release of infectious particles back into the environment (short incubation and latent periods, respectively). As a consequence, a host re-exposed to the same agent may well develop clinical symptoms and become infectious to others before the immune system has been boosted and had time to control the infection. In contrast, systemic or generalized infections such as measles and typhoid have longer incubation and latent periods, giving time for the host to mount effective immunity to reinfection (Mims, 1987). Such processes manifest their effect at the population level also. The length of the period of infectiousness is

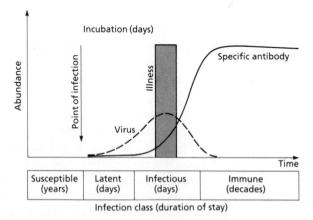

Fig. 15.2 Typical time course of an acute viral or bacterial infection in a host individual showing the corresponding progression through four infection classes or compartments. The typical duration of stay in the latent and infectious classes is transient compared with that within the susceptible and immune classes. From Nokes & Anderson (1988).

directly related to the potential of the parasite to transmit to susceptible hosts, and the time-lag induced by the latent period has an effect on the period between epidemics; the development of immunity is an important regulatory mechanism of microparasite dynamics in the host population, effectively removing individuals from the susceptible pool; and the ability of a parasite to remain in the host undetected by the immune system only to recrudesce at some future time is one answer to the problem of maintenance within a low density host population.

Looking in further detail at the population consequences of these processes, it should be apparent that the average duration of stay within each infection class shown in Fig. 15.2 determines the rate of movement or flow of individuals from one class to the next and the relative size of each class as a proportion of the total population. As illustrated in Fig. 15.3 for a typical childhood microparasite such as measles or rubella where the duration of stay in the latent and infectious classes is of the order of a few weeks, whereas the time spent in the susceptible and immune class would typically extend to years and decades respectively, the vast majority of the population at any one time would be expected to occur in the immune class and only a tiny proportion in the infected classes. This should be compared with macroparasites, whose tendency to induce chronic infection and poor host immunity typically leads to high prevalence and low herd immunity.

The process of rapid direct multiplication in the host has had considerable bearing on the theoretical study of viral and bacterial

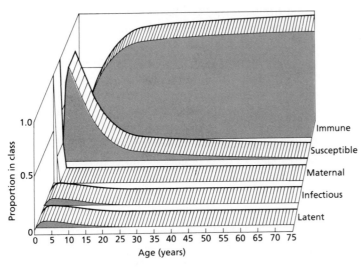

Fig. 15.3 Age-stratified proportions of a population who are susceptible, infected but not yet infectious (latent), infectious or recovered from infection for an endemic childhood infection that confers lifelong immunity (simulation based on data for rubella in the UK; proportions infected have been multiplied by 25 to make them visible). From Nokes & Anderson (1988).

population dynamics (Anderson, 1982a; Anderson & Nokes, 1990). In general, the measurement of microparasite burden is impractical, and the unit of epidemiological measurement is that of the host rather than the individual parasite. As a result the study of these infections commonly takes the form of compartmentalizing the host population according to infection status (as in Fig. 15.2) (May, 1986a). By way of contrast, for macroparasitic infections the unit of study is the parasite, and the accumulative process by which worm burden increases leads to a greater interest in the relationship between infection intensity and disease severity, and other density-dependent phenomena (Anderson, 1982b). Viral or bacterial abundance may be of some concern under circumstances where the size of infectious inoculum can affect the probability of infection and disease (certainly the case in some food- or water-borne infections such as hepatitis A, salmonella and cholera) (Mims, 1987), and where infection intensity may be related to likelihood of transmission to other hosts, as postulated for HIV (Anderson & May, 1988).

The importance of host immune response can hardly be overstated in the context of microparasite population biology. The development of an effective immunological defence is a counter to parasite regulation of host abundance. Furthermore, the development of specific, acquired immunity is a key mechanism in the regulation of parasite abundance in the host population. Not surprisingly, many viruses and bacteria have evolved methods by which to circumvent the host's defence mechanisms. For example, whereas infectious agents like the rubella and measles viruses are antigenically stable and occur as one type only, many others occur in a multitude of strains, e.g. influenza, enteroviruses and rhinoviruses, and bacteria such as *Neisseria meningitidis*, *Haemophilus influenzae* and *Salmonella spp.*, or are able to alter their antigenic coat (e.g. retroviruses and influenza viruses) in response to (and by which to evade) specific host immunity (Birkbeck & Penn, 1986; White & Fenner, 1986). Yet others seek out immunologically privileged sites where they can remain hidden from the immune system. For example, herpes viruses lie dormant with immunological impunity in the nervous system of human hosts until reactivated. The selective pressure imposed by host immunity is not peculiar to viruses and bacteria of course and is thought to have been an effective evolutionary force in many different parasites (May & Anderson, 1983a; Keymer & Read, 1990); nevertheless, the short generation time of microparasites compared to macroparasites, relative to their hosts, provides the ideal basis for developing antigenic diversity. The question then is why some viruses and bacteria *do not* present a changing 'face' to the host immune system.

Mode of transmission between hosts

Microparasites can only survive in a host population if they are able to pass from one host to the next. A variety of methods of transmission have been adopted by viruses and bacteria to achieve this goal, and are summarized in Fig. 15.4. Two main types of transmission exist: vertical, in which an infection is passed from parent to offspring (usually restricted to transmission during early development, such as germ line infection of avian oncoviruses, transplacental transmission of cytomegalovirus and perinatal infection with hepatitis B virus); and horizontal, in which the infectious agent is transferred between individuals via the environment (the majority of microparasites). Interestingly, few agents display only vertical transmission. A parasite that confers some selective disadvantage upon the host, which impairs host reproductive ability, will eventually go extinct if the sole mode of trans-

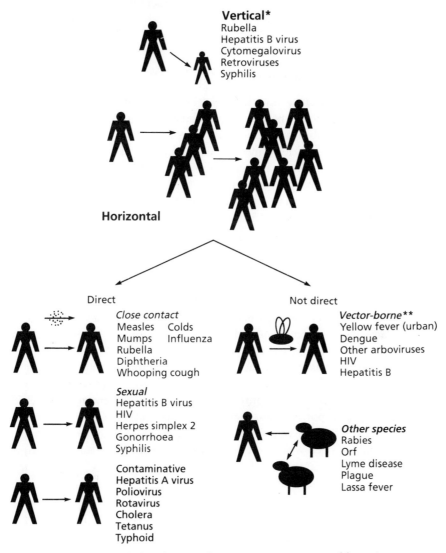

Fig. 15.4 Epidemiological classification of microparasitic viruses and bacteria.
*Vertical transmission is inclusive of transplacental and perinatal transmission, and **vector-borne includes transmission by needles. See text for further details. After Mims (1987) and White & Fenner (1986).

mission is vertical. However, microparasites which transmit in part vertically are well suited to survival in low density populations, where the chances of horizontal contact between susceptible and infectious individuals may be very low.

It is probable that each different mode of transmission described in Fig. 15.4 in part reflects a different solution to the problem of survival in a hostile heterogeneous external environment, and in part is the consequence of the natural history of an infection within the host (for example, a lung infection will favour transmission by respiratory droplets). Whatever the reasons, the resultant diversity has yielded wide variation in patterns of infection within host communities, for example in temporal incidence, prevalence, and endemic persistence. These points will be expanded upon in later sections (see also Anderson, 1982b).

Host demography and behaviour

As with all parasitic infections, the behaviour of the host individual and demographic properties of the host population, both of which affect the probability of infection transmission (i.e. the exposure of susceptibles to infection), are of fundamental importance in determining the patterns of infection in a community. Nowhere is this more clearly demonstrated than for microparasitic bacteria and viruses. The nature of this influence is related to the mode of transmission involved. Thus the population biology of directly transmitted microparasites such as measles, which require close contact between infectious and susceptible host individuals, are strongly dependent upon community size, host density and mixing rates (Kermack & McKendrick, 1927; Black, 1966; Anderson, 1982a; May & Anderson, 1985; McLean, 1986; McLean & Anderson, 1988). In the case of sexually transmitted infections, behavioural factors, such as numbers of partners per unit time and variation in this rate between individuals, predominate in determining infection patterns (Hethcote & Yorke, 1984; Dietz & Hadeler, 1988).

The very 'success' of faecal−oral contaminative infections in so many less-developed parts of the world is reliant upon behavioural and social factors that affect sanitation and hygiene and create overcrowding. Of such importance are these factors that the transition of countries from what we call developing to industrialized is in many ways heralded by social changes which act to reduce the risk of exposure to viruses and bacteria that contaminate our environment (e.g. *V. cholerae*, *S. typhi*, rotavirus and polio virus).

TOWARDS AN UNDERSTANDING OF MICROPARASITE POPULATION BIOLOGY

This section provides an outline of the transmission dynamics of what might be referred to as typical viral and bacterial infections (i.e. those directly transmitted by close contact, with short infection periods, and, following recovery, inducing lasting immunity in the host), such as the common childhood contagions like measles, whooping cough, mumps and rubella in human host populations. Much of the behaviour observed of host–microparasite population interactions in general is explicable in terms of the basic principles that follow or to some variation upon them.

Rate of transmission

The requirement of close contact (implying that the infectious particles are short lived in the environment) for the transfer of infection between infectious and susceptible host individuals suggests that the rate of transmission or spread of infection in a population will be determined by the number of individuals susceptible and infectious between whom contacts can occur and the rate of mixing of, or interaction between, individuals in the population. For a freely (randomly or homogeneously) mixing, closed (no migration) population of fixed size, the number of contacts, I_1, per unit time, between susceptible, X, and infectious, Y, individuals mixing at rate β_1, can be expressed in the form.

$$I_1 = \beta_1 XY \tag{15.1}$$

This is often referred to as the 'law of mass-action' by analogy with an ideal gas system (Hamer, 1906; Anderson & Nokes, 1990) and is a fundamental tenet in our understanding of the transmission dynamics of microparasitic agents (Anderson, 1982a). The actual number of cases of infection per unit time, I (i.e. incidence) will be dependent upon the probability, β_2, that each contact between an infectious and a susceptible individual will result in the transfer of infection, such that

$$I = \beta_1 \beta_2 XY = \beta XY \tag{15.2}$$

This probability of transfer following contact, β_2, may be dependent upon the contagiousness of the parasite, i.e. the ability of the parasite to establish within a new host, or the 'susceptibility' of the host to the infection (perhaps a genetic origin). Differences in innate transmissibility of different infections (i.e. different β_2 values) are responsible for different rates of infection of common childhood viruses and bacteria with comparable modes of transmission such as measles, mumps and rubella in the same community (see Fig. 15.5).

Clearly, for a particular value of $\beta = \beta_1 \beta_2$ (the transmission coefficient) the magnitude of transmission of the infection will be greatest within large densely packed populations. Observation supports the principle of mass-action as the major factor governing the rate of spread of viral and bacterial agents in host communities. For example, epidemics are far more explosive in the urban as opposed to rural setting. Indeed, it appears that the typical epidemic nature of measles, whooping cough, mumps and poliomyelitis have accompanied the

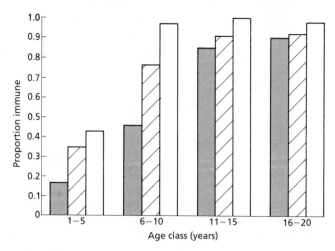

Fig. 15.5 Changes, with age, in the proportion of individuals who have antibodies to each of three childhood viruses (measles, mumps and rubella) derived from horizontal cross-sectional surveys in South Yorkshire, England. Presence of specific antibodies is a marker of past infection and subsequent immunity. Notice the markedly different rates of acquisition of immunity for the three infections. Data for measles (■) and rubella (▨) are from 1969 and for mumps (□) from 1985. (Nokes & Jennings, unpublished data.)

process of urbanization in many countries (Registrar General's Statistical Reviews for England and Wales). Community size is, in general, inversely related to the average age at infection, A, of childhood infections (note that A is inversely related to the magnitude of transmission), suggesting the rate of transmission to be dependent upon community size (Table 15.1) (Walsh, 1983). Additionally, the aggregation of children during school terms is marked by increased rates of such infections as chickenpox (varicella-zoster virus), whooping cough and measles (Fine & Clarkson, 1982; Anderson *et al.*, 1984). Black (1959) clearly showed the rate of measles acquisition to be more rapid in large families compared with those with only one or two children. However, it is unclear whether these observations are the result of increasing host density or rates of interaction; probably both contribute.

Contrary to the ideal of the mass-action principle, host populations are invariably non-homogeneous in properties like the rate of mixing, which may differ within and between age and social groups, or spatially (as in rural versus urban) (Anderson & May, 1984, 1985b). Heterogeneity may also exist in the form of differences in genetic constitution of individual hosts and parasites which may alter susceptibility and infectivity, respectively. The population biology of many microparasitic infections where transmission is not effected by short-lived particles requiring close contact between hosts (such as vector-borne and sexually transmitted diseases) will also be at odds with an explanation based upon an assumption of mass-action. These issues in addition to various other forms of heterogeneity in host and parasite populations are considered in detail in a later section.

Table 15.1 The effect of community size on the rate of infection transmission. (a) Mean age of reported cases for a variety of childhood infections in the State of New York 1918–19; (b) mean age of reported cases of measles and the basic reproductive rate, R_0, in England and Wales 1950–55. From Smith (1982)

Community size	Mean age at infection (years)			
	Measles	Whooping cough	Scarlet fever	Diphtheria
(a) New York State 1918–19				
<2500	12.9	8.2	12.3	14.2
2500–10000	10.7	6.9	11.2	12.5
10000–50000	9.0	5.7	10.2	11.5
50000–200000	9.0	6.3	10.5	10.6

	Mean age at infection (years)	R_0
(b) Measles in England and Wales 1950–55		
Rural	5.0	15.0
<50000	4.5	16.5
50000–100000	4.2	17.5
>100000	3.9	18.7

Invasion of host communities

Numerous examples of invasion by pathogens into host communities have been recorded. Notorious episodes include the epidemic of rinderpest (measles-like virus) in Africa in the late 1800s which devastated natural wild game populations (Sinclair & Norton-Griffiths, 1979), the deliberate introduction of a highly pathogenic myxoma virus into rabbit populations of Australia (Fenner, 1983), and the recent blight of the North Sea common seal populations caused by a phocine distemper virus (Dietz *et al.*, 1989). Undoubtedly there have been many other instances in which pathogens have *failed* to establish when first introduced into host populations. What factors determine the outcome of a microparasite–host population encounter?

The fate of an infection, introduced into a naive (susceptible) population of hosts, is inevitably determined by the potential for the infection to spread from host to host. Specifically, there is a requirement that an infectious individual must on average, during the period of infectiousness, pass on the infection to a susceptible individual. This concept forms the basis of an important epidemiological term known as the basic reproductive rate of a parasite, R_0. The basic reproductive rate of a microparasite is defined as the number of secondary cases which arise from the introduction of one infectious individual into a totally susceptible population. Assuming a mass-action process of infection transmission, the magnitude of R_0 will essentially be a function of the number of effective contacts (i.e. contacts that result in the transfer of

infection) per unit time between the infectious individual and susceptibles (a product of the density of susceptibles, X, and the transmission
coefficient, β) throughout the duration of the infectious period, D.
Thus,

$$R_0 = \beta X D \qquad (15.3)$$

For highly contagious pathogens, inducing a long period of infectiousness
in the host, when introduced into a rapidly mixing population of high
density, the propensity for spread will be high, i.e. a large R_0. Notice
that the factors which determine the magnitude of R_0 are of varied
origin, ranging from the characteristics of the pathogen and the biology
of the individual host, to the social/behavioural and demographic
properties of the host population.

The criterion for successful invasion of a host population is that
$R_0 \geq 1$, i.e. each infectious individual must infect, on average, at least
one other individual during the infectious period. Should R_0 be less
than unity then the infection will be unable to establish. The higher
the value of R_0 above unity the more rapid the ensuing epidemic. It is
clear to see that, other things being equal, the higher the density of
susceptible hosts the more explosive the epidemic of infection may be.
This phenomenon is well known by those in the farming world, where
unnaturally high density populations of animals are extremely susceptible to rapid epidemics of microparasite infections not normally significant in wild animal populations or in the context of low intensity
husbandry. Examples include Aujeszky's disease (pseudo rabies in pigs)
and porcine parvovirus, contagious equine enteritis and respiratory
transmitted chlamydia in sheep. It is of particular interest that outbreaks of respiratory and diarrhoea-inducing infections in cattle, like
infectious bovine rhinotracheitis and bovine virus diarrhoea (BVD), are
commonly associated with the confinement of animals during winter.

Viewing the role of host density from another angle, it may be seen
that below a critical density of hosts, the probability of contact between
an infectious and susceptible individual will be sufficiently unlikely
that the infection is unable to transmit. From equation 15.3 (setting R_0
to unity, i.e. the threshold for establishment) we find there to be a
threshold density of hosts, X_t, required for successful invasion by a
pathogen (Anderson, 1982a).

$$X_t = 1/\beta D \qquad (15.4)$$

Clearly, for low density host populations, invasion by microparasites,
which typically have short periods of infectiousness, D, will largely be
determined by the degree of parasite contagiousness, and the rate of
mixing of individuals within the population. The value of X_t is typically
high for directly transmitted microparasites like the common viral
infections of man. Such infections, which have short durations of
infection and induce lasting immunity to reinfection in those who
survive, e.g. measles, mumps, influenza and polio, may be totally

absent from small, isolated tribal communities (Black, 1962; Black *et al.*, 1974). Care should be taken to distinguish between total absence of an infectious disease and temporary extinction (reflecting an inability to persist), where invasion may occur (i.e. total population $N > X_t$) but, for reasons which are described below, infection goes extinct following the epidemic invasion and remains absent until reintroduced at a later date. Horizontal age-stratified serological surveys, which record the presence or absence of lifelong specific antibodies to viral and bacterial infections, are a record of past experience of a population to an infection, and therefore of use in differentiating between these two situations (Black, 1962; Black *et al.*, 1974).

Microparasites, other than the typical respiratory viruses and bacteria in humans, have the potential to invade host populations of low density, for example the long duration of infectiousness of the meningitis-causing bacteria (e.g. *N. meningitidis*), which establish an infectious carrier state within the host individual. Alternative modes of transmission (e.g. sexual and vector-borne) will be considered later.

Persistence within host communities

The successful invasion by a pathogen into a host population ($R_0 \geq 1$) is no assurance of persistence. Rapid multiplication of microparasites within host individuals, with possibly lethal outcome, typically leads to the development of a strong host immunological response, a concomitant short period of infectiousness, and, in those who recover, durable resistance to infection. Consequently, the host provides only a short period of sanctuary for the parasite, and the progressive loss of potential hosts through death or immunity, as an epidemic proceeds, is a self-limiting factor for pathogen perpetuation in the population. Persistence is, therefore, dependent upon replenishment of the host resource, which may arise from either births, immigration or loss of immunity in those who have previously experienced infection. If renewal of susceptibles is inadequate to ensure each primary case gives rise, on average, to at least one secondary case, then the incidence of infection may fall to very low levels and extinction may result. The major renewal process where immunity to reinfection is durable, is from births, which is itself dependent upon host demographic properties such as size and birth rate. Not surprisingly, therefore, it has been demonstrated that a critical population size is necessary for the endemic maintenance of childhood infections (Anderson, 1982c). Studies of measles incidence in human communities of different sizes show that below a threshold population size there is periodic fadeout of the infection (Fig. 15.6); the smaller the population the more probable that the infection will go extinct before susceptible numbers accumulate to epidemic proportions.

The observations in Fig. 15.6 suggest a critical community size required for measles endemicity of roughly 500 000. However, the

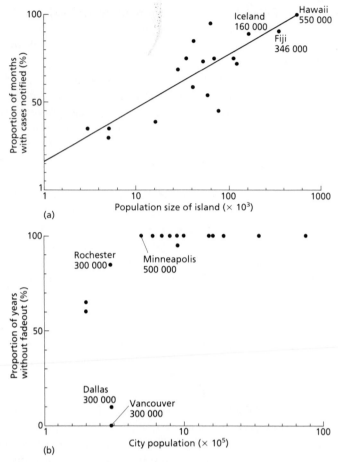

Fig. 15.6 Illustration of the dependence between persistence of a directly transmitted close-contact infection (measles) and community size, in (a) island populations and (b) North American cities over the period 1921–40. After Anderson (1982a) and Cliff *et al.* (1986).

migration of individuals between areas, locally, nationally and internationally, has never been so great as it is today, and undoubtedly this factor plays a significant part in the maintenance of close-contact infections in human communities of lower density. The Caribbean island of St Lucia for example with a population of roughly 150 000 reveals a long-term maintenance of measles infection. Records indicate that annually there is a migratory turnover of a size equal to that of the resident population. One of the key predictions arising from the mass-action assumption of transmission (βXY) is that the rate of spread of infection is strongly dependent upon community size. Analysis of observations on the incidence of infections such as measles, pertussis, scarlet fever and diphtheria, in this century, from communities differing greatly in population size suggest that the relationship is far less strong than expected (Smith, 1982, whose analysis of the data summarized in Table 15.1 showed incidence to be dependent on $N^{0.02}$

rather than N). The cause of this discrepancy is probably the degree of movement between human communities; hence rather than a number of isolated, or closed, populations of different size we are observing something approaching one large homogeneous mixing population. This idea is endorsed by measles case data for England and Wales, prior to the introduction of mass vaccination, which show cycles of incidence in different regions (each with markedly different population densities) almost completely in phase as though for a single population (Fig. 15.7a).

The average age at which individuals acquire infection, A, will be inversely related to the ability of a microparasite to spread through a population, i.e. the magnitude of R_0 (Nokes & Anderson, 1988). Figure 15.5 illustrates this point with serological profiles of measles, mumps and rubella from South Yorkshire in England. The change with age in the proportion with specific antibody (and therefore immunity) to each virus reflects the rate of acquisition of infection with increasing age. The steeper the profile (measles > mumps > rubella in Fig. 15.5) the greater the magnitude of transmission (higher R_0) and, conversely, the lower the typical age of primary infection, A (Anderson & Nokes, 1991) (Table 15.1b). Interestingly, all three profiles originate from the same host population; thus the reason for the differences in R_0 (and A) is most probably differences in the probability of transmission, β_2 in equation (15.2).

The prevalence of infection within a community will be dependent upon the magnitude of the basic reproductive rate; the higher R_0 the higher prevalence (Anderson & Nokes, 1991). It has been recognized, however, that for infections which persist more or less stably within the host population the number of new cases for each infectious individual will on average be 1, independent of R_0 (Anderson, 1982a). If this effective reproductive rate, R, were consistently greater than 1 then the infection would continue to grow in incidence which, for reasons stated above, cannot be maintained indefinitely. More often R fluctuates above and below unity as epidemics wax and wane.

Temporal incidence of infection

Regular oscillatory patterns of incidence, characteristic of the microparasitic viruses and bacteria transmitted by close contact, such as the common childhood infections (Fig. 15.7), are of two types, seasonal and longer term. Seasonality in incidence is thought to result either from climatic variation that alters microparasite survival in the external environment and, therefore, transmission potential (Giovanardi et al., 1974), or from regular changes in population mixing such as aggregation and dispersion of children in relation to school terms (Fine & Clarkson, 1982; Anderson et al., 1984).

Of major importance in generating longer term oscillations in incidence is the cycle of depletion of susceptibles during an epidemic

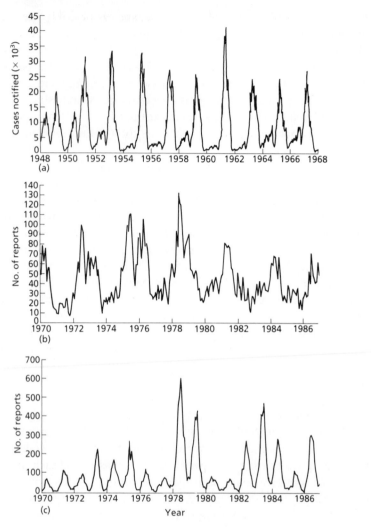

Fig. 15.7 Oscillatory behaviour of childhood viral infections in England and Wales. (a) Measles; (b) mumps; (c) rubella. The different potential for each microparasite to spread in the population (i.e. R_0 for measles > mumps > rubella) is reflected in: (i) the period between the main epidemics (roughly, 2 years for measles, 3 years for mumps and, less clearly, 4–5 years for rubella); and (ii) the synchrony of the epidemics across the country (i.e. cycles are most precise for measles, least in unison for rubella). Note that seasonal as well as longer term cycles are evident. Measles case notifications supplied by the Office of Population Censuses and Surveys, London; mumps and rubella laboratory reports obtained from CDSC, London (unpublished data).

period, removal to the immune class, and subsequent renewal of the pool of susceptibles during an interepidemic period. The period between epidemics will be directly related to the rate of transmission (i.e. R_0, which will be inversely related to the threshold number of susceptibles required to precipitate an epidemic) and the rate of renewal of susceptibles (e.g. birth rate), and inversely related to the duration of infection in the individual (the latent period conferring a time-lag on the generation of new cases, and the sum of the latent plus infectious periods dictating the rate of flow of individuals into the immune class) (Anderson & May, 1983). For these reasons infections in populations exhibiting high birth rates (typical of less-developed countries), where the susceptible pool is rapidly replenished to epidemic proportions, generally exhibit shorter periods between epidemics than the same infections in low birth rate populations (typified by developed countries). For example, the characteristic biennial measles epidemics in the UK and USA (prior to mass immunization) may be reduced to annual events in some sub-Saharan African countries (see, for example, Dabis et al., 1988). Within the same country, the periodicity of epidemics of different childhood viruses and bacteria is inversely related to the value of R_0 for each infection (Fig. 15.7). Concomitantly, there is a trend in stability of the infection within the population, infections with higher R_0 values displaying more dramatic (high amplitude) cycles as a result of the rapid depletion of susceptibles, with an increased probability of stochastic extinction during the interepidemic period. It may be presumed that as the duration of infection increases so does the period between epidemics, until depletion and replenishment of susceptibles falls out of phase and the pattern of incidence loses its periodicity. This may explain, in part, why bacterial infections able to establish long-term carriage in their human hosts (e.g. meningitis-causing bacteria, like H. influenzae and N. meningitidis) do not display regular cycles of incidence (Noah, 1987).

The maintenance of long-term cycles in incidence (i.e. not seasonal cycles) of infection appears to be dependent upon heterogeneities in host population behaviour (age-related mixing rates) and seasonality factors. However, this is an area of microparasite population biology which is poorly understood (Anderson et al., 1984).

VARIATION ON THE BASIC THEME

The preceding section, which gave an introduction to what might be called typical microparasite population behaviour, rested upon a set of basic characteristics of these infectious agents and their host populations (i.e. directly transmitted by close contact, short infectious period, lifelong specific immunity, homogeneous closed population). Such codes of behaviour are in fact commonly not adhered to. In this section the aim is to present some of the possible variations in this standard theme and indicate ways in which they cause variation in microparasite population behaviour.

Route of transmission

Whereas for close-contact infections like measles the dependence of the potential for spread R_0, and the incidence of infection, I, on host community size, N, generally holds (an important prediction of the mass-action assumption, βXY), it has long been realized that the rate of transmission of sexually transmitted infections and of those transmitted by vectors (in particular biting arthropods) is *independent* of host population size (Ross, 1911; Bailey, 1975b; Anderson, 1982c). Consider first sexually transmitted diseases (STDs). It seems reasonable to assume that the rate of sexual partner change (β_1) is a behavioural phenomenon unaffected by the number of hosts, N, in the population. It follows that for a single infectious individual, experiencing β_1 new sexual partners per unit of time, the rate at which new cases arise from the one individual will be dependent upon the proportion of partners susceptible to the infection (i.e. X/N). If β_2 is the probability that a sexual contact between an infectious and a susceptible individual will result in the transfer of infection then the incidence of new cases in a population containing Y infectious persons can be expressed as

$$I = \beta_1 \beta_2 Y(X/N) \tag{15.5}$$

(cf. equation 15.2). It follows that the number of secondary cases arising from the introduction of one infectious individual into a totally susceptible population (R_0) will be estimated by

$$R_0 = \beta_1 \beta_2 D \tag{15.6}$$

where D is the duration of infectiousness. R_0 is independent of host population size ($X = N$) (cf. equation 15.3).

These principles have foundation in observed trends. For example: (i) the persistence of STDs in populations of all sizes worldwide suggests that there is no threshold population size required for the endemic maintenance of such infections; and (ii) within highly promiscuous communities or sections of populations (such as homosexuals, prostitutes) the rate of spread of STDs is found to be more rapid than in communities with lower average rates of partner change, unless adequate precaution is taken (compare the rates of spread of HIV in homosexual males and heterosexuals in the UK (Anderson & May, 1988). Other factors being equal, for any sized population the prevalence of a particular STD is expected to be critically dependent upon the magnitude of β_1, the rate of sexual partner change. The rise in incidence of STDs such as *N. gonorrhoea* and chlamydial infections in Europe and the USA in the second half of this century are predominantly the result of changes in sexual behaviour not of increased population sizes (Cruickshank *et al.* 1973).

It is of interest that, compared with childhood viral infections, STDs commonly induce long periods of infectiousness in host individuals. For example, genital lesions resulting from infection with *Treponema pallidum* (the causative agent of syphilis) can remain infectious for a period of 5 years, and, in common with the general properties of herpes viruses, following primary infection with HSV2 virus (genital herpes) around 60% of individuals will experience recurrent episodes from their latent infections (Heath, 1987). Both examples illustrate another mechanism whereby microparasites may maintain themselves within low density populations, and suggest why regular cycles of incidence are unusual for STDs.

Many of the attributes of STDs also apply to infections transmitted by biting arthropods (Anderson, 1982b; Aron & May, 1982). In this instance we may assume that for most situations the arthropod vector has a fixed number of bites per unit time (biting rate, β_1) independent of the number of hosts available, a rate which may be dictated by biological requirements such as the time taken to digest a blood meal (horse flies) or to develop a brood of eggs (female mosquitoes). Thus the transmission rate from infected arthropods to susceptible hosts (and from infected hosts to susceptible arthropods) is likely to be proportional to the biting rate times the probability that a host is susceptible (or infected), and not simply proportional to the number of susceptible (or infected) hosts. Increasing host densities will only result in dividing the number of bites from the vector population (which will remain fixed) among more individuals thus reducing the per capita risk of infection transmission to hosts (and vectors) The magnitude of R_0 is directly proportional to the ratio of vector population size or density N_v, and host density, N_h, i.e. related to the number of vectors per host, N_v/N_h. The value of this ratio sets the critical density of vectors to hosts below which the infection cannot persist (Anderson, 1982a). The population biology of arthropod-borne viruses has many similarities to that of the vector-borne protozoans outlined in Chapter 12.

A potential source of variation in microparasite population biology is conferred upon infectious agents which remain viable in the external environment for a significant period of time (White & Fenner, 1986). The incidence of infections like the human enteroviruses, polio, HAV and rotavirus and water-borne bacteria (such as *V. cholerae* and *S. typhi*) that are spread between hosts by the faecal−oral route is inextricably linked to overcrowding, poor sanitation and hygiene, factors clearly associated with host population density. It appears that, in general, contamination of the environment by these agents may be regarded as a localized phenomenon, the population dynamics being not unlike those of typical microparasites obeying the mass-action law. At low host density however it is probable that the maintenance of infection might be attributable to long-lived environmental stages.

For other infections like the spore-bearing bacteria and viruses of forest insects (e.g. baculoviruses), where the longevity of infectious stages in the environment may be many months or years, rates of infection cannot be explained solely on the basis of host density, but are dependent also on the density of infectious stages in the habitat and the rate at which susceptibles encounter such stages. The transmission dynamics of these agents has been explored in depth elsewhere and is not considered any further here (Anderson, 1981b; Anderson & May, 1981, 1986a).

Heterogeneity in host populations

The many forms of heterogeneity in host populations derive from two main sources: (i) differences in behaviour, for example rates of mixing and sexual partner change; or (ii) genetic variability, such as suscepti- bility or resistance to infection and disease. Behavioural differences have the most obvious impact on microparasite population biology.

In the study of childhood viral and bacterial infections the question of 'who acquires infection from whom' is important in understanding the pattern of age-acquisition of infection. Analysis of horizontal age- stratified data (such as that illustrated in Fig. 15.5) for infections such as measles, mumps, rubella, pertussis and scarlet fever demonstrates a consistent age-related trend in the rate or force of infection acting on susceptible individuals. Almost invariably the force of infection (i.e. the per capita rate of susceptible infection) varies from low in young children to high in older children and teenagers, to low level in adults (Fig. 15.8). It is widely thought that such heterogeneity is a manifestation of age-dependent mixing of individuals, where maximum rates occur in school age children (Anderson & May, 1985b; Grenfell & Anderson, 1985).

Coupled with this age-dependent variation in the rate of mixing is that of age-dependent disease severity. In general, mortality due to childhood infections declines with increasing age (Mims, 1987), whereas

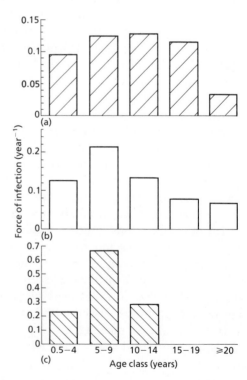

Fig. 15.8 Age-dependent forces of infection for three child- hood infections ((a) rubella; (b) mumps; (c) measles) estimated from antibody surveys in South Yorkshire, England; data which are summarized in Fig. 15.5 (for estimation procedure see Grenfell & Anderson, 1985). A similar age-dependent pattern in infection rates (from low to high to low again) is observed for each infection, although maximum forces of infection differ markedly (note scales).

morbidity increases (Anderson & Nokes, 1991). Very low average ages for measles infection in many less developed countries (commonly 1−3 years), which is attributable to, amongst other things, overcrowding and poor living conditions (Walsh, 1983), results in a severe problem of case fatalities (Morley, 1969; McLean & Anderson, 1988). Even in this, the era of vaccination, between 1 and 2 million infant deaths annually are attributable to measles (Grant, 1989). In industrialized countries measles tends to occur on average at an older age (4−5 years) when the infection is unlikely to have fatal consequences; thus morbidity becomes the significant issue (Anderson & May, 1983).

Surveys of sexual behaviour in human populations or subpopulations (e.g. heterosexuals, homosexuals and prostitutes) demonstrate marked variation between individuals in the rate of change of sexual partners (Fig. 15.9). Although, the majority of individuals have few or no new partners per unit of time, a minority will have many. The potential for spread of a sexually transmitted infection will be dependent not only upon the average sexual behaviour in the population but on the variation in sexual partner change. The few highly promiscuous individuals in an otherwise sexually reserved population have the ability to ensure infection persistence where the average behaviour would suggest otherwise (May & Anderson, 1987).

A further manifestation of variation in population behaviour is that of demographic heterogeneity, both spatial and temporal. Variation in population density has been examined in relation to its effect on the

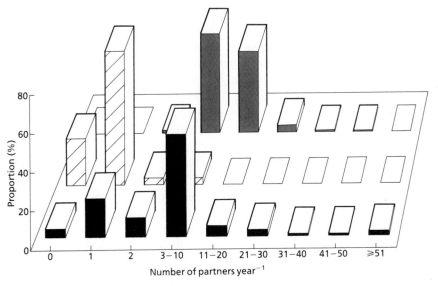

Fig. 15.9 Variation in the rate of sexual partner change in three communities, (□) heterosexuals and (■) homosexuals in the UK, 1986 (Anderson, 1988) and (▨) East African pastoralists, 1988 (Konings *et al.*, 1988). The average numbers of partners per unit time and the degree of variability indicated by the skewed distributions are both important in determining the persistence and spread of sexually transmitted infections in the population.

spread of influenza epidemics between cities (Cliff *et al.*, 1986) and the maintenance of measles in the context of marked rural and urban partitioning (Anderson & May, 1984). Considerable emphasis is now placed upon the importance of high birth rates in growing populations on the transmission dynamics of infections such as measles and HIV in developing countries (McLean, 1986; Anderson *et al.*, 1988).

Heterogeneities in parasite populations

A common feature of microparasites is their ability to present more than one antigenic profile or 'face' to the host. Specific immunity raised against one antigenically distinct profile (i.e. a variant, strain or serotype) may confer only partial or no protection to infection with other variants. As a consequence the potential for herd immunity to regulate incidence of such infections may be moderated or absent.

The population consequences of such microparasite variability are well illustrated by type A influenza in human communities. This virus possess two polymorphic, highly immunogenic, glycoprotein antigens (a haemagglutinin and a neuraminidase) on its outer surface. Marked alterations in the structure of these glycoproteins giving rise to new *subtypes* have occurred at roughly 10–12-yearly intervals between 1933 and 1968, and on each occasion have heralded pandemics of the new influenza A subtype. Each change or *shift* in surface antigen structure (the origins remain the subject of conjecture) (Hope-Simpson & Golubev, 1987; Potter, 1987) was sufficient to render the new subtype unrecognizable to host immunity raised to any previous subtype (Potter, 1987), thus rendering the whole population susceptible (naive) and making possible the widespread epidemics recorded. In addition to major shifts in surface antigen structure, subtypes of influenza A virus undergo antigenic *drift* resulting from minor changes (mutations) in the glycoproteins, resulting in new *strains* of the existing subtype. Each new strain will have a selective advantage over that which previously circulated and for which immunity in the population has built up. However, the antigenic changes between successive new strains are relatively small (in comparison to subtype shifts) and considerable cross-immunity exists. Hence epidemics resulting from antigenic drift are moderated by herd immunity to the previous strain and do not escalate to the size associated with antigenic shifts.

Host immunity is undoubtedly an important mechanism by which new microparasite variants are selected. The emergence and rapid spread of the El Tor biotype of cholera during the 1960s (Cruickshank *et al.*, 1973) is one further example of this process. Parasite-induced host mortality may also act as a potent selective mechanism, and is well illustrated by the fate of the myxoma virus when introduced into wild rabbit populations of Australia and Britain (Fenner, 1983; May & Anderson, 1983a). A number of strains of the myxoma virus exist, differing in host virulence (see Fig. 15.10). Introduced in a most virulent

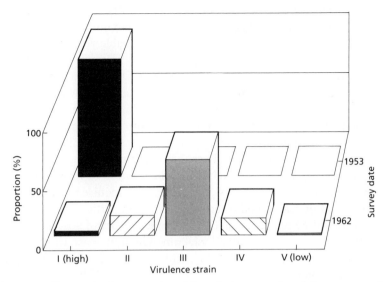

Fig. 15.10 Changes in the proportion of five virulence strains of the myxoma virus isolated from infected wild rabbits in Britain between 1953 and 1962. After May & Anderson (1983a).

form, changes were seen to occur in the strain composition of the virus population (in both Australia and the UK) such that in time a variant of moderate pathogenicity predominated in the rabbit populations (Fig. 15.10). The success of one strain in preference to any other may be explained by weighing up the various effects of virulence (i.e. on mortality, recovery rate, infectivity) on the potential of the virus to persist and spread from within the host population. Rabbits infected with the highly virulent strain were slow to recover from infection (long infectious period) and developed many open lesions suited to vectorial (mosquito or flea) transmission (high infectivity). However, they suffered a high rate of mortality, and the rabbit population plummetted. In this way the highly virulent strain effectively wiped itself out. On the other hand, rabbits infected with a strain of low virulence were less like to succumb from infection, but they also recovered rapidly and were less prone to developing lesions; factors leading to low rate of transmission. Thus it would seem that selection favoured an intermediate level of virulence maximizing transmission potential in relation to mortality costs. This observation is in contrast to the conventional wisdom that all host–parasite associations eventually settle to benign interactions.

POPULATION REGULATION

Host regulation of parasite population

It was shown earlier that the major factors regulating the microparasite population were the rate of renewal of susceptibles, in particular from births, and the development of acquired immunity (contrast with the regulatory factors in macroparasite population dynamics which tend to be density-dependent, see Chapter 14). It is probable that the classical picture of cholera occurring in explosive epidemics in non-endemic regions and as protracted epidemics in endemic regions is the result of a level of herd immunity in the latter situation (Cruickshank *et al.*, 1973). Compared with infections inducing lifelong immunity, those for which immunity is short lived or absent, such that individuals rejoin the susceptible class, may be maintained at much higher prevalence in a population (although in reality combinations of factors need to be considered). Reasons which limit the duration of immunity may be associated with the host (e.g. the absence of acquired immunity in invertebrates) or the parasite (e.g. antigenic variability of respiratory bacteria and influenza viruses). The regulatory ability of host immunity is to some extent circumvented by factors which increase the duration of infectiousness, such as the development of carrier status in individuals infected with hepatitis B virus, *H. influenzae*, or *S. typhi* ('Typhoid Mary' was a subclinical carrier of the typhoid bacillus) (Cruickshank *et al.*, 1973), or recrudescence of latent herpes viruses (White & Fenner, 1986). Such properties enhance the capacity of viruses and bacteria to persist in host populations even at low density.

Parasite regulation of host population

In the examples of microparasite invasions and epidemics we have seen the ability of these agents to inflict significant toll on host numbers. These events are, in general, transient and unlikely to have significant long-term regulatory effect on host population size. Nevertheless, they do suggest a potential for microparasites to constrain host population growth rate or depress host population levels below that which they would naturally achieve in the absence of disease, in the same way as we might view predators and parasitoids.

In population terms the essential requirement of an infectious agent in order to be able to regulate host population size is a capacity to alter host survivorship and/or reproduction, i.e. influence the natural intrinsic growth rate of the population (Anderson, 1979). Microparasites are, by their very nature, highly pathogenic organisms and their regulatory potential might, therefore, be viewed purely in terms of the direct effect of host mortality. Additionally, there are indirect consequences accompanying microparasite infection. In the developing world in particular childhood infections such as measles reduce the ability of the host to stave off opportunist infections (most often microparasites themselves, such as diarrhoea-inducing microparasites). It is also probable that during the course of an infection the host is at some competitive disadvantage to those without infection, i.e. a reduced competitive fitness.

In view of this clear potential for parasite-induced host mortality, the question then arises: why don't all microparasites that cause case fatalities regulate host population size? Theoretical and experimental studies have identified certain conditions under which microparasites may regulate host populations (Anderson, 1979; Anderson, 1981b). Disease-induced mortality must be high compared with the disease-free rate of growth (i.e. birth rate minus death rate) of the host population. However, the impact of high pathogenicity is alleviated by short duration of infectiousness (fewer individuals run the risk of infection), and long-lasting resistance to reinfection (durable immunity acts as a refuge from infection and disease for those who recover). Thus in spite of their often observed pathogenicity, the ability of the typical viruses and bacteria to regulate vertebrate host populations will be limited by transient infectiousness and long-lasting immunity. In contrast, the capacity of pathogenic viruses and bacteria to regulate invertebrate populations (the members of which are unable to develop acquired immunity) appears beyond doubt, although the degree of regulation will be moderated by the high reproductive potential of such hosts. Recent epidemiological studies have suggested that HIV, the aetiological agent of AIDS, may have the capacity to turn over positive population growth rates in developing countries (Anderson *et al.*, 1988). Factors which appear to be contribute to this capacity include the high pathogenicity of the virus and lack of immunity in the host (the infection is almost invariably fatal), high rates of sexual partner change (Fig. 15.9), long infectious period, and vertical transmission.

Understandably, the degree by which an infection depresses host population size is predicted to increase with increasing rates of transmission. It is not the case, however, that the higher the pathogenicity the greater the regulatory potential. It appears that moderate case fatality rates lead to the maximum reduction of host population size. While low pathogenicity may be insufficient for regulation, high values lead to the death of large numbers of infected host and a concomitant reduction in transmission rates. For further information on this complex issue the reader should consult Anderson and May (1991).

CONCLUSION

We have seen that a sound knowledge of the natural history of microparasites in the host, the mode of transmission between hosts and host demography and behaviour provides the basis for understanding the behaviour of microparasites in host populations.

In common with many complex natural systems it appears that a small number of factors dominate much of the typical population behaviour of microparasitic viruses and bacteria, namely their ability to multiply directly and rapidly within the host, the subsequent induction of long-term immunity to reinfection and a mass-action mode of transmission.

Thereafter, it is the considerable heterogeneity in transmission route, biological, particularly antigenic, properties of the parasite and host behaviour which are crucial to our wider understanding of the variety of patterns of microparasite infection. Undoubtedly, we do not yet know enough about these heterogeneities, particularly those related to host immunity and parasite antigenic variability, to explain many observed trends.

Direct reproduction and antigenic diversity of microparasites and specific host immunity stand out as major factors that have moulded the host—parasite interaction and distinguish microparasites from other parasitic agents. These factors in conjunction with parasite antigenic diversity play a key role in regulation and coevolution of microparasites and their hosts.

3

SYNTHESIS

16: Predator Psychology and the Evolution of Prey Coloration

TIM GUILFORD

PREDATOR PSYCHOLOGY AND PREY DEFENCE

A spectacular testimony to the influence of natural enemies in shaping the evolution of morphology and behaviour is to be found in the coloration of prey. Although animal coloration has numerous functions, such as thermoregulation (Kingsolver, 1988a) and intraspecific signalling (Butcher & Rohwer, 1989), there is little doubt that the need for defence against predators has led to a bewildering diversity of protective colour patterns. This diversity has long captured the imagination of biologists, and its incidence is well reviewed elsewhere (Poulton, 1890; Cott, 1940; Edmunds, 1974; Greene, 1988). In this chapter I want to take a different approach and view defensive colour patterns from the perspective of predator psychology. I will not attempt an exhaustive review, but will illustrate this aspect of the relationship between predator and prey by selecting a few examples about which sufficient is known to present a coherent picture. Drawing mainly on the examples of warning coloration and crypsis I will consider how knowledge about the workings of the perception and memory of one important class of predator, the actively searching vertebrate, might help us understand the nature and diversity of prey coloration. What I hope to suggest is that in many ways it is to the psychological capabilities and limitations of their enemies that prey, and their defensive colour patterns, have been forced to adapt. Because of this emphasis, I shall say little about the impact of invertebrate, sit-and-wait predators on the evolution of prey coloration (see Chapter 20).

PREDATOR PREFERENCE AND
THE PROFITABILITY SPECTRUM

Not all prey are equally good to eat. This fact depends partly on variation in the chemical constitutions of prey (leading to a 'palatability spectrum'; Brower et al., 1968; Turner, 1984b), but more generally on the sum of factors (e.g. size, energetic value, escape potential) that influence perceived profitability in the minds of potential predators. Hence each prey species, or class of individual within a species, is positioned on a 'profitability spectrum'. This position may shift with the employment of chemicals, physical structures (such as stinging hairs) or behaviours (such as the escape reaction of the spittlebug *Philaenus spumarius* (Thompson, 1973) specially adapted to reduce profitability, presumably in response to predation pressure. However, prey may also become unprofitable less directly if, having evolved an ability to feed on a toxic plant, they sequester the plant's defensive compounds for themselves. Thus monarch butterflies, *Danaus plexippus*, acquire cardenolides from many of their milkweed (*Asclepias*) food-plants as larvae, and store them as highly effective deterrents that cause emesis, and subsequent conditioned rejection, in avian predators such as the blue jay, *Cyanocitta cristata bromia* (Brower *et al.*, 1968; Brower, 1969, 1984). Not all predators will regard each prey's position on the profitability spectrum in the same way. Thus, amongst the vast concentrations of monarchs overwintering in the fir forests of Mexico, the black-headed grosbeak, *Pheucticus melanocephalus*, is able to specialize on this otherwise largely untapped food source because of its relative insensitivity to cardenolides (Brower, 1984, 1988). Brower and Calvert (1985) estimate that whilst most birds are unable to eat monarch butterflies, the black-headed grosbeaks, together with black-backed orioles (*Icterus galbula abeillei*) and a few other birds species, killed some 9% of the 22.5 million butterflies found at one overwintering site in 1978–79. A prey's position on the profitability spectrum, then, is determined not just by its defences (whether acquired through adaptation or preadaptation), and by predator counter-adaptations, but by the many factors that influence predator abundance, foraging tactics, and preferences as well (Malcolm, 1990a). Nevertheless, variation in profitability has provided a major source of variation in defensive coloration. At the profitable end, predators want to eat prey that do not want to be eaten, so prey use strategies to avoid detection in the first place. At the unprofitable end, predators do not want to eat prey that do not want to be eaten, and this closer symmetry of interests has led to the evolution of colour patterns aimed at advertising the prey's noxious qualities and enhancing detection and memorability.

PREDATOR SEARCH

Avoiding detection by predators

For profitable prey (those that are palatable, soft-bodied, and slow moving), detection is surely a dangerous event, even if other defences such as the use of startle, flash, confusion, or mimicry, make death less likely. So how do prey avoid detection? Prey do of course employ many behavioural mechanisms, such as hiding, or keeping still, but perhaps the most striking has been the way in which the visual hunting skills of predators have been foiled by the evolution of cryptic coloration.

It is the match between animal and background that reduces detectability, and so makes a prey cryptic. Naturalists have been fascinated by the elaborate ways in which cryptic prey appear to resemble their backgrounds, in grain (or patch size distribution), colour, reflectance, contrast, and geometry (Cott, 1940; Wickler, 1968; Endler, 1978). Some may even change colour to match different backgrounds as they change seasonally (as in arctic birds), or more rapidly (as in the cuttlefish, *Sepia officionalis* (Edmunds, 1974)). Many limitations in the predator can increase effective crypsis, such as a large visual acuity angle (the smallest angle to the eye between two resolvable points), poor colour vision, or a large normal viewing distance. But if predators can perceive discrepancies, then there is likely to be much more to crypsis than just a general resemblance to the background. In particular, we need to take account of predators' abilities to detect the individual elements within a prey's pattern. Thus Endler has defined a pattern as cryptic '... if it resembles a random sample of the background perceived by predators at the time and age, and in the microhabitat where the prey is most vulnerable to visually hunting predators' (Endler, 1978, p. 321).

Matching a random sample of the background

If we can photograph an animal's natural background at the appropriate time and place, we can estimate the frequency distributions of the size, shape, colour and brightness of the elements that make up the background, and compare them with those of the animal itself. Then we can quantify the probability that distributions characteristic of the prey could have been drawn at random from the distributions characteristic of the background, and we have a more objective measure of crypsis. This is the method pioneered by Endler in a study of 321 species of North American moths trapped within their natural forest habitats (Endler, 1984). Endler measured, along a series of transect lines across each moth's forewing, the distances between each patch boundary in the pattern, the average colour of each patch (compared to known colour standards), and each patch's average brightness. He made similar measurements along transects across photographs of forest backgrounds. Comparisons could then be made between the characteristic distributions of given moths and given backgrounds. Colour frequency distributions, for example, were made by calculating the percentage total transect length that fell into each colour class, and were compared by calculating the product-moment correlation coefficient between the pairs of numbers in each class (see Endler, 1984, for details). Endler showed that moths vary considerably in their degree of match to the background. Sweet (1985) used similar methods to show that western rattlesnakes (*Crotalus viridis*) and gopher snakes (*Pituophis melanoleucus*) appear to have convergently cryptic patterns in various microhabitats across western North America. In more anecdotal fashion, Godfrey *et al.* (1987) showed something similar when they demonstrated, using Fourier analysis, that whilst a zebra (*Equus grevyi*) stripe pattern contains spatial frequencies uncommon in a sample background habitat, the stripes of a tiger (*Panthera tigris*) do not (although these conclusions are based on single photographs). So there is quantitative confirmation of the naturalists' view that some animals blend into their background (Poulton, 1890; Cott, 1940). But is this sufficient evidence that colour patterns have been selected to be cryptic?

The evidence that crypsis works

Endler (1984) found that moths' crypsis values were greatest against those backgrounds characteristic of the time of year at which the adult insects normally flew. Those species with known resting places or microhabitats were more cryptic on the appropriate background than on others; (indeed, some moths are able to choose appropriately coloured backgrounds, and to orient themselves so as to increase their match (Kettlewell, 1955; Sargent, 1966, 1968, 1969a,b, 1973)). Background generalists scored lower than specialists, whilst species that hide on the undersides of leaves, presumably avoiding the selection pressure imposed by visually searching predators overhead, also had low values. These findings support the general idea that visually hunting predators may select for prey colour patterns that display a random sample of background constituents, but the relationship is rather crude and there are exceptions to it (Endler, 1984). One problem is that we know too little about the way predators actually search for and detect prey to be very specific in predicting how prey should look to avoid detection (see below). Nevertheless, in a more direct test of Endler's random sample theory of crypsis, Endler and Kamil (Endler, 1986b) showed that general mismatch between prey and background was a good predictor of how quickly a real predator detected prey. In an operant set-up, blue jays were shown slide-projected images of natural backgrounds that either did or did not contain a *Catocala* moth (details in Pietrewicz & Kamil, 1981). By rewarding birds for pecking the screen only when it contained a moth, Endler and Kamil were able to measure the time between presentation and peck for moth/background combinations of varying degrees of crypsis, as measured by Endler's random sample method (Fig. 16.1). Time to detection correlated well with crypsis values.

Male guppies (*Poecilia reticulata*) are also cryptic against their natural gravel-bottomed stream backgrounds, but they are faced with the additional problem of having to be conspicuous in courtship. Endler (1978) found that in Trinidadian streams populated by especially dangerous, visually hunting predators, the selection balance seemed to have shifted towards greater crypsis because fish had smaller pigmented spots, and fewer of the bright spots produced by structural colours. But the most convincing evidence that crypsis works comes from experimental studies in which both the size of the gravel and the intensity of predation were manipulated. Endler (1980) showed that in the presence of predators there was selection for smaller spots on small gravel backgrounds, and larger spots on large gravel backgrounds, whilst bright

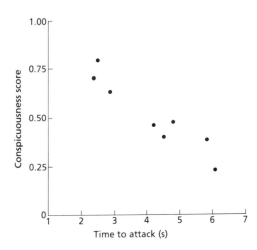

Fig. 16.1 Relationship between the time it takes the blue jays presented with slide-projected images to find and attack *Catocala* moths on various tree trunks and the moth's crypsis, as measured by the similarity between the prey and background with respect to patch size and colour frequency distributions (Endler, 1986b, p. 119). More cryptic moths have a lower 'conspicuousness' score.

colours diminished in both cases. In the absence of predation, the opposite changes occurred, driven by sexual selection for increased conspicuousness.

SOPHISTICATED SEARCH AND ENHANCED DETECTION

The limits to simple search

Treating visual predators as broad band discrepancy analysers, which impose selection on the patterns of palatable prey in proportion to their overall mismatch with the environment, has allowed us to explain the general form of cryptic colour patterns rather well. But there must be more to the design of crypsis, because predators are surely more sophisticated in their tactics than this. Predators (and parasitoids) often use area-restricted search after discovering one prey item in the hope of discovering more, and some cryptic prey may space out as an effective defence against this kind of behaviour (Tinbergen *et al.*, 1976). Predators may specialize on a subset of available cues. For example, amphibian predators may employ specialized shape detectors for enhanching detection, or recognition, of appropriate prey, and may be specially tuned to detect jerky movements as well (see, for example, Roth, 1986). Predators are sophisticated in the ways they analyse shape, movement and location in prey detection, so we might expect them to be sophisticated in their analysis of coloration too. Perhaps we may interpret aspects of defensive coloration in terms of specific adaptations to search by sophisticated predators. What if, for example, predators specialize on a subset of pattern constituents, such as outline, which may normally be good indicators of a prey's presence amongst the background noise? This may have encouraged the evolution of 'disruptive coloration' (Poulton, 1890; Cott, 1940; Edmunds, 1974), in which animals avoid detection by employing blocks or stripes of colour, or highly contrasting adjacent tones, to break up their outline, as perhaps in the butterfly *Anartia fatima* (Silberglied *et al.*, 1980), or the outline of some prominent feature such as the eye, as in the plover *Charadrius alexandrinus* (Edmunds, 1974). Although disruptive coloration has not been widely demonstrated experimentally (Silberglied *et al.*, 1980; Endler, 1984; Kingsolver, 1988a) it is one way in which prey may have outwitted a more sophisticated aspect of search by the actively hunting predator. As Cott (1940) realized, the effectiveness of disruptive patterns must stem from the way that predators normally search for and recognize individual objects by shape (using outline and gradual shading), for example, and it is by exploiting weaknesses in the receiver's psychology (Guilford & Dawkins, 1991) by presenting patterns that draw attention only to portions of the object, so deterring perception of outline, that protection is attained.

Another way in which predator psychology may be exploited is through the use of 'flash colours', in which prey (such as the grass-hopper *Trilophidia tenuicornis*) that have been disturbed by a predator display highly detectable visual features (or other cues such as clicking sounds) only during a short escape flight, turning them off on landing (Cott, 1940; Edmunds, 1974). It seems paradoxical that prey should use conspicuous cues to evade detection, and yet it is quite conceivable that a predator's attention may be drawn towards the obvious cues that are informative (in indicating the prey's position during flight), away from the less detectable cues that will be essential for location once the prey lands, and that may in fact have given the prey away in the first place. If predators adjust their search strategy to suit the particular detection task at hand, perhaps through selective attention to relevant cues, or through the adjustment of viewing time to that optimal for cues of a particular degree of detectability (for possible mechanisms see Dawkins, 1969; Bond, 1983), then perhaps flash coloration could be understood as a mechanism for fooling the predator into thinking it should search in a way appropriate only for conspicuous colours. Perhaps by being duped into searching only for easy cues, the predator then overlooks the difficult ones which are essential for detecting the prey after it has landed. Much work still needs to be done if we are to understand the psychological principles behind defensive strategies such as disruptive and flash coloration. There is, however, one area where the relationship between cryptic coloration and the psychology of the actively searching predator has been better explored. It concerns the relationship between polymorphism and mechanisms of enhanced prey detection.

Searching for variably abundant cryptic prey

Many predators, especially those that use active search to catch prey, are well designed to extract subtle prey signals from the noise of a visually complex world; signals made subtler still by the evolution of sophisticated crypsis. Yet the predator's capacity to perceive and process information is not infinite, so how might we expect predators to make the best use of the capacities they have for detecting prey? One possibility is that predators concentrate their detection systems towards abundant prey types, switching targets as abundances vary. This efficient allocation of limited resources is already seen in predators who concentrate their hunting with specific techniques, or in specific locations (Lawrence & Allen, 1983), but flexible mechanisms for enhancing the detectability of the most abundant prey have caused more controversy. Yet they are important because they may hold the key to explaining both the coexistence of multiple cryptic species, and the maintenance of colour pattern polymorphisms.

Search image formation

One solution to the problem of detecting variably abundant cryptic prey is Tinbergen's specific searching image hypothesis (Tinbergen, 1960). There is still much debate about what exactly constitutes search image formation (Pietrewicz & Kamil, 1981; Lawrence & Allen, 1983; Lawrence, 1986, 1989; Guilford & Dawkins, 1987, 1989a,b; Hollis, 1989; Kamil, 1989), but Tinbergen envisaged some kind of temporary perceptual change occurring in the predator, in response to recent encounters with a particular prey type, that brought about transitory enhanced detection accuracy for that prey type. Essentially the idea is that after encountering a few prey items of a particular morph the predator learns to attend selectively (Bond, 1983) to the particular configuration of stimuli that make that morph detectable. This kind of perceptual specialization brings about temporarily enhanced detection accuracy for that particular prey type, thus allowing the predator to make the best use of its perceptual capacities by directing them selectively towards currently relevant stimuli. If the stimuli to which the animal is attending selectively now fail to be informative (the prey has dropped in abundance), then it cannot pay to maintain reduced processing capacity for other stimuli, so the search image should be lost. In other words, because of this selective attention the search image causes 'interference', so that alternative prey types which are differently patterned but otherwise equally cryptic become less readily detected by the predator. Enhanced detection accuracy for a single, or small set of similar prey comes at the expense of detection accuracy for other prey types.

If they really exist, search images could provide an efficient way to hunt for a variety of variably abundant cryptic prey, but by generating interference they provide prey with a psychological loop-hole for evading detection. Imagine two different morphs that are equally cryptic but at different frequencies. The rarer is less likely to be encountered by predators with the appropriate search image, because predators are themselves less likely to have recently encountered sufficient similar prey to have formed the search image. Hence, both inherent degree of crypsis (see pp. 378–381) and rarity could contribute to evading detection (Endler, 1986b). Even marginally less cryptic morphs might be maintained, but at frequencies lower than their more cryptic alternatives. Understanding both the inherent degrees of crypsis, and the dynamics of search image formation, might help us to predict the stable frequencies of polymorphic prey (Endler, 1986b; Guilford, 1989a). However, as yet we do not understand search images. We do not know what sort of encounter schedule is required to produce a search image, how long it lasts, how effective it can be at enhancing detection, or in exactly what ways other prey have to be different if their detectability is to be interfered with rather than enhanced (i.e. what sort of changes in the pattern will successfully divide the predator's attention). Indeed, and despite their common acceptance, we are still not sure that search images exist at all because the direct evidence is often equivocal (Guilford & Dawkins, 1987, 1989a,b; but see Kamil, 1989). Many field experiments have shown that visual, actively hunting predators such as birds can impose frequency-dependent selection (known as 'apostatic selection') on visually polymorphic prey (Greenwood, 1984; Allen, 1988; Endler, 1988). Apostatic predation is not in doubt, but it could have a number of causes, of which search image formation is only one. As Gendron and Staddon (1983) first noted, predators may also alter their search rate appropriately to allow enhanced detection of relatively abundant cryptic prey (Guilford & Dawkins, 1987). So how does search rate adjustment work as a mechanism for dealing with variably abundant cryptic prey?

Search rate adjustment

The search rate adjustment hypothesis (Guilford & Dawkins, 1987) suggests that predators might achieve temporarily enhanced detection not through the perceptual specialization of search image formation, but simply by adjusting the length of time they spend viewing each piece of the environment before deciding that there is no food present and moving on to the next. Perhaps it should really be called the 'stare duration hypothesis' to distinguish it from other effects on search rate such as the speed of travelling between patches. If more cryptic prey require a longer look before they are detected, then a predator might increase its capture rate when cryptic prey are common by slowing down its search rate to allow a longer stare at each piece of background. Conversely, when cryptic prey are rare, a slow search rate will waste valuable time and bring little extra reward, so should be adjusted in favour of shorter stare durations (faster search) allowing a higher encounter rate with more conspicuous prey. The sensible predator may make the best use of a limitation in its capacity to process information (the fact that detection takes time) by trading off detection accuracy against the time allowed for visual stimuli from each portion of the environment to impinge on the retina in a way that depends on the abundance of prey of different degrees of crypsis. Prey cannot foil the search rate adjuster simply by donning rare patterns, since the appropriate stare duration for one pattern is the same for all other patterns of the same inherent degree of crypsis. In this case relative rarity does not help. Of course, the vulnerability of a cryptic prey will depend on its density in the environment, and on the density of other more, less, or equally cryptic prey (since these all affect the predator's best search rate). However, it will not depend on the relative frequency of equally cryptic morphs (or coexisting species), because there is no interference.

So, is stare duration important to the sophisticated, actively searching predator? Some predators do seem to slow down as they hunt for more cryptic prey (Gendron & Staddon, 1983, 1984; Gendron, 1986), but it is often difficult to determine whether this is because they are staring for longer at each piece of the background, or because they are searching more individual pieces within a restricted area so that it merely appears as if they are slowing down. What we do know is that more conspicuous prey take less time to detect (Endler, 1986b) and are captured faster even by animals trained to search for cryptic prey (Dawkins, 1971; Guilford & Dawkins, 1987), in line with the search rate adjustment model. Yet attempts to demonstrate enhanced detection without decreases (or with increases) in search rate have been hampered by an inability to measure stare durations sufficiently accurately (Lawrence, 1985a,b; Kamil, 1989). Perhaps the best data so far are those reported by Lawrence (1989) who showed that blackbirds (*Turdus merula*) appeared to reduce their scan times from about 400 ms to about 300 ms between cryptic prey captures with experience. However, since birds may make use of viewing times much shorter than this (Friedman, 1975, found actively foraging house sparrows (*Passer domesticus*) moved their gaze about every 100 ms) we cannot be sure that Lawrence's blackbird scans do not in fact contain a series of views of neighbouring pieces of the background, and that it is increases in individual viewing times that both enhance prey detection and decrease scan durations by replacing many short but unsuccessful views with a few long but successful views (Guilford & Dawkins, 1989a). Kamil (1989) suggests that the original experiments of Pietrewicz and Kamil (1979) show a similar (though non-significant) increased speed of response in birds that have developed greater improved detection accuracy with recent experience of cryptic prey. In this case, though, the birds, blue jays, were searching for moths hidden in slide-projected images presented in an operant set-up. Trained birds shown picture series containing just one prey type compared with those shown series containing a mixture of two different types not only performed with greater detection accuracy (more correct decisions), but may also have responded faster. Some may regard search image formation as the most parsimonious explanation of this result (Kamil, 1989), but technically the stare duration hypothesis could account for the same result if searching a slide involved a series of multiple views. The problem is that it is difficult to know exactly what an animal is looking at even when its head is stationary, so only very careful measurements of search rate (taking account of eye movements) can be used to distinguish the search image and search rate adjustment hypotheses, and these have not yet been made.

The search image prediction of interference should provide a clearer distinction between the two mechanisms, but those studies that purport to show interference either fail to control adequately for preference effects (see, for example, Bond, 1983) which can also change with recent experience of a particular prey type (and possibly even in a background or context specific way), or they do not use equally cryptic prey (Pietrewicz & Kamil, 1979, 1981; for discussions see Guilford & Dawkins, 1987; Kamil, 1989). If prey are not equally cryptic, then what looks like impaired detection accuracy caused by interference of search images could actually be caused by the adoption of the appropriate stare duration for only the less cryptic of the prey types, causing the predator to overlook the other more cryptic ones. These two explanations are clearly distinguishable given the right kind of data. Unfortunately, it is just the kind that still seems to be lacking.

The sophisticated predator seems to use some mechanism of transitorily enhanced prey detection to make the best use of its perceptual abilities, or to save time. Perceptual hypotheses like the search image prove to be to difficult to verify, especially since the underlying mechanisms of attention and pattern recognition are poorly understood (Guilford & Dawkins, 1987; Lamb, 1988; Hollis, 1989). Motor change hypotheses like search rate adjustment are in principle easier to verify, but the tests may require sophisticated measurement. And yet, behavioural mechanisms like these may be an important component in understanding the structure of cryptic polymorphisms, and the coexistence of many cryptic species.

PREDATOR LEARNING AND MEMORY

Beyond detection

Although we can learn much about the design of prey colour patterns by treating the predator as nothing more than a detecting device, albeit a sophisticated one, the predator's brain is also designed to gather, store and use information about the outside world to guide future behaviour. These processes are important aspects of the natural enemies to which prey colour patterns have had to adapt. Herons (*Butorides striatus*) and egrets (*Casmerodius albus* and *Egretta thula*) enter their world already equipped with inherited knowledge that black and yellow banded sea snakes are dangerous (Caldwell & Rubinoff, 1983; see also Smith, 1975, 1977; Schuler & Hesse, 1985). Nonetheless, and for good reason, most knowledge about the value of potential prey, about their position on the profitability spectrum, is acquired during the animal's own lifetime, through learning. Although the principles of animal learning have been extensively studied (Mackintosh, 1974, 1983; Staddon, 1983; Rescorla, 1988), their relevance to the interactions between prey and their natural enemies has received far less attention. By extracting a few key examples I hope to show how understanding this aspect of the predator's mind can provide insights into the evolution of prey colour patterns.

Exploiting generalization

Learning to associate the colour pattern of a prey individual with its perceived quality is beneficial to the predator only if it can use this knowledge to predict the quality of similarly coloured prey in the future. This usually works, of course, because similarly coloured prey are generally of the same species, and so of similar quality. However, it does not always work, and the process of generalization by predators has provided another psychological loop-hole adeptly exploited by prey that use deceptive colour patterns in disguise. Young larvae of the moth *Trilocha kolga* bear an uncanny resemblance to bird droppings (Edmunds, 1974), whilst lizards, many insects and even fish (Cott, 1940) have both colour pattern and morphological adaptations designed to make them look like leaves. Such animals rely for their defence on fooling the predator into recalling the wrong association when they are detected, and generalizing from previous experience with (or perhaps inherited expectations about) real bird droppings and real leaves. And because predators generalize across similar objects even when they are not identical, resemblance does not necessarily have to be perfect to provide protection.

Most striking, however, is the evolution of mimicry in which one prey species (the mimic) has come to resemble a second (the model) that is regarded as unprofitable by predators (Wickler, 1968; Edmunds, 1974; Pasteur, 1982; Turner, 1984b; Pough, 1988a,b). The mimic may itself be either profitable (Batesian) or unprofitable (Müllerian), and this can affect the evolutionary dynamics in interesting ways (see, for example, Turner, 1984b). At both ends of the profitability spectrum mimetic individuals reduce their probability of being attacked, by exploiting the propensity of experienced predators to generalize from their previously acquired association between an unprofitable model and its colour pattern. Many factors affect how far predators will generalize around an unprofitable pattern, of which the degree of unprofitability is the most obvious. The more dangerous or toxic the model, the more likely it is that even a partial resemblance will afford a mimic protection (Pough, 1988b).

Mimics benefit because the predator generalizes to patterns similar to those that it has learnt to avoid, but some fear or avoidance responses are, to a greater or lesser degree, inherited by predators, and do not require that the unprofitable nature of their objects be learnt. Although there is some evidence of inherited aversions (Smith, 1975, 1977; Caldwell & Rubinoff, 1983) or hesitations (Schuler & Hesse, 1985) to potential prey animals, innate reactions seem most likely to be directed at extremely dangerous objects, such as other predators, about which there might be little chance to learn. Perhaps the extreme danger associated with the sudden appearance of a pair of eyes forces animals to generalize widely to react to all sorts of eye-like stimuli (Blest, 1957), even if they vary in size (Jones, 1980; Inglis *et al.*, 1983) and are not in their appropriate context (Scaife, 1976a,b). Coupled with the need for a swift response, this may have allowed the evolution of sometimes quite rudimentary eyespot patterns in a wide variety of prey, and perhaps other deimatic (frightening) displays as well (Cott, 1940; Edmunds, 1974). We do not know what psychological process is being exploited (Scaife, 1976a,b; Coss, 1978; Jones, 1980; Inglis *et al.*, 1983), but it has clearly been influential in the evolution of defensive colour patterns, and warrants further investigation.

In contrast to crypsis, which is designed to minimize the probability of detection, mimicry, special resemblance, and perhaps even deimatic displays rely on the predator's propensity to generalize from previous experience with (or prior expectations about) more or less similar patterns to deter attack even after detection. However, predators also have reasons to discriminate.

Discrimination and enhanced learning

Generalization and discrimination are two parts of the same process whereby predators recognize objects as belonging to categories, with which they may have associated values. The propensity of predators to discriminate presents problems for prey, and these can be seen reflected in their colour patterns. As predators encounter unprofitable prey they will learn to associate the prey's colour pattern with unpleasant experience, and it is in the interests of similar prey to give as clear an indication as possible that they belong to the same category. This need to avoid predator discrimination, and so enhance recognition, produces normalizing selection on the 'warning colours' of unprofitable prey, and is probably the reason why warning colours tend to be monomorphic (for reviews of warning colours see Cott, 1940; Edmunds, 1974; Guilford, 1990a), though polymorphism in spittlebugs may be an exception (Thompson, 1973). Rare or variant forms are penalized by anti-apostatic (negative frequency-dependent) selection (Greenwood, 1984; Endler, 1988), because they have a reduced chance of being recognized as belonging to a class of prey that is already ranked as unprofitable, and yet they may still have a high chance of being detected (for discussions of the problems associated with the initial evolution of warning colours see Guilford, 1985, 1988; Engen *et al.*, 1986; Leimar *et al.*, 1986; Mallet & Singer, 1987). It is the same process that encourages the evolution of Müllerian mimicry, and there is a symmetry of interests between predator and prey.

Recognition errors are a nuisance to unprofitable prey, but profitable prey might in fact benefit from a predator's failure to recognize them as profitable. This raises another potential source of apostatic selection on profitable prey, in addition to those associated with detectability effects (see pp. 381–386), because a rare morph may be less likely to be recognized as profitable than a common one with which predators are more likely to have had experience. Again, too little is known about how predators learn to recognize profitable prey to make predictions about the sort of polymoprhism this would generate. However, it is conceivable that quite transitory effects on perceived profitability might be caused by recent encounters with a cryptic morph, because a predator's certainty about the prey's value will decay with time since last encounter, and may even depend on the context (or background) in which prey are found (see, for example, Batson & Best, 1979). The problem of distinguishing the attentional effects inherent in the search image hypothesis from the associative effects of transitory changes in preference has hardly been considered in the context of predator–prey interactions, but it is well known in the field of animal learning (see, for example, Lamb, 1988). The problem is that we now have two separate ways in which a predator's performance at picking up a particular cryptic profitable prey type can be altered transitorily by very recent experience, and they are not likely to be easy to distinguish experimentally. First, enhanced detection might be caused by the animal's shifting attention to the relevant cues presented by that prey type (a search image). Second, enhanced recognition that the cues presented by that prey type are associated with reward might also occur as a result of recent experience with similar prey (a change in preference). Perhaps we are further from understanding the basis of apostatic predation than we thought (Allen, 1988). Nevertheless, attention and association are separate processes, so it should in principle be possible to distinguish them. Perhaps, for example, the kind of polymorphic patterns that might benefit by dividing the predator's attention would be different from those that benefit by appearing not to be from an easily generalized category associated with food reward. The two psychological processes might, if we knew enough about them, predict different kinds of polymorphism.

Recognition errors and conspicuousness

So far I have indicated why warning colours tend to be monomorphic, but that is all. Yet their most striking characteristic is that they are generally very conspicuous, at least to the human eye. In some cases this may be simply because unprofitable prey are freed from the behavioural constraint of having to be cryptic, or because being conspicuous (for some function unrelated to defence) has brought with it selection for unprofitability to counter the effects of increased predation risk (Guilford, 1988), but there seems to be more to it than this. Why does the South American arrow poison frog *Dendrobates histrionicus* sport such bold black and red stripes when it would seem to invite detection? One suggestion (Grafen, 1990) is that this may be a strategy for ensuring the honesty, and therefore trustworthiness, of the warning signal. Such a 'handicap' interpretation (Zahavi, 1987) argues that predators will treat conspicuous prey with caution precisely because on the whole only genuinely unprofitable prey can afford the consequences of inviting detection. While this suggests that warning colours are conspicuous in order to ensure that the signal is honest, most hypotheses assume that the role of conspicuousness is in enhancing the effectiveness of the warning signal at conveying its honest message (Guilford & Dawkins, 1991). For example, the detection distance hypothesis (Guilford, 1986) argues that conspicuousness may ensure that predators easily reject the prey at a distance and are not forced to detect prey at close quarters when they may have to make a snap decision to attack. Delaying attack wastes time, and may often be unnecessary, so there are reasons why predators might be expected to regard some categories of prey as innocent until proved guilty. Delay may be particularly costly when the predator's energy reserves are very low (Stephens & Krebs, 1986), if feeding itself brings risk of predation (Dill & Fraser, 1984), where there are competitors for the same resource (Barnard *et al.*, 1983; Dill & Fraser, 1984), or where prey are mobile or have escape reactions (Thompson, 1973; Gibson, 1974, 1980). Because accurate identification is a psychological process that takes time (Guilford & Dawkins, 1987, and references therein), familiar unprofitable prey are less likely to be attacked by mistake if they are always detected from a distance when even an approaching predator may be forced to scrutinize the prey for longer. Though not directly manipulating the detection distance, Guilford (1986, 1989b) provided experimental support for this hypothesis using artificial prey presented on a conveyor belt, by showing that experienced chicks were less likely to make recognition errors to unprofitable prey if they had more time to view, but not more time to peck the prey. Of course, such an effect of viewing time may be restricted to certain predator types (those that are fast moving perhaps), and certain habitats (viewing conditions may be so restricted in turbid water, for instance, that conspicuousness can have little impact on detection distance), but this is an empirical question that presently remains unresolved.

One function of conspicuousness, then, might be to reduce recognition errors in experienced predators, by increasing the time between detection and potential capture. However, it is through enhancing the acquisition and maintenance of learnt aversions to unprofitable prey (rather than in reducing recognition errors in experienced predators) that conspicuousness is more commonly thought to function. The reasons for this fall into three broad categories (Turner, 1984b; Guilford,

1990a): (i) contrast enhances learning; (ii) particular colours or patterns enhance learning; (iii) novel colours or patterns enhance learning.

[391]
Chapter 16
Prey Coloration

CONTRAST ENHANCES LEARNING

The first possibility (Rettenmeyer, 1970; Turner, 1975; Guilford, 1990a), that conspicuousness in an animal's colour pattern enables predators to learn more readily to associate that pattern with unpalatability, or perhaps with any level of profitability, is supported by experiments demonstrating that chicks learn more effectively to avoid unpalatable items if these are presented so as to contrast with the background (Gittleman & Harvey, 1980; Roper & Wistow, 1986; Roper & Redston, 1987). In fact a number of processes may contribute to the success of conspicuous warning colours here. Contrast with the background may increase the predator's encounter rate, allowing higher initial ingestion rates and the consequent build-up of delayed action toxins (see, for example, Brower & Calvert, 1985; Malcolm, 1989), giving the predators a nastier experience when it finally takes effect (Gittleman & Harvey, 1980). It is a general property of animal learning (see, for example, Mackintosh, 1983) that the stronger punishment should then make the association more durable. Conspicuousness might also help draw the gaze of other predators nearby, and enhance observational learning (Harvey et al., 1982; Guilford, 1990a). Mason and Reidinger (1982) demonstrated that red-winged blackbirds (Agelaius phoeniceus) could learn to avoid food associated with the appropriate colour through watching others being sickened after eating, but we do not yet know whether this commonly happens in nature, or whether conspicuousness will enhance it. However, none of these effects can account for the striking result observed by Roper and Redston (1987) that even if they are allowed to peck at a coloured, noxious bead only once, naive domestic chicks learn to avoid beads of the same colour more strongly, and perhaps for longer, if the colour contrasts with the background. Perhaps there is something inherently memorable about the contrast of conspicuous warning patterns.

SOME COLOURS OR PATTERNS MAY BE ESPECIALLY MEMORABLE

A second possibility is that predators more readily associate unprofitability with specific colours or patterns which just happen usually to be conspicuous (Harvey & Paxton, 1981; Rothschild, 1984, 1985; Turner, 1984b; Sillén-Tullberg, 1985; Guilford, 1990a). It is obvious that if a predator cannot perceive a colour or pattern then it cannot provide a useful cue in avoidance learning, so perhaps amongst those that a predator can perceive some may be inherently more memorable than others. Animals may have prior expectations about the value of some colours or patterns (Roper & Cook, 1989; Roper, 1990). Thus Schuler and Hesse (1985) found that domestic chicks that had never eaten before were more hesitant to eat (but not to peck) black and yellow banded prey than olive coloured controls, and that this predisposition was quickly stabilized if the prey were made unpalatable. Although it is difficult to control for the effects of conspicuousness generated by the internal colour boundaries of striped patterns (indeed, this may be part of their function), perhaps Schuler (1989) is right to claim that conspicuousness is a necessary but not sufficient component of an effective warning colour pattern. Simpler patterns may be easier to learn than more complex ones. It is also possible that repeated stripes or spots present a 'consistent signal' to predators, so that from many angles of attack, or even if partially hidden by foliage, the predator will always be able to learn or recognize the appropriate pattern. Or perhaps some patterns, such as larval stripes, are largely the consequence of development expedience. There is no doubt that the challenge of understanding the significance of aposematic patterns still remains.

NOVELTY

A third possibility is that predators more readily associate unpalatability with coloration that is different from that for which they normally hunt (Wallace, 1867; Shettleworth, 1972; Edmunds, 1974; Turner, 1975; Greenwood, 1978; Guilford, 1990a). This could be because predators already associate inconspicuous coloration with profitable prey (something likely to be fairly general to actively searching predators as a group), and hence unprofitable prey evolve conspicuousness in order to minimize the retarding effect that generalization from such associations have on learning (Guilford, 1990a, 'novelty of association'). Paradoxically, profitable prey may themselves wish to avoid this generalization, so that, within the limitations of having to be cryptic, they too may attempt to look different to those patterns that predators already associate with profitability. This may be another cause of polymorphism in cryptic species. Alternatively, it may simply be that conspicuous colours are relatively rare in the environment generally, and so are looked at more thoroughly when they are detected (Guilford, 1990a, 'novelty of detection'). Though we do not know exactly why, it is clear that novelty can enhance aversion learning under artificial conditions (see, for example, Shettleworth, 1972; Mackintosh, 1983; for discussion see Guilford, 1990a), so it may contribute to (though it cannot account for all of) the effectiveness of conspicuous warning coloration in unprofitable prey.

Potentiation and the aposematic ensemble

I have suggested that defensive colour patterns of prey can, at least in part, be understood as adaptations designed to exploit the psychological apparatus of their natural enemies, and that understanding this apparatus, its complexities, subtleties and weaknesses, will help us to explain why prey defences are the way they are. But so far, attempts to understand the function of, say, conspicuous colour patterns have looked at the effect the pattern has on predator detection and memory in isolation from other aspects of the prey's behaviour. Whilst this may be an essential first step, it is now clear that stimuli in different modalities can interact in unexpected and significant ways that may limit the usefulness of piecemeal investigation. Vertebrate brains do not always process stimuli in isolation from each other (see, for example, Lett, 1980).

For example, when a naive predator attacks an aposematic prey it is often faced not just with a conspicuous colour pattern which it can associate with the punishing experience, but a whole battery of defensive reactions. The monarch butterfly is not just warningly coloured and toxic, its cardenolide toxins have a bitter taste, and its tissues contain a distinctive pyrazine odour (found in many aposematic insects; Rothschild et al., 1984; Guilford et al., 1987; Kaye et al., 1989), whilst other species may employ distinctive sounds and movements when attacked. What is the function of this 'aposematic ensemble' (Rothschild, 1985)? Each reaction may of course have its own separate effect in deterring predation, and we know, for example, that predators can learn to avoid odour cues associated with noxious experience (Mason & Silver, 1983; Whitman et al., 1986; Guilford et al., 1987). But one intriguing possibility is that a distinctive odour, rather than acting independently of other signals, might enhance learning of the association between noxiousness and the visual characteristics of the prey (Rothschild et al., 1984; Rothschild & Moore, 1986; Guilford et al., 1987; Guilford, 1989a,b, 1990b; Avery & Nelms, 1990). Such potentiation effects are quite common in aversion learning (Clarke et al., 1979; Lett, 1980; Palmerino et al., 1980; Best & Meachum, 1986), and Kaye et al. (1989) demonstrated that pyrazine odour can potentiate a rat's ability to associate contextual cues with unpalatability. Since pyrazines have not yet been shown to be noxious in themselves (at least to birds; Guilford et al., 1987; Avery & Nelms, 1990), it is possible that they have the curious function in aposematic insects of increasing the effectiveness of warning coloration, without necessarily becoming associated with unprofitability in the predator's mind. This means that a potentiating odour would be under quite different evolutionary pressures from a warning odour (or any similar defence), because it is individuals with the warning signal, not the potentiating odour, that benefit from the adaptation (Guilford, 1989a,b, 1990b, for discussions). A warning odour helps itself, but a potentiating odour really only helps others.

Not all cues presented in combination cause potentiation: they can also block the formation of associations between other cues and punishment, particularly if they are highly salient (an effect known to psychologists as 'overshadowing'; Mackintosh, 1974, 1983). Avery & Nelms (1990), for example, demonstrated that in food aversion learning experiments with red-winged blackbirds, whilst odour and colour cues appeared to have a synergistic effect on learnt aversion, colour seems to have blocked aversion learning to taste. We do not yet know when to expect potentiation rather than overshadowing when cues are presented together, or, indeed, how exactly potentiation works (Lett, 1980; Palmerino *et al.*, 1980; Best & Meachum, 1986; Westbrook & Brookes, 1988). Nevertheless, in order to understand the function of the aposematic ensemble, and in order to predict under what conditions compound defensive stimuli will evolve, we will ultimately need answers to such questions about the psychology of learning in different predators.

CONCLUSION

Whilst some anti-predator colour patterns work by evading detection (e.g. general resemblance), others work by passing information to the predator (e.g. warning and mimetic signals). The arguments presented in this chapter suggest that in either case much can be learnt about the design of a defensive colour pattern by considering the psychological processes of some of the natural enemies ranged against the prey. This is not a complete picture, because many factors affect the variety and design of prey defences. The actively searching vertebrate predator is only one kind of natural enemy: trappers, sit-and-wait predators, and enemies without the kinds of complex psychological apparatus that make the selective forces I have discussed relevant will also contribute to defensive design. Colour pattern may be compromised by functions and costs unrelated to predation. Furthermore, in many cases the prey's natural enemies may be sufficiently disparate in their psychological abilities and limitations that the prey is forced to compromise in its anti-predator strategies in ways that are not easy to predict. And finally, the signalling properties of defensive colour patterns have to accommodate factors other than predator psychology. In particular, signals, such as warning signals, that are designed to transmit information effectively have to deal with the transmission properties of the medium through which they signal (warning under water may be different to warning in air), and the sensory properties of their receivers as well as the effect they have once they reach the receiver's brain (Guilford & Dawkins, 1991). Nevertheless, it is clear that predator psychology has played an important role as a designing force in the evolution of prey defensive coloration.

ACKNOWLEDGEMENTS

I thank Marian Stamp Dawkins, John Endler, Sue Healy, Sean Nee, Christine Nicol, Andrew Read and Steve Malcolm for helpful comments on this article, and Reg for ceaseless criticism and inspiration. I am funded by a Royal Society University Research Fellowship, and by St John's College, Oxford.

17: Natural Enemies and Community Dynamics

ANDREW REDFEARN & STUART L. PIMM

INTRODUCTION

Hierarchical explanations

What determines how population densities change over time? This simple question does not have a simple answer, because built into the question is a problem of scale. The kinds of answers depend on how long a time period we are looking at those changes. If we are considering very short time periods, then features of the individuals' behaviour, such as how they aggregate, will be important. Over longer time periods, such features as the population's birth and death rates will be important.

However, no species exists in isolation from the other species in its community. There will be interactions with the species that are the prey, predators, competitors, and mutualists of our species of interest. These species also interact with other species, that interact with still other species, some of which may interact with our species of interest. The change in the density of one species will cause changes in densities of others and the effects will ripple through the community like the waves in a pond. Over yet longer periods of time, species will invade or be lost from communities, so the patterns of interactions will change. Population changes will also follow slowly changing features of the planet's climate.

In short, it is possible to answer the simple question by studying behaviour, population dynamics, community interactions, or ecosystem processes. All provide part of the answer. The part of the answer we address in this chapter involves the role of community interactions, in particular, how predators affect the population dynamics of their prey. We will primarily discuss these effects in herbivorous insect communities.

Community structure and population dynamics

Ecologists have long used the idea of 'stability' as a means to characterize the temporal changes in populations and communities. The problem has been that ecologists have many meanings of 'stability' (Pimm, 1982, 1984a). In this chapter, we examine two of these, resilience and variability, and look at how community-wide patterns of predation affect them.

Resilience is measured by how fast a system returns to equilibrium once displaced from it. Resilience could be estimated by a decay time, the amount of time taken for the displacement to decay to some specified fraction of its initial value. (There is an obvious analogy to the half life of, say, a radioactive decay.) Long decay times mean low resilience and vice versa. Resilience is measured as a rate of change. High resilience is usually equated with a high degree of stability.

The variance of a variable over time, or allied measures, such as the standard deviation, or coefficient of variation, are obvious measures of *variability*. Variability is equated with instability.

RESILIENCE

How fast a population can grow when it is below equilibrium must depend, in part, on fecundity and other properties of the species itself (Southwood, 1981; Pimm, 1991). Resilience must also depend on how the species interacts with other species, for predation and competition must slow down the potential growth. Here we will be concerned with how predation modifies resilience.

Food chain length and resilience

THEORY

When we disturb the densities of a set of interacting species, no species density can return to equilibrium until all the others have done so. Each species is only as resilient as the least resilient species in their food chain or community. Consider this example. Suppose we reduce the density of phytoplankton in a lake. This will reduce the density of their zooplankton predators, and, eventually, the fish that feed on the zooplankton. The phytoplankton might have the capacity to return quickly to their former densities. But while the fish densities are recovering (slowly), zooplankton will be unusually abundant and this will keep the density of the phytoplankton down. In this example, the systems resilience is limited by the resilience of the predatory fish. This points to the possibility that food chain length may have a strong effect on resilience.

To examine these effects of food chain length on resilience, Pimm and Lawton (1977) modelled a variety of food chains and food-webs. They used Lotka−Volterra models to construct food chains of varying length and investigated how long the constituent populations took to return to equilibrium following a disturbance.

For each of the three food-webs shown in Fig. 17.1, they created 2000 models. The parameters describing the interactions between species, were chosen randomly over intervals that seemed ecologically

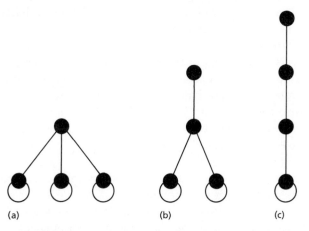

Fig. 17.1 Model food chains with four species differing in food chain length from two trophic levels (a) to four (c). Each food-web was simulated 2000 times to obtain the results shown in Fig. 17.2.

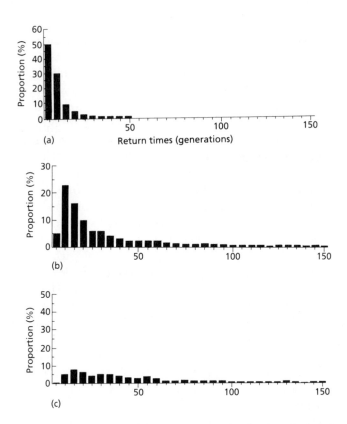

Fig. 17.2 Effect of food chain length on return time for the three food chains shown in Fig. 17.1. Results based on 2000 simulations of each food-web show that return time increases as the number of trophic levels increases. See text for details.

sensible. The return times shown in Fig. 17.2 were calculated by a well-known mathematical result detailed by Pimm (1982). As the number of trophic levels increases, so do the return times. For two trophic levels, nearly 50% of the models had return times less than or equal to 5 time units. The comparable percentage falls to 5% for three trophic levels and to less than 1% for four. The percentages of models where return times exceed 150 time units increases from 0.1% for the two trophic level models to 9% and 34% for the three and four trophic level models respectively.

An alternative approach is to have only one species per trophic level (as in the case of the four-species, four trophic level model). For a two-species model with two trophic levels, about 70% of the models had return times of less than 5 time units. Hence, models with different numbers of species overall and the same number of species per level, as well as models with the same number of species overall, lead to the same conclusions: as the number of trophic levels increases so do the return times.

TESTING THE THEORY: RESILIENCE IN ARTHROPOD PESTS AFTER PESTICIDE USE

How might we test the idea that long food chains cause the population densities of the individual species to be less resilient? Consider a herbivorous insect pest population increasing toward high density. A species with a resilient population will reach unacceptably high levels faster than a species that is not resilient. Population resilience can thus be a measure of *instability* and also a measure of how often we may have to control the pest by chemical means. There is an interesting paradox here. For species that we deem desirable, a high resilience means a quick return following some disturbance. For these desirable species high resilience would be equated with *stability*!

The preceding theory suggests that the longer the food chain, the lower is the resilience of the system's constituent species. So removing predators should mean that pest outbreaks will occur faster. A way to evaluate this idea may also give some insight into why insecticide applications often induce pest outbreaks. Insecticides clearly kill many pests, but many predators are often killed as well. The predators, moreover, cannot recover until the pests come back. This suggests that there will be a 'time window' after insecticide application in which pests have very few predators. Food chain lengths are effectively shortened and so, according to the theory, populations should increase more rapidly. Redfearn and Pimm (1987) asked: at comparable densities, do pests increase more rapidly after insecticide applications than do untreated controls?

It is not a new idea that insect pests can reach higher densities after spraying with pesticides, and that reduced predator populations may be a cause of this (Pimentel, 1961a; Debach, 1974). This does not necessarily mean, however, that the pest populations are increasing faster at any particular density. Although this seems reasonable, it had not been tested experimentally. Here, we summarize the essential points of our earlier analysis (Redfearn & Pimm, 1987).

Data collection

We restricted our attention to aphids and mites on cereal, vegetables, and orchard crops. We grouped chemicals according to whether they were applied as granules or liquids into the soil and then covered ('soil-applied' pesticides) or whether they were applied as sprays, dusts, or foliage granules above ground ('aerial' pesticides). It was assumed that pesticides in both categories would reduce the densities of herbivorous species, whereas those in the second category would also severely affect predatory species. Systemic pesticides applied above ground were assigned to the aerial category.

The data were counts of individual species, groups of species, or all mites or aphids encountered. The data were extracted from articles appearing in the *Journal of Economic Entomology* during the years 1957−76. Some articles presented more than one set of data, for example, when more than one pest species, area, or year was studied. All data sets included insect counts in one untreated plot or field plus separate measurements from one or more treated plot or field, each receiving a particular amount or type of chemical. In total, we made 112 comparisons between treated and untreated populations from 34 data sets and from 23 articles.

Data analysis

We used two methods to analyse the data. The first method assumes exponential population growth and suggests the following simple linear regression model:

$$\text{log density} = \beta_0 + \beta_1(\text{time}) \qquad (17.1)$$

where β_0 is a constant and β_1 is an estimate of r, the growth rate of the population, which we considered to be its resilience. An example of this method of comparing a treated and an untreated population is provided by Fig. 17.3a.

From 60 comparisons, 88% of populations treated with aerial pesticides had higher growth rates than the corresponding untreated population (Table 17.1). For populations treated with soil-applied pesticides, 67% of 52 comparisons had greater growth rates than the corresponding untreated populations (Table 17.1).

At least some of the populations were increasing progressively more slowly at higher densities, i.e. the plots of log(densities) versus time are non-linear. Populations may behave in this fashion if, for example, resource limitation were restricting population growth at

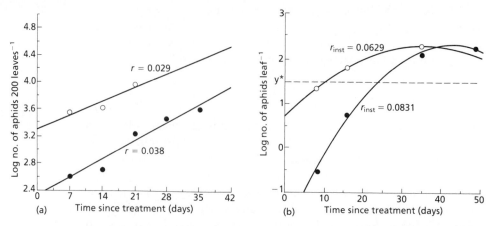

Fig. 17.3 The logarithm of aphid numbers as a function of days since treatment for treated (●) and control (○) populations. (a) The linear (exponential) model typified by green peach aphids on peppers; and (b) the quadratic model, typified by apple aphids on apple. Slope values (r) are adjacent to each line. See text for details. From Hale and Shorey (1971) and Madsen *et al.* (1961).

high densities. Non-exponential growth raises the following possibility. Our previous estimates of growth rates for treated populations (from equation 17.1) may have been higher than those for untreated populations due to sampling at different points along a common curve of log density against time, namely at lower densities. After all, the purpose of the treatment is to reduce the pests' densities!

If population growth is not simply exponential, i.e. log density does not increase linearly with time, then we must use a model that also includes a quadratic term in time:

$$\text{log density} = \beta_0 + \beta_1(\text{time}) + \beta_2(\text{time})^2 \tag{17.2}$$

where β_0, β_1 and β_2 are constants. To test whether our results are altered if one assumes that growth is not exponential, we found the values of β_0, β_1 and β_2 that best describe the data for each population. We need to compare growth rates of treated and untreated populations at equal densities. We compared instantaneous growth rates (r_{inst}) at a density (y^*) midway between the mean log densities of each corresponding treated and control population (see Fig. 17.3b).

Not all of the data could be analysed in this way. From 55 comparisons, however, 82% of populations treated with aerial pesticides had higher instantaneous growth rates than the corresponding untreated populations (see Table 17.1). For populations treated with soil-applied pesticides, only 41% of 51 comparisons had greater instantaneous growth rates than the control (Table 17.1). These results parallel those from the exponential model. For both models, when mite populations were analysed separately from aphid populations, their growth rates were not significantly increased by either category of pesticide (Table 17.1).

Table 17.1 The effect of soil-applied and aerial pesticides on the population growth rates of aphids and mites. Results of paired t-tests on the differences between mean growth rates for treated and untreated populations are given in Redfearn and Pimm (1987). For both models, aerial pesticides caused significant increases in population growth for all species combined, and for aphids alone, but not mites alone. Soil-applied pesticides had no significant effects on population growth in all cases. After Redfearn & Pimm (1987). Model (a) assumes exponential population growth, while model (b) includes a quadratic term and so population growth is density-dependent

Model used to calculate population growth rates	Soil-applied pesticides			Aerial pesticides		
	Overall	Aphids	Mites	Overall	Aphids	Mites
(a) log density = $\beta_0 + \beta_1$(time)	$\frac{35}{52} = 67\%$ *	$\frac{31}{37} = 84\%$	$\frac{4}{15} = 27\%$	$\frac{53}{60} = 88\%$	$\frac{44}{46} = 96\%$	$\frac{9}{14} = 64\%$
(b) log density = $\beta_0 + \beta_1$(time) $+ \beta_2$(time)^2	$\frac{21}{51} = 41\%$ +	$\frac{16}{36} = 44\%$	$\frac{5}{15} = 33\%$	$\frac{45}{55} = 82\%$	$\frac{41}{45} = 91\%$	$\frac{4}{10} = 40\%$

*Thus, 67% of the treated populations had higher r values than the corresponding untreated population.
+ Thus, 41% of the treated populations had higher r_{inst} values than the corresponding untreated population.

Summary

Aerial pesticides accelerate population growth of aphids but not mites. This accelerated growth is not due entirely to the fact that these insecticides might release the aphid populations from some density-dependent factor, such as resource limitation, which affects the untreated populations at high densities. We reach this conclusion because our results were not altered by assuming density-dependent growth and comparing estimates of the instantaneous growth rates of treated and untreated populations at equal densities.

We suggest that these data support the theoretical prediction that population resilience increases as predators are removed from a community. The fact that soil-applied pesticides did not have a significant effect on population growth rates is consistent with this hypothesis. These chemicals are expected to kill fewer predators than those applied aerially. There are, however, other possible reasons why soil-applied pesticides had no effects on population growth. Soil treatments may be chronic rather than acute in their effects on pest populations, because release of the toxins may be a gradual process, which means that these treatments may cause mortality long after the original application. (Perhaps because of this, populations that were treated with soil-applied pesticides were generally monitored for longer periods of time than those treated with aerial pesticides.) This chronic mortality may compensate for effects due to any reduced predation.

VARIABILITY

Introduction

Population variability is a measure of population instability, which we define as the degree to which the density or number of individuals in a given population fluctuates from year to year. It can be measured by such statistics as the standard deviation of the logarithmically transformed yearly abundances or the coefficient of variation of the yearly abundances (Williamson, 1984). Like resilience, the temporal variability of populations depends on many factors including features of the species themselves (Glazier, 1986; Taylor, 1986; Gaston & Lawton, 1988; Spitzer & Leps, 1988) and the variability of the physical environment in which the species are found (Malicky, 1976; McGowan & Walker, 1985). However, here we will focus on the effects on variability of the patterns of species interactions.

Studies of the causes of variability must inevitably compare populations. Since different systems are characterized by particular patterns of species interactions, we can compare populations in different systems (e.g. agricultural versus natural, tropical versus temperate versus arctic) with respect to the variabilities of the component species (see, for example, Elton, 1958; Krebs & Myers, 1974; Murdoch, 1975; Wolda,

1983). Systems differ in many ways, however, and the set of species in each system is likely to be uniquely adapted to the special selective pressures imposed by their respective environments. An alternative approach is to compare species rather than systems. For example, within a given system we can look at effects on population variability of interactions with prey, interactions with potential competitors, and interactions with predators.

MacArthur (1955) was one of the first to argue that interactions among species would modify temporal variability. Species that fed on few species would be more likely to become very rare following the loss of single prey species than species that were more polyphagous. Watt (1964), in contrast, argued that polyphagous species would be able to increase faster, and become more abundant, than monophagous species following the loss of the predators that controlled their densities. Both arguments are plausible, but they are not truly alternatives. MacArthur concentrated on food-limited species, and the reliability of that food supply. Watt focussed on predator-limited species and the consequences of the failure of that limitation. A comprehensive theory of food-web structure and population variability must take into account the arguments of both MacArthur *and* Watt. Redfearn and Pimm (1987, 1988) and others (Watt, 1964, 1965; Rejmanek & Spitzer, 1982) have empirically shown that herbivorous insect species with many host plants are in some cases more, and in other cases less, variable than species with more restricted diets.

These arguments are about the variability of the predator (or herbivore) and how it is modified by interactions with prey (or host). We begin by asking how predators affect the variability of their prey. Experimental populations of the Azuki bean weevil, *Callosobruchus chinensis* L., were less variable in cultures with two parasitoid species than in those with one (Utida, 1957; den Boer, 1971). For voles in northern Europe, there is also evidence that increased predator diversity reduces variability (Hanski, 1987a; Hansson, 1987). In the far north, where there is only one important vole predator, vole densities cycle and, as a consequence, are relatively more variable. Further south, there is a diversity of relatively more polyphagous predators, so vole densities do not cycle and are less variable.

Here we consider insects again and ask: are populations of herbivorous insect species with many parasitoid species more, or less, variable through time than those with few parasitoid species? Why study effects of parasitoids rather than those of classical predators? There is a great deal of information in the literature on insect parasitoids (see Chapter 11). Classical predation is relatively difficult to observe or quantify, whereas parasitism rates can be estimated and parasitoid species identified in the laboratory 'after the fact'. We analyse data on the forest Lepidoptera of southern Ontario and their parasitoids (Raizenne, 1952).

Variability and the diversity of parasitoids

Parasitoids and other natural enemies are important forces in the population dynamics of herbivorous insects (Beddington *et al.*, 1978; Hassell, 1978; Lawton & Strong, 1981; Strong *et al.*, 1984; Waage & Greathead, 1986). This suggests that the diversity of parasitoids may be at least as important as the diversity of host plants in determining the population variability of herbivorous insects.

There are at least three hypotheses that predict that the population variabilities of the host species decrease with an increase in the diversity of the parasitoids that attack them. (These arguments may apply equally to the effects of classical predators on their prey.)

1 A large number of species of parasitoids may be able to impose higher mean mortality rates and thus reduce the host's population growth rates more than can a small set of parasitoid species (see, for example, Huffaker & Kennett, 1966; Huffaker *et al.*, 1986). Low rates of population growth may lead to high variability in birds (Pimm, 1982, 1984b) and in fish (May *et al.*, 1978), but in moths, species with low rates of population growth are those with low population variability (Spitzer *et al.*, 1984; Spitzer & Leps, 1988). Thus, for insects we postulate that high parasitoid diversity leads to low population growth rates, and low growth rates to low population variability.

These relationships are not inevitable, however, even for insects. Competition amongst the many parasitoid species may mean that overall host mortality rates are not significantly higher than for few parasitoids (Zwolfer, 1963, 1971; Force, 1974; Ables & Shepard, 1976). Many species may 'spoil the broth' rather than 'make light work' (Varley, 1959; Turnbull & Chant, 1961; Turnbull, 1967; Ehler & Hall, 1982). Dean and Ricklefs (1979, 1980), however, found no evidence for such competition amongst the parasitoid species attacking the forest Lepidoptera of southern Ontario (but see Bouton *et al.*, 1980; Force, 1980).

2 A diverse suite of parasitoids may provide more reliable control of a host species than just a few species of parasitoid (Smith, 1929; den Boer, 1971). The total density of a few parasitoid species might fluctuate considerably, so allowing the host species to attain high densities on some occasions. If there are many species of parasitoid, there may always be at least one parasitoid species that is sufficiently abundant to keep the host population in check, and so host population variability would be low.

Watt (1965), on the other hand, argued that a high degree of population stability in diverse parasitoid complexes themselves would mean high host population variability. According to Watt, the diverse parasitoid complex would not be able to respond fast enough to prevent population outbreaks of the host.

3 The first two hypotheses consider how the parasitoid complex might influence the population dynamics of the host. The direction of causation may be the reverse, however. The host's variability may limit how many parasitoid species can successfully attack it. Unpredictable, variable populations may support few species of parasitoids for the same reasons that unpredictable environments support few species of all kinds (MacArthur, 1975).

METHODS

Data on the forest Lepidoptera of southern Ontario were collected by the Canadian Forest Insect Survey (Raizenne, 1952). The insects were collected as larvae by hand-picking and hand-beating trees. We restricted our attention to the 38 species of Lepidoptera (from 13 families) for which there were at least 10 years of data. There were too many zero years for the remaining species for reliable estimates of population variability.

The population estimates from the data source showed slight trends. Population counts tended to increase after 1943 (Fig. 17.4). This was probably due to an increase in the intensiveness of the sampling programme. We used the total number of collections of all the species analysed to obtain a correction factor for the individual species' abundances in each year. We calculated variability and mean abundance as the standard deviation and mean, respectively, of the logarithmically transformed index of abundance. Therefore, zero years did not enter into these calculations.

For each Lepidoptera species and for each year, Raizenne (1952) provides the number of specimens of each of the species of parasitoid reared from the host Lepidoptera collections. For each Lepidoptera

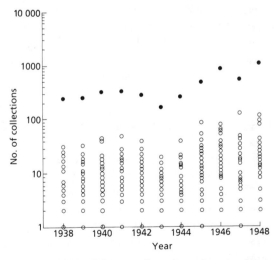

Fig. 17.4 Data on the forest Lepidoptera of southern Ontario showing a slight upward trend after 1943 due to increased intensiveness of sampling. ●, the total number of collections of all species analysed each year. ○, collections of individual species. Data from Raizenne (1952).

species, we used the total number of parasitoid species reared over the entire census period as our measure of parasitoid diversity. Hyperparasitoids are also included. This is unavoidable since many species of parasitoid are facultatively hyperparasitic and so we cannot distinguish between cases of direct and hyperparasitism (see Chapter 11).

We also recorded the number of counties from which each species of host Lepidoptera was collected, as a measure of the regional distribution of the species in southern Ontario. We used logarithmically transformed indices of parasitoid divesity and regional distribution throughout.

CONFOUNDING FACTORS

Before we describe the results, we need to discuss the various factors that may mask or confound the relationship of interest. Species differ in many ways that have nothing directly to do with how the diversity of parasitoids might affect population variability. We must attempt to account for these differences before we come to any conclusions about possible causes of variability.

Large collections of host insects are likely to yield many species of parasitoids. Thus, abundant hosts may have a higher number of parasitoid species simply because they are more intensively sampled.

Host insect species collected from a wide geographic range may appear to be exposed to a wide variety of parasitoid species simply due to passive sampling (Karban & Ricklefs, 1983; Hawkins & Lawton, 1987; Hawkins, 1988). If different parasitoid species are found in different parts of the geographic range of a host insect, the diversity of parasitoids at any one locality may still be quite low, even though the host may have many parasitoid species over the entire range.

We can examine these possible confounding effects of abundance or regional distribution on parasitoid diversity using regression analysis. If the effect on parasitoid diversity is significant we then take the residuals around the regression line as our corrected measure of parasitoid diversity. However, abundant or widespread species may genuinely have more parasitoid species at a given locality (Dritschilo *et al.*, 1975) and so we also used uncorrected measures of parasitoid diversity for all analyses.

The Lepidoptera species analysed exhibit a wide range of feeding habits as larvae. Most live and feed externally on the host plant ('open' species), but some build simple structures or are in some other way more concealed or restricted in their movements ('concealed' species, e.g. leaf rollers, leaf tie-ers, case bearers, web makers, leaf miners, wood borers). We categorized Lepidoptera as being 'open' or 'concealed' in habit according to descriptions by McGugan (1958), Prentice (1962, 1963, 1965) and Baker (1972). Of the species categorized 22 were open and 10 were concealed. We were unable to find information about the feeding habits of six of the species.

species with greater mobility have fewer parasitoid species and, amongst relatively immobile species, more concealed species have fewer species of parasitoid (see Askew & Shaw, 1986). The species within our 'concealed' category are both more concealed and more restricted in their movements than open species and so the effect on parasitoid species diversity may go either way. Whether they have few or many species of parasitoid, the open species may be more exposed to the vagaries of the environment than are concealed species, and may thus be more variable.

RESULTS

There was a negative but non-significant relationship between population variability and parasitoid diversity in Lepidoptera (Table 17.2). However, more species of parasitoid were reared from highly abundant Lepidoptera ($P = 0.0001$). Population variability was significantly negatively correlated to parasitoid diversity once we corrected for parasitoid abundance (see Table 17.2).

More parasitoid species were reared from Lepidoptera species with wide regional distribution than from species with restricted distributions ($P = 0.0001$). Even when we accounted for differences in abundance, more widely distributed species had significantly more species of parasitoid ($P = 0.01$). We used the residuals from a model that simultaneously regressed Lepidoptera abundance and distribution against parasitoid diversity to obtain a measure of parasitoid diversity that is corrected for both abundance and distribution. Lepidoptera population variability was also negatively correlated with this measure of parasitoid diversity. Similar results were obtained when we corrected parasitoid diversity solely for geographical distribution (see Table 17.2).

Open versus concealed species of Lepidoptera

Open species were attacked by fewer species of parasitoid than concealed species ($P = 0.006$). This was also true when we accounted for differences in abundance ($P = 0.05$). It seems that the effect on parasitoid diversity of increased mobility overrides the effect of decreased concealment (Hawkins & Lawton, 1987; Hawkins, 1988) in these Lepidoptera.

We regressed population variability against parasitoid diversity, treating the two groups (open, concealed) separately in an analysis of covariance. The slopes were not significantly different for the two groups (Table 17.2). The 'same slopes' model suggested a significant negative relationship between population variability and parasitoid diversity for the two groups (Table 17.2, Fig. 17.5a).

The results were not significantly changed when we initially correct our measure of parasitoid diversity for abundance and/or regional distribution (Table 17.2, Fig. 17.5b).

Table 17.2 Results of regression analyses and analyses of covariance to show relationships between population variability and parasitoid diversity in Lepidoptera. Our measure of parasitoid diversity is the logarithm of the total number of parasitoid species reared from the host insect collections. The significance level for the relationship between parasitoid diversity and mean log abundance was $P = 0.0001$. We used the residuals around this relationship as our corrected measures of parasitoid diversity. In order to correct for abundance and regional distribution, we used the residuals around a model that simultaneously regressed parasitoid diversity against mean log abundance and the logarithm of the number of counties from which the host insect was collected. The overall significance level for that model was $P = 0.0001$

	Parasitoid diversity	Parasitoid diversity corrected for abundance ($P = 0.0001$)	Parasitoid diversity corrected for regional distribution ($P = 0.0001$)	Parasitoid diversity corrected for abundance and regional distribution ($P = 0.0001$)
All species[*]	$P = 0.13$ (−)	$P = 0.04$ (−)	$P = 0.05$ (−)	$P = 0.03$ (−)
Open versus concealed				
Are slopes similar?[†]	$P = 0.69$	$P = 0.41$	$P = 0.37$	$P = 0.21$
Are intercepts similar?[‡]	$P = 0.16$	$P = 0.18$	$P = 0.15$	$P = 0.17$
Overall slope[§]	$P = 0.03$ (−)	$P = 0.01$ (−)	$P = 0.02$ (−)	$P = 0.01$ (−)

[*] Significance levels (and the direction of the relationship) between variability and parasitoid diversity, from simple linear regression analyses.
[†] Results from an analysis of covariance 'different slopes' model with open and concealed species treated as separate classes. In all cases, open and concealed species showed similar slopes.
[‡] Results from an analysis of covariance 'same slopes' model in which open and concealed species differ only in terms of the intercept.
[§] Significance level and direction of the relationship between variability and parasitoid diversity, from the 'same slopes, different intercepts' model.

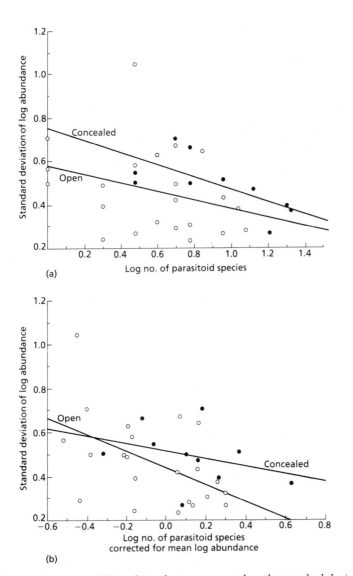

Fig. 17.5 Population variability of Lepidoptera measured as the standard deviation of log abundance plotted against: (a) parasitoid species diversity (log scale); or (b) parasitoid species diversity corrected for mean log abundance (log scale). Notice how population variability declines with increasing parasitoid species diversity for both 'concealed' host populations (●) and 'open' host populations (○).

MECHANISMS

Recall that there are three hypotheses that might explain a negative relationship between host population variability and parasitoid diversity. Hypothesis 1 emphasizes mean parasitism rates and their effects on the rates of population increase of the host and, indirectly, on the variability of the host. Hypothesis 2 emphasizes the temporal variability of those parasitism rates, and their direct influence on the variability of the host. Hypothesis 3 considers the consequences that the variability of the host might have on the parasitoid complex.

To separate the first two hypotheses we would need estimates of the mean yearly parasitism rates as well as the variability of those rates from year to year. The data are not adequate to provide reliable estimates of variability in parasitism rates.

It is difficult to see how one could truly evaluate the relative merit of the third hypothesis (that high host variability leads to reduced parasitoid diversity) without a controlled experiment or a comparison of insect species in different systems.

An analysis of the skewness of the population data provides limited support for hypotheses 1 and 2, however. In the Lepidoptera, species with more negatively (or less positively) skewed populations tended to have marginally more species of parasitoid ($P = 0.11$, Fig. 17.6). This means one or both of two things. Host species characterized by occasional population crashes were more likely to be parasitized by many species, or species characterized by population outbreaks were more likely to be parasitized by few species. Neither interpretation seems compatible with the notion that parasitoid diversity is determined by population variability. If it were, then we would expect the population crashes of the host to cause a *reduced* number of parasitoid species. This is not the case.

IMPLICATIONS

In Lepidoptera, species with many species of parasitoid tend to have less variable populations than species with few parasitoids. The population variability of herbivorous insects does appear to be linked to the structure of their parasitoid complexes.

This relationship between population variability and the diversity of parasitoids has implications about how best to introduce biological agents to control insect pests (Turnbull & Chant, 1961; Watt, 1965; Turnbull, 1967; Hassell & Varley, 1969). Insect species with many parasitoid species may be less variable and less likely to achieve those high densities that have economic impacts. This does not necessarily mean that we should always introduce many parasitoid species in order to control pest populations. Low mean density rather than low variability may be the goal of pest management programmes and low variability may not be associated with low density. Furthermore, it may be more difficult to successfully introduce a large number of parasitoid species (Ehler & Hall, 1982; but see Keller, 1984).

SUMMARY

Resilience and variability are two definitions of stability which permit comparisons of populations. Resilience measures how fast a population

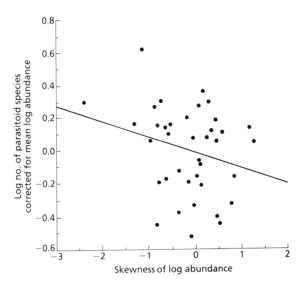

Fig. 17.6 The relationship between parasitoid species diversity (log scale corrected for log abundance) and the skewness of log abundance, showing the suggestion of a negative association. Negatively skewed populations of Lepidoptera may support marginally more species of parasitoid.

recovers to some equilibrium level following a disturbance. We define variability as the degree to which a population's densities fluctuate from one year to the next.

We have argued that these measures of stability may be related not just to the intrinsic properties of the species, but to interactions with the other species in the community to which the species belongs. Specifically, we have discussed the possible effects on stability of interactions with predators (or parasitoids), focusing primarily on insect herbivore communities.

We have not considered interactions involving more than two trophic levels (Faeth, 1985; Price 1988). For example, the nature of the predator complex attacking a particular herbivore may not be independent of the host plants on which the herbivore feeds (Freeland, 1983; Askew & Shaw, 1986; Hawkins & Lawton, 1987; Hawkins, 1988; Price, 1988). Also the effects on herbivore population dynamics of predators may depend on aspects of the spatial arrangement of host plants and the movements amongst those plants of the herbivore and its predators (Root, 1973; Risch et al., 1983; Andow & Risch, 1985; Karieva, 1986; Sheehan, 1986; Letourneau, 1987). A challenge for ecologists now is to understand the combined effects on population dynamics of interactions with species at higher trophic levels and lower trophic levels.

18: Biological Control

JEFF K. WAAGE & NICK J. MILLS

INTRODUCTION

Few activities demonstrate more clearly the effect of natural enemies on animal populations than the practice of biological control, the use of living organisms as agents of pest control. By manipulating indigenous natural enemies or introducing new species into ecosystems, biological control seeks to increase the mortality inflicted on pests and thereby to reduce mean population levels and population fluctuations.

Disastrous forays in the eighteenth and nineteenth centuries into the control of vertebrate pests using vertebrate predators (e.g. the notorious introduction of mongooses onto islands to control rats) proved that biological control would only be effective and safe where the natural enemies used were relatively specific to the pest and would not, therefore, become pests themselves. Insect pests provided the first opportunity to demonstrate this, and the use of specific predators, parasitoids and pathogens against insects has become the major focus of biological control today. Similar specificity is found amongst the insect herbivores and pathogens of plants, and this has led to their widespread use in the successful control of noxious weeds.

While biological control is an essentially empirical discipline, there is considerable potential for improvement through basic research on the ecology of interactions between natural enemies and pests. This potential is largely unrealized as yet. It is the aim of this chapter to draw together the practice and theory in biological control, and to indicate where improving our understanding of natural enemies may improve our ability to use them.

APPROACHES TO BIOLOGICAL CONTROL

If all pests have natural enemies, why do we have pests? Clearly, the impact of natural enemies is often insufficient to prevent pests from causing losses to crop production. This may be simply because agricultural losses occur at pest densities well below those which natural enemies can maintain in the field.

On the other hand, natural enemies may not be realizing their full potential to depress crop pest populations. Modern crop monocultures do not provide the range of alternative foods and the variety of microhabitats which many natural enemies depend on in more diverse, natural vegetation. Consequently, natural enemy numbers may be low in crops, or late to arrive during a cropping season. To rectify this, efforts can be made to conserve natural enemies by modifying cropping practices, e.g. by increasing vegetational diversity near or in the crop (Altieri & Letourneau, 1984; van Emden, 1990) or by modifying pesticide use (Waage et al., 1985; Waage, 1989).

Where conservation is insufficient, natural enemies may need to be added to the crop system. This might involve the augmentation of field populations of natural enemies with laboratory-reared individuals released early in a season, to establish populations which can check the pest before it reaches damaging levels. Alternatively, it may involve massive and repeated applications of natural enemies as biopesticides through a crop season. The chosen method depends largely on the ecology of the natural enemy. For instance, insect pathogens, with their relatively poor capacity for transmission, are often best used as inundative biopesticides.

Perhaps the best known method of biological control is that used to combat exotic pests. Exotic pests frequently invade regions without their adapted natural enemy complex and, in the absence of effective local natural enemies, they can reach very high population levels. Their control involves the introduction and establishment of effective natural enemies imported from the pest's area of origin. This method is frequently called *classical biological control* in recognition of its relatively early first use in the 1800s.

All of the biological control methods pose interesting questions about the ecology of natural enemies and how it might be manipulated. However, classical biological control has a particular appeal to ecologists, as it represents a unique opportunity to experiment with natural enemy populations. Classical biological control lets us compare under field conditions the dynamics of pest species with and without (i.e. before and after) natural enemies, and therefore to make and test predictions about the role of natural enemies in pest population depression and regulation. Further, the exotic nature of both pest and introduced natural enemy tends to isolate their interaction in their new environment. This facilitates the application to classical biological control of fairly simple two-species predator—prey models as a means to understand processes and generate predictions.

STEPS IN A CLASSICAL BIOLOGICAL CONTROL PROGRAMME

The aim of classical biological control is to reduce the population density of a pest, through the importation and establishment of natural enemies, to a level at which it no longer causes economic damage. Ecologists have tended to equate this aim with a necessity for the natural enemy to depress pest population densities to a very low stable level. This depression is often represented by the ratio q, the equilibrium population size of the host in the presence (N) and absence (K) of the natural enemy ($q = N/K$) (Beddington *et al.*, 1978). However, since economic thresholds for pests are determined by pest impact, crop value and socio-political value as well as pest density, the degree of depression of pest abundance required for successful biological control will vary from case to case.

The first step in a classical biological control project is the *evaluation* of the pest problem in the target region for the programme. This establishes the correct taxonomic identity of the pest and determines its probable area of origin. In addition, information must be gained on its ecology in its exotic range and the natural enemies that may be associated with it there.

The second step in a programme is the *foreign exploration* for natural enemies of the pest in its area of origin. Foreign exploration should involve quantitative surveys to assess the complex of natural enemies that attack the pest, their impact and their degree of specialization. This begins a process of *selection* of biological control agents from this complex for importation and establishment in the target region. This step is perhaps the most controversial aspect of biological control and is the one which relies most heavily on ecological principles.

Once natural enemies have been selected they must go through the processes of *quarantine*. For insect natural enemies, this means removing any hyperparasitoids, plant pathogens and insect pathogens from the culture. Natural enemies cleared through quarantine are candidates for *release* in the target region. If establishment is achieved, the final step in a biological control programme is the evaluation of the impact of the released agents and the *monitoring* of the change of the natural enemy and pest population.

PATTERNS OF CLASSICAL BIOLOGICAL CONTROL SUCCESS

Much has been made of the historical data on biological control, both to provide guidelines for future practice and ecological insight. This is a risky exercise at best, because the record of classical biological control is troubled by erroneous identifications of pests and natural enemies, occasional errors in dates and places and the arbitrary (and sometimes entirely incorrect) interpretation of success.

A database of introductions recently developed at the International Institute of Biological Control (IIBC), called BIOCAT, has attempted to obtain accurate information on all introductions to date of insect natural enemies against arthropod pests. It provides an interesting resource for examining patterns.

According to BIOCAT, using records up to 1988 (Waage & Greathead, 1989), there have been over 4000 separate introductions of insect biological control agents against insect pests. These have involved the introduction of 563 different species of natural enemies against 292 pest species in 168 countries. About 25% of introduced agents became established. Of these, about 49% contributed to substantial control of the pest.

Figure 18.1 presents some of these data in a different way, to illustrate patterns in the control of different groups of insect pests. Here we see the number of distinct programmes which have been mounted against pests belonging to different orders, and the number of establishments associated with those programmes. The roughly equal numbers of programmes and agents hides the fact that many successful programmes have involved the establishment of more than one agent, while many past programmes have failed with no establishment. In the pie diagram (Fig. 18.1), the success achieved by established agents is broken down according to the order of the target pest.

It is clear from this summary that the Homoptera have been both the most targeted and most successfully controlled group of exotic pests (Greathead, 1989). Their prevalence as exotic targets for biological control may reflect their feeding niche on the woody stems of perennial crops. This greatly enhances their potential to be transported around the tropics and subtropics on propagative plant material.

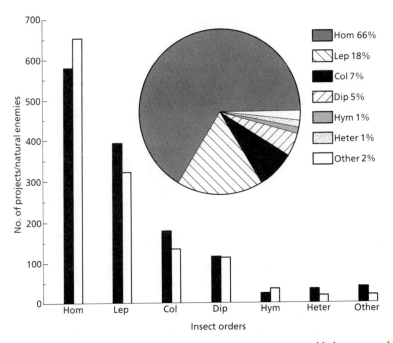

Fig. 18.1 Patterns of classical biological control programmes, establishments and success on different orders of insect pests. Histogram give number of programmes (■) and established natural enemies (□) for each insect order. Pie chart gives distribution of all successful introductions: i.e. where substantial control of the pest was achieved) across different orders of target pest. Hom, Homoptera; Lep, Lepidoptera; Col, Coleoptera; Dip, Diptera; Hym, Hymenoptera; Heter, Heteroptera. Data from BIOCAT.

The successful control of Homoptera, largely by the introduction of parasitoids, probably reflects the considerable impact that a guild of highly mobile and efficient natural enemies can have on relatively sedentary and exposed hosts. Recent theoretical studies have suggested that Homoptera may be particularly amenable to stable, successful biocontrol because of the large proportion of the life cycle which is invulnerable to parasitism (Murdoch *et al.*, 1987), the long reproductive life of adults (Godfray & Waage, 1991) and their colonial habits, which may facilitate stabilizing numerical responses by the parasitoids to their patchy distribution (Waage & Greathead, 1989).

Classical biological control agents, like the indigenous natural enemies discussed earlier, are likely to be influenced by cropping systems. This is borne out in an analysis of the success of biological control in crop systems of differing stability (ranging from unstable, seasonally planted field crops through to highly stable, long-term forest crops). From a theoretical standpoint, Southwood (1977) has argued that natural enemies are likely to be more effective in stable than unstable habitats. The record from classical biological control tends to support this, with greater success in orchard crop systems than in seasonal field crops (Hall & Ehler, 1979; Hall *et al.*, 1980; Greathead & Waage, 1983)

DECISION MAKING IN CLASSICAL BIOLOGICAL CONTROL

Classical biological control is often thought of as a process of reconstructing the natural enemy complex of a pest in its exotic range. In practice, only a small fraction of the natural enemy complex of a pest is introduced into its exotic range in the process of classical biological control. This is illustrated in Fig. 18.2 for insect parasitoids. The solid histograms indicate the distribution of the number of parasitoids introduced against a particular exotic pest in a particular country. The data come from BIOCAT and indicate that the average number of parasitoid species introduced in a particular programme is 3.4, with a range from one to 53.

These biological control records can be compared with the size distribution of parasitoid complexes from British insects, derived from Hawkins (1988) (Fig. 18.2, open histograms). While this is not the best comparison, since the parasitoids used for biological control come from all over the world, it does give a (probably conservative) impression of what a small fraction of the natural enemy complex of a pest is used in a typical classical biological control programme.

The reasons for the limited use of potential natural enemy complexes in classical biological control are various, and include practical constraints as well as ecological considerations.

1 We must be careful not to introduce natural enemies which have unwanted side effects. This applies particularly to generalist natural

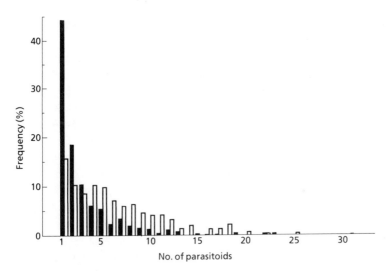

Fig. 18.2 Comparison of the number of parasitoids released per pest with the size of natural parasitoid complexes. The histogram gives the frequency distribution of numbers of parasitoids released in classical biocontrol programmes (■) (from BIOCAT), and the number of parasitoid species associated with a selection of British insects (□). From Hawkins (1988) as a representative sample of natural parasitoid complexes.

enemies from the pest's area of origin, which may have an unpredictable host range in its area of introduction. Agents should be sufficiently specific so as not to put at risk populations of insects of economic importance (e.g. bees, silkworms), insects used for the biological control of weeds (Goeden & Louda, 1986) or insects of conservation value, particularly in highly indigenous island faunas (Howarth, 1983).

2 When carried out with due care, classical biological control is a lengthy and expensive process, and programmes are usually constrained by limited funds and limited time. As a result, it is rarely possible to introduce more than a few of the candidate natural enemies identified during the exploration programme.

3 Some potential agents may be detrimental to overall control, by reducing the efficiency of superior parasitoids. A commonly cited example is that of a facultative hyperparasitoid, which may raise pest populations by attacking an effective primary parasitoid more than it lowers them by attacking the pest itself. An example of another kind of negative interference concerns the exotic pest *Rhyacionia buoliana* in North America (see below). As Ehler (1990) points out, it is by no means obvious that the natural structure of a parasitoid complex is compatible with the practical goals of biological control.

4 Many efforts to establish agents fail, and some of these failures probably reflect the difficulty of establishing new agent species on pest populations which have already been reduced in density by agent species introduced earlier. There is some evidence for this from analyses of the record of biological control (Ehler & Hall, 1982). Thus, inferior

control agents, if introduced first, may prevent the subsequent introduction of better species.

For all of the above reasons, classical biological control involves
decisions about which part of a natural enemy complex to introduce
against an exotic pest. The need for such decisions is the strongest
argument for an ecologically based approach to biological control. If
we can only introduce a fraction of the natural enemy complex, it is
critical that we introduce the most important species. This is even
more true if initial decisions limit future options, as they might where
one introduction interferes with or limits the chances of establishing a
second one. The entire process, its constraints and possibilities, is well
illustrated by the example of a recent successful biological control
programme against the exotic mealybug, *Rastrococcus invadens*, in
West Africa (Box 18.1).

In the sections which follow, we consider how natural enemy
ecology influences the stages, and decisions, which lead to the introduction of a particular natural enemy in a classical biological control
programme.

EXPLORATION FOR
POTENTIAL BIOLOGICAL CONTROL AGENTS

Selecting the target for exploration

Exploration for the natural enemy of an exotic pest usually focuses on
the pest in its area of origin, where there is the potential to find the
greatest diversity of natural enemies. Even at this early stage, however,
theoretical arguments come to bear on the practical decision of where
to look for agents.

Hokkanen and Pimentel (1984) have argued that better success may
be achieved by seeking exotic natural enemies from species related to
the target pest, so that natural enemies are less 'coevolved' with the
target pest and therefore more virulent. There are clear instances
where such 'new associations' have led to very successful biological
control.

Taking this advice might broaden an exploratory programme, both
in terms of species and geographical areas examined. Clearly, the need
to select reasonably specific natural enemies will constrain the range
of species, and the increased effort needed for an expanded exploration
programme may be a practical constraint. However, what about the
ecological basis of the argument that coevolution leads to reduced
virulence?

The Hokkanen and Pimentel argument was supported by an analysis of the record of past biological control. Their methods and conclusions have engendered some controversy and subsequent analyses
have given conflicting results. Part of the problem may be due to
differences in interpretation of the data (Waage & Greathead, 1989).

Box 18.1 How errors in exploratory research can affect success

This example describes a biological control programme against the pine shoot moth, *Rhyacionia buoliana*, a European insect which became an exotic pest in Canada early in this century. In Europe, where an exploratory programme was mounted, the larvae of this tortricid moth are attacked by a complex of 13 parasitoids, most of which are relatively uncommon (Schroeder, 1974). A foreign explorer, collecting parasitoids by rearing samples of host larvae, would have found results as shown in Table B18.1, comparable to those of Schindler (1965). The four most frequent parasitoids were assumed to be the best for biological control introductions, due to their apparent impact on the host population, and were introduced into Canada between 1928 and 1967 (McGugan & Coppel, 1962).

Later, more detailed studies of pine shoot moth parasitism (Arthur *et al.*, 1964; Schroeder, 1974), in which host larvae were dissected to look for parasitoid recruitment to the host population as well as their eventual emergence from the host, clearly indicated that the three ichneumonid species *Pristomerus* sp.nr. *orbitalis*, *Sinophorus rufifemur* and *Temelucha interruptor* are facultative cleptoparasitoids, preferentially attacking hosts previously parasitized by the braconid *Orgilus obscurator*. Levels of parasitoid recruitment (measured above from dissection) show that in fact *O. obscurator* is the most effective parasitoid, in terms of its ability to locate and attack hosts, but that it is a poor competitor in relation to the three ichneumonid competitors.

Table B 18.1 Parasitism of the European pine shoot moth by the most frequent larval parasitoids at Meppen in northern Germany

Parasitoid	% Apparent parasitism (emergence)	% Hosts attacked (dissection)	% Apparent parasitism (dissection)
Lypha dubia	23	7	7
Orgilus obscurator	5	37	12
Pristomerus vulnerator	4	19	13
Sinophorus rufifemur	5	7	5
Temelucha interruptor	21	25	17
Year	1958	1964	
Source	Schindler (1965)	Schroeder (1974)	

While *O. obscurator* is today widely established and successful in much of Canada, in some areas it is still affected by *T. interruptor*, which can cause up to 50% mortality of *O. obscurator* larvae in hosts, and thereby reduce the local impact of *O. obscurator* on *R. buoliana* (Syme, 1981).

Recent refinements of these analyses (Waage, 1990; Hokkanen *et al.*, unpublished data) leave little doubt that new associations can be as effective as old ones once the natural enemy is established. However, establishment of a natural enemy on a species it has not previously encountered is very difficult. This is understandable, given the high specificity of many natural enemies. Thus in Box 18.1, we found that parasitoids of the closely related *Rastrococcus spinosus* did not attack the target pest, *R. invadens*.

In practice, therefore, studies on the target pest in its area of origin remain the most promising approach to finding an effective control agent quickly, but the potential for new associations should also be explored.

Maximizing coverage of the natural enemy complex

The aim of foreign exploration is to determine the full species richness of its natural enemy complex, the potential impact of its component species and their host specialization. The species richness of a natural enemy complex varies enormously between host species and can be affected by a wide range of factors associated with the habitat, host plant, taxonomy and abundance of the host (Askew & Shaw, 1985).

The richness of the natural enemy complex, varies with the area covered during the survey (Hawkins & Lawton, 1987). For example, Hawkins (1988) shows that the mean number of parasitoids associated with a host insect in Great Britain increases from 5 for local sites to 8.6 on a national scale. Clearly this indicates that foreign exploration should be extended beyond a localized survey but is it necessary to cover the full geographic range of a host? In fact, geographic variability in the composition of natural enemy complexes appears generally to be small (Askew & Shaw, 1985), and a small series of well separated locations can often provide the full complement of natural enemies.

Host plant factors can also affect natural enemy complexes and the success of exploration programmes. Natural enemy complexes can be reduced where the range of a host plant has been artificially extended by man in recent centuries. An example of this is seen with the larch casebearer, *Coleophora laricella*, which supports 20 parasitoid species in the Alps (primary distribution of larch) but only 9–10 species in northern Europe (secondary distribution of larch). The parasitoid complex of a pest frequently varies on different crops, and records for *Helicoverpa armigera* in Africa (van den Berg *et al.*, 1990) clearly demonstrate this, with cotton and sorghum supporting almost mutually exclusive assemblages of parasitoid species. Since many parasitoids locate hosts by response to chemical cues derived from the host's food-plant (Vinson, 1986), it is not surprising to find such different parasitoid faunas in different crops.

The composition of a natural enemy complex also varies with the density of the host population (Price, 1973; Mills, 1990). The European larch bud moth, *Zeiraphera diniana*, exhibits cyclical outbreaks on larch across the Alps. During periods when the bud moth is at low density the most frequent parasitoids are three eulophid species, but these species are totally absent from the parasitoid complex of outbreak populations (Delucchi, 1982). These ectoparasitoids attack mid-stage host larvae, which construct characteristic feeding tubes from the needles of the host tree. As host density and level of defoliation increase, feeding tubes become less obvious and the eulophid parasitoids are probably excluded by loss of visual host searching cues and lack of protection for their exposed larval development.

This aspect of natural enemy ecology has important implications for the success of classical biological control which have only recently been appreciated. For ease of collection, most foreign exploration programmes have sought local outbreaks of the target pest in its region of origin. However, the natural enemies collected from outbreaks may not be those most suited to maintain the pest at low densities in its exotic range. The gypsy moth, *Lymantria dispar*, a European pest introduced accidentally into North America, provides an illustrative example.

Foreign exploration for natural enemies of gypsy moth in Europe was confined to high host density populations with the result that the majority of the 50 or more agents released were unspecialized parasitoids that were merely taking advantage of the abundance of the gypsy moth to exploit a marginal host (Dahlsten & Mills, 1992). More recently, the use of experimental populations of gypsy moth to expose cohorts of larvae to endemic parasitism has revealed the occurrence of a previously unknown and specialized parasitoid that shows much promise as a biological control agent (Mills, 1990). The use of experimental host cohorts placed in the field provides a useful and often neglected technique for the survey and collection of natural enemies from low host density populations.

Interpreting the abundance of natural enemies

In survey work, the importance and impact of natural enemies is generally assessed by their relative abundance at the various survey locations. Abundance and perhaps size, used as a correlate of feeding potential, are often used to rank the relative importance of predators, while the proportion of hosts killed is used to rank different species of parasitoids.

These methods of assessing the importance of particular natural enemies are less rigorous than one would like. Even the simplest predator–prey models demonstrate clearly that ineffective natural enemies will be abundant if the pest which they attack is abundant. Furthermore, the tendency to collect from outbreak populations of the pest (see above) may create unwarranted interest in a common predator or parasitoid. Care must also be taken in interpreting impact of natural enemies from limited sampling of the pest. Van Driesche (1983) has shown how poorly single point measures of percentage parasitism may estimate the true generational mortality imposed by a natural enemy.

When a misleading measure of field parasitism is used to select natural enemies, the consequences for biological control can be dramatic. A famous example involves the classical biological control of European pine shoot moth, *Rhyacionia buoliana*, in Canada (McGugan & Coppel, 1962), which is described in Box 18.1. Here, introductions were made on the basis of estimated percentage parasitism, and not impact on the host population. As a result, cleptoparasitoids were inadvertently introduced which reduced the effectiveness of the biological control effort.

This example gives an extreme case where some agents were not only inferior, but actually interfered directly with the more important species. A less severe case of this phenomenon may occur with natural enemy complexes which exhibit 'counterbalanced' competition, where parasitoids show different and complementary advantages in finding hosts and competing in hosts (Zwölfer 1971). This is a frequent occurrence, and it is conceivable that a failure to appreciate how these species affect each other's impact might lead to the selection of the inferior species for introduction.

The logical alternative to superficial interpretation of natural enemy abundance, as a basis for selecting the most important species, is the collection and analysis of detailed life table data in the area of origin of a pest. Life table studies of this kind have been done for two biological control programmes, one against the winter moth, *Operophtera brumata*, and the other against the larch casebearer, *Coleophora laricella*. Both life table studies carried out in Europe (Jagsch, 1972; Varley *et al.*, 1973) failed to detect any significant role of parasitoids in the dynamics of the host populations, although density-dependent predation of the winter moth pupae was revealed.

In both studies, however, the introduced parasitoids have played a major role in the decline of pest control either by direct (Embree, 1966; Ryan, 1986) or possibly indirect (Roland, 1988) processes. Thus the life table data from Europe proved not to be predictive of the outcome of parasitoid introductions in North America. It should also be observed that both life table studies followed the first programmes of parasitoid introduction against these pests — they were retrospective analyses. Indeed, few classical biological control programmes run with the time and funding to undertake detailed ecological studies in the area of origin of the pest before introduction.

There is no question that life tables give a more accurate impression of the importance of natural enemies than point samples of predator abundance or parasitism. Their failure to identify the potential of a particular natural enemy reflects in part the different situations in which natural enemies act in their areas of origin and introduction. Thus, for winter moth, the relative unimportance of the parasitoid,

Cyzenis albicans, in Britain was attributed to the subsequent action of pupal predators which killed the parasitoid as well. Release from such predation may have allowed it to have a greater impact in North America. Besides such constraints on the survival of natural enemies in their area of origin, measurements of their impact on small pest populations cannot predict their capacity to respond to the larger populations which they may encounter in their area of introduction.

Overall, our understanding of the importance of natural enemies in a pest's indigenous range would seem to be maximized by a combination of extensive studies, to encompass both variation in the natural enemy complex, the pest's ecology and its density, and intensive studies, such as life table studies, which provide more accurate measures of how particular natural enemies affect generational mortality. No simple scoring system for 'good agents' will emerge from such studies. Rather, we gain a better knowledge of a natural enemy which we can then use to generate hypotheses on how it will behave on the pest in a new region, where both its survival and phenology and that of the pest's may be different.

SELECTION OF AGENTS FOR INTRODUCTION

As the result of an exploration programme, we arrive at a partial understanding of a natural enemy complex, its characteristics and dynamics. We use this information to begin that draconian process by which a diverse natural enemy complex is narrowed to a few priority species for introduction.

Clearly, most interest will be directed at species which appear to have a substantial impact on the pest population, or those which might have such an impact were other constraining factors removed (e.g. *Orgilus obscurator* free of its cleptoparasitoids).

Often, generalist natural enemies will be identified as important mortality factors, frequently in the form of a complex of species acting on a particular life stage. Both theory (Hassell, 1985) and practice have demonstrated the importance of generalists to biological control of indigenous pests. However, it is also possible, as in the case of the gypsy moth, that generalists may be opportunists exploiting pest outbreaks. Further, as *introduced* agents their impact on the pest equilibrium will always be constrained by the vagaries of the abundance of other host or prey species. But the deciding argument which excludes most generalist natural enemies from selection is the potential risk which they pose to non-target species in the area of introduction.

Left with a list of promising, specialist natural enemies, a number of factors will influence the priority that these are given for introduction. These include very practical constraints, such as our ability to rear the natural enemy for quarantine and release.

Very often, precedent from previous programmes is used in the selection of agents. Hence, encyrtid wasp parasitoids have a better record for control of mealybugs than the coccinellid beetle predators which specialize on these pests, and therefore parasitoids are usually preferred in new programmes (Moore, 1988, and see Box 18.2).

Finally, there is an opportunity for the application of ecological theory to agent selection. Theory has been used to identify criteria for 'good agents', and these criteria tent to fall into two broad groups (Waage, 1990). *Holistic criteria* are derived from concepts of how a particular natural enemy fits into the ecology of its pest and other mortality factors acting on it. *Reductionist criteria*, by contrast, are derived from theoretical views on which specific characteristics of an individual natural enemy species embodies its potential as a biological control agent. These criteria often arise from the dissection of a natural enemy into a set of component characteristics, for instance, search efficiency, aggregation and handling time.

Holistic criteria for agent selection

Holistic criteria for selection can be applied at the earliest stage in exploration, as when we choose to seek natural enemies in low density rather than outbreak pest populations, or when we investigate the structure and dynamics of a natural enemy complex to see how one natural enemy species may be influenced by another.

One of the most contentious issues within this holistic approach is

---

Box 18.2 A classical biological control example

The mango mealybug, *Rastrococcus invadens*, was first discovered in Ghana and Togo around 1981 (Agounké *et al.*, 1988) causing damage to fruit trees, particularly mango and citrus. It feeds by sucking sap from the undersurface of the leaves and its honeydew leads to the growth of sooty moulds on the foliage. The net result is a marked reduction in mango production, a fruit that provides an essential source of both energy and vitamins in the diet of the West African people. Apart from being a staple food it has been estimated that the annual value of the mango crop in Togo is £1.95 million (J.M. Voegele, unpublished data). The extent of economic losses are difficult to assess, given the traditional nature of the horticultural systems in the region, and remain largely unquantified although crop losses of up to 80% from mangoes have been noted in Ghana (Willink & Moore, 1988). Thus potential annual losses exceeding £1.5 million for individual nations indicate that the problem was sufficiently severe to warrant a classical biological control programme.

At first, the mealybug was confused with another, Asian species, but subsequently proved to be new to science. It was described in 1986 (Williams, 1986) and a reassignment of museum material revealed that its probable area of origin was South and South East Asia. Accordingly, IIBC initiated surveys for natural enemies in India and

continued on p. 425

Box 18.2 *contd*

Malaysia in July 1986. From a series of three *Rastrococcus* species found feeding on mangoes in India, about 15 species of parasitoids and predators were recorded (Narasimham & Chacko, 1988), most of them new to science. These included several encyrtid parasitoids, which have proved to be the most successful agents used against mealybugs in classical biological control (Moore, 1988). Host specificity of some encyrtid parasitoids of other *Rastrococcus* species revealed that they did not attack *R. invadens*. Therefore, further work concentrated on the parasitoids of *R. invadens*.

These initial surveys indicated that *R. invadens* has two encyrtid parasitoids, *Gyranusoidea tebygi* and *Anagyrus* sp. Field observations in India, where *R. invadens* was not abundant at any localities, suggested that parasitism seldom exceeded 20% and that *Anagyrus* sp. was the dominant parasitoid (Narasimham & Chacko, unpublished data). Further studies of the biology of *R. invadens* and its parasitoids at IIBC in the UK (Willink & Moore, 1988; Moore, unpublished results) showed that both *G. tebygi* and *Anagyrus* sp. in addition to being very host specific also have a generation time less than that of their host. *Gyranusoidea tebygi* proved to be more easily reared than *Anagyrus* sp. and so was selected as the first candidate for introduction.

The first releases of *G. tebygi* were made near Lomé in Togo in November 1987. The parasitoid readily established from releases of as few as 20–30 individuals in some sites and after only 1 year the parasitoids had migrated up to 100 km from the original release sites (Agricola *et al.*, 1989). Releases of *G. tebygi* have now been extended to Benin, Ghana, Nigeria, Zaire and Gabon and in all cases the parasitoids have become established. The current indications are that good control of the mealybug has been achieved throughout the areas colonized and monitoring of the impact of the releases is in progress.

In 1987, work began on a population model to predict the potential impact of different natural enemies of *R. invadens*. The model was constructed for *G. tebygi* and *Anagyrus* sp., using basic data collected in India and UK, with some information on survivorship of the mealybug in the field in Togo.

Due to the urgency of the problem, *G. tebygi* was released before the model was complete, but subsequent analysis predicted that *G. tebygi* would cause greater depression of mealybug populations than *Anagyrus* sp., and that the latter would not contribute to substantial additional depression if both parasitoids were used (Godfray & Waage, 1991). Subsequent laboratory population studies in the UK supported these findings: *Anagyrus* sp. was less effective at suppressing colonies of *R. invadens* over 10 generations and was driven to extinction in the presence of the superior *G. tebygi* (D. Moore & A. Cross, unpublished data).

As a result of these studies, any new introductions in affected countries will focus on *G. tebygi*, but opportunities will be sought to test *Anagyrus* sp. in combination with it.

the decision to introduce more than one agent (multiple introduction) in preference to a single best agent. Ignoring the question of whether a single best agent can even be identified, it has been argued that several introduced species may compete with one another in such a way that the overall levels of pest depression will be less than might be achieved with one agent alone (see, for example, Turnbull, 1967). It is possible to find examples where natural enemies have interacted in this way, but these examples often involved quite particular, even peculiar, interactions between agents. Thus an agent may be introduced which causes a pest population with overlapping generations to become discrete, thereby eliminating another agent whose developmental period depended on the continuous presence of a particular host stage (Godfray & Hassell, 1987). Alternatively, an agent may be introduced which is facultatively hyperparasitic or cleptoparasitic on another, as described above for *Rhyacionia buoliana*.

In an important paper, May and Hassell (1981) used simple analytical population models to show that two parasitoids attacking a single host may either coexist (by virtue of the 'balanced competition' mentioned above) or one will be eliminated, but that equilibrium population levels will not be higher than they would have been with either species acting on its own. Later, Kakehashi *et al.* (1984) observed that this held true where the parasitoids distributed their attacks on the host independently of each other. If attacks were distributed on a similar subset of hosts then it was possible to have lower depression of pest numbers with both natural enemies than with one on its own.

It remains to be investigated how important the situation described by Kakehashi *et al.* (1984) might be in real biological control programmes. The extreme case of their model, where a less efficient parasitoid distributes its eggs on precisely the same hosts as a more efficient species, and always wins in competition, is very similar to the action of cleptoparasitoids or hyperparasitoids. These kinds of natural enemies would be eliminated from consideration at an early stage of the selection process.

In planning multiple introductions, practitioners of biological control often seek natural enemies which occupy different feeding niches on the host. The strategy here derives from concepts of community ecology, and the desire to minimize competition and maximize impact on the host. The conclusion of Kakehashi *et al.* (1984) that low niche overlap in parasitoids promotes pest depression is relevant to this argument. In a somewhat different vein, Murdoch (1990) suggested that pest depression is encouraged by maximizing the proportion of a pest's life cycle which is vulnerable to natural enemies.

In any discussion of this kind, we need to bear in mind the dynamical nature of natural enemy complexes and the possibility that natural enemies acting at different stages of the pest's life cycle can influence one another strongly. Thus, Hill (1988) suggested that the impact of the introduced larval parasitoid, *Apanteles rufricrus*, on populations of the moth, *Mythimna seperata*, may have been reduced by density-dependent parasitism by a later-acting, indigenous pupal parasitoid. Similarly, the contribution of egg parasitoids may be reduced by density-dependent mortalities acting on young larval stages of their moth hosts (van Hamburg & Hassell, 1984). In general, strongly density-dependent mortalities will tend to have substantial negative effects on the impact of natural enemies which precede them (May et al., 1981).

Interactions between natural enemies attacking the same and different stages of a pest are particularly relevant when considering the introduction of exotic agents to control native pests, a kind of 'new association' which has much promise. Here, practitioners often seek to 'fill the empty niche', finding an exotic species which attacks a pest stage not attacked by any members of the indigenous natural enemy complex.

The subject of multiple introductions is more complex than previously thought and deserves more scrutiny. However, negative effects appear so rarely in careful biological control programmes that the benefits of multiple introductions far outweigh the alternative of trying to introduce the best agent the first time. Stopping after the first introduction just in case it *was* the best species, and just in case the subsequent introduction would reduce the efficacy of control, is not justified on field experience.

Reductionist criteria for agent selection

The development of simple analytical models for predator–prey interactions lies at the heart of the reductionist approach to selecting biocontrol agents. Reductionist criteria are derived from the parameters of these models, particularly those which appear important to pest depression and regulation. Searching efficiency (or area of discovery), handling time, aggregation, mutual interference have all been suggested as criteria for selecting agents. An example of how such criteria might be used would be a functional response experiment to calculate and compare the parameters of three candidate natural enemies, so as to select the agent with the highest searching efficiency.

In practice, reductionist criteria have rarely been used in selecting agents. This is partly due to the difficulty of estimating the parameters. Laboratory measures of searching efficiency and aggregation are unlikely to reflect the real efficiency in the field, where a more complex environment is searched and patches of hosts or prey may be of different sizes and distributions. To measure such parameters in the field is possible, but difficult, particularly if different species are to be compared.

Beyond these practical constraints, the breakdown of a natural enemy's ecology into component attributes gives the misleading impression that these attributes can be compared independently. In fact, natural enemies' attributes are often associated in particular patterns. Thus fecundity or efficiency in parasitoids is often traded off against the intrinsic competitive ability once in a host (May *et al.*, 1981), and other distinctive suites of attributes have been identified (Waage, 1990).

It is the whole organism, not its component parts, which forms the basis for prediction of success, and this realization has led to the exploration of population models as a means of predicting success. In a model, attributes such as searching efficiency, the stage of the host attacked, longevity and larval survival, can be assembled in a realistic manner to characterize a natural enemy and to examine how it might influence depression of a pest population. Mathematical models of this kind have been built to examine the biological control of the winter moth (Hassell, 1980), red scale (Murdoch *et al.*, 1987), cassava mealybug (Gutierrez *et al.*, 1988) and mango mealybug (Godfray & Waage, 1991). While most of these models have been retrospective, they reveal the potential for using these models in a predictive manner.

The model for mango mealybug (see Box 18.1) was developed pro-spectively as an exercise in agent selection. As such, it emphasized the incorporation of important but *easily measurable* life history par-ameters, which could be collected in a few months of field and labora-tory study. Such parameters included the stage of host attacked by the different parasitoids, age-specific development rates for hosts and para-sitoids, age-specific survivorship of hosts in the field, and adult long-evities and daily oviposition rates. Difficult parameters, such as searching efficiency, were treated as variables. This model predicted the superiority of one of two potential control agents, a prediction later supported by laboratory studies on long-term dynamics of the host with one or both parasitoids (A. Cross & D. Moore, personal communication).

The aim of prospective modelling is to assist in decision making, particularly when many interacting elements are involved and where the outcome may not be easy to visualize. While they aim to be as accurate as possible, prospective models cannot be the products of many years of careful study and must be built and used quickly and cheaply.

RELEASE AND ASSESSMENT OF SELECTED AGENTS

When an agent is finally selected for release, it is important that the investment in its selection is matched by a similar commitment to its establishment. In fact, this is not always the case, and some of the 75% of failed establishments for arthropod agents is probably attribu-table to insufficient effort. Analyses of past programmes indicates that greater effort at establishment leads to greater success (Beirne, 1980; Hoy, 1985).

Overall, there is a considerable and unrealized opportunity to examine the ecological basis of natural enemy establishment. While research on agent selection has been directed at properties which affect ultimate, equilibrium levels of a pest, research on establishment needs to be focused on short-term population trajectories, the spatial dispersion and distribution of pests and agents, and the dispersal of agents away from the release point.

Once established, it is important to assess the impact of a natural enemy. Again, this is rarely done in practice. In classical biological control, successes and failures are often quite dramatic and obvious, and this perhaps inhibits efforts to understand them better. It must also be said that our ability to quantify the impact of natural enemies, particularly in tropical agricultural systems, is still rudimentary, and methods are limited and often approximate (see Luck *et al.*, 1988 for a review).

A good example of the value of a well-quantified assessment of biological control is that of the larch casebearer in the Pacific northwest of the USA (Ryan, 1990). Detailed monitoring of the host population both before and after the release of parasitoids was combined with a life table analysis of the role of the introduced parasitoids. The study demonstrates clearly that parasitoids were the key factors in the decline of casebearer densities to a new lower equilibrium level.

Assessment of biological control programmes sometimes reveals quite complex and unexpected interactions. The much-quoted success against the winter moth in Canada has for a long time been attributed to the direct action of introduced parasitoids (Hassell, 1980), as was the case with the larch casebearer (above). However, the presumed role of the introduced larval parasitoid, *Cyzenis albicans*, has been recently reassessed by Roland (1988), who argues that the additional mortality directly attributable to the parasitic fly was quite low. He suggests that control was actually achieved largely by an increase in pupal mortality caused by indigenous predators following establishment of the parasitoid. A possible explanation for this curious finding is the tendency of parasitized hosts to remain longer in the pupal stage in the soil, thereby permitting the build-up and continuity of indigenous pupal predators.

Only with the accumulation of detailed ecological studies which identify the specific impact of introduced natural enemies on pest population change can we begin to understand whether our methods for selecting natural enemies are useful and correct.

CONCLUSIONS

We hope that the examples presented in this chapter have left little doubt of the important contribution which basic research on the ecology of natural enemies can make to practical biological control. More importantly, perhaps, we hope that they have shown the constraints under which this contribution is made.

Perhaps the most frustrating aspect of the development of scientific biological control are the constraints on testing hypotheses. This problem is particularly acute for classical biological control, where it would be most desirable to test selection procedures by releasing and comparing selected and rejected agents in different but comparable parts of a pest's exotic range. On the other hand, this kind of grand ecological experiment does not take into account the sensibilities of an affected country, where a rapid solution to a serious agricultural problem is required, and where one would be reluctant to risk further losses in order that the investigation be a guinea pig for the refinement of scientific methods.

Nonetheless, a way must be found to test hypotheses regarding exploration, selection and release of biological control agents. This challenge, and of course the generation of the hypotheses themselves, will guarantee many years of fruitful collaboration between biological control practitioners and students of natural enemy ecology.

19: The Dynamics of Predator–Prey and Resource–Harvester Systems

ROBERT M. MAY & CHARLOTTE H. WATTS

INTRODUCTION

... In that Empire, the craft of Cartography attained such Perfection that the Map of a Single province covered the space of an entire City, and the Map of the Empire itself an entire Province. In the course of Time, these Extensive maps were found somehow wanting, and so the College of Cartographers evolved a Map of the Empire that was of the same Scale as the Empire and that coincided with it point for point. Less attentive to the Study of Cartography, succeeding Generations came to judge a map of such Magnitude cumbersome, and, not without Irreverence, they abandoned it to the Rigours of sun and Rain. In the western Deserts, tattered Fragments of the Map are still to be found, Sheltering an occasional Beast or beggar; in the whole Nation, no other relic is left of the Discipline of Geography
[Suarez Miranda, 1658]

The fate of Suarez Miranda's imaginary cartographers seems to us to sum up the case for simplified models — maps on the scale of 1 inch to 1 mile, as it were — more crisply and vividly than conventional scholarly justifications. This book in general, and our chapter in particular, aims to present a broad and deliberately oversimplified framework for thinking about the dynamical properties of prey–predator systems, be they plants and herbivores, animal prey and four- or six-legged predators, insects and parasitoids, hosts and helminth or arthropod parasites ('macroparasites'), hosts and viral, bacterial, protozoan or fungal infections ('microparasites'), or plant and animal resources that are harvested by humans. All such systems share certain general properties; against this background significant differences emerge as we focus on the details of any one association.

Thus the approach in this chapter is, as it were, on the scale of a road atlas, expanding to city-map scale for some illustrative examples. Exhaustive studies of specific situations — maps on the scale of 1 inch to 1 inch — of course have their place, but not in this overview chapter.

Our chapter begins by recapitulating the range of dynamical behaviour, from stable points to stable cycles to chaos, that predator–prey interactions can exhibit (see Chapter 3). We suggest some general rules

[431]

about the kinds of biological factors that push a system towards steadiness or towards self-sustained oscillation, emphasizing that exceptions to virtually all such rules can be found. We then sketch some applications of these ideas, giving special emphasis to systems where humans are the 'predators' as they harvest plant or animal populations, or where humans seek to intervene (by immunization or other means) in interactions between parasites and human hosts.

BASIC DYNAMICS OF
ONE PREY−ONE PREDATOR ASSOCIATIONS

Lotka−Volterra and other neutrally stable models

As explained in Chapter 3, predator−prey associations have an inherent propensity to oscillatory behaviour. If there are initially few predators, the prey population tends to flourish. This creates a situation where life is good for the predators, and their population will increase. The rising numbers of predators will then begin to have an impact on prey numbers, causing the prey population to decline. In the wake of falling prey density, predator populations will then drop. But as predator populations decline to low levels, the stage is set for a relative increase in the prey population, starting the cycle all over again.

This intuitive argument contains the gist of the more analytic discussion given in earlier chapters, backed by phase-plane diagrams and explicit mathematical models. Whether the innate tendency to oscillatory behaviour results in persistent, self-sustaining cycles or is damped toward constant population densities depends on the details of the specific predator−prey interaction in question.

The standard Lotka−Volterra model, which is the simplest imaginable mathematical metaphor for a predator−prey system, has dynamics that are balanced, with exquisite and improbable delicacy, exactly on the razor's edge between damped oscillation and self-sustaining cycles ('stable limit cycles'). Like the fictitious frictionless pendulum beloved of elementary physics courses (and found nowhere in the world outside them), the Lotka−Volterra model possesses 'neutral stability' in the sense that the system oscillates endlessly in cycles whose periods are indeed determined by the parameters of the model, but whose amplitudes are determined solely by the initial conditions. Such neutrally stable cycles are to be sharply distinguished from stable limit cycles, which are unique cycles (with both period and amplitude determined by the biological and physical parameters characterizing the association) to which the system will settle no matter what the initial conditions.

Specifically, the Lotka−Volterra model for interactions between prey and predator populations of magnitudes $H(t)$ and $P(t)$, respectively, at time t has the form:

$$\frac{dH}{dt} = H(a - \alpha P) \qquad (19.1)$$

$$\frac{dP}{dt} = P(-b + \beta H) \qquad (19.2)$$

where a represents the prey population growth rate in the absence of predators, and b the rate at which the predator population declines in the absence of its sustaining resource of prey. The parameters α and β measure the negative impact of predators on prey, and the positive impact of prey on predators, respectively. As indicated in earlier chapters, graphical or algebraic analysis shows that the equilibrium point at $H^* = b/\beta$, $P^* = a/\alpha$ is neutrally stable, and that the system oscillates in cycles whose period, T, is approximately the geometric mean of the intrinsic time scales for prey and for predator populations:

$$T = 2\pi(ab)^{-0.5} \qquad (19.3)$$

This approximation is accurate if the initial conditions produce cycles of relatively small amplitude.

The Lotka–Volterra metaphor derives from studies, in the 1920s and 1930s, of animal predator–prey systems, with population growth being a continuous process (hence differential equations). More recent studies of the interactions between disease organisms and their host populations have produced other neutrally stable models. One example is Anderson and May's (1978) so-called 'Basic Model' for a host population whose density is regulated by helminth or arthropod parasites ('macroparasites', *sensu* Anderson & May, 1979), which are distributed among the hosts in a random way (that is, according to a Poisson distribution). Another example is for a host population interacting with a viral, bacterial or other infection such that hosts can be partitioned into a handful of categories, such as susceptible, infected-and-infectious, recovered-and-immune ('microparasites', in contrast to the macroparasites where pathogenetic effects, transmission efficiencies and other things depend on a host's parasite burden, which varies from host to host). If these host–microparasite systems are modelled in discrete time (as difference equations rather than the more familiar differential equations), then again we find neutrally stable oscillations in, for example, the fraction of hosts infected (May, 1986b). The period of these neutrally stable cycles is the appropriately altered analogue of equation 19.3:

$$T = 2\pi(A\tau)^{0.5} \qquad (19.4)$$

where A is the average age at infection (the 'characteristic lifetime' of the uninfected hosts or 'prey'), and τ is the average duration of infection (the 'characteristic lifetime' of the infectious agent or 'predator').

All such neutrally stable models are in one sense grossly unrealistic, because in the real world there are no frictionless pendulums; there are no systems balanced on the razor's edge between damped oscillation and stable limit cycles, forever oscillating, with an amplitude that remembers how things started off in the distant past. However, in another sense such models are useful both in clarifying the oscillatory tendency of prey–predator systems (complete with an estimate of the inherent period of oscillation), and as a point of departure for adding realistic refinements.

Stable points, stable cycles, chaos

All manner of refinements may be grafted onto the basic Lotka–Volterra model of equations 19.1 and 19.2, and its related neutrally stable models for host–macroparasite and host–microparasite associations. Some such refinements will tend to tip the system toward a stable point, in the sense of constant prey and predator population values; any disturbance from such a state will tend to show damped oscillations back toward the stable point. Other refinements tend to precipitate the system into stable limit cycles, uniquely defined oscillatory states, to which the system will return following a perturbation. A partial cata-logue of such refinements has been given by May (1981), and in the next subsection we will discuss the kinds of biological effects that make for stable point, versus stable cycle, behaviour.

Before doing this, however, we emphasize that those systems that are tipped off the watershed of neutral stability into sustained cyclic behaviour do not always exhibit orderly, regular cycles. Simple and purely deterministic models for prey–predator interactions, with no random elements whatsoever, can also exhibit the kinds of irregular and apparently random fluctuations that have come to be called 'chaotic' (May, 1976, Gleick, 1987).

Such chaotic phenomena have only recently been recognized. Partly this is because chaos cannot arise in differential equation (continuous time) models for one prey–one predator interactions, with no time delays in the interaction effects and with constant coefficients. Such two-dimensional (one prey, one predator; H, P) dynamical systems cannot do anything very complicated, because the trajectories in the two-dimensional $H–P$ phase plane cannot ever cross. If they did cross, then we would have two alternative trajectories from a single point, which would violate the basic result that systems of first-order differ-ential equations with constant coefficients have unique solutions, that is have a unique trajectory from each point. There are, however, at least three ways in which simple one prey–one predator systems can escape this constraint and exhibit chaotic behaviour.

One way is to have time delays in one or more of the interaction terms. Once explicit time lags are present, so that expressions like $H(t - T)$ or $P(t - T)$ are present in the differential equations at time t, then we no longer are dealing with a set of first-order equations, and trajectories can cross. The mathematical models used by the Inter-national Whaling Commission in its analyses of harvesting baleen and sperm whale populations — a prey–predator system of an un-conscionably bloody kind — have the propensity to exhibit cascades of period-doubling bifurcations and chaotic dynamics (although not for parameter values that pertain to real whale populations).

Another way to get chaotic dynamics from one prey–one predator models with continuous time is to allow one or more of the interaction parameters to vary over time; trajectories can now cross, without violating uniqueness of solutions, because the same point in H–P space can correspond to different values of the interaction parameters (and hence different trajectories). This situation arises in host–microparasite models in which the transmission efficiency of the infectious agent exhibits annual variation (resulting from climate conditions, or the opening and closing of schools). Such annual variability can interact with the underlying propensity to prey–predator oscillation, characterized by the Lotka–Volterra period of equation 19.4, to produce chaotic fluctuations in the incidence of infection. We will return to this case when discussing some examples of these ideas (see p. 451).

A third way for chaos to arise in one prey–one predator models is when generations are discrete and non-overlapping, so that time is a discrete rather than a continuous variable. We then deal with difference equations, not differential equations. This situation is rather like having time delays not merely in an interaction term, but built into the basic fabric of the model.

An illuminating example of this third route to chaos in one prey–one predator systems is provided by a simple and a natural model for an insect population (with discrete, non-overlapping generations) regulated by a lethal pathogen that sweeps through each generation in epidemic fashion before the reproductive stage is reached. For such a situation, the number of insects in generation $t + 1$, N_{t+1}, is related to the number in generation t, N_t, by

$$N_{t+1} = \lambda N_t[1 - f(N_t)] \tag{19.5}$$

where λ is the 'finite rate of increase' of this population in the absence of the pathogen, and f represents the fraction of hosts that are killed by the pathogen (so that a fraction $1 - f$ survive to reproduce). The fraction, f, infected over the course of the epidemic is given by the Kermack and McKendrick (1927) equation,

$$1 - f = \exp(- fN_t/N_T) \tag{19.6}$$

This non-linear equation gives f in terms of the ratio between the host population and some threshold density, N_T, which depends on the virulence and transmissibility of the pathogen. If N_t is below threshold in generation t, $N_t < N_T$, the epidemic does not spread and all hosts survive to reproduce ($f = 0$). But if $N_t > N_T$, then equation 19.6 has a non-trivial solution, $f \neq 0$, and a larger and larger fraction of hosts are killed as N_t increases above N_T. Figure 19.1 shows this relation between N_{t+1} and N_t, for a representative value of the single parameter, λ, that characterizes the dynamics of this system (N_t is plotted in units of N_T, that is as $X_t = N_t/N_T$). Interestingly, this system has *only*

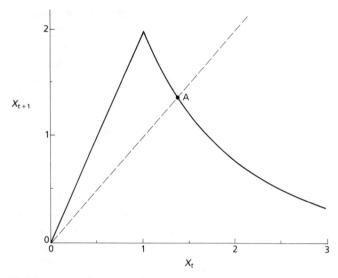

Fig. 19.1 The solid line depicts the functional relationship between the (dimensionless) population density in generation $t + 1$, X_{t+1}, in terms of that in generation t, X_t, for a population regulated by a pathogen. Specifically, the relationship is for equations 19.5 and 19.6 with $\lambda = 2$. The spikey hump in the curve derives from the discontinuous nature of the threshold phenomenon, whereby a deterministic epidemic spreads if, and only if, the host population exceeds a critical value. The dashed line corresponds to an unchanging population, $X_{t+1} = X_t$, so that the point A represents a possible equilibrium point. After May (1985).

chaotic solutions, for all values of λ (of course, $\lambda > 1$). Figures 19.2 and 19.3 illustrate this rather surprising fact. It will be seen that the system alternates between relatively high and relatively low host population values in a fairly regular manner for values of λ only slightly in excess of unity. For values of λ above 4 or so, however, this simple and natural model exhibits conspicuously chaotic behaviour, with a purely deterministic equation yielding trajectories that look random.

The existence of apparently random or chaotic trajectories that are produced by simple predator—prey models containing no random elements whatsoever raises serious questions about how ecologists interpret and analyse apparently randomly fluctuating time-series of field data. We shall return to this.

Factors predisposing toward stable points versus cycles

A few tentative generalizations can be made about which factors tend to tip predator—prey systems from the watershed of neutral stability toward stable point behaviour, or conversely toward stable cycles or chaos. What follows is largely an abridged version of May (1981).

In the Lotka—Volterra model, equations 19.1 and 19.2, the prey population undergoes unbounded exponential growth if no predators are present ($dH/dt = aH$ if $P = 0$). Clearly other density-dependent

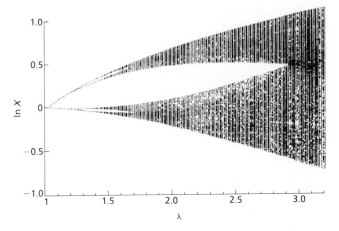

Fig. 19.2 This figure is constructed by plotting the population values (on a logarithmic scale, ln$X$) generated by iterating the difference equations 19.5 and 19.6 many times for each of a sequence of λ-values. The figure gives an impression of the probability distribution of population values generated by this purely deterministic difference equation; for a more full discussion, see May (1985).

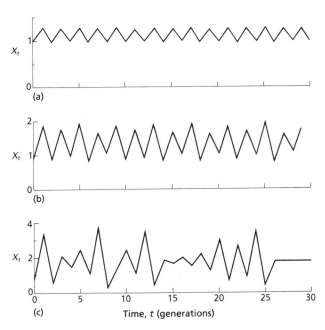

Fig. 19.3 This figure shows the variation in the population, X, with time, t, as described by equations 19.5 and 19.6, for three particular values of λ. These three population trajectories further illustrate the character of the chaotic dynamical behaviour noted in Fig. 19.2, and discussed more generally in the text: (a) for λ = 1.3, the trajectory, although strictly aperiodic, is close to a 2-point cycle; (b) for λ = 2, the trajectory is obviously aperiodic, but still alternates up and down in successive generations; (c) for λ = 4, the trajectory is quite ragged.

effects will ultimately govern the host population. Thus, in simple terms, one class of refinements is that dealing with the environmental 'carrying capacity', K say, for the prey in the absence of predators. Such density-dependent effects tend to stabilize predator–prey interactions. But if K is significantly larger than the average prey density in the presence of predators, then the stabilizing elements contributed to the dynamics of the prey–predator association by the prey's density-dependence will be relatively weak. This may underlie the 'paradox of enrichment', whereby increasing K makes for lower stability, and eventually for stable limit cycle behaviour. The 'paradox' was identified by Rosenzweig (1971) on the basis of numerical studies. It is possibly exemplified by natural populations such as the larch bud moth in Switzerland (whose population cycles in the optimum part of its range, between 1700 and 1900 m, but is relatively steady in marginal habitats), and by laboratory studies on *Paramecium* and its predator *Didinium* (Luckinbill, 1973). Other laboratory studies of predatory freshwater zooplankter *Daphnia* and their algal prey have, however, found no evidence of this effect (McCaughley & Murdoch, 1990).

The way in which predator attack rates respond to changing host density can tend to stabilize, or to destabilize, depending on the details. Switching behaviour, whereby predators tend to concentrate on those prey that are locally most abundant, can produce upward-curving predator attack rates, making predation relatively more intense as host densities rise. Clearly such effects tend to stabilize the system. Conversely, saturation effects that derive from finite handling times or satiation result in predation becoming relatively less important as host densities rise; such effects are likely to favour stable limit cycle behaviour (compared with the strictly linear relation between predator attack rates and host densities, αN per predator, in the Lotka–Volterra model).

Most kinds of time lags in the effects of interactions between prey and predators tend to push the system off the Lotka–Volterra knife-edge into stable limit cycle behaviour. Moreover, if the intrinsic growth rate of the prey population (r) significantly exceeds that of its predators (s), then, in effect, the predators find it harder to keep pace with the prey dynamics, tending to produce positive feedback and sustained cycles. Tanner's (1975) necessarily crude analysis of eight vertebrate prey–predator associations suggested that the seven stable ones had either a relatively small value of K or s significantly greater than r, while the one stably cyclic example had the combination of relatively large K and relatively high $r:s$ ratio that would predispose to instability.

For viral, bacterial, and protozoan (broadly, microparasite) or helminth and arthropod (broadly, macroparasite) infections, the characteristic time scales for 'predators' tend to be faster than for hosts. Partly for this reason, this special class of prey–predator models tends to show stable point rather than stable limit cycle behaviour, except when extrinsic time dependence in parameters (such as seasonality in transmission probabilities) or other complications intervene. Cycles also arise for systems involving some forest insects and their viral or protozoan parasites, where the hosts can have high population growth rates whereas long-lived transmission stages effectively mean the 'predator's' characteristic lifetime is long, resulting in high '$r:s$' ratios.

Generalizations about what promotes stable point behaviour and what promotes stable cycles or chaos must always be treated with caution. Deceptively simple models can have complex dynamics that provide counter-examples to many tempting generalizations. One early example is due to May (1973), who analysed a prey–predator model in which the main refinement was a density-dependent limitation on the growth rate of the prey, of logistic form (stabilizing) but operating with a time delay (destabilizing). For smooth changes in parameters such as the intrinsic growth rate of the prey, r, this system can unfold a sequence of changes in dynamical behaviour, from stable point, to stable cycle, to stable point, to stable cycle, and so on. Other examples of simple predator–prey systems in which monotonic changes in a parameter produce confusing and kaleidoscopic changes in the dynamics have subsequently been found (see, for example, Murdoch *et al.*, 1987).

Other complications

For one prey–one predator systems where population growth is continuous and where there are no time lags in the interaction terms (ordinary first-order differential equations), Kolmogorov (1936) showed that, provided the interaction terms satisfy a particular set of mathematical conditions, there will be either a stable point or a stable cycle. This set of mathematical conditions, moreover, is fulfilled by most of the prey–predator models to be found in the ecological literature (May, 1973). In essentials, Kolmogorov's result is again a consequence of the fact that trajectories cannot cross in two-dimensional systems of ordinary differential equations.

However, there do exist some ecologically interesting prey–predator models where the interaction terms do not satisfy the Kolmogorov conditions, and where two (and possibly more) alternative states can arise for some ranges of parameter values. In particular, Wollkind *et al.* (1988) have studied a simple and sensible model for an insect prey–predator association where, for a biologically interesting although narrow range of parameter values, there is both a stable point and a stable limit cycle; the system settles to one or the other, depending on the initial conditions. More generally, it is possible in principle to design pairs of coupled first-order differential equations where two or more stable limit cycles are nested inside each other in the H–P phase plane, in between unstable limit cycles which divide the two-dimensional phase plane into annular bands of points that are attracted to the stable limit cycle within each band. The extent to which these mathematical curiosities are generally relevant to real predator–prey systems is debatable.

Once we start dealing with more elaborate one prey–one predator models, with various refinements describing density-dependent effects on prey density and with non-linearities in the predator's functional and numerical responses, alternative states can arise easily. For example, Noy-Meir (1975) has described how plant–herbivore systems can exhibit two alternative stable states, one essentially where plant density is governed by nutrients or other resource considerations, and a second governed by herbivory; he also suggests possible examples of this phenomenon in grazing systems. Several authors have shown how insect predator–prey systems may possess two alternative stable states, one where the prey are limited essentially by the predators, and a second where the prey escape the predators (owing to saturation effects) to be controlled by other mechanisms (disease or food supplies); again, specific instances have been suggested (Southwood & Comins, 1976; Hochberg *et al.*, 1990).

If more than one predator species is involved, such multiple stable states arise even when no regulatory effects other than predation exist. Focusing on arthropod host–parasitoid systems, Hassell and May (1986) have demonstrated such effects both with two specialist parasitoids and with systems of one specialist and one generalist. The alternative states may both be stable points, or both cycles (periodic or chaotic), or a mixture.

A different kind of complication arises in models of one prey–one predator systems that take explicit account of age structure. Of special interest is the finding that models in which population growth is continuous (age-structured differential equations), or where there are many distinct but overlapping age classes (coupled difference equations), often have instabilities that result in one particular age class dominating the dynamics. That is, these age-structured models behave as if they had discrete, non-overlapping generations. This observation, made independently in different contexts by several authors (Oster & Takahashi, 1974; Gurney & Nisbet, 1985; Godfray & Hassell, 1989), may explain why many arthropod predator–prey or host–parasitoid associations can be successfully modelled using generalized Nicholson–Bailey models with discrete non-overlapping generations, even though the animals in question (often tropical insects) can breed throughout the year.

Spatial heterogeneity

Another class of complications, deserving a subsection of their own, are those produced by spatial patchiness. As emphasized in many of the earlier chapters, most prey are scattered in subpopulations in many different patches, among which predators seek for them. Conventional animal predator–prey and insect host–parasitoid studies have only relatively recently begun to deal with the dynamical consequences of such heterogeneity; the Lotka–Volterra model and its direct heirs assume a homogeneous world.

Other kinds of 'prey−predator' studies have implicitly used meta-population models from the start, describing the dynamics of populations of predators that seek their prey in a patchy world. For instance, the very first model for host−macroparasite systems recognized that the way in which the helminth or arthropod macroparasites were distributed among their host-islands of habitat was central to the dynamics (Crofton, 1971a,b). If the helminths were distributed independently randomly (a Poisson distribution), Crofton's discrete-generation model gave diverging oscillations. But if the macroparasite distribution was a clumped or 'overdispersed' one, with most hosts having relatively few parasites and a few having many, then Crofton found that steady levels of both parasite populations and parasite-regulated host populations could ensue. A similar clumped distribution for the macroparasites was found to be important to producing steady, rather than oscillatory, host−parasite associations in Anderson and May's (1978) more realistic models with continuously overlapping generations of hosts and macroparasites. These authors also have documented the ubiquity of significant such clumping in the distribution of many different helminth and arthropod parasites of human and other animal populations (Anderson & May, 1978, 1985a), although the underlying reasons are not yet fully understood.

This earlier work on host−macroparasite systems in some ways anticipates Pacala *et al.*'s (1990) study of the relation between variability in parasitoid attack rates among patches and stability in arthropod host-parasitoid systems. In Chapter 11, Hassell and Godfray discuss the approximate rule '$CV^2 > 1$': host populations that are regulated mainly by a parasitoid will tend to be roughly steady if the square of the coefficient of variation of parasitoid attack rates among patches, CV^2, exceeds unity, and oscillatory or chaotic otherwise. In retrospect, Anderson and May's criterion that macroparasite distributions have to be 'sufficiently clumped' to produce stable associations can similarly be recast as an approximate '$CV^2 > 1$' rule, with CV here interpreted as the coefficient of variation of the distribution of macroparasites among hosts.

Studies of the effects of different patterns of movement and distribution of predators (herbivores, vertebrate or invertebrate predators, parasitoids, macroparasites or microparasites) among patches are still, in our opinion, at an early stage. We see this as a major growth area for theoretical and empirical studies.

Predator−prey systems and 'single-species systems'

We conclude this breathless gallop through a maze of ideas with a general remark on the relation between predator−prey models and models for a single species. The remark will be illustrated by an example in the applications section, below.

In some formal sense, we could take any predator−prey model and reduce it to a single-species equation for the prey population only.

Often, indeed, our data will be such that we only have information about the prey population in what we believe to be intrinsically a two-species, predator–prey, system. More generally, this is a particular example of the problem of 'embedding' a time series for one variable in what is really an n-dimensional dynamical system (Takens, 1981; for reviews see Sugihara *et al.*, 1990, and Godfray & Blythe, 1990).

In order to replace the pair of coupled equations for prey and predator populations by a single equation for the prey population, we need to integrate the predator equation to get an expression for $P(t)$. In the particular case where predators do not interfere with each other, so that predator population growth rates depend only on host density, we can do this explicitly:

$$\text{if} \quad dP/dt = PF(H)$$

$$\text{then} \quad P(t) = P(0) \exp\left\{ \int_0^t F[H(t')]dt' \right\}$$

In this event, we see explicitly that the predator population at time t, $P(t)$, involves an integral over past values of the prey population. More generally, the value of $P(t)$ will always depend on past values of H. Substituting the explicit expression for $P(t)$ above, into the equation for the prey dynamics, gives an expression of the form

$$\frac{dH}{dt} = \text{function of} \left\{ H(t) \quad \text{and} \quad \int_0^t F[H(t')] \, dt' \right\} \quad (19.7)$$

That is, we have a single-species equation for $H(t)$, with a time-dependent regulatory mechanism. Even when we cannot solve explicitly for $P(t)$ in terms of past values of H, the same general result will be formally valid: the rate of change in $H(t)$, dH/dt, will involve not just $H(t)$ but past values of H, so that the regulatory effects will operate with time-delays.

In this sense, a predator–prey system is equivalent to a single-species system with time-lags.

APPLICATIONS TO HARVESTING AND DISEASE CONTROL

The concept of maximum sustainable yield (MSY)

Suppose we are presented with a resource, in the form of a plant or animal population, that we wish to harvest in a sustainable way. Suppose also that before we introduce ourselves as predators upon this resource, the population on average (with environmental fluctuations smoothed out) obeys a non-linear differential equation of the form

$$\frac{dN}{dt} = rNf(N). \quad (19.8)$$

Here $f(0) = 1$, so that r represents the intrinsic rate of population growth, in the absence of the density-dependent effects that are encapsulated in $f(N)$. If $f(N)$ is a monotone decreasing function of N, the population will settle to an equilibrium value at $N^* = K$, where $f(K) = 0$.

Let us harvest this population with a 'constant effort', h, such that the yield per unit time, Y, is

$$Y = h N \qquad (19.9)$$

Other harvesting strategies are, of course, possible. But the constant effort strategy is representative, and some specificity is required lest this chapter balloon to book-length (for a more detailed discussion, see Clark, 1990).

Our harvest represents an additional mortality on the resource population, whose dynamics now obey

$$\frac{dN}{dt} = rNf(N) - Y \qquad (19.10)$$

Such a harvest is sustainable (so long as $h < r$), and the population will settle to a new equilibrium value, $N^*(Y)$, determined by $rN^*f(N^*) = Y$; this new equilibrium population will, of course, be less than K (see Fig. 3.5, p. 57).

The question now arises, at what level should the effort, h, be set in order to maximize the sustainable yield? If h is set too low, N^* will be close to K but, from equation 19.9, Y will be small because h itself is small. If h is set too large (close to r), then N^* will be close to zero ($f(N^*) = h/r$, and as $h \to r$, $N^* \to 0$), and now Y will be small because N^* is small. The maximum sustainable yield, Y_{MSY}, lies between these extremes, and can be found either numerically or analytically by calculating the value of N that maximizes $Y = rNf(N)$. That is, as common sense suggests, the MSY is attained by holding the population at the point where its overall growth rate is maximal, and taking this maximal excess of births over deaths as the harvest.

As an explicit example, suppose the pristine population exhibits a logistic growth pattern. That is, suppose $f(N) = 1 - N/K$ (whereupon equation 19.8 becomes the familiar $dN/dt = rN(1 - N/K)$). Then the equilibrium population under harvesting effort h is

$$N^*(h) = K (1 - h/r) \qquad (19.11)$$

The corresponding yield is

$$Y = Kh (1 - h/r) \qquad (19.12)$$

Clearly the MSY is attained for $h = r/2$, giving $N_{MSY} = K/2$ and $Y_{MSY} = rK/4$.

All this is fairly routine stuff, although it is not always thought of as an aspect of predator–prey studies.

So far, we have dealt only with the statics of harvesting. Dynamical studies go beyond this, to ask how the level of harvesting affects both the dynamics of the resource population, and the stability of the yield itself. Emphasis on such questions led in the 1970s to a recognition that increasing harvesting levels often produced a concomitant length-ening in the characteristic time for the harvested population to re-cover from environmentally imposed fluctuations (Doubleday, 1976; Beddington & May, 1977; Sissenwine, 1977). Thus, as harvesting levels and yields increase from zero toward the MSY point, if often happens that the relative magnitude of fluctuations in the harvested population and in yields also increases. These and other considerations have since led to less emphasis on MSY as such, and more emphasis on designing harvesting policies that are responsive to fluctuations and uncertainty in resource levels, and that seek to avoid excessive variability in yields. Such policies will typically require a yield somewhat below the MSY level, in order to have significantly less variability in yield and less risk of driving the harvested population to dangerously low levels. Needless to say, while the policies may be comparatively clear, their implementation is often difficult. This is particularly the case when the harvesters are not really interested in sustainable yields, but rather in liquidating the capital stock and then moving on. This situation tends to arise when the resource generates maximum sustainable yield rates that are less than inflation-corrected interest rates (as is the case, for instance, for whales or redwoods; for a more formal discussion, see Clark, 1990).

Against this background, we can now pursue two points, which help illustrate some of the themes of this chapter.

Do natural predators harvest at MSY?

Much has been written about whether natural selection is likely to cause predators to be 'prudent', with prudent usually interpreted as exploiting their prey at around MSY levels. Some conservation and wildlife management literature takes it for granted that predators are likely to have evolved to be prudent in this sense. However, natural selection usually works on differential survival among individuals, and so a 'cheat' who increases its own reproductive success by doing things that act against the interests of the group as a whole (for example, consuming resources at an unsustainable rate) will prosper; natural selection acts on individuals, not on groups as such. So, more careful authors have noted that arguments about 'prudent predation' appeal, explicitly or implicitly, to group selection, and have sought to under-stand how predator behaviour is in fact likely to evolve (see, for example, Slobodkin, 1974; Schaffer & Rosenzweig, 1978; Roughgarden, 1979).

For some host–parasitoid associations, however, a purely empirical statement can be made about whether or not the parasitoids are ex-ploiting their host populations at MSY. Following the general guidelines laid down in Chapter 11 on host–parasitoid interactions, we consider a host population, N, whose dynamics in the absence of parasitoids obeys the discrete analogue of equation 19.8:

$$N_{t+1} = \lambda\, N_t g(N_t) \qquad (19.13)$$

Here λ is the intrinsic finite rate of increase, and $g(N)$ represents density-dependent effects that may regulate the population in the absence of parasitoids. This population will have an equilibrium density, N^*, determined from the relation

$$\lambda\, g(N^*) = 1 \qquad (19.14)$$

We leave open the question of whether the population is steady at this level, or fluctuates cyclically or chaotically around it.

Suppose now that a population of parasitoids exploits this population, at MSY. The techniques already discussed (see pp. 442–443) can be used to calculate the MSY level of the exploited host population, N_{MSY}. This level depends only on λ and on the functional form of $g(N)$; the details of this calculation are given by Hassell and May (1985). Figure 19.4 shows the degree to which the host population is depressed below its parasitoid-free level, $q_{MSY} = N_{MSY}/N^*$, as a function of λ, for two opposite extreme assumptions about the functional form $g(N)$: curve (A) is for $g(N) = \exp(-aN)$, representative of 'scramble' competition; curve (B) is for $g(N) = 1/(1 + aN)$, representative of 'contest' competition. In either case, we see that MSY levels of exploitation by the parasitoids are unlikely to depress the host population much below roughly half its unexploited equilibrium value (that is, $q_{MSY} \simeq 1/2$), unless λ takes extremely high values.

In a study altogether independent of the above considerations, Beddington *et al.* (1978) have estimated the parasitoid reduction of host density below the parasitoid-free value (the q-value) for 10 host–parasitoid associations in the field and laboratory. These observed q-values are indicated by the horizontal lines to the left in Fig. 19.4. The six solid lines come from field studies, and the four dashed lines from laboratory studies; in all cases the λ-values are uncertain (which is why only the q-values are shown). Figure 19.4 shows that the q-values for the four laboratory situations could be consistent with the parasitoids exploiting their hosts at an MSY level, but the six field q-values are grossly inconsistent with MSY exploitation.

This population-level approach, based on ideas about maximum sustainable harvesting of a resource, is quite different from studies that seek to determine whether predators are prudent on the basis of the evolution of individual behaviour. Although the answers suggested by Fig. 19.4 are very crude, they appear to say that, in natural settings, at least six species of parasitoids exploit their host populations to densities well below MSY. It could be that a more refined analysis, that explicitly allowed for the effects of spatial patchiness upon the host–parasitoid interactions, alters this conclusion; for the six studies in question, the available data do not permit such refinement. It seems to us more likely that the coevolutionary trajectories of host–parasitoid associations depend on many aspects of the life histories of both

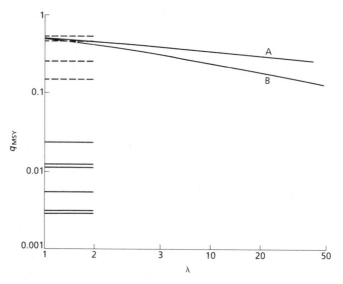

Fig. 19.4 The degree to which host density is depressed below its pristine value by a parasitoid that 'harvests' the host population for maximum sustainable yield, q_{MSY}, is shown as a function of the intrinsic growth rate of the host population, λ. Curve A is for the case where the parasitoid-free host population is regulated by 'scramble' competition and curve B is for the opposite extreme case of 'contest' competition. The solid horizontal lines to the left show estimated q-values for six host–parasitoid associations in the field, and the dashed horizontal lines are similar estimates for four laboratory populations (from Beddington *et al.*, 1978). As discussed in the text, it does not appear likely that the six field populations of parasitoids exploit their host populations at MSY.

species, in ways that are unlikely to lead to MSY as a general outcome. There is room for much more work here.

How does harvesting affect predator–prey cycles?

A basic theme throughout this chapter has been that predation, whether from natural enemies or from harvesting by humans, tends to make the prey population more prone to oscillatory behaviour or erratic fluctuations than it might otherwise be, essentially because such regulatory effects as are exerted by the predators tend to act with a time lag. But if the prey population already exhibits cyclic changes in density in the absence of predation or harvesting, then harvesting may act to reduce the severity of the oscillations or even to damp them out altogether. We can get an intuitive feeling for this possibility by looking back to p. 438, where one of the factors predisposing a predator–prey system to stable limit cycle behaviour was high intrinsic growth rate of the prey population relative to that of the predators (relatively high $r:s$). Additional harvesting of the prey population has the effect of imposing further mortality, and thus reducing the prey population's intrinsic growth rate and thereby reducing the ratio $r:s$.

Suppose we are seeking to harvest a population, such as red grouse in Britain, which appears to exhibit pronounced and fairly regular oscillations. What is the best sustainable policy? If taking the MSY also happens to damp out the oscillations, resulting in grouse populations that are roughly steady from year to year, then there is little problem. But suppose that harvesting at MSY levels leaves the population still cycling (albeit with reduced amplitude), whereas harvesting at higher, but nonetheless sustainable, rates results in a steady prey population and a steady yield. Should we opt for the yield that fluctuates but on average is higher, or for a somewhat lower but steady yield? Such questions ultimately involve social and economic choices. However, these choices are not clear unless the underlying biology is understood.

To illustrate these ideas, we go back to the single-species model of equation 19.8, and let the density-dependent function $f(N)$ have the explicit form $f = 1 - N(t - T)/K$:

$$\frac{dN}{dt} = r N [1 - N(t - T)/K] \tag{19.15}$$

This is the so-called time-delayed logistic equation. Its dynamic properties are well known: if $rT < \pi/2$, there is stable equilibrium point at $N^* = K$; if $rT > \pi/2$, the population exhibits stable limit cycles, whose period is roughly $4T$ and whose amplitude increases steeply as the product rT increases above the threshold value of $\pi/2$ (May, 1973).

Now add a constant-effort harvesting term to equation 19.15, as discussed above and defined in equations 19.9 and 19.10:

$$\frac{dN}{dt} = r N [1 - N(t - T)/K] - hN \tag{19.16}$$

As before, an analysis of the static, equilibrium properties of this system is carried out by putting $dN/dt = 0$, to arrive at equation 19.12 for the relation between yield, Y, and harvesting effort h. This $Y–h$ relation is illustrated in Fig. 19.5. As before the MSY, $Y_{MSY} = rK/4$, is attained for $h = r/2$.

However, if $rT > \pi/2$ so that the pristine system oscillates, then the harvested system can exhibit interesting dynamical complications. Specifically, the harvested system will continue to exhibit stable limit cycle behaviour if, and only if, $(r - h)T > \pi/2$ [this result can be obtained by the simple trick of writing $r - h = r'$ and $K(1 - h/r) = K'$ in equation 19.16, which transforms it identically into equation 19.14 but with r replaced by $(r - h)$]. There is thus a critical level of harvesting effort, h_c, given by

$$h_c/r = 1 - \pi/(2rT) \tag{19.17}$$

above which harvesting will produce a stable point, and constant (rather than cycling) yields. Because the MSY harvesting level is $h_{MSY} = r/2$, we see that the system will continue to cycle under MSY

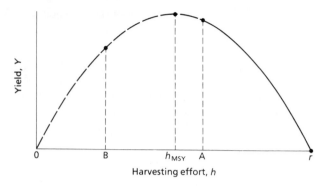

Fig. 19.5 Showing the equilibrium yield, Y, as a function of the (fixed) harvesting effort, h, for a population whose dynamics obeys the time-delayed logistic, equation 19.16. Where the relation is shown as a dashed line, the equilibrium state is not steady, but instead exhibits stable limit cycles. The solid part of the $Y–h$ curve corresponds to stable equilibrium yields. Specifically, this curve is drawn for $rT = 3.6$ ($>\pi$), so that the population (and the yield) exhibits cycles under MSY harvesting levels; yields become steady at harvesting levels in excess of that labelled A ($h > 0.56r$). Alternatively, for $rT = 2\pi/3$ ($<\pi$), yields are stable for harvesting effort greater than that indicated by the point B; in this case, MSY harvesting gives stable yields.

harvesting levels if $rT > \pi$. Figure 19.5 illustrates these dynamical aspects of harvesting by showing the $Y–h$ curve as a dashed line when the system is cycling, and as a solid line when it has a stable equilibrium point. The figure is drawn for the specific parameter choice $rT = 3.6$: the system exhibits cycles at MSY harvesting levels, and does not give a steady yield until h_c exceeds the level denoted by the point A. The point B in Fig. 19.5 shows h_c for the case $rT = 2\pi/3$; here the system settles to stable point behaviour at harvesting below the MSY level.

Figure 19.6 further illustrates these points, showing the yields, as functions of time, obtained under two different harvesting levels (specified in the legend). Again this example has $rT > \pi$, so that the system cycles under MSY harvesting levels. Figure 19.6 makes plain the choices facing a manager. The oscillating yields of curve A correspond to MSY harvesting levels, and give, on average, the maximum sustainable yield. But the higher harvesting effort of curve B gives unfluctuating steady yields that are not a lot less than the larger averages generated by curve A. The choice between these two strategies will be dictated by economic considerations, made more complicated by the fact that prices may well be lower in years when the cycling yields are at their peaks.

We now pursue this example a bit further, partly to show it is robust and partly to illustrate the point made on pp. 441–442 about the equivalence between predator–prey models and single-species models with time-lags.

Suppose our harvested population is the prey, $H(t)$, in some pristine

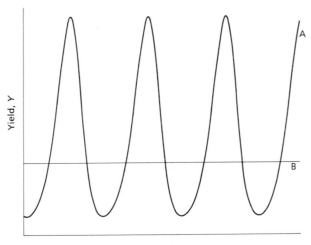

Fig. 19.6 Further illustrating the points seen in Fig. 19.5, this figure shows the equilibrium yield as a function of time, for the system described by equation (19.16) with $rT = 3.6$. The cycling curve A is for MSY harvesting ($h = 0.5r$), whereas curve B is for a harvesting effort ($h = 0.75r$) that gives a constant, although on average somewhat lower, yield.

predator−prey association. Specifically, suppose this predator−prey system can be described by the pair of equations

$$\frac{dH}{dt} = rH - \alpha\, PH/(H + A) - hH \qquad (19.18)$$

$$\frac{dP}{dt} = \rho P\, (1 - P/H). \qquad (19.19)$$

Here H and P denote the prey and predator populations, respectively, and r and ρ are their intrinsic rates of population growth. The term $\alpha PH/(H + A)$ describes the effects of predation upon the prey dynamics; the predators have a 'Type II' functional response, with saturation at prey densities significantly above A. The predator dynamics obey a logistic, with a 'carrying capacity' determined by the prey density. As before, harvesting is at constant effort, h, and the resulting yield is $Y = hH$.

In the absence of harvesting, $h = 0$, this prey−predator system has an equilibrium point at $H^* = P^* = rA/(\alpha - r)$. This equilibrium point is stable if $r < (\alpha\rho)^{0.5}$. If not, the system exhibits cycles with ever-increasing amplitude in this simplest model; for inclusion of intrinsic density-dependence for the prey in equation 19.18 and other discussion, see May (1973). Thus, broadly speaking, this prey−predator system has dynamics that are similar to the earlier single-species system of equation 19.14, with $(\alpha\rho)^{-0.5}$ playing the role of the time-lag, T.

Now suppose we harvest this system. The equilibrium yield is found by putting $dH/dt = dP/dt = 0$ in the usual way, and is given in terms of h by

$$Y = Ah(r - h)/(\alpha - r - h) \tag{19.20}$$

This yield—effort, Y–h, relation is illustrated in Fig. 19.7. The curve is similar to the earlier one in Fig. 19.5, except the peak is shifted to the left [the MSY harvesting level can be seen from equation 19.20 to be attained for $h_{MSY} = [\alpha(\alpha - r)]^{0.5} - (\alpha - r)$, which is always less than $r/2$]. As before, the dynamics of the harvested system are the same as those of the pristine system, with the change that $r-h$ replaces r in the criterion for stable point behaviour (this can be shown by rescaling equations 19.18 and 19.19, as spelled out explicitly above for equation 19.16 in relation to equation 19.15). That is, the harvested system will continue to exhibit diverging oscillations until harvesting levels exceed a critical value, h_c, given by

$$h_c/r = 1 - (\alpha\rho)^{0.5}/r \tag{19.21}$$

This is illustrated in Fig. 19.7, where (as in Fig. 19.5) the dashed part of the Y–h curve corresponds to stable limit cycle behaviour, and the solid part to a stable equilibrium point and steady yields.

The similarities between Figs 19.5 and 19.7 bear out our earlier general remarks about the broad equivalence between prey—predator models and single-species models in which regulatory effects operate with time delays. Whether the oscillatory dynamics of the pristine prey population is seen as deriving from time lags in regulatory mechanisms or from a prey—predator cycle, it can pose complicated questions

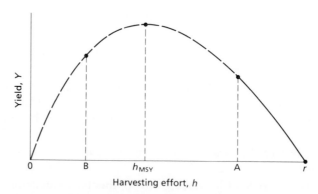

Fig. 19.7 This figure shows the equilibrium yield, Y, as a function of harvesting effort, h, as in Fig. 19.5 except that now the harvested population is the prey in the prey—predator system described by equations 19.18 and 19.19. Specifically, the figure is for these equations with $\alpha = 2r$, so that $h_{MSY} = (2^{0.5} - 1)r$. If $\rho = r/32$, then (as indicated by the dashed line) yields exhibit stable limit cycle behaviour until harvesting efforts exceed the level denoted by the point A ($h = 0.75r$). Alternatively, if $\rho = 0.32r$, the yield is stable for harvesting efforts in excess of that denoted by B, so that MSY yields are steady. For a fuller discussion of this figure, and of the parallels with Figs 19.5 and 19.6, see the text.

for harvesters. As Figs 19.5–19.7 clearly show, there can be choices between sustainable yields that oscillate about relatively high average values, and stable yields at somewhat lower, but steady, levels. Such choices have received surprisingly little attention in the context of harvesting cycling populations, such as red grouse (Krebs & May, 1990). As noted earlier, these choices will ultimately be constrained by sociopolitical factors, but a lucid understanding of the underlying dynamics is a prerequisite to informed choice.

Host–parasite dynamics and chaos

As discussed in Chapters 14 and 15, the interactions between human and other animal hosts and their viral, bacterial, protozoan, fungal or helminth infections represent one particular class of predator–prey association. Moreover, for infectious childhood diseases, such as measles, rubella, chickenpox and others, there are long runs of data of a quality and duration rarely found elsewhere in the ecological sciences. Building on the work presented in Chapters 14 and 15 we focus on two topics that illustrate the basic themes of the present chapter.

As mentioned above (pp. 432–433) host–microparasite models have a propensity to oscillate, with a period given in terms of epidemiological variables by equation 19.4. If time is a continuous variable (as seems sensible in most contexts), then the basic model possesses a stable point, with perturbations exhibiting damped oscillations back to equilibrium (Soper, 1929). These oscillations are, however, very weakly damped (provided infectiousness is short compared with other relevant time scales, and provided that recovery is into a class with lifelong immunity and no subsequent episodes of infectiousness; Anderson & May, 1991).

There are, moreover, time series of case notifications of measles, rubella (German measles), pertussis (whooping cough), mumps, and other childhood infections in developed countries, before the advent of vaccination, that exhibit rough periodicities at around the periods given by the approximate formula of equation 19.4, namely about 2 years for measles, 3 years for pertussis and mumps, and 5 years for rubella (see Anderson *et al.*, 1984). Figure 19.8 gives one such time series, for weekly case notifications of measles in England and Wales.

At one level, we could say that such data provide the best examples we have of prey–predator cycles whose periods accord with rough Lotka–Volterra-like theoretical estimates. At a more technical level, we have the difficulty that these 'interepidemic' periodicities are persistent phenomena, whereas the basic Soper model (and its age-structured refinements; Anderson & May, 1985b) exhibit damped — albeit weakly damped — oscillations to a stable point. This has prompted a quest for mechanisms that can 'pump' the intrinsic propensity to predator–prey oscillations, producing the sustained cycles that are observed.

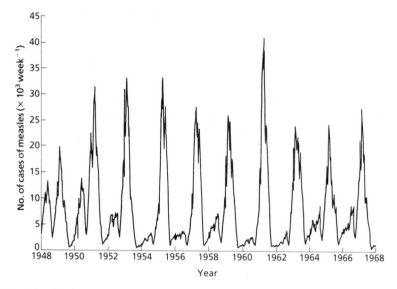

Fig. 19.8 This figure shows weekly case notifications of measles in England and Wales for the period 1948 to 1968, before the introduction of mass vaccination.

One suggestion, due to Bartlett (1957), is that stochastic fluctuations in the input of new susceptibles by births can produce sustained oscillations. In this view, we expect noisy cycles, with the periods given roughly by equation 19.4 and the unevenness (as seen, for example, in Fig. 19.8) deriving from the stochastic effects that pump the cycle. Another suggestion, due to London and Yorke (1973) and later elaborated by Schwartz (1985) and others, is that annual periodicities in trans- mission efficiency — deriving from school calendars and/or the effects of temperature and humidity on the survival of viral droplets — can be the pumping mechanism. Such an interplay between annual and 'Lotka— Volterra' periods, within a simple and fully deterministic model, can produce chaotic dynamics; in this view, both the interepidemic period and the unevenness in the time series, such as that shown in Fig. 19.8, derive from the deterministic (but chaotic) dynamics.

These ideas have in recent years given us a chance to add some flesh to the bare bones of the ideas about chaos in prey—predator systems that were sketched earlier (see pp. 434—437).

Given a noisy-looking time series, such as that in Fig. 19.8, can we distinguish between 'real noise' (sampling error or other sources of externally caused, unpredictable fluctuations) and chaotic trajectories that are generated by simple and deterministic rules? Various techniques for answering this question are under active development, and Schaffer (1987) has pioneered their application to epidemiological and other ecological time series. One difficulty, however, is that most such biological time series, long though they may be by ecological standards, are too short for reliable analysis by the methods the physicists are developing. Farmer and Sidorowich (1989) and Sugihara and May (1990) have, however, begun to develop methods that are applicable to relatively short time series. These methods are based on forecasting: one assembles a library of past patterns (from the first half of the series, say), and uses this as a basis for projecting forward from each point in the second half of the series; one then computes the conventional statistical correlation, ρ, between predicted and observed outcomes as a function of the number of time steps into the future, T_p, for which the projection is made. For 'real noise', past spins of the roulette wheel tell us as much or as little about the next throw as the next hundredth throw, and ρ shows no significant dependence on T_p. For a chaotic time series, the extreme sensitivity to initial conditions that is characteristic of chaos means that long-term forecasting will be impossible, but short-term forecasts based on past patterns are possible; a $\rho-T_p$ curve that declines from relatively high to low values as T_p increases may be interpreted as a characteristic signature of a chaotic time series. For a fuller discussion of these ideas, see Sugihara and May (1990).

Figure 19.9 shows an analysis of this kind, for data from case notifications of measles (1928–1963) and for chickenpox (1928–1972) in New York City. The correlation, ρ, between predicted and observed values for measles declines as the prediction interval, T_p, lengthens, suggesting the time series, with its erratic and rough 2-year cycle, is indeed the outcome of some deterministic mechanism (thus supporting

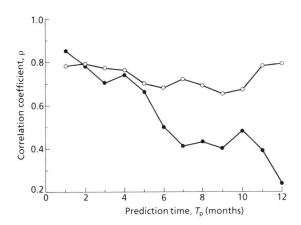

Fig. 19.9 This figure shows the statistical correlation, ρ, between predicted and observed numbers of cases of measles and chickenpox each month in New York City, as a function of the prediction interval, T_p (measured in months). The predictions are based on the forecasting technique sketched in the text, and described more fully in Sugihara and May (1990). The data for measles ($\bullet$) (from 1928 to 1963, before vaccination was introduced) show the decline in ρ with increasing T_p characteristic of chaotic dynamics. In contrast, the chickenpox data ($\circ$) (1928–1972) show a flat $\rho-T_p$ relation, characteristic of some regular pattern (here, annual periodicity) with additive noise.

the idea that seasonality, rather than stochastic fluctuations in births, pumps the inherent cycle). Conversely, the $\rho-T_p$ curve for chickenpox is flat, suggesting an annual cycle (which explains roughly 70% of the variance, and which is readily found by the forecasting technique), with random noise superimposed.

The corresponding story for case notifications of measles in England and Wales prior to vaccination is even more interesting and informative. If we apply the above-described technique to these data, the $\rho-T_p$ curve looks more like that for chickenpox in Fig. 19.9 than that for measles, suggesting a Lotka−Volterra cycle pumped by stochastic effects. Why should measles in England and Wales differ from measles in New York City? Second thoughts suggest that one should tease apart the aggregated data for England and Wales, and look separately at the larger cities, e.g. London, Birmingham, Manchester, Liverpool, Sheffield, etc., if one wants a valid comparison with New York City. When this is done, the British cities indeed look just like New York City, with the $\rho-T_p$ curves exhibiting the decline in ρ with increasing T_p that is characteristic of chaotic dynamics (Sugihara *et al.*, 1990).

It thus appears that time series for measles cases in individual cities, both in the UK and the USA, exhibit irregular but roughly periodic fluctuations that are chaotic dynamics resulting from simple and deterministic mechanisms (most likely the interplay between annual periodicities in transmission and the Lotka−Volterra interepidemic period). When these data are aggregated over large regions, however, the chaotic structure is averaged out, and we end up with something that looks like the Lotka−Volterra period plus additive noise. This example provides striking testimony to the fact that, in ecology, the search for understanding depends on looking at the data at an appropriate scale of aggregation.

Host−parasite dynamics and the control of infection

The inherent propensity toward oscillatory behaviour in host−microparasite systems can have important implications for programmes that aim to reduce or eradicate infection. What follows is a sketchy abbreviation of work presented in more detail elsewhere (Anderson & May, 1983, 1985b, 1991).

Suppose we have some microparasitic infection that is endemic in a given population, either at steady levels or fluctuating in 'interepidemic' fashion as discussed above. Suppose we begin a programme of immunization that will eventually, when a new steady state is reached, eradicate the infection or reduce its incidence to some lower level. Uninformed intuition might suggest a smooth decline in the incidence of infection as we move to the new state. However, an appreciation of the oscillatory nature of host−microparasite associations suggests, to the contrary, that such a perturbation is likely to excite pronounced oscillations *en route* to the new state. An individual who anticipated a smooth transition could be disconcerted by the manifestation of such oscillations.

Figure 19.10 provides a concrete example. This figure provides a measure of the number of cases of congenital rubella syndrome (CRS) following the introduction of a programme of vaccination against rubella that was begun in Britain around 1973. CRS occurs in the offspring of women infected with rubella in the first trimester of pregnancy, and the curve in Fig. 19.10 can therefore be taken as reflecting the incidence of rubella among pregnant woman, year by year, following the initiation of the vaccination programme (which was aimed primarily at successive cohorts of 13-year-old girls, with coverage of around 60%). The noticeable oscillation, with a period of around 8 years, is close to what is predicted from relatively simple age-structured models (Anderson & May, 1983).

To summarize the last two sections, we see that the interactions between human hosts and microparasitic infections provide an especially rich collection of quantitative data. These interactions seem to illustrate ideas about chaotic dynamics in relatively simple systems (provided we look at the data at the appropriate level of aggregation). More than this, an understanding of the basic dynamical properties of host–microparasite systems can have important applications in the design and implementation of public health programmes.

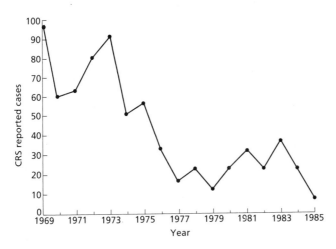

Fig. 19.10 The annual number of reported cases of congenital rubella syndrome (CRS) in the years 1969–85 in Britain. Vaccination, mainly of 13-year-old girls, was begun around 1971–73; for a given year the total number of cases is based on diagnoses up to 4 years after the birth of the child, so that the figures for 1981 onwards are incomplete. The statistically significant upward swing in the incidence of CRS about 8 years after the initiation of the vaccination programme is as predicted by appropriately age-structured host–microparasite models. From Anderson & May (1991).

Table 19.1 Comparisons of some characteristics of the life histories of microparasites, macroparasites, parasitoids and predators. (After Dobson 1982)

Life history characteristic or other property	Microparasite	Macroparasites	Parasitoid	Predator
Ratio of average life span to that of host or prey	$\ll 1$ (very small)	<1 (fairly small, $1-10^{-2}$)	$\simeq 1$ (usually about equal)	>1 (usually live longer)
Ratio of body sizes	Much smaller than hosts	Smaller than hosts	Mature stages often are of similar size to host	Usually larger than their prey
Intrinsic growth rate of population	Much faster than hosts	Faster than hosts	Comparable, but usually slightly slower than hosts	Usually slower than prey
Interaction with individual hosts, as observed in natural populations	One host usually supports a number of populations of different species	One host usually supports from a few to many individuals of different species	One host can support one or several individuals of one (or rarely two) species	One prey item can feed one or a few individuals of the same predator species, but many individual prey are required during a predator's lifespan
Effect of the above interaction on the host individual	Mildly to fairly deleterious	Variable: not usually too virulent in definitive host; can be very virulent in intermediate host	Eventually fatal	Usually fatal immediately
Ratio between numbers of species, at the population level	Many species of parasites recorded from each member of the host population	Many species of parasites recorded from each population of hosts	Most host species support several parasitoid species, both specialist and generalist (but only a proportion of hosts are actually attacked)	Individual species of predators tend to use more than one prey species
Degree of overlap of the ranges of the two species	Occur as diffuse foci throughout host's range	Occur as diffuse foci throughout host's range	Usually present throughout host's range	Range is usually greater than that of host
Genotypes per host or prey	Single or multiple	Multiple	Single or multiple sibships	Single

MORE GENERAL ASPECTS OF
PREDATOR–PREY ASSOCIATIONS

This chapter has focused on the dynamics of predator–prey associations. It has drawn some rough parallels between different kinds of such associations, and has illustrated the points with some particular examples pursued in greater depth.

The chapter has, however, neglected many other aspects of predator–prey interactions, where interesting similarities and differences among host–microparasite, host–macroparasite, host–parasitoid, and conventional predator–prey interactions can be explored. For example, we could have asked whether this continuum of types of prey–predator association has interesting correlations with patterns in the body sizes of the constituent individuals, or with their relative abundance, or with geographical ranges, or with genetic diversity. Table 19.1 (after Dobson, 1982) summarizes some tentative ideas about such comparisons. This table is presented essentially as a genuflection in the direction of paths not taken.

20: Prey Defence and Predator Foraging

STEPHEN B. MALCOLM

INTRODUCTION

It may seem simplistic and obvious to define predation as the process that describes the interactions between prey defences and the foraging tactics of predators (Fig. 20.1). However, the notion prevails that predation is either what individual predators do to catch and eat their prey (Curio, 1976), or is a description of the dynamics of interacting predator and prey populations (Taylor, 1984). Since prey defence is implicit, but rarely invoked explicitly, in the extensive literatures on predator–prey population dynamics (Hassell, 1978; Hassell & May, 1985; Hassell & Anderson, 1989), and predator foraging behaviour (Schoener, 1971; Hassell & Southwood, 1978; Kamil & Sargent, 1981; Stephens & Krebs, 1986; Kamil et al., 1987), it is worth emphasizing the significance of the defences of prey against predators as a primary selective determinant of how prey and predators interact (Jeffries & Lawton, 1984).

Herbivory is a similar process to predation and describes the interactions between plant defences and herbivore foraging (Fig. 20.1), although herbivores exploit their food resources in ways more like parasites than predators (Price, 1980; Janzen, 1981). Despite the similarity, we understand the process of herbivory better than that of predation, because equal attention has been given to the operation of plant defences and the tactics of herbivore foraging. Consequently, several theories of herbivory have been developed to describe interactions between plant defences and herbivore foraging (Dethier, 1954; Fraenkel, 1959; Ehrlich & Raven, 1964; Freeland & Janzen, 1974; Janzen, 1974; Feeny, 1976; Futuyma, 1976; Rhoades & Cates, 1976; Rhoades, 1979, 1983, 1985; Fox, 1981; Crawley, 1983, 1989; Jermy, 1984; McKey, 1984; Coley et al., 1985).

The behavioural and mathematical tools used to describe predation vary according to the level of organization, from individuals, through populations, to communities. These tools have been effectively used to describe predator foraging (Kamil & Sargent, 1981; Stephens & Krebs, 1986; Kamil et al., 1987), and the dynamics of interacting predator and prey populations (Hassell, 1978; Taylor, 1984; Comins & Hassell, 1987; Hassell & Anderson, 1989). Both approaches have been integrated to some extent by Hassell and Southwood (1978), Comins

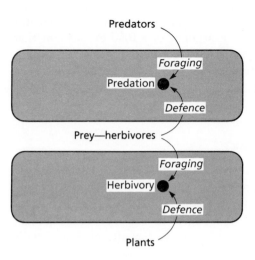

Fig. 20.1 Interactions between prey defence and predator foraging describe the process of *predation*, and similar interactions between plant defence and herbivore foraging describe the process of *herbivory*. The primary distinction between the processes is that herbivores act more like parasites than predators.

and Hassell (1978) and Hassell and May (1985) to show how foraging behaviour can influence the population dynamics of predators and their prey. However, none of these approaches have been used to anything like the same extent to examine the functioning of prey defences. Thus, although Tinbergen (1960), Holling (1965, 1966) and Tostowaryk (1972) have shown that predator functional responses are influenced by prey defences, and Kidd and Mayer (1983) have examined the influence of host escape defences on the stability of host–parasite dynamics, we still know little about how prey defences influence both predator foraging behaviour and the dynamics of predator–prey interactions.

Despite the fact that the experimental, descriptive and theoretical tools currently available are sufficient to assess the role of prey defences in predation, we continue to ignore a general synthesis of predator–prey interactions that includes prey defences. The purposes of this chapter are (i) to offer a simple framework for the consideration of predation, based on interactions between prey defence and predator foraging; (ii) to explore how prey defences operate; and (iii) to investigate some of the reasons why the role of prey defences has been largely ignored.

INTERACTIONS BETWEEN PREY DEFENCE AND PREDATOR FORAGING

Predation is considered, in this chapter, to be an inextricably linked interaction between prey defence and predator foraging. A simple categorization of prey defences, as 'the missing link' to understanding predation, is followed by an equally simple categorization of the basic diversity of predator foraging tactics. Interactions between these two sets of categories are then described and the importance of predator guild structure is illustrated with some laboratory experiments. This restricted account of interactions between prey defences and predator foraging tactics is followed by an account of a field experiment that shows the relevance of predator guild structure to understanding variation in the defence of a single species against predators. Finally, the points raised in these four sections are discussed in a concluding section.

Prey defences: the missing link

Prey defences have long interested biologists due to the extraordinary diversity of ways in which animals reduce the impact of predation on their fitness (Cott, 1940; Edmunds, 1974). Defences operate against all the known sensory modes of foraging predators. The seemingly bewildering array of defensive ploys employed has attracted an equally diverse array of subjective, intuitive interpretations of their functions from biologists. The sheer inventiveness of some of the teleological interpretations, divorced from efforts to observe predator selection (e.g. Hinton, 1974; Stradling, 1976), may have even hindered understanding of defence in the predation process, because the whole subject of defence has become associated with subjectivity and a lack of rigour.

In contrast, Endler (1986a,b, 1988) champions an objective approach to the functioning of different defences, based on how they operate at various stages of predator foraging. Using the *search, pursuit* and *subduing* phases of predation introduced by Holling (1966), MacArthur and Pianka (1966), Griffiths (1980a), and Vermeij (1982), he categorized defences by their effectiveness in interrupting the behavioural sequence of a foraging predator through at least five stages of prey *detection, identification, approach, subjugation* and *consumption*. Although different prey defences will operate most effectively at one or more particular stages in the foraging sequence, the categories are dependent upon predator foraging for their definition. For example, the marked influence of an aposematic defence on *consumption* of distasteful aphid prey after rapid and easy *detection, approach* and *subjugation* by spiders and birds (Malcolm, 1986). This means that independent predictions as to the structure and dynamics of predator–prey communities are precluded. Essentially, this is the same kind of sophism that Rhoades (1979) considered inherent in Feeny's (1976) 'apparency' rationale for classifying plant defences based on the behaviour of foraging herbivores.

As an alternative to Endler's (1986b) account of defence, it is clear from the thorough natural history surveys by Cott (1940) and Edmunds (1974), that defences at the same time and place can be grouped into three basic, functional categories: *'fight', 'run'* and *'hide'* (Fig. 20.2; Malcolm, 1990a); much like the prey responses of *'resistance', 'escape'* and *'undetectability'* of Vermeij (1987). In addition, variations in spatial and temporal distributions of prey that are adaptively defensive (Taylor, 1976; Waldbauer *et al.*, 1977; Treherne & Foster, 1980, 1981; Kidd, 1982; Sweeney & Vannote, 1982; Kidd & Mayer, 1983; Price, 1984;

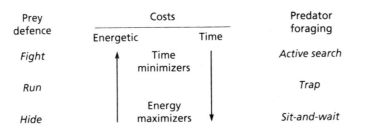

Fig. 20.2 Basic tactics of prey defence and predator foraging when prey and predator distributions in space and time are congruent, ranked according to energetic and time costs of implementation. These tactics are not exclusive and most prey or predators employ various combinations. When prey that employ any of these defence tactics also use distributional tactics to minimize temporal and spatial overlap with foraging predators, the predator's foraging tactics may also include a behavioural or physiological component to increase overlap with prey (e.g. migration and diapause). From Malcolm (1990a).

Waldbauer, 1988a) could be grouped as distributional or *'life history'* defences. Such defences could also include other life-history responses by prey to predator foraging, such as changes in the timing and amount of reproductive investment in guppies, because their predators alter age-specific survival through size-specific prey choice (Reznick *et al.*, 1990).

Thus, the four basic categories would include the defences described by Edmunds (1974), plus distributional defences, as follows:

1 *Fight.* Aposematism (including Müllerian mimicry), deimatic (frightening) behaviour, retaliation, group behaviour or mutualism.

2 *Run.* Flight, attack deflection or withdrawal to a prepared retreat.

3 *Hide.* Anachoresis (living in holes), crypsis, Batesian mimicry, thanatosis (feigning death).

4 *Life history.* Temporal or spatial distributions that minimize prey overlap with predator distributions.

When distributions of prey and predator coincide, energetic costs will be least for *hide* defences and greatest for *fight* defences. Conversely, time costs will be greatest for prey that hide from predators and least for prey that fight. Prey that hide could therefore be viewed as energy maximizers, and prey that fight as time minimizers (Fig. 20.2). Thus, prey that fight can engage in activities such as mate and food location without incurring the time costs of prey that need to hide from predators. Of course, there are ways to minimize these time costs, such as maintaining concealment while foraging. Fight defences incur energetic costs in various ways, such as metabolic energy expenditure in interactions with naive or persistent predators, or the resources incorporated in armour, or stored toxins.

Other predictions that might stem from these different defensive categories include the possibility that the diet breadth of time-minimizing prey with fight defences could be narrower than that of prey that hide, because they can exercise food choice, comparatively unconstrained by foraging predators. This may be especially true for aposematic, specialist herbivores that rely on plant toxins for their defences against predators (Malcolm, 1989; Malcolm & Brower, 1989). Cryptic, energy-maximizing, palatable prey may be forced by their crypsis and time constraints to exploit a wide range of food types. Moreover, when the distributions of prey and predators are variable, it is likely that prey with fight defences will have more predictable distributions in space and time than prey that hide (see also Price, 1984).

Many authors have stressed that no defence is absolute and it is likely that most prey species will employ combinations of defences that will vary in effectiveness according to the nature of predator attention (Pearson, 1989). Thus, many aposematic prey are also well armoured (Wiklund & Järvi, 1982; Malcolm, 1986; Vermeij, 1987), and many prey species that engage in group defensive behaviours are also adept at rapid flight defences. Seasonal or diurnal migrations may also have a defensive component, forcing predators to maintain similar distributions (Kruuk, 1972; Wilson & Clark, 1977; Malcolm & Brower, 1989).

Although prey defences are undoubtedly more complex than the simple categories shown in Fig. 20.2, two further factors cloud our perception of prey defence.

1 Considerations of the responses of single, somewhat arbitrarily chosen, predator species to the defences of prey species are unlikely to cast much light on how defences operate. Blue jays make excellent laboratory predators of chemically defended monarch butterflies (Brower, 1969, 1984), but their relevance to monarch population biology has never been established. Instead, a suite of different predator species has been observed attacking these aposematic butterflies in their over-wintering aggregations (Calvert *et al.*, 1979; Fink & Brower, 1981; Fink *et al.*, 1983; Brower & Calvert, 1985; Brower & Fink, 1985; Brower *et al.*, 1985, 1988; Glendinning *et al.*, 1988; Glendinning & Brower, 1990; Arellano-G. *et al.*, 1992; Glendinning, 1992).

2 Field observations on cryptic polymorphisms of the peppered moth, *Biston betularia* and the snail, *Cepaea nemoralis*, have been related to environmental changes and apostatic selection by bird predators (Kettlewell, 1973; Bishop *et al.*, 1978a,b; Brakefield, 1987). Similarly, butterfly and syrphid mimicry have been attributed to selection by bird predation (Sternburg *et al.*, 1977; Waldbauer *et al.*, 1977; Jeffords *et al.*, 1979; Waldbauer, 1988a,b). However, the process of predation has not been established in these observations and so alternative explanations are not eliminated and cause and effect can only be inferred indirectly.

As a solution to these problems, perhaps the best way to understand prey defences is to attempt the difficult task of determining the real diversity of natural enemies and their responses to the defences of a prey species. Just as Endler (1978, 1980, 1982) showed the importance of different predator species in maintaining a balance between predator— and mate-selected colour polymorphisms in tropical stream fish, the diversity and dynamics of different predator foraging tactics must be the key to understanding the operation and diversity of prey defences.

Predator foraging: diversity of tactics

The simplest description of predator foraging, when prey and predator distributions are congruent, is to include tactics within an energy expenditure continuum from energy consuming *active search*, through less energetic *trap* tactics, to energy conservative *sit-and-wait* foraging (Fig. 20.2; Huey & Pianka, 1981). Trap and sit-and-wait tactics are different ways to ambush prey. In concert with these decreasing energetic costs, time costs are likely to increase with increasing inactivity of foraging tactics, because predators become more dependent upon prey activity to maintain encounters rates than on their own mobility (Fig. 20.2). Although these foraging behaviours are more tactical than strategic (*sensu* Holling, 1968), active search predators could be described as time minimizers in the strategic sense of Schoener (1971), because they may forage unconstrained by time (Fig. 20.2). At the other end of the spectrum, sit-and-wait foragers could be viewed as *energy maximizers*, constrained by time but investing minimal energy in foraging. Trap foragers occupy an intermediate position, with some energetic costs associated with trap construction (e.g. spider webs, Prestwich (1977); caddisfly nets, Wallace *et al.* (1977); and antlion pits, Griffiths (1980b), Lucas (1985a)) and less time constraints than sit-and-wait foragers, because the trap may permit time for non-foraging activities such as prey consumption, mating or parental care.

Although Schoener (1971) defined time minimizers and energy maximizers differently, in terms of the energy gained from food, they are defined here in terms of the time and energy constraints associated with foraging. This distinction is made to avoid using prey to define different foraging tactics, because prey defences also influence the energetic return (or currency) accruing to successfully foraging predators (Stephens & Krebs, 1986).

In order to exploit prey that are spatially or temporally separated, predators also use *life history*, or distributional, foraging tactics to locate and track their prey (Hassell & Southwood, 1978). These foraging movements in space or time could include both small-scale dispersals within habitats and large-scale migratory and diapausing behaviours between habitats. Such distributional responses of predators to the density and distribution of their prey are among some of the best understood facets of the interaction between prey and predators and it is clear that prey patchiness and spatial scale have important consequences for the dynamics of interacting predator—prey populations (Taylor, 1976; Hassell, 1978; Comins & Hassell, 1979; Kidd & Mayer,

1983; Hassell & May, 1985; Comins & Hassell, 1987; Hassell & Anderson, 1989).

Interactions: prey defence and predator foraging

Interactions between prey and predators are likely to partition prey among predators according to the effectiveness of foraging tactics at dealing with various prey defences. For example, Huey and Pianka (1981) found that active search (or 'widely foraging') lizard predators, that incurred high foraging energy expenditures, were able to exploit more sedentary and unpredictably distributed prey than less energy-consuming sit-and-wait lizards that ate active prey. We might also expect active search predators to be more selective than sit-and-wait, or trap, predators because they can afford to exploit sedentary prey with high energetic returns and use search images (see Chapter 16 to circumvent the cryptic defences of these prey. Conversely, sit-and-wait and trap predators may exploit wider spectra of prey types, since they are constrained by encounter rates and these are determined by prey activity. An indication of these differences was shown by an active search bird predator that learned rapidly to discriminate palatable aphid prey from unpalatable, aposematic aphids (fight defence tactic) and prey exclusively on the palatable aphids (Malcolm, 1986). In contrast, a web-building spider (trap foraging tactic) continued to attack and kill the same species of aposematic aphid, despite very low food returns from the prey (Malcolm, 1986, 1989), and a cost resulting from seriously disrupted web structures after feeding on the toxic steroids present in the aphid (Malcolm, 1989, 1990b). Spider foraging on palatable aphid prey was also disrupted immediately after an encounter with the aposematic aphid. Thus, the most effective foraging option available to the spider is to relocate its trap to an area with a higher profitable prey encounter rate.

These kinds of interactions point towards the need to understand the diversity of foraging and defensive interactions that occur in natural communities before we can appreciate how defences operate and influence predator foraging and community structure. Unfortunately, we do not know much more about these interactions now than we did almost 60 years ago, at the time of the lively McAtee controversy.

In 1932 McAtee published a vituperative attack on what he considered to be the extant 'selectionist' teleological interpretation of prey defences against bird predation. McAtee found that 237 399 prey identifications from 80 000 stomachs of more than 300 bird species in North America included many aposematic insects and he concluded that defensive appearances were a figment of selectionist imaginations. Instead, McAtee's thesis was that insect prey were eaten by birds in direct proportion to their availability, and that their different appearances, which the selectionists assumed to be adaptively protective, had nothing to do with defence. This heresy was received with a torrent of articulate protest, organized by E.B. Poulton, the leading selectionist proponent, at that time, of protective adaptations (at least 33 publications between 1932 and 1935 rebuffed McAtee to support the defensive interpretations of cryptic, mimetic and aposematic appearances, e.g. 'various authors' (1932, 1934), including such great names as Fisher, Huxley and Nicholson). This important and lively debate illustrates much of the confusion which still exists in attempts to understand prey defences (see Malcolm, 1990a).

Nicholson (1927) was one of the few to recognize that McAtee's conclusion, that all birds in a large geographical area prey on almost all insects in the same area, regardless of their defences, is exactly what we should expect if birds are generally representative of the full spectrum of predator foraging tactics (Malcolm, 1990a). McAtee observed the product of predation and not the process. If he had observed both he might have found that certain defences were effective against some bird species, but not against others. This means that the only way we will understand prey defences is to appreciate the interaction between particular prey defences and a diversity of predator foraging tactics.

Interactions between prey defences and predator foraging tactics are often discussed using the rhetorical analogy of competitive human 'arms races' to describe the assumed coevolution between defence and foraging (Cott, 1940; Edmunds, 1974; Dawkins & Krebs, 1979; Malcolm 1990a). Dawkins and Krebs distinguished between competitive and predator—prey arms races by labelling them symmetric and asymmetric, respectively. Although the arms race metaphor may be an inappropriate description of predator—prey interactions (Vermeij, 1982; Malcolm, 1990a), it is useful for explaining why Poulton and his fellow selectionists were just as wrong as McAtee in adhering so unrelentingly to their ideas. It is very unlikely that predator—prey species interactions will show the coevolutionary reciprocity that is implied by an asymmetrical arms race, because prey occur in habitats with many potential predator species, and predators occur in habitats with many potential prey species. If McAtee and Poulton had appreciated prey and predator species diversity they might have found a compromise that at best would invoke the diffuse or multispecies coevolution discussed by Gould (1988), rather than reciprocal, pairwise adaptation and counteradaptation (Vermeij, 1982).

A CASE STUDY

A solution to the McAtee—Poulton impasse is only likely to result from tackling the difficult task of determining the real and potential predators of prey species and examining both the process and product of interactions between prey and their guilds of natural enemies. For example, the developmental responses of nine species of predator to a toxic, chemically defended aphid *Aphis nerii*, show considerable variation (Table 20.1; Fig. 20.3). Such interactions between aphid

Table 20.1 Four life history measurements of nine species of aphid predator when fed either the toxic, aposematic oleander aphid, *Aphis nerii* or the palatable pea aphid, *Acythosiphon pisum*, compared by one-way analyses of variance. Predators are separated into *included*, *peripheral* and *excluded* species according to their ability to exploit *A. nerii* as a prey resource. The larval predator weight changes with time are shown graphically in Fig. 20.3.

	Prey							
	A. nerii			*A. pisum*				
Life-history measurement	Mean	SE	N	Mean	SE	N	F^*	P
Included predator species								
Cycloneda sanguinea (L), Coleoptera, Coccinellidae								
Max. larval weight (mg)	15.75	0.61	28	16.46	0.49	28	0.78	>0.25 NS
Development time** (days)	15.77	0.22	30	15.53	0.24	30	0.48	>0.25 NS
Development rate (mg day^{-1})†	1.60	0.06	28	1.67	0.05	28	0.61	>0.25 NS
Survivorship‡ (%)	75.00		40	75.00		40		NS
Cheilomenes lunata (F), Coleoptera, Coccinellidae								
Max. larval weight (mg)	26.00	1.37	8	27.07	0.99	9	0.36	>0.50 NS
Development time** (days)	21.63	0.53	8	21.00	0.38	9	0.83	>0.25 NS
Development rate (mg day^{-1})†	1.93	0.14	8	2.08	0.09	9	0.72	>0.25 NS
Survivorship‡ (%)	36.36		22	40.91		22	0.31*	>0.50 NS
Adonia variegata (Goeze), Coleoptera, Coccinellidae								
Max. larval weight (mg)	9.72	0.30	30	9.89	0.29	17	0.13	>0.50 NS
Development time** (days)	15.07	0.25	41	15.37	0.32	19	0.46	>0.50 NS
Development rate (mg day^{-1})†	1.02	0.04	30	1.06	0.05	17	0.43	>0.50 NS
Survivorship‡ (%)	93.18		44	90.48		21	0.38*	>0.50 NS
Peripheral predator species								
Chrysopa carnea Stephens, Neuroptera, Chrysopidae								
Max. larval weight (mg)	11.00	0.22	5	10.30	0.45	9	1.06	>0.25 NS
Development time** (days)	28.60	1.61	5	22.33	0.69	15	15.36	<0.005
Development rate (mg day^{-1})†	0.72	0.05	5	1.02	0.09	9	5.12	<0.05
Survivorship† (%)	26.32		19	84.21		19	3.59*	<0.005

Coccinella septempunctata L, Coleoptera, Coccinellidae								
Max. larval weight (mg)	24.49	1.20	17	38.06	1.15	19	62.69	<0.001
Development time** (days)	19.48	0.73	23	14.91	0.18	22	34.15	<0.001
Development rate (mg day^{-1})†	1.70	0.06	17	3.95	0.14	18	195.73	<0.001
Survivorship‡ (%)	74.19		31	100.00		22	3.82	<0.001
Ischiodon aegyptius (Wied.), Diptera, Syrphidae								
Max. larval weight (mg)	20.14	0.19	3	25.68	1.45	5	6.51	<0.05
Development time** (days)	17.67	0.67	3	14.25	0.48	4	4.30	<0.01
Development rate (mg day^{-1})†	2.26	0.14	3	3.19	0.23	5	6.35	<0.05
Survivorship‡ (%)	100.00		3	80.00		5	1.27*	>0.20 NS
Excluded predator species								
Syrphus ribesii (L.), Diptera, Syrphidae								
Max. larval weight (mg)	—	—	—	70.89	1.51	16	—	—
Development time** (days)	—	—	—	22.42	0.60	12	—	—
Development rate (mg day^{-1})†	—	—	—	6.91	0.17	16	—	—
Survivorship‡ (%)	0.00		35	80.95		21	—	—
Sphaerophoria sp., Diptera, Syrphidae								
Max. larval weight (mg)	—	—	—	13.83	1.60	3	—	—
Development time** (days)	—	—	—	18.67	0.98	3	—	—
Development rate (mg day^{-1})†	—	—	—	1.22	0.12	3	—	—
Survivorship‡ (%)	0.00		2	75.00		4	—	—
Adalia bipunctata L., Coleoptera, Coccinellidae								
Max. larval weight (mg)	—	—	—	12.69	0.63	7	—	—
Development time** (days)	—	—	—	15.00	0.40	7	—	—
Development rate (mg day^{-1})†	—	—	—	1.54	0.13	7	—	—
Survivorship‡ (%)	0.00		11	77.78		9	—	—

* Percentage survivorships are compared by *d*-tests on arcsined data.
** Larval emergence from egg to adult emergence from pupa.
† Days to reach maximum larval weight.
‡ To live adult emergence.

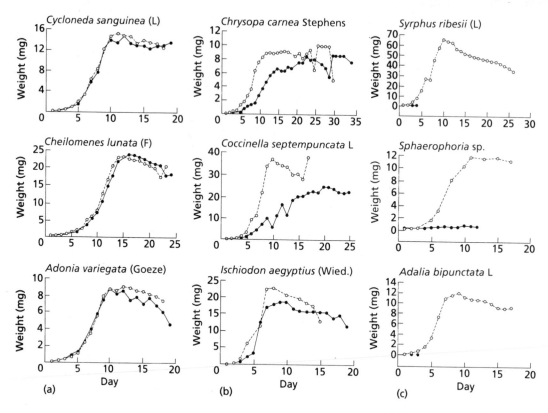

Fig. 20.3 Weight change with time of the larvae of nine species of aphid predator when fed either the toxic, aposematic oleander aphid, *Aphis nerii*, or the palatable pea aphid, *Acyrthosiphon pisum*. As in Table 20.1, the predators are separated into *included* (a), *peripheral* (b) and *excluded* (c) species according to their ability to exploit *A. nerii* as a prey resource. Sample sizes, significances of differences and the taxonomic affiliations of the predators are given in Table 20.1. −●−, fed *A. nerii*; −○−, fed *A. pisum*.

defences and foraging by aphidophagous insect predators make them useful candidates for research on defence, because both prey and predators are easier to observe than the species chosen by McAtee and Poulton, which interact over much greater spatial scales.

A. nerii is conspicuously coloured yellow and black and it sequesters toxic cardenolide steroids from its milkweed host plants (Malcolm, 1990b) that are an effective fight defence against bird and spider predators (Malcolm, 1986, 1989). Any of the nine insect predator species (Table 20.1; Fig. 20.3) could feed on *A. nerii* naturally: I collected *Cycloneda sanguinea* feeding on *A. nerii* in Venezuela; *Cheilomenes lunata*, *Adonia variegata* and *Ischiodon aegyptius* feeding on *A. nerii* in South Africa (*A. variegata* was also recorded as a predator of *A. nerii* in France and Greece by Iperti (1966) and Pasteels (1978)). *Chrysopa carnea* and *Coccinella septempunctata* were observed feeding on *A. nerii* in Israel by Rothschild (1973); and *Syrphus ribesii*, the *Sphaero-*

phoria species, and *Adalia bipunctata* are common and widely distributed aphid predators in Europe. Larvae of the nine predators were reared from egg emergence on either *A. nerii* or a non-toxic palatable control aphid, *Acyrthosiphon pisum*. *A. pisum* was used as a control aphid because it employs run and hide defences (it falls rapidly from host plants when disturbed and is very mobile, and its green colour is cryptic on host plants (the broad bean, *Vicia faba*)). Aphids were fed to the larval predators on cut leaves placed in Petri dishes. Thus *A. nerii* remained aposematic and chemically defended to predator larvae, but the defences of *A. pisum* were neutralized because it could not escape or hide from the predators.

The results show that the nine predator species could be separated into three categories, based on their comparative ability to exploit aposematic *A. nerii* as a food resource (Table 20.1, Fig. 20.3). *Included* predators could exploit either aphid prey equally successfully with no significant differences in life-history measurements. *Peripheral* predators showed significantly reduced survivorship, smaller sizes and extended development times when reared on the aposematic aphid. *Excluded* predators were unable to exploit the aposematic aphid as prey and all larvae died.

Although additional natural enemies, including hymenopteran parasitoids and other dipteran, neuropteran and coleopteran predators can attack *A. nerii* naturally, the results point to what is likely to be a universal generalization; the defences of prey species will be completely effective against excluded predators, partially effective against peripheral predators and ineffective against included predators. All prey are likely to encounter natural enemies that could lie at any point on the continuum from included to excluded predator categories. These categories could be distributed among many different guilds of natural enemies, including pathogens, parasitoids and predators, or distributed within a single guild. Most importantly, such a diversity of defence effectiveness suggests that peripheral guild members are likely to be the pivotal or *key* species in determining the success of particular defences.

Variation in the effectiveness of defences, whether physiological, morphological or behavioural, will have its greatest impact on peripheral key predators. A small reduction in the effectiveness of a defence is likely to result in an increase in the ability of a predator to exploit the prey as food. Ageing and pathogen attack can reduce the effectiveness of fight and run defences, with the result that large mammalian predators attack old and sick prey more frequently than younger, healthy prey (Kruuk, 1972). Similarly, many insect herbivores are specialists (Bernays & Graham, 1988), and many also rely on their host plants for their own chemical defences. This means that variation in host biochemistry, either within or among plant species, can have profound effects on the chemical defences of these herbivores (Brower, 1969; Malcolm & Brower, 1989). Such plant variation can often have a marked indirect impact on the peripheral predators of chemically defended prey; much like many of the examples discussed in Kerfoot and Sih (1987).

Three trophic level interactions

The current interest in three trophic level interactions shows how pervasive plant influences on the defences of herbivores against natural enemies can be (Barbosa & Letourneau, 1988; Bernays & Graham, 1988). For example, like many aposematic insect herbivores, *A. nerii* depends upon its host plants as a source of defensive chemicals sequestered for use against natural enemies (Malcolm, 1986, 1989, 1990b). The effectiveness of chemical defence in *A. nerii* thus depends on the aphid's ability to sequester plant-derived chemicals from milkweed phloem and on the variability of plant chemical content. Plant cardenolides may vary qualitatively and quantitatively, both within and among species in the milkweed genus *Asclepias* (Malcolm & Brower, 1989; Malcolm *et al.*, 1989), and they may vary both spatially and temporally (Malcolm *et al.*, 1989, 1992). Thus the defences of *A. nerii* will co-vary with their host plants despite the aphid's ability to sequester cardenolides effectively (Malcolm, 1990b).

In a simple field experiment, Barbara Cockrell and I compared the population fluctuations of *A. nerii* and its natural enemies colonizing two milkweed species growing on the edge of a longleaf pine forest in Gainesville, north central Florida. One milkweed species, *Asclepias amplexicaulis*, contained very low quantities of cardenolide (pooled samples contained 5–6 µg cardenolide per 0.1 g dry leaf weight, measured by spectrophotometry (Roeske *et al.*, 1976; Malcolm *et al.*, 1989)); and the other species, *A. curassavica*, contained very high cardenolide concentrations (797–864 µg per 0.1 g dry weight). The purpose of the experiment was to observe the influence of plant-determined chemical defences of the aphid against the aphidophagous predator guild. On the high cardenolide *A. curassavica*, the aphid contained 81–160 µg cardenolide per 0.1 g dry aphid (see also Malcolm, 1990b), and on the low cardenolide *A. amplexicaulis* the aphid contained 15–21 µg cardenolide per 0.1 g dry aphid.

The experiment was set up by planting a patch of 100 single-stemmed *A. curassavica* (collected from a field in south Florida) immediately adjacent to a patch of *A. amplexicaulis* that was also maintained at 100 single stems by removing new shoots. Densities of plant stems in the two patches were the same. *A. nerii* was allowed to colonize the plants naturally, as was the suite of natural enemies that followed (this included two syrphid species, one hemerobiid, one chrysopid and two coccinellids). Aphids were also placed in natural enemy exclosures on an additional 10 stems of each milkweed species and the intact aphid colonies counted. Observations over 44 days (Fig. 20.4) show that aphid populations were smaller and more influenced by predators on the low cardenolide milkweed than on the high car-

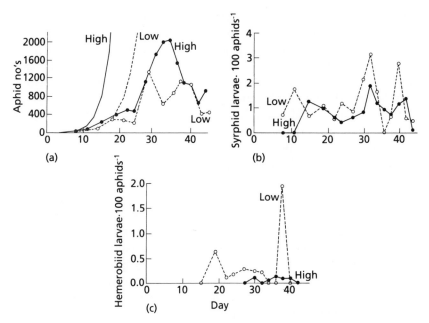

Fig. 20.4 Field observations over 44 days of colonization by the aposematic oleander aphid, *Aphis nerii*, and its predators, on the milkweeds *Asclepias curassavica* with high cardenolide content (——), and *A. amplexicaulis* with low cardenolide content (– – –). (a) The exponentially increasing aphid numbers observed in controls that excluded natural enemies (lines with data points removed), and the total aphid numbers on 100 stems of each host plant species after attack by natural enemies. The numbers of predator larvae per 100 aphids of (b) the syrphid *Ocyptamus fuscipennis* and (c) the brown lacewing (hemerobiid) *Micromus posticus* on the two milkweeds. Observations were made in June and July at the Devil's Millhopper State Geological Site in NW Gainesville, Florida.

denolide milkweed. This difference may have been a result of the presence of higher predator numbers per aphid on low rather than high cardenolide milkweeds. Most of the difference was attributable to one key predator species, rather than being caused by predators in general. When the two dominant predator species were compared, an interesting difference between the plants emerged. The two predator species that accounted for most of the predator numbers were the syrphid *Ocyptamus fuscipennis* and the hemerobiid *Micromus posticus*. The syrphid was equally abundant on both milkweed species throughout the experiment (Fig. 20.4), but the hemerobiid larvae were considerably more common on the low cardenolide milkweed than on the high cardenolide species — despite adults laying approximately equal numbers of eggs on both plant species. This field result suggests to us that *M. posticus* is a peripheral key predator of *A. nerii* and that its ability to exploit the aphid is influenced by host plant chemical defences.

We also took the hemerobiid into the laboratory and compared its larval development on the aphid feeding on the same two milkweed species with high or low cardenolide contents. The results (Table 20.2) show that *M. posticus* was unable to survive on *A. nerii* from *A. curassavica* with high cardenolide, but 70% survived on *A. nerii* from *A. amplexicaulis* with low cardenolide. Larval weights and development rates of the hemerobiid were also significantly higher on aphids from the low cardenolide milkweed.

This kind of field and laboratory experimentation highlights the need to describe both the process and product of predation before we can appreciate how defence operates in interactions between prey and a diversity of foraging predators.

CONCLUSION

Variation in the effectiveness of prey defences against foraging by peripheral predators is likely to be the key to understanding the ecology and evolution of interactions among prey and their predators. Increases in defence effectiveness will expand enemy-free space (Jeffries & Lawton, 1984) at the expense of peripheral predators, and will have little effect on either included or excluded predators (Fig. 20.5). Similarly, decreases in defence effectiveness will reduce enemy-free space to the benefit of peripheral predators, but will leave both included and excluded predators unaffected. Defence effectiveness can vary in many different ways. The example used here of variation in chemical host plant influences on the effectiveness of a fight defence against predators is just one

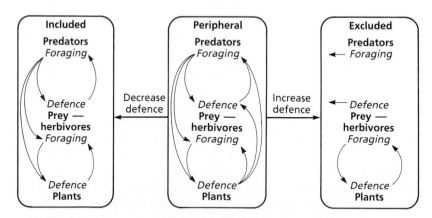

Fig. 20.5 Three trophic level interactions separated by predator ability to exploit defended prey. Interactions with *included* predators are symmetrical because foraging predators directly influence the foraging of their prey. Interactions with *peripheral* predators are symmetrical because plants commonly influence the defences of their herbivores against predator foraging. *Excluded* predators are eliminated from interactions and must forage elsewhere. Decreases in the effectiveness of prey defences attacked by *peripheral* predators may result in their becoming *included* predators. Alternatively, increases in defence effectiveness may *exclude* peripheral predators from foraging.

Table 20.2 Five life-history measurements of the hemerobiid, *Micromus posticus* fed *A. nerii* on either the high cardenolide milkweed *Asclepias curassavica*, or the low cardenolide milkweed, *A. amplexicaulis*, compared by one-way analyses of variance.

		Aphid host plant								
		A. curassavica			*A. amplexicaulis*					
Life-history measurement		Mean	SE	N	Mean	SE	N	F	P
Max. larval weight	(mg)	2.82	0.15	12	5.73	0.23	24	71.56	0.0001
Development time*	(days)	–	–	0	14.82	0.19	19	–	–
Time to pupation	(days)	9.90	0.90	10	9.36	0.20	22	0.65	0.43 NS
Development rate	(mg.day^{-1})**	0.38	0.04	12	0.84	0.03	24	76.10	0.0001
Survivorship^{+}	(%)	0.00		32	70.37		29	–	–

* Larval emergence from egg to adult emergence from pupa.

** Days to reach maximum larval weight.

$^{+}$ To live adult emergence.

commonly encountered way in which variation occurs. Other ways to vary defence effectiveness could include changes in the effectiveness of hide defences, such as crypsis, with different light conditions at different times of day, or with different positions in a forest (Endler, 1988). Other fight defences could vary in effectiveness over time, for example, when an armoured crab moults its exoskeleton and is soft and vulnerable, for a period of time, to attack by a wider range of predators.

That peripheral predators are an abundant and potent evolutionary influence on prey defences is shown by Vermeij (1982) who surveyed 60 predator species and found that only 19% of their prey were attacked with an efficiency (after detection) of more than 90%. In most cases however, predator detection, pursuit and subjugation of prey was considerably less successful. It is likely that a predator's success will depend on whether it is an included, peripheral or excluded guild member attacking a particular prey species.

An important caveat in descriptions of prey defence is to avoid the teleological idea that particular defences convey an advantage; a warning that Nicholson (1927) stressed so eloquently. For example, analyses of both the process and result of field predation within mimic-model-predator systems still do not exist, despite the fact that Nicholson called for exactly this kind of information on defensively mimetic systems more than 60 years ago. Although excellent field experiments are available (Brower *et al.*, 1964, 1967; Jeffords *et al.*, 1979; Waldbauer, 1988a), they all examine the result, or product, of predation and not the process, and are thus open to alternative explanation. Waldbauer's defensively adaptive explanation of the temporal distribution of Batesian syrphid mimics of hymenoptera, in relation to bird predation (for summary see Waldbauer, 1988a), has an alternative interpretation. The mimetic syrphids show a unimodal summer distribution sandwiched between a bimodal wasp distribution of abundance in spring and late summer. Waldbauer and colleagues suggest that syrphid mimics appear after their wasp models in order to avoid naïve fledgling bird predators learning that wasps sting. However, since the syrphid mimics are mostly univoltine larval detritivores that overwinter as immatures, and the hymenopteran models are partially social predators that overwinter as adults, these marked life-history differences (unrelated to defence) could also explain the different temporal distributions of models and mimics. This does not preclude an effective Batesian mimicry from protecting the syrphids, but mimicry would exist in spite of the different temporal distributions of model and mimic, and predation would not be a cause of the distribution. In this case it is probably also important to compare the life histories and temporal distributions of non-mimetic, closely related syrphids before Waldbauer's conclusion can be applied. An adaptive explanation for the different temporal distributions could be invoked with more confidence if the difficult task of examining the process of predation, via the behavioural responses of each potential bird predator species to models and mimics throughout the season was attempted.

We need to understand prey defences in a way that relates to the conceptual ideas of how predators forage (Schoener, 1971), and to the extensive literature on optimal foraging (MacArthur, 1972; Pyke *et al.*, 1977; Pyke, 1984; Stephens & Krebs, 1986; Kamil *et al.*, 1987). Rhoades (1979) used a similar cost–benefit, optimal foraging, fitness-maximization argument to develop his theory of *optimal defense*, with an emphasis on plants and their defences against herbivores. Apart from differences in predatory versus parasitic foraging of most herbivores, prey defences function much like plant defences. The primary distinction is that plants are sessile (although the rapid responses of the leaves of legumes, such as *Mimosa* and *Schrankia*, to herbivore contact are akin to run defences from foraging) and their defences are characterized by fight and hide tactics against actively searching herbivores. Exceptions would include aquatic filter-feeders (trap or sit-and-wait foragers) exploiting mobile, water-borne plants. Thus most plants invest in a fight against herbivores with diverse arrays of chemical defences, spines and hairs, etc., or hide by being cryptic, mimetic or part of plant defence guilds (Atsatt & O'Dowd, 1976). Distributional, life-history defences are also very important to plants in avoiding herbivore foraging.

Although defence is implicit in all considerations of predation, we need to understand how the defences of single prey operate against suites of predators, each foraging in a different way. Similarly, we need to know how single predators forage among suites of prey that defend themselves in quite different ways. The most likely reason that progress in understanding prey defence has been so slow in comparison with predator foraging and predator–prey population dynamics, is that theoretical work has offered such a wealth of testable hypotheses in foraging and dynamics (Kareiva, 1989). The study of defence has remained anecdotal and subjective, and has been largely the realm of the enthusiastic natural historian. Given equal emphasis on the roles of prey defence and predator foraging, a much clearer understanding of predation is likely to emerge, that will rival our understanding of herbivory.

21: Overview

This brief summary attempts an overview of some of the general patterns that emerge from the foregoing chapters. Given the vast array of taxonomic groups represented, and the variety of biological processes involved in the study of the ecology and evolution of natural enemies, you will not be surprised that the number of generalizations is small and that, as ever, caveats abound.

NATURAL ENEMIES AND PREY ABUNDANCE

When prey populations are low, mortality inflicted by natural enemies is bound to lead to reductions in prey population density. This is because compensation can only occur when the prey population is subject to density-dependence, and this, of course, happens only at relatively high densities. Thus, to the extent that prey populations are typically scarce, then natural enemies will generally reduce prey numbers.

The problem, of course, lies in the definition of 'scarce'. In the present context, 'scarce' means *free from intraspecific density-dependence*. But unless we know about the details of the resources for which the prey might be competing (and usually we know next to nothing about them), then we cannot say whether the prey are scarce, and so we cannot predict whether or not attack by natural enemies will cause reduced prey population densities. The fact that absolute prey densities are low does *not* mean that prey are scarce in the sense we mean here; it may simply be that the resources for which the prey are competing are present at low absolute densities.

The subtleties are introduced when prey are subject to density-dependence as well as natural enemy attack. Then, if the prey contest for resources, natural enemy attack may have no impact at all on prey densities. Instead, enemies simply alter the identity of the survivors. If the prey scramble for resources, then predation might even lead to increased prey density, by relaxing the intensity of intraspecific competition. In either case, the medium-term change in prey population density is substantially less than the mortality inflicted by natural enemy attack.

It is worth noting in passing that most commercial fish stocks are scarce, and that increased fishing pressure almost always leads to

[476]

reduced fish stocks (Pauly, 1987). Indeed, it may be a general characteristic of populations harvested by humans that exploitation will always tend to drive them into the region in which continued harvesting leads to population decline.

NATURAL ENEMIES AND POPULATION REGULATION

Natural enemies can only regulate prey population if their attack is density-dependent. That is to say, the percentage of the prey population killed by natural enemies must increase as prey population density rises. This could come about through short-term changes in the behaviour of generalist predators (e.g. switching behaviour or aggregation in local patches of relatively high host density), or through longer term (but inevitably time-lagged) numerical responses by specialist natural enemies. It is reasonable to suppose that numerical responses leading to density-dependent attack rates will be more prevalent with specialist natural enemies than with generalists, although some generalists can, of course, exhibit pronounced numerical responses if their preferred prey become sufficiently abundant, e.g. vertebrate predators of cyclic small mammals; Angelstam *et al.* (1984), Erlinge *et al.* (1983, 1984), Hansson & Hettonen (1988), Ims & Steen (1990).

The evidence on population regulation by natural enemies is equivocal. Most natural enemy exclosure experiments have demonstrated concomitant increases in prey numbers (Table 21.1), but the studies were often short term and the changes in prey numbers may have been only transitory. It is often difficult to separate temporary increases in breeding success from genuine, long-term increases in prey population density. Several detailed studies fail to show any response of prey density to predator exclusion. Again, the fact that prey density increases following predator exclusion does not prove that natural enemies were responsible for prey population regulation, but simply that predation was an additive (i.e. non-compensated) mortality factor. Proof of regulation requires the demonstration that attack is density-dependent at least at some times and/or some places. Thus, while predation by one corvid like the carrion crow may have a significant impact in reducing the densities of ground-nesting birds (Reynolds, Angelstam & Redpath, 1988), predation by another like the magpie, may have no effect at all on the density of nesting songbirds (Gooch, Baillie & Birkhead, 1991).

The jury is still out. Of course natural enemies *could* be important in prey population regulation, but there is little direct experimental evidence that they are important. Many theoretical models show that natural enemies could regulate prey densities at levels well below those at which they would be constrained by other factors like competition for food. But other models can equally well be used to show that realistic-looking patterns of herbivore and predator dynamics can

Table 21.1 Predator exclusion experiments in terrestrial habitats. Predator exclusion almost always led to increases in prey breeding success, but, since most studies were short term, these dynamics were almost certainly transitional. Note also that many of the studies alleging a major impact of predator removal were anecdotal, often referring to alien herbivores and/or alien predators, and some of the experimental studies are poorly replicated

Example	Reference
Predation on white-winged doves reduced by shooting predatory grackles; nest numbers doubled in response	Blankinship (1966)
Providing milk for cats reduces rat populations in buildings	Elton (1953)
Hand removal of coccinelids, hoverflies and lacewings caused 30% increase in peak aphid densities	Edson (1985)
Ant exclusion from bracken fronds led to an increase in density of only one of the herbivore species, suggesting that ant predation was feeble relative to other factors	Heads & Lawton (1984)
Shooting woodpigeons reduced overwinter mortality rather than increasing it and shooting caused no decline in pigeon numbers	Murton *et al.* (1974)
Small carnivore removal led to double the rate of chick production compared to control areas with similar breeding density	Chesness *et al.* (1968)
Nesting densities of upland ducks were threefold higher and percentage nests with eggs that hatched were 50% greater on areas from which red foxes, racoons, striped skunks and badgers were removed	Duebbert & Kantrud (1974)
Fox, stoat, cat and carrion crow removal led to a twofold increase in nest density and increased nesting success in partridges	Potts (1980)
Racoon, skunk and fox removal led to 60% increase in duckling production	Balser *et al.* (1968)
Jackrabbits increased three-fold on fox removal plots but not at all on small-carnivore removal plots; pheasants increased by 75% on small-carnivore removal plots, but not at all on fox removal plots	Trautman *et al.* (1973)
Caribou calf survival at 6 months of age increased from 2.7% to 16.3% following removal of lynx from the breeding ground	Bergerud (1971)
On the island of Guam, the introduced brown tree snake caused the extinction or range contraction of 10 native bird species	Savidge (1987)

Increase in sparrowhawk predation from virtually nothing to 18–34% following the banning of toxic pesticides did not lead to reduced tit numbers in Wytham Wood, England	Perrins & Geer (1980)
Experimental reduction of prey density led to reduced badger predation on ground squirrels	Slade & Balph (1974)
Experimental removal of rabbit predators caused fourfold increase in rabbit numbers over 3 years in New South Wales, Australia	Newsome *et al.* (1989)
In Australia, erection of a 9660 km dingo fence led to increases in populations of 11 medium-sized mammal species on the dingo-free side of the fence	Newsome & Corbett (1975)
Lion shooting 1903–60 led to such a large increase in wildebeest numbers that culling became necessary	Walker & Noy Meir (1982)
Bird exclusion reduced mortality on overwintering larvae of codling moth by 95%	Solomon *et al.* (1976)
Bird exclusion from bilberry stands led to increased damage from insects	Atlegrim (1989)
Blattidae reduced by 50% in 6 weeks when exposed to predation by ant-wren	Gradwohl & Greenberg (1982)
Bird exclusion reduced larval mortality by 50% in spruce budworm	Torgersen & Campbell (1982)
Dingo control in Queensland led to outbreak of feral pigs, while there were no pigs at all in two National Parks where there were no dingo control measures	Letts (1979), Newsome *et al.* (1983)

be generated when different processes are assumed to limit prey numbers (e.g. plant–herbivore interactions). Again, most of the observational evidence for the importance of natural enemies in prey population regulation comes from managed systems in which man is the predator (e.g. fisheries), from systems involving the biological control of exotic pest species using introduced insect parasitoids (see Chapter 18), or from cases where alien herbivorous insects have become much more abundant when freed from their native natural enemies (McClure, 1980, 1986; Hails & Crawley, 1991).

GENERALISTS AND SPECIALISTS

Generalist natural enemies tend to exhibit more or less pronounced preferences for their various prey species. If the enemy is capable of mounting a pronounced functional or numerical response to increases in the abundance of one of its preferred prey species, then the population density of the preferred prey species may be regulated by that generalist natural enemy. Generalist natural enemies are unlikely to regulate the population density of less preferred prey species.

Specialist natural enemies are capable, at least in theory, of regulating the abundance of their prey, but the time-lags involved in their numerical responses will often mean that they do not regulate prey numbers. More normally, changes in the abundance of specialist natural enemies will be driven by extrinsic changes in the abundance of their prey. An example might be a monophagous insect parasitoid feeding on a food-limited caterpillar, where changes in the abundance of the food plant were driven by changes in spring rainfall. Most of the examples of prey population regulation by specialist natural enemies come from managed systems in applied ecology (e.g. biological control of exotic pest species).

HETEROGENEITY IN THE RISK OF DEATH

Refuges are ubiquitous. Individual prey almost always differ in the probability of their being attacked by a given natural enemy. This heterogeneity may come about as a result of prey phenotype (size or body condition), or through the existence of physical refugia in which some prey are safe from natural enemy attack. Further, the foraging behaviour of the natural enemies (e.g. habitat selection or oviposition behaviour) adds to the heterogeneity between prey individuals in the probability of being attacked. The existence of refugia will often stabilize population dynamics by providing an unexploited (or under-exploited) reserve which buffers the prey from local extinction.

Prey population density will tend to be higher as a result of the existence of refuges simply because only a fraction of the prey population is susceptible to predation at any given moment. This will be a benefit when the prey is a valued resource and a problem when the prey is a pest. Murphy's Law indicates that pest populations can be harvested indefinitely at high intensities, while resource populations decline to extinction under even the mildest exploitation.

The existence of refugia does not guarantee stability. Refugia may be too small to stabilize a predator–prey interaction, while too large a refuge will lead to the prey population escaping control.

THE DEVIL AND THE DEEP BLUE SEA

The question of whether or not natural enemies can or do regulate prey numbers is intimately connected with other elements of prey demography. As Lawton and McNeill (1979) pointed out, the ability of natural enemies to regulate prey populations depends upon the prey's rate of increase. If the net rate of increase is too great (perhaps because temperatures are optimal and food quality is high), then the prey population can escape control; no matter what functional and numerical responses are mounted by the natural enemies, they simply cannot keep up with prey reproduction. Only if the net population increase of the prey is relatively low (e.g. as a result of low food quality, suboptimal temperatures, or attack by generalist predators) can the natural enemies mount a density-dependent attack that is capable of regulating prey numbers.

The ability of predators to control prey populations may also depend upon prey density, especially as this is influenced by other, extrinsic environmental factors like drought or fire. Thus, vertebrate predators in Australia appear to be able to regulate rabbit numbers once severe drought has reduced rabbit densities to within range the of $8-15$ rabbits km^{-1} of transect. If prolonged rains allow repeated successful breeding, then rabbits escape control and increase to very high densities (e.g. $\geqslant 50\,km^{-1}$; Newsome, 1990).

Thus, asking whether predation regulates a given prey population may be posing a question that is just too simple to have a sensible answer. The prey's rate of increase will almost always vary from year to year as weather and food conditions change. There may be a sequence of years in which conditions for the prey are relatively bad, and during this period prey numbers may be regulated by predation. But then a bumper year for the prey will occur in which prey reproduction swamps the natural enemies, and the prey population escapes control. It will then increase until some new density-dependent check on its numbers, like an outbreak of epidemic disease or mass starvation intervenes to reduce prey density. In this case, natural enemies can regulate prey numbers but their regulating ability is limited. Natural enemy attack may be the only source of density-dependence acting on the prey in most years, and may maintain prey numbers at such low levels that food competition and disease have no influence on prey demography. The regulating ability of the natural enemies is limited, however, because an increase in the net reproductive rate of the prey will cause the prey to escape control.

WHAT REGULATES
NATURAL ENEMY POPULATIONS?

We tend to make the tacit assumption that natural enemy numbers are food-limited (i.e. enemy populations are regulated by competition for prey in the manner advocated by Hairston et al., 1960). However, this book has provided several examples of natural enemy populations that are regulated by the availability of safe nest sites, or of suitable sites in which to build traps, and these densities are well below the level at which the animals would compete for food.

Now, unless natural enemies are food-limited, it is most unlikely that they will regulate prey numbers (except, perhaps, in the case of generalists which might limit the abundance of their most attractive prey species; the 'ice cream principle'; see Crawley, 1987). It is simply not good enough to *assume* that natural enemies are food-limited; it is necessary to demonstrate experimentally that natural enemies compete for food. If the predators are food-limited, then prey supplementation experiments should lead to increased natural enemy densities, and experimental prey reductions should lead to reduced enemy densities.

SYMMETRY OR ASYMMETRY IN REGULATION?

It is most unlikely that prey populations are regulated all of the time. Furthermore, Murphy's Law states that if we are interested in regulation, then our study will take place during those periods when the population is unregulated. In populations which are regulated, the factors that are responsible for regulation may not always be the same, and may differ from year to year or from place to place. Thus, the ability to say something unambiguous like 'this population is food-limited', or 'that population is predator-limited' will be an unusual luxury. More typically, we will need to be more circumspect and to say such things as 'this population wanders around with the vagaries of the weather for 6 years out of 7, but is food-limited in the seventh year and the population seems never to be regulated by natural enemies'. Alternatively, there will be cases when we would prefer to say that 'this species is sufficiently scarce, and suffers such high rates of parasitism, that no hypothesis other than enemy-limitation is at all plausible'. It is most unlikely that we shall ever be able to convince the extreme sceptic who demands the evidence of long-term, large-scale, well-replicated manipulative field experiments before being persuaded that natural enemies are responsible for prey population regulation.

NATURAL ENEMIES AND SPECIES RICHNESS

In principle, natural enemies could have any one of three kinds of impact on prey species richness: they could increase it, decrease it, or have no effect at all. What do the data suggest? In the case of keystone species, the result is clear. By killing what would otherwise be the competitive dominant, a keystone natural enemy allows an increase in prey species richness, as a result of competitor release amongst what would otherwise be excluded prey species. You will notice, however, that this is yet another tautology; keystone predators affect prey diversity because predators that affect prey diversity are called 'keystone species'. While there are several textbook examples of keystone predators (starfish feeding on mussels, Paine, 1974; sea otters feeding on urchins, Estes *et al.*, 1978), there are no examples of keystone parasitoids. Whether this means that predators are more important than parasites in affecting prey species richness is a moot point. It may simply mean that parasitoid exclusion experiments are harder to carry out than predator exclusions. Also, a number of the examples of so-called key-stone predators have turned out on critical re-examination to be nothing more than anecdotes (and wrong anecdotes at that; e.g. the lobsters and urchins of the north-western Atlantic), so we need to be 'cautious about allowing paradigms to become established without adequate scrutiny' (Elner & Vadas, 1990).

Gamekeepers have traditionally believed that vermin (large, mobile generalist predators like foxes, crows and magpies) were responsible for reducing the diversity of 'desirable' prey species. The view was that these animals were sufficiently abundant that they could seek out and destroy the nests and young of songbirds and game-birds. While this view is still controversial, it remains plausible that abundant generalist predators could have severe depressive effects on the abundance of highly preferred (or highly vulnerable) prey species (see Table 21.1).

In so far as local extinction is more likely when population density is lower, then natural enemies which reduce prey abundance would increase the probability of local extinction, and hence tend to reduce local prey species richness.

TOP-DOWN OR BOTTOM-UP DYNAMICS

It is 30 years since Hairston et al. (1960) put the cat amongst the pigeons with their provocative essay on trophic regulation. Their argument (paraphrased by generations of students as 'the world is green and we're not making coal') was that all trophic levels were regulated by competition for resources, except for the herbivores which were regulated by natural enemies. In his recent book, Hairston (1989) reviews the experimental evidence that has been amassed during the intervening 30 years, and concludes that the hypothesis has stood the test of time. There is nothing in the present book to refute Hairston's assessment. (See also Oksanen et al., 1981; Yodzis, 1984; DeAngelis, 1992.)

This is largely because the focus of the present book has been on species and populations and not on trophic levels as a whole. Hairston can argue that plants as a group are not herbivore-limited because the grazing-tolerant, or herbivore-avoided species tend to increase to dominance in places where herbivory is intense. Even if it is true that interspecific competition is the most important process affecting plant community dynamics (Crawley, 1990), small changes in the levels of herbivory could lead to sufficiently great changes in competitive ability as to change the identity of the eventual dominant plant (Pacala & Crawley, 1992). At the scale of the whole trophic level, herbivory has changed nothing (the world is still covered with plants), but in terms of plant population dynamics, everything has changed, and the reason it has changed is herbivory.

Similar reasoning could apply at the level of carnivory. Many (perhaps even most) carnivores are food-limited. If the supply of prey were increased, the long-term average population density of the carnivores would increase. But this does not mean that most carnivores are involved in symmetrical interactions with their prey species. It is entirely reasonable that the carnivores are food-limited, but the prey are not enemy-limited. Thus, the observation that the carnivore trophic level is food-limited cannot be used as evidence that the herbivore trophic level is enemy-regulated. An alternative explanation is that the herbivores are limited by competition for resources (despite the fact that the world may appear to us to be green). Prey regulation might occur only infrequently; most of the time, the prey populations might be recovering from density-independent catastrophes caused by drought, storm, fire or community succession.

BIG FLEAS HAVE LITTLE FLEAS

We know a great deal about the population biology of carnivorous animals and the dynamics of their interactions with prey populations. However, what we know has only served to highlight the extent of our ignorance about the precise way that natural enemy systems work in the field. Theory is developing steadily, and observational and comparative studies continue to distinguish the actual from the possible. It is to be hoped that the coming decades will see a flourishing of field experiments, involving large-scale, long-term manipulations of prey and enemy densities, linked to sharply focused theoretical models and supported by detailed observational studies. Then we may find out what is really going on.

The experimental demonstration that natural enemies are food-limited is not enough. We need also to determine whether changes in prey numbers are driven by natural enemy attack. A given prey population could be regulated by a natural enemy that was not food-limited. Similarly, a food-limited natural enemy might have no impact on the abundance of prey in the next generation. It will probably turn out that real examples of the text-book symmetrical predator–prey interaction where each species influences the abundance of the other are extremely unusual. Asymmetric interactions appear to be the norm in predator–prey interactions just as they are in competitive (Lawton & Hassell, 1981) and plant–herbivore interactions (Crawley, 1983).

It is important to remember that real world predator–prey interactions are imbedded between two other trophic levels (Table 21.2), and that the outcome of a predator–prey interaction can be affected by changes in the plants that form the food of the prey, and in the diseases and parasites that attack the predator. Interactions with competing prey and competing natural enemies can have profound effects on population dynamics (Table 21.3).

Table 21.2 Multitrophic level processes involving predator–prey interactions

Interaction	Example and Reference
Plant quality affects prey rate of increase	Natural enemies can regulate prey feeding on low quality host plants, but increases in plant food quality lead to insect outbreaks; Lawton & McNeill (1979)
Plant spatial pattern affects prey distribution	Crawley (1983)
Plant chemicals attract predators directly	Parasites attracted by plant rather than host cues, Ayal (1987)
Plants of other species mask host plant detection cues causing reduced predation	Reduced natural enemy efficiency in weedy crops
Plants of other species are a source of generalist natural enemies	Increased natural enemy efficiency in weedy crops
Pathogens and microorganisms compete with predators and parasites for prey	Hochberg & Lawton (1990)
Parasites and diseases restrict habitat use by natural enemies	Prey range is greater than natural enemy range, providing a stabilizing refuge for the prey in the disease-infested areas
Plant diseases reduce host plant availability for specialist herbivores, and hence specialist natural enemies	Chestnut blight and Dutch elm disease reduce specialist predator densities on the tree-feeding herbivores of these plants
Plant–herbivore–predator interactions	Generation of 10-year-cycles Akcakaya (1992)
Predators and nutrient cycling	DeAngelis (1992)
Predator density is reduced by human habitat interference	Scrub clearance, etc. may have more effect on predators than on prey
Abiotic factors reduce refugia	Fire or storm destroys cover leading to increased natural enemy attack
Changed management practice alters degree of cover for prey	Reduced livestock grazing increases perennial grass cover, thus reducing avian predation on voles

EVOLUTION AND THE DETECTION OF SELECTIVE PREDATION IN THE FIELD

Natural enemies will influence prey evolution so long as they inflict genotype-specific mortality and there is no subsequent compensatory selection against the surviving genotypes. As to the problems of studying the evolutionary aspects of natural enemy attack under field conditions I

Table 21.3 Theoretical models of multispecies natural-enemy–prey interactions. Many of these systems can exhibit more complex dynamics than are displayed by any of the pair-wise interactions

Interaction	Reference
One predator, several prey	Parrish & Saila (1970)
	Cramer & May (1972)
	Roughgarden & Feldman (1975)
	Comins & Hassell (1976)
	Hanski (1981)
Several predators, one prey	Nicholson & Bailey (1935)
	Hassell (1978)
	May & Hassell (1981)
Specialist and generalist predators attack one prey	Hassell & May (1985)
Several parasites attack one host	Dobson (1985)
	Hochberg & Holt (1990)
Host, parasite, hyperparasite interactions	Beddington & Hammond (1977)
	Hassell (1978)
	May & Hassell (1981)
Pathogen and parasitoid competition for one prey	Hochberg *et al.* (1990)
	Hochberg & Lawton (1990)

can do no better than to quote from Endler (1986a). 'There are many difficulties in detecting natural selection: (i) lack of detection of natural selection when it exists; (ii) apparent detection when it does not exist; and (iii) misleading detection of natural selection. The most important problems in detecting natural selection arise out of incomplete knowledge of the ecology and general biology of the species studied, and from incomplete or poorly designed sampling. There is no substitute for careful and intensive field work if one wants to find out what is happening in natural populations.'

SUMMARY

There is widespread agreement about the important elements which affect the dynamics of the interaction between natural enemies and their prey: net reproductive rates, degree of polyphagy, apparent competition, spatial refuges, temporal asynchrony, co-variance in spatial aggregation, time lags, threshold densities, and the timing and nature of prey density dependence have all been shown to be influential in observational, theoretical and experimental studies. The major difficulty is that there is not a single case for which we have all of this information, and for many of the best-known predator–prey systems, we understand only one or two of the component elements.

It is important to know how the costs and benefits of *real* foraging tactics compare with those of optimal foraging behaviour under field conditions. Detailed, long-term field observations are needed to understand the way that prey animals perceive and respond to the risks of predation (Abrahams & Dill, 1989; Brown, 1988; Lima & Dill, 1990; Nonacs & Dill, 1990; Bouskila & Blumstein, 1992). Likewise, we still know very little about the details of real encounters between natural enemies and their prey under field conditions; how often do they happen, who encounters whom, and what determines the outcome? Modern methods of molecular genetics are sure to play a major role in highlighting the importance of individual differences; which prey are most likely to get eaten, to what extent do differences in the likelihood of predation reflect differences in prey genotype, and so on? Fingerprinting may also allow us to determine the identity of those who survive and those who die from predation under realistic conditions in the field.

We need to know much more about the spatial co-variance of the distributions of natural enemies and other species: between predators and their preferred prey species; between prey and their competitors (especially those that are also preyed upon by shared natural enemies); and between predators and their own natural enemies. It is clear that spatially explicit predator–prey models will exhibit at least as great a range of dynamic behaviour as spatially explicit competition models. Field work on the relative mobility of natural enemies and their prey holds great promise for determining the correct spatial scale at which to study their dynamics (e.g. do predators always exhibit higher mean dispersal distances than their prey?). There has been considerable interest in the dynamics of fragmented populations (meta-population dynamics); this may form a basis for unifying several presently disparate strands of work on local extinctions caused by predation, relative mobility of predators and prey, patch structure, isolation distances, patch-leaving rules and so forth. The importance of spatial refuges was realized by the pioneering laboratory ecologists (Gause, 1934; Crombie, 1947; Ivlev, 1961; Huffaker et al., 1963) who were attempting to maintain long-term predator–prey cultures. (One or both of the species tended to go extinct rapidly in uniform culture vessels.) The addition of spatial heterogeneity (and other between-individual differences in the risk of predation) to theoretical models has developed rapidly over the past 20 years; its potential importance for population dynamics is now universally admitted. What we do not know, aside from a handful of medical, fisheries and pest population examples, is how these heterogeneities in the risk of attack work under field conditions.

Large, mobile, generalist predators may have a profound effect on the distribution and abundance of their most highly preferred prey species. In the limit, preferred prey species may be excluded altogether from certain habitats (the 'ice-cream principle' (Crawley, 1989) states that there is no levelling-off of the functional response equation at high prey densities, and prey are consumed every time they are encountered). For predators which feed on a succession of seasonally available prey types (e.g. many insectivores), it is unlikely that we shall find pronounced numerical responses to changes in the abundance of any one prey species. Here, behavioural flexibility, functional responses and rapid spatial aggregation in regions of ephemeral prey abundance are likely to be the norm. Specialist natural enemies will inevitably suffer time lags in any numerical responses which they might exhibit, and hence might be expected to have rather less impact on average prey abundance. The paradox of enrichment is a central theme in much of the theoretical work on predation in multiple trophic level dynamics (e.g. DeAngelis, 1992). For example, increasing the productivity of prey can reduce the stability of a predator–prey interaction by relaxing the intensity of competition between the enemy-limited prey (see Crawley, 1983, p. 254). It is not clear whether specialist or generalist predators are more or less likely to exhibit the paradox of enrichment.

It is important not to underestimate the importance of parasitic infection as distinct from predation. Many of the most important effects of parasite attack act through reduced fecundity rather than through a direct effect on increased death rates. In addition, parasite attack may expose individuals to increased risks of death from other causes, or reduce the probability of successful mating. These indirect effects could have important population dynamic and evolutionary consequences.

We need more and better manipulative field experiments. They need to be carried out on a bigger spatial scale and over a longer time period. There is a wide variety of different ideas that need to be tested: for example, what are the consequences for prey dynamics of increasing plant productivity by nutrient-addition, experimental increases and reductions of prey densities, predator-exclusion trials, and so forth? In the past, these experiments have been carried out by small groups, working on different species in different places. The next generation of experiments should seek to combine these questions in larger, inter-disciplinary projects. The main difficulty with previous predator exclusion experiments has been their short duration (this has meant that it is difficult or impossible to separate short term effects on breeding success from long-term effects on prey density), small scale, inadequate replication (many just had two areas, one predator-free and another 'control') and uncritical analysis (e.g. a chronic inability to spot pseudo-replication).

As to the top-down, bottom-up debate, the jury is still out. The notion of alternating regulation in alternating trophic levels has a certain aesthetic appeal. In one kind of scenario, a top-predator regulates predators at such low densities that herbivores become sufficiently abundant as to be food-limited, and their food-plants are herbivore-regulated at low densities. In another scenario there is no top predator, so predators regulate herbivores at such low densities that their food plants are abundant and limited by competition with other plants. So, unless we know that there is a keystone top predator, we shall not know whether to expect the herbivore trophic level to be enemy-regulated or food-limited. Further, it is not at all obvious how arguments like this that refer to *chains* of species extend to more realistic *webs* (or networks) of species.

The reader will judge the degree to which the questions posed in the Preface have been successfully answered by the various contributors. There is considerable agreement about the basics; about the theoretical underpinning and about the kinds of work that produce useful results. The disagreement centres on the details, but as we have seen amply well demonstrated, the details usually matter. Let us hope that this book helps to clarify some of the patterns, and that the next generation of students of natural-enemy ecology will tackle the questions that we have not thought of yet.

References

Ables, J.R. & Shepard, M. (1976)
Seasonal abundance and activity of
indigenous hymenopterous
parasitoids attacking the house fly
(Diptera: Muscidae). *Can. Entomol.*
180: 841–844.

Abrahams, M. & Dill, L.M. (1989) A
determination of the energetic
equivalence of the risk of predation.
Ecology **79**: 999–1007.

Abrams, P.A. (1986) Adaptive responses
of predators to prey and prey to
predators: the failure of the arms-race
analogy. *Evolution* **40**: 1229–1247.

Abramsky, Z. & Rosenzweig, M.L.
(1984) Tilman's predicted
productivity–diversity relationship
shown by desert rodents. *Nature*, **309**:
150–151.

Addicott, J.F. (1974) Predation and prey
community structure: an
experimental study on mosquito
larvae and protozoans. *Ecology* **55**:
475–492.

Agounké, D., Agricola, U. & Bokonon-
Ganta, H.A. (1988) *Rastrococcus
invadens* Williams (Hemiptera:
Pseudococcidae), a serious exotic pest
of fruit trees and other plants in West
Africa. *Bull. Entomol. Res.* **78**:
695–702.

Agricola, U., Agounké, D., Fischer, H.U.
& Moore, D. (1989) The control of
Rastrococcus invadens (Hemiptera:
Pseudococcidae) in Togo by the
introduction of *Gyranusoidea tebygi*
Noyes. *Bull. Entomol. Res.* **79**:
671–678.

Aitchison, C.W. (1987) Review of winter
trophic relations of soricine shrews.
Mammal Rev. **17**: 1–24.

Akcakaya, H.R. (1992) Population cycles
of mammals: evidence for a ratio-
dependent predation hypothesis. *Ecol.
Monogr.* **62**: 119–142.

Albers, G.A.A. & Gray, G.D. (1986)
Breeding for worm resistance: a
perspective. In: Howell, M.J. (ed.)

Parasitology — Quo Vadit?,
pp. 559–567. Proceedings of the 6th
International Congress of
Parasitology, Australian Academy of
Science, Canberra.

Alexander, R. McN., Jayes, A.S., Maloiy,
G.M.O. & Wathuta, E.M. (1979)
Allometry of limb bones from shrews
(*Sorex*) to elephants (*Loxodonta*). *J.
Zool. Lond.* **189**: 305–314.

Alexander, R. McN., Jayes, A.S., Maloiy,
G.M.O. & Wathuta, E.M. (1981)
Allometry of the leg muscles of
mammals. *J. Zool. Lond.* **194**:
539–552.

Allen, J.A. (1988) Frequency-dependent
selection by predators. *Phil. Trans. R.
Soc. Lond. B* **319**: 485–503.

Allen, K.R. (1985) Preliminary notes on
impact on fish stocks and catches
(Task Force 2). MS submitted to
Canadian Royal Commission on Seals
and Sealing.

Alstad, D.N. & Corbin, K.W. (1990)
Scale insect allozyme differentiation
within and between hosts. *Evol. Ecol.*
4: 43–56.

Alstad, D.N. & Edmunds, G.F. (1983)
Selection, outbreeding depression and
the sex ratio of scale insects. *Science*
220: 93–95.

Alstad, D.N. & Edmunds, G.F. (1989)
Haploid and diploid survival
differences demonstrate selection in
scale insect demes. *Evol. Ecol.* **3**:
253–263.

Altieri, M.A. & Letourneau, D.K. (1984)
Vegetation diversity and insect pest
outbreaks. *CRC Crit. Rev. Plant Sci.*
2: 131–169.

Andersen, S.H. (1982) Change in
occurrence of the harbour porpoise,
Phocoena phocoena, in Danish waters
as illustrated from catch statistics
from 1834 to 1970. In: *Mammals in
the Seas*, FAO Fish. Series V, Vol. 3,
pp. 131–134. FAO, Rome.

Anderson, D.T. (1981a) Cirral activity

and feeding in the barnacle *Balanus perforatus* Bruguiere (Balanidae), with comments on the evolution of feeding mechanisms in thoracician cirripedes. *Phil. Trans. R. Soc. London. B* **291**: 412–449.

Anderson, J.F. (1974) Responses to starvation in the spiders *Lycosa lenta* Hentz and *Filistata hibernalis* (Hentz). *Ecology* **55**: 576–585.

Anderson, R.M. (1979) The influence of parasitic infection on the dynamics of host population growth. In: Anderson, R.M., Turner, B.D. & Taylor, L.R. (eds) *Population Dynamics*, pp. 245–281. Blackwell Scientific Publications, Oxford.

Anderson, R.M. (1981b) Population ecology of infectious disease agents. In: May, R.M. (ed.) *Theoretical Ecology*, pp. 318–355. Blackwell Scientific Publications, Oxford.

Anderson, R.M. (1982a). Directly transmitted viral and bacterial infections of man. In: Anderson, R.M. (ed.) *Population Dynamics of Infectious Diseases: Theory and Applications*, pp. 1–37. Chapman and Hall, London.

Anderson, R.M. (1982b). Epidemiology. In: Cox, F.E.G. (ed.) *Modern Parasitology*, pp. 204–251. Blackwell Scientific Publications, London.

Anderson, R.M. (1982c). Transmission dynamics and control of infectious disease agents. In: Anderson, R.M. & May, R.M. (eds) *Population Biology of Infectious Diseases*, pp. 149–176. Springer-Verlag, Berlin.

Anderson, R.M. (1988) The epidemiology of HIV infection: variable incubation plus infectious periods and heterogeneity in sexual activity. *J.R. Stat. Soc. A* **151**: 66–98.

Anderson, R.M. (1992) Reproductive strategies of trematodes. In: Adiyodi, K.G. & Adiyodi, R.G. (ed.) *Reproductive Biology of Invertebrates*, Vol. 6. John Wiley & Sons, New York (in press).

Anderson, R.M. & Crombie, J.A. (1985) Experimental studies of age-intensity and age-prevalence profiles of infection: *Schistosoma mansoni* in snails and mice: In: Rollinson, D. & Anderson, R.M. (eds) *Ecology and Genetics of Host–Parasite Interactions*, pp. 111–145. John Wiley & Sons, New York.

Anderson, R.M. & Gordon, D.M. (1982) Processes influencing the distribution of parasite numbers within host populations with special emphasis on parasite-induced host mortalities. *Parasitology* **85**: 373–398.

Anderson, R.M., Grenfell, B.T. & May, R.M. (1984) Oscillatory fluctuations in the incidence of infectious disease and the impact of vaccination: time series analysis. *J. Hyg.* **93**: 587–608.

Anderson, R.M. & May, R.M. (1978) Regulation and stability of host–parasite population interactions. I. Regulatory processes. *J. Anim. Ecol.* **47**: 219–247, 249–267.

Anderson, R.M. & May, R.M. (1979) Population biology of infectious diseases. *Nature* **280**: 361–367, 455–461.

Anderson, R.M. & May, R.M. (1980) Infectious diseases and population cycles of forest insects. *Science* **210**: 658–661.

Anderson, R.M. & May, R.M. (1981) The population dynamics of microparasites and their invertebrate hosts. *Phil. Trans. R. Soc. Lond. B* **291**: 451–524.

Anderson, R.M. & May, R.M. (1983) Vaccination against rubella and measles: quantitative investigations of different policies. *J. Hyg.* **90**: 259–325.

Anderson, R.M. & May, R.M. (1984) Spatial, temporal and genetic heterogeneity in host populations and the design of immunization programmes. *IMA J. Math. Appl. Med. Sci.* **1**: 233–266.

Anderson, R.M. & May, R.M. (1985a) Helminth infections of humans: mathematical models, population dynamics and control. *Adv. Parasitol* **24**: 1–101.

Anderson, R.M. & May, R.M. (1985b) Age-related changes in the rate of disease transmission: implications for the design of vaccination programmes. *J. Hyg.* **94**: 365–436.

Anderson, R.M. & May, R.M. (1985c) Herd immunity to helminth infection and implications for parasite control. *Nature* **315**: 493–496.

Anderson, R.M. & May, R.M. (1986) The invasion, persistence and spread of infectious diseases within animal and plant communities. *Phil. Trans. R. Soc. Lond. B* **314**: 533–570.

Anderson, R.M. & May, R.M. (1988) Epidemiological parameters of HIV transmission. *Nature* **333**: 514–518.

Anderson, R.M. & May, R.M. (1991)

Infectious Diseases of Humans: Dynamics and Control. Oxford University Press, Oxford.

Anderson, R.M., May, R.M. & McLean, A.R. (1988) Possible demographic consequences of AIDS in developing countries. *Nature* **332**: 228–234.

Anderson, R.M. & Nokes, D.J. (1991) Mathematical models of transmission and control. In: Holland, W.W., Detels, R.A. & Knox E.G. (eds) *Oxford Textbook of Public Health*, 2nd edn, vol. 2, pp. 225–252. Oxford Medical Publications, Oxford.

Anderson, S.S. & Fedak, M.A. (1985) Grey seal males: energetic and behavioural links between size and sexual success. *Anim. Behav.* **33**: 829–838.

Andow, D.A. & Risch, S.J. (1985) Predation in diversified agroecosystems: relations between a coccinellid predator *Coleomegilla maculata* and its food. *J. Anim. Ecol.* **22**: 357–372.

Angelstam, P., Lindstrom, E. & Widen, P. (1984) Role of predation in short-term population fluctuations of some birds and mammals in Fennoscandia. *Oecologia* **62**: 199–208.

Antonovics, J. & Ellstrand, N.C. (1984) Experimental studies on the evolutionary significance of sexual reproduction. I. A test of the frequency-dependent selection hypothesis. *Evolution* **38**: 103–115.

Arditi, R. (1982) Quelques difficultés de la détermination expérimentale des trois types de réponse fonctionelle des prédateurs et des parasitoides: consequences sur la distribution des réponses. *Mitteilungen der Schweitzerischen Entomologischen Gesellschaft* **55**: 151–168.

Arditi, R. & d'Acorogna, B. (1988) Optimal foraging on arbitrary food distributions and the definition of habitat patches. *Am. Nat.* **131**: 837–846.

Arellano-G., A., Glendinning, J.I. & Brower, L.P. (1992) Interspecific comparisons of the foraging dynamics of black-backed orioles and black-headed grosbeaks on overwintering monarch butterflies in Mexico. In Malcolm, S.B. & Zalucki, M.P. (eds) *Biology and Conservation of the Monarch Butterfly*. Los Angeles County Natural History Museum, Los Angeles (in press).

Aron, J.L. & May, R.M. (1982) The population dynamics of malaria. In: Anderson, R.M. (ed.) *The Population Dynamics of Infectious Diseases: Theory and Applications*, pp. 139–179. Chapman and Hall, London.

Aronson, R.B. & Givinish, T.I. (1983) Optimal central-place foragers: a comparison with null hypotheses. *Ecology* **64**: 395–399.

Arthur, A.P. (1962) Influence of host tree on abundance of *Itoplectis conquisitor* (Say) (Hymenoptera: Ichneumonidae), a polyphagous parasite of the European pine shoot moth, Rhyacionia buoliana (Schiff) (Lepidoptera: Oleuthreutidae). *Can. Entomol.* **94**: 337–347.

Arthur, A.P. (1966) Associative learning in *Itoplectis conquisitor* (Say) (Hymenoptera: Ichneumonidae). *Can. Entomol.* **98**: 213–223.

Arthur, A.P., Stainer, J.E.R. & Turnbull, A.L. (1964) The interaction between *Orgilus obscurator* (Nees) (Hymenoptera: Braconidae) and *Temelucha interruptor* (Grav.) (Hymenoptera: Ichneumonidae), parasites of the pine shoot moth, *Rhyacionia buoliana* (Schiff.) (Lepidoptera: Olethreutidae). *Can. Entomol.* **96**: 1030–1034.

Askew, R.R. (1961) On the biology of the inhabitants of oak galls of Cynipidae (Hymenoptera) in Britain. *Trans. Soc. Br. Entomol.* **14**: 237–268.

Askew, R.R. (1971) *Parasitic Insects.* Heinemann, London.

Askew, R.R. & Shaw, M.R. (1985) Parasitoid communities: their size, structure and development. In: Waage, J.K. & Greathead, D.J. (eds) *Insect Parasitoids* pp. 225–264. Academic Press, London.

Atlegrim, O. (1989) Exclusion of birds from bilberry stands: impact on insect larval density and damage to bilberry. *Oecologia* **79**: 136–139.

Atsatt, P.R. & O'Dowd, D.J. (1976) Plant defense guilds. *Science* **193**: 24–29.

Aukema, B. (1986) Winglength determination in relation to dispersal by flight in two wing dimorphic species of *Calathus* Bonelli (Coleoptera, Carabidae). In: Den Boer, P.J., Luff, M.L., Mossakowski, D. & Weber, F. (eds) *Carabid Beetles, their Adaptations, Dynamics and Evolution* pp. 91–99. Fischer, Stuttgart.

Aukema, B. (1987) Differences in egg production and egg-laying period between long and short-winged *Calathus erythroderus* (Coleoptera, Carabidae) in relation to wingmorph frequencies in natural populations. *Acta Phytopathol. Entomol. Hung.* **22**: 45–56.

Aumann, T. (1986). *Aspects of the biology of the brown goshawk* Accipiter fasciatus fasciatus *in southeastern Australia*. MSc thesis, Monash University, Melbourne. (Cited in Olsen & Cockburn, 1991).

Auslander, D., Guckenheimer, J. & Oster, G. (1978) Random evolutionarily stable strategies. *Theor. Pop. Biol.* **13**: 276–293.

Avery, M.L. & Nelms, C.O. (1990) Food avoidance by red-winged blackbirds conditioned with a pyrazine odor. *Auk* **107**: 544–549.

Ayal, Y. (1987) The foraging strategy of *Diaeretiella rapae*. I. The concept of the elementary unit of foraging. *J. Anim. Ecol.* **56**: 1057–1068.

Baars, M.A. (1979) Patterns of movement of radioactive beetles. *Oecologia* **44**: 125–140.

Baars, M.A. & Van Dijk, T.S. (1984) Population dynamics of two carabid beetles at a Dutch heathland. II. Egg production and survival in relation to density. *J. Anim. Ecol.* **53**: 389–400.

Bailey, G.N.A. (1975a) Energetics of a host–parasite system: a preliminary report. *Int. J. Parasitol.* **5**: 609–613.

Bailey, N.T.J. (1975b) *The Mathematical Theory of Infectious Diseases and its Applications*. 2nd Edn. Hafner Press, New York.

Bailey, P.C.E. (1986) The feeding behaviour of a sit-and-wait-predator, *Ranatra dispar* (Heteroptera: Nepidae): optimal foraging and feeding dynamics. *Oecologia* **68**: 291–297.

Bailey, V.A., Nicholson, A.J. & Williams, E.J. (1962) Interactions between hosts and parasites when some individuals are more difficult to find than others. *J. Theor. Biol.* **3**: 1–18.

Baker, W.L. (1972) *Eastern Forest Insects*. Misc. Publ. 1175. USDA, Forest Service.

Bakker, F.M. & Sabelis, M.W. (1989) How larvae of *Thrips tabaci* reduce the attack success of phytoseiid predators. *Entomologia Experimentalis et Applicata* **50**: 47–51.

Bakker, R.T. (1983) The deer flees, the wolf pursues: incongruencies in predator–prey coevolution. In: Futuyma, D.J. & Slatkin, M. (eds) *Coevolution*, pp. 350–382. Sinauer, Sunderland, MA.

Balashov, Y.S. (1984) Interaction between bloodsucking arthropods and their hosts, and its influence on vector potential. *Ann. Rev. Entomol.* **29**: 137–156.

Balfour, E. & Cadbury, J.C. (1979) Polygyny, spacing and sex ratio among hen harriers *Circus cyaneus* (L.) in Orkney, Scotland. *Ornis Scand.* **10**: 133–141.

Balgooyen, T.G. (1976) Behaviour and ecology of the American kestrel (*Falco sparverius* L.). *Univ. Calif. Publ. Zool.* **105**: 1–85.

Ballard, W.B., Whitman, J.S. & Gardner, C.L. (1987) Ecology of an exploited wolf population in south-central Alaska. *Wildlife Monogr.* **98**: 1–54.

Balser, D.S., Dill, H.H. & Nelson, H.K. (1968) Effect of predator reduction on waterfowl nesting success. *J. Wildlife Manage.* **32**: 669–682.

Barbosa, P. & Letourneau, D.K. (eds) (1988) *Novel Aspects of Insect–Plant Interactions*. John Wiley & Sons, New York.

Barkan, C.P.L. & Withiam, M.L. (1989) Profitability, rate maximization, and reward delay: a test of the simultaneous-encounter model of prey choice with *Parus atricapillus*. *Am. Nat.* **134**: 254–272.

Barker, M.F. (1977) Observations on the settlement of the brachiolaria larvae of *Stichaster australis* (Verrill) and *Coscinasterias calamaria* (Gray) (Echinodermata: Asteroidea) in the laboratory and on the shore. *J. Exp. Marine Biol. Ecol.* **30**: 95–108.

Barker, M.F. (1979) Breeding and recruitment in a population of the New Zealand starfish *Stichaster australis* (Verrill). *J. Exp. Marine Biol. Ecol.* **41**: 195–211.

Barnard, C.J. (1980) Flock feeding and time budgets in the house sparrow (*Passer domesticus* L.) *Anim. Behav.* **28**: 295–309.

Barnard, C.J. & Brown, C.A.J. (1981) Prey size selection and competition in the common shrew (*Sorex araneus* L.). *Behav. Ecol. Sociobiol.* **8**: 239–243.

Barnard, C.J., Brown, C.A.J. & Gray-

Wallis, J. (1983) Time and energy budgets and competition in the common shrew (*Sorex araneus* L.). *Behav. Ecol. Sociobiol.* **13**: 13–18.

Barrett, J.A. (1984) The genetics of host–parasite interaction. In: Shorrocks, B. (ed.) *Evolutionary Ecology*, pp. 275–294. Blackwell Scientific Publications, Oxford.

Barrett, J.A. (1985) The gene-for-gene hypothesis: parable or paradigm. In: Rollinson, D. & Anderson, R.M. (eds) *Ecology and Genetics of Host–Parasite Interactions*, pp. 215–225. Academic Press, London.

Barrett, J.A. (1988) Frequency-dependent selection in plant–fungal interactions. *Phil. Trans. R. Soc. Lond. B* **319**: 473–483.

Bartlett, M.S. (1957) Measles periodicity and community size. *J. Roy. Stat. Soc. Ser. A* **120**: 48–70.

Basson, M. (1989) *Population dynamics of krill interactions with its major predators*. PhD thesis, University of London.

Batson, J.D. & Best, P.J. (1979) Drug-preexposure effects in flavor-aversion learning: associative interference by conditioned environmental stimuli. *J. Exp. Psychol: Anim. Behav. Proc.* **5**: 273–283.

Baum, W.M. (1974) On two types of deviation from the matching laws: bias and under-matching. *J. Exp. Anal. Behav.* **22**: 231–242.

Baynes, A. (1975) The distribution of the native mammals of south-western Australia. *Aust. Mamm.* **1**: 404–405.

Beattie, A.J. (1985) *The Evolutionary Ecology of Ant–Plant Mutualisms*. Cambridge University Press, Cambridge.

Beddington, J.R., Beverton, R.J.H. & Lavigne, D.M. (eds) (1985) *Marine Mammals and Fisheries*. George Allen & Unwin, London.

Beddington, J.R., Free, C.A. & Lawton, J.H. (1978) Characteristics of successful natural enemies in models of biological control of insect pests. *Nature* **273**: 513–519.

Beddington, J.R. & Hammond, P.S. (1977) On the dynamics of host–parasite–hyperparasite interactions. *J. Anim. Ecol.* **46**: 811–821.

Beddington, J.R., Hassell, M.P. & Lawton, J.H. (1976) The components of arthropod predation. II. The predator rate of increase. *J. Anim. Ecol.* **45**: 165–185.

Beddington, J.R. & May, R.M. (1977) Harvesting natural populations in a randomly fluctuating environment. *Science* **197**: 463–465.

Bednarz, J.C. (1988) Co-operative hunting in Harris hawks (*Parabuteo unicinctus*). *Science* **239**: 1525–1527.

Begon, M., Harper, J.L. & Townsend, C.R. (1986) *Ecology: Individuals, Populations and Communities*. Blackwell Scientific Publications, Oxford.

Beirne, B.P. (1980) Biological control: benefits and opportunities. In: *Perspectives in World Agriculture*, pp. 307–321. Commonwealth Agricultural Bureaux, Farnham Royal, Slough.

Beissinger, S.R. & Snyder, N.F.R. (1987) Mate desertion in the snail kite. *Anim. Behav.* **55**: 477–487.

Bell, G. (1982) *The Masterpiece of Nature*. Croom Helm, London.

Bell, G. & Maynard Smith, J. (1987) Short-term selection for recombination among mutually antagonistic species. *Nature* **328**: 66–68.

Bellows, T.S. & Hassell, M.P. (1988) The dynamics of age-structure host–parasitoid interactions. *J. Anim. Ecol.* **57**: 259–268.

Bergerud, A.T. (1971) The population dynamics of Newfoundland caribou. *Wildlife Monogr.* **25**: 1–55.

Bergerud, A.T. (1974a) Decline of caribou in North America following settlement. *J. Wildlife Manage.* **38**: 757–770.

Bergerud, A.T. (1974b) The role of the environment in the aggregation, movement and disturbance behaviour of caribou. In: Geist, V. & Walther, F. (eds) *The Behaviour of Ungulates and its Relation to Management. IUCN Publ. New Ser.* **24**: 552–584.

Bergerud, A.T. (1988) Caribou, wolves and man. *Trends Ecol. Evolut.* **3**: 68–72.

Bergerud, A.T. & Elliot, J.P. (1986) Dynamics of caribou and wolves in northern British Columbia. *Can. J. Zool.* **64**: 1515–1529.

Bergerud, A.T. & Page, R.E. (1987) Displacement and dispersion of parturient caribou at calving as antipredator tactics. *Can. J. Zool.* **65**: 1597–1606.

Bernays, E. & Graham, M. (1988) On the evolution of host specificity in phytophagous arthropods. *Ecology* **69**:

886–892.

Bernstein, C. (1984) Prey and predator emigration responses in the acrine system *Tetranychus urticae–Phytoseiulus persimilis*. *Oecologia* **61**: 135–142.

Bernstein, C., Kacelnik, A. & Krebs, J.R. (1988) Individual decisions and the distribution of predators in a patchy environment. *J. Anim. Ecol.* **57**: 1007–1026.

Bernstein, C., Kacelnik, A. & Krebs, J.R. (1991). Individual decisions and the distribution of predators in a patchy environment. II. The influence of travel costs and structure of the environment. *J. Anim. Ecol.* **60**: 205–225.

Bertram, B.C.R. (1978a) Living in groups: predators and prey. In: Krebs, J.R. & Davies, N.B. (eds) *Behavioural Ecology: An Evolutionary Approach*, pp. 64–96. Blackwell Scientific Publications, Oxford.

Bertram, B.C.R. (1978b) *Pride of Lions*. Dent, London.

Bertram, B.C.R. (1979) Serengeti predators and their social systems. In: Sinclair, A.R.E. & Norton-Griffiths, M. (eds) *Serengeti: Dynamics of an Ecosystem*, pp. 221–248. Chicago University Press, Chicago.

Bess, H.A., Spurr, G.H. & Littlefield, E.W. (1947) Forest site conditions and the gypsy moth. *Harvard For. Bull.* **22**: 56pp.

Best, M.R. & Meachum, C.L. (1986) The effects of stimulus pre-exposure on taste-mediated environmental conditioning: potentiation and overshadowing. *Anim. Learning Behav.* **14**: 1–5.

Beverton, R.J.H. (1985) Analysis of marine mammal–fisheries interactions. In: Beddington, J.R., Beverton, R.J.H. & Lavigne, D.M. (eds) *Marine Mammals and Fisheries*, pp. 3–33. George Allen & Unwin, London.

Biere, J.M. & Uetz, G.W. (1981) Web orientation in the spider *Micrathena gracilis* (Araneae: Araneidae). *Ecology* **62**: 336–344.

Bildstein, K.L. (1983) Why white-tailed deer flag their tails. *Am. Nat.* **121**: 709–715.

Bilsing, W.S. (1920) Quantitative studies in the food of spiders. *Ohio J. Sci.* **20**: 215–260.

Birch, L.C. (1948) The intrinsic rate of natural increase of an insect population. *J. Anim. Ecol.* **17**: 15–26.

Birkbeck, T.H. & Penn, C.W. (1986) *Antigenic Variation in Infectious Diseases*. Special Publications of the Society for General Microbiology, volume 19. IRL Press, Oxford.

Bishop, J.A., Cook, L.M. & Muggleton, J. (1978a) The response of two species of moths to industrialization in northwest England. I. Polymorphisms for melanism. *Phil. Trans. R. Soc. Lond. B.* **281**: 489–515.

Bishop, J.A., Cook, L.M. & Muggleton, J. (1978b) The response of two species of moths to industrialization in northwest England. II. Relative fitness of morphs and population size. *Phil. Trans. R. Soc. Lond. B.* **281**: 517–540.

Black, F.L. (1959) Measles antibodies in the population of New Haven, Connecticut. *J. Immunol.* **82**: 74–83.

Black, F.L. (1962). Measles antibody prevalence in diverse populations. *Am. J. Dis. Child.* **103**: 242–249.

Black, F.L. (1966). Measles endemicity in insular populations: critical community size and its evolutionary implications. *J. Theor. Biol.* **11**: 207–211.

Black, F.L., Hierholzer, W.J., Pinheiro, F.P., *et al.* (1974) Evidence for persistence of infectious agents in isolated human populations. *Am. J. Epidemiol.* **100**: 230–250.

Black, R. (1978) Tactics of whelks preying on limpets. *Marine Biol.* **46**: 157–162.

Blankinship, D.R. (1966) The relationship of white-winged dove production to control of great-tailed grackles in the lower Rio Grande Valley of Texas, *Transactions of the North American Wildlife and Natural Resources Conference* **31**: 45–58.

Blest, A.D. (1957) The function of eyespot patterns in the Lepidoptera. *Behaviour* **11**: 209–256.

Blum, M.S. (1981) *Chemical Defenses of Arthropods*. Academic Press, London.

Boertje, R.D., Gasaway, W.C., Grangaard, D.V. & Kellyhouse, D.G. (1988) Predation on moose and caribou by radio-collared grizzly bears in east central Alaska. *Can. J. Zool.* **66**: 2492–2499.

Boethel, D.J. & Eikenbarry, R.D. (1986) *Interactions of Plant Resistance and Parasitoids and Predators of Insects*. Ellis Horwood, Chichester.

Bond, A.B. (1980) Optimal foraging in a uniform habitat: the search mechanism of the green lacewing. *Anim. Behav.* **28**: 10–19.

Bond, A.B. (1983) Visual search and selection of natural stimuli in the pigeon: the attention threshold hypothesis. *J. Exp. Psychol: Anim. Behav. Proc.* **9**: 292–306.

Bortolotti, G.R. (1986) Influence of sibling competition on nestling sex ratios of sexually dimorphic birds. *Am. Nat.* **127**: 495–507.

Boucher, D.H. (1985) *The Biology of Mutualism: Ecology and Evolution.* Croom Helm, London.

Bouskila, A. & Blumstein, D.T. (1992) Rules of thumb for predation hazard assessment: predictions from a dynamic model. *Am. Nat.* **139**: 161–176.

Boutin, S., Krebs, C.J., Sinclair, A.R.E. & Smith, J.M.N. (1986) Proximate causes of losses in a snowshoe hare population. *Can. J. Zool.* **64**: 606–610.

Bouton, C.E., McPheron, B.A. & Weis, A.E. (1980) Parasitoids and competition. *Am. Nat.* **116**: 876–881.

Boyd, I.L. (1984) The relationship between body condition and the timing of implantation in pregnant grey seals (*Halichoerus grypus*). *J. Zool. Lond.* **203**: 113–123.

Braendegaard, J. (1937) Observations on spiders starting off on 'ballooning excursions'. *Vidensk Meddrdansk Naturh Foren KBH* **101**: 115–117.

Braithwaite, R.W. & Lee, A.K. (1979) A mammalian example of semelparity. *Am. Nat.* **113**: 151–156.

Brakefield, P.M. (1985) Polymorphic Müllerian mimicry and interactions with thermal melanism in ladybirds and a soldier beetle: a hypothesis. *Biol. J. Linnean Soc.* **26**: 243–267.

Brakefield, P.M. (1987) Industrial melanism: do we have the answers? *Trends Evol. Ecol.* **2**: 117–122.

Braune, H.J. (1982) Effect of the structure of the host egg-mass on the effectiveness of an egg parasite of *Spodoptera litura* (F.) (Lepidoptera, Noctuidae). *Drosera* **1**: 7–16.

Brayley, E.W. (ed.) (1722) '*A Journal of the Plague Year; or Memorials of the Great Pestilence in London, in 1665, by Daniel Defoe*'. Thomas Tegg, London.

Bremermann, H.J. (1980) Sex and polymorphism as strategies in host–pathogen interactions. *J. Theor. Biol.* **87**: 671–702.

Bremermann, H.J. & Fiedler, B. (1985) On the stability of polymorphic host–pathogen populations. *J. Theor. Biol.* **117**: 621–631.

Brodie, P.F. (1975) Cetacean energetics, an overview of intraspecific size variation. *Ecology* **56**: 152–161.

Brooks, D.R. (1988) Macroevolutionary comparisons of host and parasite phylogenies. *Ann. Rev. Ecol. Syst.* **19**: 235–259.

Brooks, D.R. & Glen, D.R. (1982) Pinworms and primates: a case study in coevolution. *Proc. Helminthol. Soc. Wash.* **49**: 76–85.

Brooks, D.R., O'Grady, R.T. & Glen, D.R. (1985) Phylogenetic analysis of the Digenea (Platyhelminthes: Cercomeria) with comments on their adaptive radiation. *Can. J. Zool.* **63**: 411–443.

Brooks, J.L. & Dodson, S.I. (1965) Predation, body size and composition of plankton. *Science* **150**: 28–35.

Brower, L.P. (1969) Ecological chemistry. *Sci. Am.* **220**: 22–29.

Brower, L.P. (1984) Chemical defence in butterflies. In: Vane-Wright, R.I. & Ackery, P.R. (ed.) *The Biology of Butterflies*, pp. 109–134. Academic Press, London.

Brower, L.P. (1988) Avain predation on the monarch butterfly and its implications for mimicry theory. *Am. Nat.* **131**: S4–S6.

Brower, L.P., Brower, J.V.Z., Stiles, F.G., Croze, H.J. & Hower, A.S. (1964) Mimicry: differential advantage of color patterns in the natural environment. *Science* **144**: 183–185.

Brower, L.P. & Calvert, W.H. (1985) Foraging dynamics of bird predators on overwintering monarch butterflies in Mexico. *Evolution* **39**: 852–868.

Brower, L.P., Cook, L.M. & Croze, H.J. (1967) Predator responses to artificial Batesian mimics released in a neotropical environment. *Evolution* **21**: 11–23.

Brower, L.P. & Fink, L.S. (1985) A natural toxic defense system: cardenolides in butterflies versus birds. *Ann. NY Acad. Sci.* **443**: 171–186.

Brower, L.P., Horner, B.E., Marty, M.A., Moffitt, C.M. & Villa-R.B. (1985) Mice (*Peromyscus maniculatus, P. spicilegus,* and *Microtus mexicanus*) as predators of overwintering monarch butterflies (*Danaus plexippus*) in Mexico. *Biotropica* **17**:

89–99.

Brower, L.P., Nelson, C.J., Seiber, J.N., Fink, L.S. & Bond, C. (1988) Exaptation as an alternative to coevolution in the cardenolide-based chemical defense of monarch butterflies (*Danaus plexippus* L.) against avian predators. In: Spencer, K.C. (ed.) *Chemical Mediation of Coevolution*, pp. 447–475. Academic Press, San Diego.

Brower, L.P., Ryerson, W.N., Coppinger, L.L. & Glazier, S.C. (1968) Ecological chemistry and the palatability spectrum. *Science* **161**: 1349–1351.

Brown, C.R. (1986) Cliff swallow colonies as information centers. *Science* **234**: 83–85.

Brown, C.R. & Brown, M.B. (1986) Ectoparasitism as a cost of coloniality in cliff swallows (*Hirundo pyrrhonota*). *Ecology* **67**: 1206–1218.

Brown, J.S. (1988) Patch use as an indicator of habitat preference, predation risk, and competition. *Behav. Ecol. Sociobiol.* **22**: 37–47.

Brown, M.W. & Cameron, E.A. (1979) Effects of dispalure and egg mass size on parasitism by the gypsy moth egg parasite, *Ooencyrtus kuwani*. *Environ. Entomol.* **8**: 77–80.

Bryan, K.M. & Wratten, S.D. (1984) The responses of polyphagous predators to prey spatial heterogeneity: aggregation by carabid and staphylinid beetles to cereal aphid prey. *Ecol. Entomol.* **9**: 251–259.

Bryden, M.M. (1972) Growth and development of marine mammals. In: Harrison, R.J. (ed.) *Functional Anatomy of Marine Mammals*, Vol. 1, pp. 1–80.

Buchler, E.R. (1976) The use of echolocation by the wandering shrew *Sorex vagrans. Anim. Behav.* **24**: 858–873.

Buckner, C.H. (1955) Small mammals as predators of sawflies. *Can. Entomol.* **87**: 121–123.

Buckner, C.H. (1958) Mammalian predators of the larch sawfly in eastern Manitoba. *Proceedings of the International Congress of Entomology, Montreal* **4**: 353–361.

Buckner, C.H. (1959) Mortality of cocoons of the larch sawfly, *Pristiphora erichsonii* (Htg.), in relation to distance from small-mammal tunnels. *Can Entomol.* **91**: 535–542.

Buckner, C.H. (1964) Metabolism, food

consumption, and feeding behaviour in four species of shrews. *Can. J. Zool.* **42**: 259–279.

Buckner, C.H. (1966) The role of vertebrate predators in the biological control of forest insects. *Ann. Rev. Entomol.* **11**: 449–470.

Buckner, C.H. (1969) Some aspects of the population ecology of the common shrew *Sorex araneus* near Oxford, England. *J. Mammal.* **50**: 326–332.

Bullock, T.H. (1953) Predator-recognition and escape responses of some intertidal gastropods in the presence of starfish. *Behaviour* **5**: 130–140.

Bulmer, M.G. (1986) Sex ratios in geographically structured populations. *Trends Ecol. Evol.* **1**: 35–38.

Burkot, T.R., Narara, A., Paru, R., Graves, P.M. & Garner, P. (1989) Human host selection by anophelines: no evidence for preferential selection of malaria or microfilariae infected individuals in a hyperendemic area. *Parasitology* **98**: 337–342.

Burnett, T. (1958) A model of host–parasite interaction. *Proceedings of the 10th International Congress of Entomology* **2**: 679–686.

Bush, R.R. & Mosteller, F. (1951) A mathematical model for simple learning. *Psychol. Rev.* **68**: 313–323.

Bushman, L.L., Whitcomb, W.H., Hemenway, R.C. *et al.* (1976) Predators of the velvetbean caterpillar eggs in Florida soybeans. *Environ. Entomol.* **6**: 403–407.

Butcher, G.S. & Rohwer, S. (1989) The evolution of conspicuous and distinctive coloration for communication in birds. *Curr. Ornithol.* **6**: 51–108.

Butterfield, J., Coulson, J.C. & Wanless, G. (1981) Studies on the distribution, food, breeding biology and relative abundance of the pygmy and common shrews (*Sorex minutus* and *S. araneus*) in upland areas of northern England. *J. Zool. Lond.* **195**: 169–180.

Butterworth, D.S. (1988) A simulation study of krill fishing by an individual Japanese trawler. Submitted to the 1988 CCAMLR Scientific Committee meeting.

Byers, J.A. & Byers, K.Z. (1983) Do pronghorn mothers reveal the locations of their hidden fawns?

Behav. Ecol. Sociobiol. **13**: 147–156.

Caldwell, G.S. & Rubinoff, R.W. (1983) Avoidance of venomous sea snakes by naive herons and egrets. *Auk* **100**: 195–198.

Callan, E.McC. (1944) A note on *Phanuropsis semiflaviventris* Girault (Hym: Scelionidae), an egg parasite of cacao stink-bugs. *Proc. R. Ent. Soc.* (*Lond.*) Series A **19**: 48–49.

Calow, P. (1981) *Invertebrate Biology*. Croom Helm, London.

Calow, P. & Jennings, J.B. (1974) Calorific values in the phylum Platyhelminthes: the relationship between potential energy, mode of life and the evolution of entoparasitism. *Biol. Bull.* **147**: 81–94.

Caltagirone, L.E. (1981) Landmark examples in classical biological control. *Ann. Rev. Entomol.* **26**: 213–232.

Caltagirone, L.E. & Doutt, R.L. (1989) The history of the *Vedalia* beetle importation to California and its impact on the development of biological control. *Ann. Rev. Entomol.* **34**: 1–16.

Calvert, W.H., Hedrick, L.E. & Brower, L.P. (1979) Mortality of the monarch butterfly (*Danaus plexippus* L.): avian predation at five overwintering sites in Mexico. *Science* **204**: 847–851.

Campbell, R.W. & Sloan, R.J. (1977a) Release of gypsy moth populations from innocuous levels. *Environ. Entomol.* **6**: 323–330.

Campbell, R.W. & Sloan, R.J. (1977b) Natural regulation of innocuous gypsy moth populations. *Environ. Entomol.* **6**: 315–322.

Caraco, T., Blanckenhorn, W.U., Gregory, G.M., Newman, J.A., Recer, G.M. & Zwicker, S.M. (1990) Risk-sensitivity: ambient temperature affects foraging choice. *Anim. Behav.* **39**: 338–345.

Caraco, T., Martindale, S. & Whittam, T.S. (1980) An empirical demonstration of risk-sensitive foraging preferences. *Anim. Behav.* **28**: 820–830.

Caraco, T. & Wolf, L.L. (1975) Ecological determinants of group sizes in foraging lions. *Am. Nat.* **109**: 343–352.

Caro, T.M. (1980) Effects of experience on the predatory patterns of cats. *Behav. Neural Biol.* **23**: 1–28.

Caro, T.M. (1986a) The functions of stotting: a review of the hypotheses.

Anim. Behav. **34**: 649–662.

Caro, T.M. (1986b) The functions of stotting in Thomson's gazelles: some tests of the predictions. *Anim. Behav.* **34**: 663–684.

Caro, T.M. (1987a) Indirect costs of play: cheetah cubs reduce maternal hunting success. *Anim. Behav.* **35**: 295–297.

Caro, T.M. (1987b) Cheetah mothers' vigilance: looking out for prey or for predators? *Behav. Ecol. Sociobiol.* **20**: 351–360.

Caro, T.M. (1989) Determinants of asociality in felids. In: Standen, V. & Foley, R.A. (eds) *Comparative Socioecology: The Behavioural Ecology of Humans and Other Mammals*, pp. 41–74. Blackwell Scientific Publications, Oxford.

Caro, T.M. (under review) *Cheetahs of the Serengeti Plains: Grouping in an Asocial Species*. Chicago University Press, Chicago.

Caro, T.M. & Collins, D.A. (1987) Male cheetah social organization and territoriality. *Ethology* **74**: 52–64.

Carrel, J.E. (1978) Behavioral thermoregulation during winter in an orb-weaving spider. *Symp. Zool. Soc. Lond.* **2**: 41–50.

Carroll, W.J. (1964) Predation of larch sawfly cocoons by the introduced shrew, *Sorex cinereus cinereus* (Kerr.), in Newfoundland. *Bi-monthly Progress Report, Dept of Forestry* **20**: 1–2.

Cartar, R.V. & Dill, L.M. (1990) Why are bumble bees risk-sensitise foragers? *Behav. Ecol. Sociobiol.* **26**: 121–127.

Carter, M.C. & Dixon, A.F.G. (1984) Honeydew: an arrestant stimulus for coccinellids. *Ecol. Entomol.* **9**: 383–387.

Cavé, A.J. (1968) The breeding of the kestrel, *Falco tinnunculus* L., in the reclaimed area Oostelijk Flevoland. *Netherlands J. Zool.* **18**: 313–407.

Chabaud, A.G. (1950) L'infestation par des Ixodines provoque-t-elles une immunité chez l'hote? *Ann. Parasitol.* **25**: 474–479.

Charnov, E.L. (1976) Optimal foraging: the marginal value theorem. *Theor. Pop. Biol.* **9**: 129–136.

Charnov, E.L. (1979) The genetical evolution of patterns of sexuality: Darwinian fitness. *Am. Nat.* **113**: 465–480.

Charnov, E.L., Los-den Hartogh, R.L., Jones, W.T. & van den Assem, J.

(1981) Sex ratio evolution in a variable environment. *Nature* **289**: 27–33.

Charnov, E.L. & Skinner, S.W. (1984) Evolution of host selection and clutch size in parasitic wasps. *Florida Entomol.* **67**: 5–21.

Charnov, E.L. & Stephens, D.W. (1988) On the evolution of host selection in solitary parasitoids. *Am. Nat.* **132**: 707–722.

Cheira, J.W., Newson, R.M. & Cunningham, M.P. (1985) Cumulative effect of host resistance on *Rhipicephalus appendiculatus* Neumann (Acarini, Ixodidae) in the laboratory. *Parasitology*, **90**: 401–408.

Chesness, R.A., Nelson, M.M. & Longley, W.H. (1968) The effect of predator removal on pheasant reproductive success. *J. Wildlife Manage.* **32**: 683–697.

Chesson, J. (1978) Measuring preference in selective predation. *Ecology* **59**: 211–215.

Chesson, P.L. (1985) Coexistence of competitors in spatially and temporally varying environments: a look at the combined effects of different sorts of variability. *Theor. Pop. Biol.* **28**: 263–287.

Chesson, P.L. & Murdoch, W.W. (1986) Aggregation of risk: relationships among host–parasitoid models. *Am. Nat.* **127**: 696–715.

Cheverton, J., Kacelnik, A. & Krebs, J.R. (1985) Optimal foraging: constraints and currencies. In: Hölldobler, B. & Lindauer, M. (eds) *Experimental Behavioural Ecology*, pp. 109–126. G. Fischer-Verlag, Stuttgart.

Chitty, H. (1950) Canadian Arctic wildlife enquiry, 1943–49: with a summary of results since 1933. *J. Anim. Ecol.* **19**: 180–193.

Chivers, D. & Hladik, C.M. (1980) Morphology of the gastrointestinal tract in primates: comparisons with other mammals in relation to diet. *J. Morphol.* **166**: 337–386.

Chiverton, P.A. (1986) Predator density manipulation and its effects on populations of *Rhopalosiphum padi* (Hom.: Aphididae) in spring barley. *Ann. Appl. Biol.* **109**: 49–60.

Christophers, S.R. (1960) *Aedes aegypti (L.), The Yellow Fever Mosquito.* Cambridge University Press, Cambridge.

Churchfield, S. (1982) Food availability and the diet of the common shrew *Sorex araneus* in Britain. *J. Anim. Ecol.* **51**: 15–28.

Clark, C.W. (1990) *Mathematical Bioeconomics*, 2nd edn. John Wiley & Sons, New York.

Clarke, B. (1976) The ecological genetics of host–parasite relationships. In: Taylor, A.E.R. & Muller, R. (eds) *Genetic Aspects of Host–Parasite Relationships. Symposia of the British Society for Parasitology* **14**: 87–103. Blackwell Scientific Publications, Oxford.

Clarke, B.C. (1979) The evolution of genetic diversity. *Proc. R. Soc. Lond. B* **205**: 453–474.

Clarke, J.C., Westbrook, R.F. & Irwin, J. (1979) Potentiation instead of overshadowing in the pigeon. *Behav. Neural Biol.* **25**: 18–29.

Clarke, R.D. & Grant, P.R. (1968) An experimental study of the role of spiders as predators in a forest litter community. Part 1. *Ecology* **49**: 1152–1154.

Clausen, C.P. (1940) *Entomophagous Insects.* Hafner, New York.

Clausen, C.P. (1972) *Entomophagous Insects.* Hafner Publishing Company, New York.

Cliff, A.D., Haggett, P. & Ord, J.K. (1986) *Spatial Aspects of Influenza Epidemics.* Pion Ltd, London.

Clutton-Brock, T.H., Albon, S.D. & Guinness, F.E. (1988) Reproductive success in male and female red deer. In: Clutton-Brock, T.H. (ed.) *Reproductive Success: Studies of Individual Variation in Contrasting Breeding Systems*, pp. 325–343. Chicago University Press, Chicago.

Clutton-Brock, T.H. & Guinness, F.E. (1975) Behaviour of red deer (*Cervus elaphus*) at calving time. *Behaviour* **55**: 287–300.

Cock, M.J.W. (1978) The assessment of preference. *J. Anim. Ecol.* **47**: 805–816.

Cockburn, A., Scott, M.P. & Dickman, C.R. (1985) Sex ratio and intrasexual kin competition in mammals. *Oecologia* **66** 427–429.

Coddington, J.A. (1986a) The genera of the spider family Theridiosomatidae. *Smithsonian Contributions to Zoology* **422**: 1–96.

Coddington, J.A. (1986b) The monophyletic origin of the orb web. In: Shear, W.A. (ed.) *Spider Webs and Spider Behavior*, pp. 319–363.

Stanford University Press, Palo Alto.

Cole, L.C. (1954) The population consequences of life history phenomena. *Q. Rev. Biol.* **29**: 103–137.

Coley, P.D., Bryant, J.P. & Chapin, F.S. III (1985) Resource availability and plant antiherbivore defence. *Science* **230**: 895–899.

Comins, H.N. & Blatt, D.W.E. (1974) Prey–predator models in spatially heterogeneous environments. *J. Theor. Biol.* **48**: 75–83.

Comins, H.N. & Hassell, M.P. (1976) Predation in multiprey communities. *J. Theor. Biol.* **62**: 93–114.

Comins, H.N. & Hassell, M.P. (1979) The dynamics of optimally foraging predators and parasitoids. *J. Anim. Ecol.* **48**: 335–351.

Comins, H.N. & Hassell, M.P. (1987) The dynamics of predation and competition in patchy environments. *Theor. Pop. Biol.* **31**: 393–421.

Commons, M.L., Kacelnik, A. & Shettleworth, S.J. (eds) (1987) *Quantitative Analyses of Behavior, Vol. 6: Foraging.* L. Erlbaum, Hillsdale.

Connell, J.H. (1961) Effects of competition, predation by *Thais lapillus* and other factors on natural populations of the barnacle *Balanus balanoides*. *Ecol. Monogr.* **31**: 61–104.

Connell, J.H. (1970) A predator–prey system in the marine intertidal region. I. *Balanus glandula* and several predatory species of *Thais*. *Ecol. Monogr.* **40**: 49–78.

Connell, J.H. (1972) Community interactions on marine rocky intertidal shores. *Ann. Rev. Ecol. Systemat.* **3**: 169–192.

Connell, J.H. (1975) Some mechanisms producing structure in natural communities: a model and evidence from field experiments. In: Cody, M.L. & Diamond, J.M. (eds) *Ecology and Evolution of Communities*, pp. 460–490. Harvard University Press, Cambridge, MA.

Conway, G.R. (1973) Ecological aspects of pest control in Malaysia. In: Farvar, J.T. & Milton, J.P. (eds) *The Careless Technology*, pp. 467–488. Tom Stacey Ltd, London.

Cook, R.M. & Cockrell, B.J. (1978) Predator ingestion rate and its bearing on feeding time and the theory of optimal diets. *J. Anim. Ecol.* **47**: 529–547.

Cook, R.M. & Hubbard, S.F. (1977) Adaptive searching strategies in insect parasites. *J. Anim. Ecol.* **46**: 115–125.

Cook, R.S., White, M., Trainer, D.O. & Glazener, W.C. (1971) Mortality of young white-tailed deer fawns in south Texas. *J. Wildlife Manage.* **35**: 47–56.

Cope, E.D. (1885) On the evolution of the vertebrata, progressive and retrogressive. *Am. Nat.* **19**: 140–353.

Corbet, G.B. & Hill, J.E. (1980) *A World List of Mammalian Species*. British Museum (Natural History), London.

Corbet, S.A. (1973) Concentration effects and the response of *Nemeritis canescens* to a secretion of its host. *J. Insect Physiol.* **19**: 2119–2128.

Corbett, L.K. & Newsome, A.E. (1987) The feeding ecology of the dingo. III. Dietary relationships with widely fluctuating prey populations in arid Australia: an hypothesis of alternation of predation. *Oecologia* **74**: 215–227.

Cornelius, M. & Barlow, C.A. (1980) Effect of aphid consumption by larvae on development and reproductive efficiency of a flower fly, *Syrphus corollae* (Diptera: Syrphidae). *Can. Entomol.* **112**: 989–992.

Cott, H.B. (1940) *Adaptive Coloration in Animals*. Methuen, London.

Cowan, I. McT. (1947) The timber wolf in the Rocky Mountain national parks of Canada. *Can. J. Res.* **25**: 139–174.

Cragg, J.B. (1961) Some aspects of the ecology of moorland animals. *J. Anim. Ecol.* **30**: 205–234.

Cramer, N.F. & May, R.M. (1972) Interspecific competition, predation and species diversity: a comment. *J. Theor. Biol.* **34**: 289–290.

Crawley, M.J. (1975) The numerical responses of insect predators to changes in prey density. *J. Anim. Ecol.* **44**: 877–892.

Crawley, M.J. (1983) *Herbivory. The Dynamics of Animal–Plant Interactions*. Blackwell Scientific Publications, Oxford.

Crawley, M.J. (1989) The relative importance of vertebrate and invertebrate herbivores in plant population dynamics. In: Bernays, E.A. (ed.) *Insect–Plant Interactions 1*, pp. 45–71. CRC Press, Boca Raton.

Crawley, M.J. (1989) Insect herbivores and plant population dynamics. *Ann.*

Rev. Ent. **34**, 531–564.

Crawley, M.J. (1990) Plant population dynamics. *Phil. Trans. R. Soc. Lond. B* **330**: 125–140.

Crawley, M.J. & Akhteruzzaman, M. (1988) Individual variation in the phenology of oak trees and its consequences for herbivorous insects. *Functional Ecol.* **2**: 409–415.

Crawley, M.J. & Pattrasudhi, R. (1988) Interspecific competition between insect herbivores: asymmetric competition between cinnabar moth and the ragwort seedhead fly. *Ecol. Entomol.* **13**: 243–249.

Crisler, L. (1956) Observations of wolves hunting caribou. *J. Mammal.* **37**: 337–346.

Crisp, D.J. (1964) An assessment of plankton grazing by barnacles. In: Crisp, D.J. (ed.) *Grazing in Terrestrial and Marine Environments*, pp. 251–264. Blackwell Scientific Publications, Oxford.

Crisp, M. (1969) Studies on the behavior of *Nassarius obsoletus* (Say.) (Mollusca: Gastropoda). *Biol. Bull.* **136**: 355–373.

Crofton, H.D. (1971a) A quantitative approach to parasitism. *Parasitology* **63**: 179–193.

Crofton, H.D. (1971b) A model of host–parasite relationships. *Parasitology* **63**: 343–364.

Croin Michielsen, N. (1966) Intraspecific and interspecific competition in the shrews *Sorex araneus* L. and *S. minutus* L. *Arch. Neerl. Zool.* **17**: 73–174.

Crombie, A.C. (1945) On competition between different species of graminivorous insects. *Proc. R. Soc. Lond. B* **132**: 362–395.

Crombie, J.A. & Anderson, R.M. (1985) Population dynamics of *Schistosoma mansoni* in mice repeatedly exposed to infection. *Nature* **315**: 491–493.

Crompton, D.W.T. (1985) Reproduction. In: Crompton, D.W.T. & Nickol, B.B. (eds) *Biology of Acanthocephala*, pp. 213–272. Cambridge University Press, Cambridge.

Crompton, D.W.T. & Harrison, J.G. (1965) Observations on *Polymorphus minutus* (Goeze, 1782) (Acanthocephala) from a wildfowl reserve in Kent. *Parasitology* **55**: 345–355.

Crompton, D.W. & Joyner, S.M. (1980) *Parasitic Worms*. Wykeham Publications Ltd, London.

Crowley, P.H. (1981) Dispersal and the stability of predator–prey interactions. *Am. Nat.* **118**: 673–701.

Crowley, P.H., Nisbet, R.M., Gurney, W.S.C. & Lawton, J.H. (1987) Population regulation in animals with complex life histories: formulation and analysis of a damselfly model. *Adv. Ecol. Res.* **17**: 1–59.

Cruickshank, R., Duguid, J.P., Marmion, B.P. & Swain, R.H.A. (1973) *Medical microbiology*, 12th edn, vol. 1. Churchill Livingstone, London.

Cunningham, P.N. & Hughes, R.N. (1984) Learning of predatory skills by shorecrabs *Carcinus maenas* feeding on mussels and dogwhelks. *Marine Ecol. Prog. Ser.* **16**: 21–26.

Curio, E. (1976) *The Ethology of Predation*. Springer-Verlag, Berlin.

Curry, G.L. & DeMichele, D.W. (1977) Stochastic analysis for the description and synthesis of predator–prey systems. *Can. Entomol.* **109**: 1167–1174.

Cuthill, I.C. (1985) *Experimental studies of optimal foraging theory*. Unpublished D. Phil. thesis, Oxford University.

Cuthill, I.C. & Kacelnick, A. (1991) Central place foraging: a re-appraisal of the 'loading effect'. *Anim. Behav.* **40**: 1087–1101.

Cuthill, I.C. & Kacelnik, A., Krebs, J.R., Haccou, P. & Iwasa, Y. (1991) Patch use by startlings: the effect of recent experience on foraging decisions. *Anim. Behav.* (in press).

Czaika, M. (1963) Unknown facts of the biology of the spider *Ero furcata* (Villers) (Mimetidae, Araneae). *Pol. Pismo Entomol.* **33**: 229–231.

Dabis, F., Abdourahmane, S., Waldman, R.J., *et al.* (1988) The epidemiology of measles in a partially vaccinated population in an african city: implications for immunization programs. *Am. J. Epidemiol.* **127**: 171–178.

Dabrowska-Prot, E. & Luczak J. (1968) Studies on the incidence of mosquitos in the food of *Tetragnatha montana* Simon and its food activity in the natural habitat. *Ekol. Pol. Ser. A* **16**: 843–853.

Dahlsten, D.L. & Mills, N.J. (1992) Biological control of forest insects. In: Fisher, T.W., Bellows, T. & Gordh, G. (eds) *Principles and Applications of Biological Control*. University of California Press (in press).

Daniel, T.L. & Kingsolver, J.G. (1983) Feeding strategy and the mechanics of blood sucking in insects. *J. Theor. Biol.* **105**: 661–672.

Danks, H.V. (1987) *Insect Dormancy: An Ecological Perspective*, p. 125. Biological Survey of Canada Monograph Series no. 1, Entomological Society of Canada, Ottawa.

Dare, P. (1961) *Ecological observations on a breeding population of the common buzzard* Buteo buteo. PhD thesis, Exeter University.

Darwin, C. (1859) *On the Origin of Species*. John Murray, London. Facsimile of 1st edn, with an introduction by E. Mayr: Harvard University Press, Cambridge, MA, 1964.

Davidson, D.W., Samson, D.A. & Inouye, R.S. (1985) Granivore in the Chihuahuan Desert: interactions within and between trophic levels. *Ecology* **66**: 486–502.

Davies, N.B. & Houston, A.I. (1984) Territorial economics. In: Krebs, J.R. & Davies, N.B. (eds) *Behavioural Ecology: An Evolutionary Approach*, pp. 148–169. Blackwell Scientific Publications, Oxford.

Davies, N.B. & Lundberg, A. (1984) Food distribution and a variable mating system in the Dunnock *Prunella modularis. J. Anim. Ecol.* **53**: 895–912.

Davis, R.B., Herreid, II. C.F. & Short, H.L. (1962) Mexican free-tailed bats in Texas. *Ecol. Monogr.* **32**: 311–346.

Dawbin, W.H. (1966) The seasonal migratory cycle of humpback whales. In: Norris, K.S. (ed.) *Whales, Dolphins and Porpoises*. University of California Press, Berkeley.

Dawkins, M. (1971) Perceptual changes in chicks: another look at the 'search image' concept. *Anim. Behav.* **19**: 566–574.

Dawkins, R. (1982) *The Extended Phenotype*. W.H. Freeman & Co., Oxford.

Dawkins, R. & Krebs, J.R. (1979) Arms races between and within species. *Proc. R. Soc. Lond. B* **205**: 489–511.

Dawkins, R.D. (1969) The attention threshold model. *Anim. Behav.* **17**: 134–141.

Day, J.F. & Edman, J.D. (1983) Malaria renders mice susceptible to mosquito feeding when gametocytes are most infective. *J. Parasitol.* **69**: 163–170.

Day, P.R. (1974) *Genetics of Host–Parasite Interactions*. Freeman, San Francisco.

Day, P.R. & Jellis, G.J. (eds) (1987) *Genetics and Plant Pathogenesis*. Blackwell Scientific Publications, Oxford.

Dayton, P.K. (1971) Competition, disturbance and community organization: the provision and subsequent utilization of space in a rocky intertidal community. *Ecol. Monogr.* **41**: 351–389.

Dayton, P.K. (1973) Two cases of resource partitioning in an intertidal community: making the right predictions for the wrong reason. *Am. Nat.* **107**: 662–670.

DeAngelis, D.L. (1992) *Dynamics of Nutrient Cycling and Food Webs*. Chapman & Hall, London.

Dean, J.M. & Ricklefs, R.E. (1979) Do parasites of Lepidoptera larvae compete for hosts? No! *Am. Nat.* **113**: 302–306.

Dean, J.M. & Ricklefs, R.E. (1980) Do parasites of Lepidoptera larvae compete for hosts? No evidence. *Am. Nat.* **116**: 882–884.

DeAngelis, D.L. (1975) Stability and connectance in food web models. *Ecology* **56**: 238–243.

Debach, P. (1974) The role of weather and entomophagous species in the natural control of insect populations. *J. Econ. Entomol.* **51**: 474–484.

Deevey, E.S. (1947) Life tables for natural populations of animals. *Q. Rev. Biol.* **22**: 283–314.

Delucchi, V. (1982) Parasitoids and hyperparasitoids of *Zeiraphera diniana* (Lep. Tortricidae) and their role in population control in outbreak areas. *Entomophaga* **27**: 77–92.

Dempster, J.P. (1975) *Animal Population Ecology*. Academic Press, London.

Dempster, J.P. (1983) The natural control of populations of butterflies and moths. *Biol. Rev.* **58**: 461–481.

Dempster, J.P. & Pollard, E. (1986) Spatial heterogeneity, stochasticity and the detection of density dependence in animal populations. *Oikos* **46**: 413–416.

Den Boer, P.J. (1971) Stabilization of animal numbers and the heterogeneity of the environment: the problem of the persistence of sparse populations. In: den Boer, P.J. & Gradwell, G.R. (eds) *Proceedings of* the Advanced Study Institute on Dynamics of Numbers in Populations, pp. 77–97. Oosterbeek, The Netherlands.

Den Boer, P.J. (1990) Density limits and survival of local populations in 64 carabid species with different powers of dispersal. *J. Evol. Biol.* **3**: 19–48.

Den Boer, P.J., Van Huizen, T.H.P., Den Boer-Daanje, W., Aukema, B. & Den Bieman, C.F.M. (1980) Wing polymorphism and dimorphism as stages in an evolutionary process (Coleoptera: Carabidae). *Entomologia Generalis* **6**: 107–134.

Desender, K. (1989) Heritability of wing development and body size in a carabid beetle, *Pogonus chalceus* Marsham, and its evolutionary significance. *Oecologia* **78**: 513–520.

Dethier, V.G. (1954) Evolution of feeding preferences in phytophagous insects. *Evolution* **8**: 33–54.

De Vries, P.J. (1990) Enhancement of symbioses between butterfly caterpillars and ants by vibrational communication. *Science* **248**: 1104–1106.

De Vries, Tj. (1975) The breeding biology of the Galapagos hawk *Buteo galapagoensis*. *Le Gerfant* **65**: 29–57.

Diamond, J. & Case, T.J. (eds) (1986) *Community Ecology*. Harper & Row, New York.

Dice, R.L. (1947) Effectiveness of selection by owls on deer mice which contrast in colour with their background. *Laboratory of Vertebrate Biology of the University of Michigan* **34**: 1–20.

Dicke, M. & Sabelis, M.W. (1988) How plants obtain predatory mites as bodyguards. *Netherlands J. Zool.* **38**: 148–165.

Dicke, M. & Sabelis, M.W. (1989) Does it pay plants to advertize for bodyguards? Towards a cost–benefit analysis of induced synomone production. In: Lambers, H., Konings, H., Cambridge, M.L. & Pons, Th.L. (eds) *Consequences of Variation in Growth Rate and Productivity of Higher Plants*, pp. 341–358. SPB Academic Publishing BV, The Hague.

Dicke, M. & Sabelis, M.W. (1992) Costs and benefits of chemical information conveyance: proximate and ultimate aspects. In: Roitberg, B.D. & Isman, M. (eds) *Evolutionary Perspectives in Insect Chemical Ecology*. Chapman and Hall, London.

Dicke, M., Sabelis, M.W. & De Jong, M. (1988) Analysis of prey preference in phytoseiid mites by using an olfactometer, predation models and electrophoresis. *Exp. Appl. Acarol.* **5**: 225–241.

Dicke, M., Sabelis, M.W., De Jong, M. & Alers, M.P.T. (1990) Do phytoseiid mites select the best prey species in terms of reproductive success? *Exp. Appl. Acarol.* **8**: 161–173.

Dicke, M., Sabelis, M.W., Takabayashi, J., Bruin, J. & Posthumus, M.A. (1991) Plant strategies of manipulating predator–prey interactions through allelochemicals: prospects for application in pest control. *J. Chem. Ecol.* **16**: 3091–3118.

Dicke, M., Sabelis, M.W. & Van de Berg, H. (1989) Does prey preference change as a result of prey species being presented together? Analysis of prey selection by the predatory mite *Typhlodromus pyri* (Acarina: Phytoseiidae). *Oecologia* **81**: 302–309.

Dickman, C.R. (1986a) An experimental study of competition between two species of dasyurid marsupials. *Ecol. Monogr.* **56**: 221–241.

Dickman, C.R. (1986b) An experimental manipulation of the intensity of interspecific competition: effects on a small marsupial. *Oecologia* **70**: 536–543.

Dickman, C.R. (1988a) Body size, prey size, and community structure in insectivorous mammals. *Ecology* **69**: 569–580.

Dickman, C.R. (1988b) Sex-ratio variation in response to interspecific competition. *Am. Nat.* **132**: 289–297.

Dickman, C.R. (1989) Patterns in the structure and diversity of marsupial carnivore communities. In: Abramsky, Z., Fox, B.J., Morris, D.W. & Willig, M. (eds) *Patterns in the Structure of Mammalian Communities*. Texas Technical University Press, Texas. (In press).

Dickman, C.R., Green, K., Carron, P.L., Happold, D.C.D. & Osborne, W.S. (1983) Coexistence, convergence and competition among *Antechinus* (Marsupialia) in the Australian high country. *Proc. Ecol. Soc. Austr.* **12**: 79–99.

Diekmann, O., Metz, J.A.J. & Sabelis, M.W. (1988) Mathematical models of predator–prey–plant interactions in a patchy environment. *Exp. Appl.*

Acarol. **5**: 319–342.

Diekmann, O., Metz, J.A.J. & Sabelis, M.W. (1989) Reflections and calculations on a prey–predator-patch problem. *Acta Applicandae Mathematicae* **14**: 23–25.

Dietz, K. & Hadeler, K.P. (1988) Epidemiological models for sexually transmitted infections. *J. Mathemat. Biol.* **26**: 1–25.

Dietz, R., Heide-Jorgensen, P. & Harkonen, T. (1989). Mass death of harbour seals (*Phoca vitulina*) in Europe. *AMBIO* **18**: 258–64.

Dijkstra C. (1988) *Reproductive Tactics in the kestrel* Falco tinnunculus. Drukkerij van Denderen BV, Groningen.

Dill, L.M. (1987) Animal decision making and its ecological consequences: the future of aquatic ecology and behaviour. *Can. J. Zool.* **65**: 803–811.

Dill, L.M. & Fraser, A.H.G. (1984) Risk of predation and the feeding behaviour of juvenile coho salmon (*Oncorhynchus kisutch*). *Behav. Ecol. Sociobiol.* **16**: 65–71.

Dixon, A.F.G. (1970) Factors limiting the effectiveness of the coccinellid beetle, *Adalia bipunctata* (L.), as a predator of the sycamore aphid, *Drepanosiphum platanoides* (Schr.) *J. Anim. Ecol.* **39**: 739–751

Dobson, A.P. (1982) Comparisons of some characteristics of the life histories of microparasites, macroparasites, parasitoids, and predators. In: Anderson, R.M. & May, R.M. (eds) *Population Biology of Infectious Diseases*, p. 5. Springer-Verlag, New York.

Dobson, A.P. (1985) The population dynamics of competition between parasites. *Parasitology* **91**: 317–347.

Dobson, A.P. (1988a) The population biology of parasite-induced changes in host behavior. *Q. Rev. Biol.* **63**: 139–165.

Dobson, A.P. (1988b) Restoring island ecosystems: The potential of parasites to control introduced mammals. *Conservation Biol.* **2**: 31–39.

Dobson, A.P. & Hudson, P.J. (1992) The population dynamics and control of the parasitic nematode *Trichostrongylus tenuis* in red grouse in the North of England. In: Jain, S. & Botsford, L. (eds) *Applied Population Biology*. Springer-Verlag, Berlin. (In press.)

Dobson, A.P. & Keymer, A.E. (1985) Life history models. In: Crompton, D.W.T. & Nickol, B.B. (eds) *Biology of Acanthocephala*, pp. 347–384. Cambridge University Press, Cambridge.

Dodson, S. (1989a) Predator-induced reaction norms. *Bioscience* **39**: 447–452.

Dodson, S.I. (1989b) The ecological role of chemical stimuli for the zooplankton: predator-induced morphology in *Daphnia*. *Oecologia* **78**: 361–367.

Doubleday, W.G. (1976) Environmental fluctuations and fisheries management. *Int. Comm. Northw. Atlant. Fish. Sel. Pap.* **1**: 141–150.

Dreisig, H. (1981) The rate of predation and its temperature dependence in a tiger beetle, *Cicindela hybrida*. *Oikos* **36**: 196–202.

Driessen, G.J.J., Van Raalte, A.Th. & Bruyn, G.J. (1984) Cannibalism in the red wood ant, *Formica polyctena* (Hymenoptera: Formicidae). *Oecologia* **63**: 13–22.

Dritschilo, W., Cornell, H., Nafus, D. & O'Connor, B. (1975) Insular biogeography: of mice and mites. *Science* **190**: 467–469.

Duebbert, H.F. & Kantrud, H.A. (1974) Upland duck nesting related to land use and predator reduction. *J. Wildlife Manage.* **38**: 257–265.

Duelli, P. (1984) Flight, dispersal, migration. In: Canard, M., Semaria, Y. & New, T.R. (eds) *Biology of the Chrysopidae*, pp. 110–116. Junk, The Hague.

Duffey, E. (1956) Aerial dispersal in a known spider population. *J. Anim. Ecol.* **25**: 85–111.

Duncan, P. & Cowton, P. (1980) An unusual choice of habitat helps Camargue horses to avoid blood-sucking horseflies. *Biol. Behav.* **5**: 55–60.

Duncan, P. & Vigne, N. (1979) The effect of group size in horses on the rate of attacks by bloodsucking flies. *Anim. Behav.* **27**: 623–625.

Dunham, K.M. & Murray, M.G. (1982) The fat reserves of impala, *Aepyceros melampus*. *Afr. J. Ecol.* **20**: 81–87.

Dunkin, S. de B. & Hughes, R.N. (1984) Behavioural components of prey selection by dogwhelks, *Nucella lapillus* (L.) feeding on barnacles *Semibalanus balanoides* (L.) in the laboratory. *J. Exp. Marine Biol. Ecol.*

79: 91–103.

Dupré, J. (ed.) (1987) *The Latest on the Best. Essays on Evolution and Optimality*. MIT Press, Cambridge, MA.

Dye, C. (1990) Epidemiological significance of vector-parasite interactions. *Parasitology* **101**: 409–415.

Dye, C. (1992) The analysis of parasite transmission by bloodsucking insects. *Ann. Rev. Entomol.* **37**: 1–19.

Dye, C. & Hasibeder, G. (1986) Population dynamics of mosquito-borne disease: effects of flies which bite some people more frequently than others. *Trans. R. Soc. Trop. Med. Hyg.* **80**: 69–77.

Dye, C. & Wolpert, D.M. (1988) Earthquakes, influenza and cycles of Indian kala-azar. *Trans. R. Soc. Trop. Med. Hyg.* **82**: 843–850.

Eberhard, W.G. (1971) The ecology of the web of *Uloborus diversus* (Araneae: Uloboridae). *Oecologia* **6**: 328–342.

Eberhard, W.G. (1975) The inverted ladder orb web of *Scloloderus* sp. and the intermediate orb of *Eustala* (?) sp. (Araneae: Araneidae). *J. Nat. Hist.* **9**: 93–106.

Eberhard, W.G. (1977) Aggressive chemical mimicry by a bolas spider. *Science* **98**: 1173–1175.

Eberhard, W.G. (1980) The natural history and behavior of the bolas spider *Mastophora dizzydeani* sp.n. (Araneidae). *Psyche* **87**: 143–169.

Edgar, W.D. (1969) Prey and predators of the wolf spider *Lycosa lugubris*. *J. Zool. Lond.* **159**: 405–411.

Edgar, W.D. (1970) Prey and feeding behavior of adult females of the spider *Pardosa amentata* (Clerck). *Netherlands J. Zool.* **20**: 487–491.

Edmunds, G.F. & Alstad, D.N. (1978) Coevolution in insect herbivores and conifers. *Science* **199**: 941–945.

Edmunds, G.F. & Alstad, D.N. (1981) Reponses of black pine leaf scales to host plant variability. In: Denno, R.F. & Dingle, H. (eds) *Insect Life History Patterns: Habitat and Geographic Variation*, pp. 29–38. Springer-Verlag, New York.

Edmunds, M. (1974) *Defence in Animals. A survey of Anti-predator Defences*. Longman, Essex.

Edson, J.L. (1985) The influences of predation and resource sub-division on the coexistence of goldenrod aphids. *Ecology* **66**: 1736–1743.

Edson, K.M., Vinson, S.B., Stoltz, D.B. & Summers, M.D. (1981) Virus in a parasitoid wasp: suppression of the cellular immune response in the parasitoid's host. *Science* **211**: 582–583.

Edwards, C.A., Sunderland, K.D. & George, K.S. (1979) Studies on polyphagous predators of cereal aphids. *J. Appl. Ecol.* **16**: 811–823.

Edwards, D.C., Conover, D.O. & Sutter, F. (1982) Mobile predators and the structure of marine intertidal communities. *Ecology* **63**: 1175–1180.

Edwards, G.B., Carroll, J.F. & Whitcomb, W.H. (1974) *Stoidis aurata* (Araneae: Salticidae) a spider predator of ants. *Florida Entomol.* **57**: 337–345.

Edwards, J. (1983) Diet shifts in moose due to predator avoidance. *Oecologia* **60**: 185–189.

Eggleston, D.B. (1990) Behavioural mechanisms underlying variable functional responses of blue crabs, *Callinectes sapidus*, feeding on juvenile oysters, *Crassostrea virginica*. *J. Anim. Ecol.* **59**: 615–630.

Ehler, L.E. (1986) Distribution of progeny in two ineffective parasites of a gall midge (Diptera: Cecidomyiidae). *Environ. Entomol.* **15**: 1268–1271.

Ehler, L.E. (1990) Introduction strategies in biological control of insects. In: Mackauer, M., Ehler, L.E. & Roland, J. (eds) *Critical Issues in Biological Control*, pp. 111–134. Intercept Ltd, Andover.

Ehler, L.E. & Hall, R.W. (1982) Evidence for competitive exclusion of introduced natural enemies in biological control. *Environ. Entomol.* **11**: 1–4.

Ehler, L.E. & Miller, J.C. (1978) Biological control in temporary agroecosystems. *Entomophaga* **23**: 207–212.

Ehrlich, P.R. & Raven, P.H. (1964) Butterflies and plants: a study in coevolution. *Evolution* **18**: 586–608.

Eisenberg, J.F. (1981) *The Mammalian Radiations. An Analysis of Trends in Evolution, Adaptation, and Behavior*. University of Chicago Press, Chicago.

Elgar, M.A. (1989) Predator vigilance and group size in mammals and birds: a critical review of the empirical evidence. *Biol. Rev.* **64**: 13–33.

Elgar, M.A. & Harvey, P.H. (1987) Basal

metabolic rates in mammals: allometry, phylogeny and ecology. *Funct. Ecol.* **1**: 25–36.

Elner, R.W. & Hughes, R.N. (1978) Energy maximization in the diet of the shore crab, *Carcinus maenas. J. Anim. Ecol.* **47**: 103–116.

Elner, R.W. & Vadas, R.L. (1990) Inference in ecology: the sea urchin phenomenon in the northwestern Atlantic. *Am. Nat.* **136**: 108–125.

Elton, C. (1927) *Animal Ecology*. Sidgwick & Jackson, London.

Elton, C.S. (1942) *Voles, Mice and Lemmings*. Oxford University Press, Oxford.

Elton, C. (1953) Use of cats in farm rat control. *Br. J. Anim. Behav.* **1**: 151–155.

Elton, C.S. (1958) *The Ecology of Invasions by Animals and Plants*. Chapman and Hall, London.

Embree, D.G. (1966) The role of introduced parasites in the control of the winter moth in Nova Scotia. *Can. Entomol.* **98**: 1159–1168.

Emlen, J.M. (1973) *Ecology: an Evolutionary Approach*. Addison-Wesley, Reading, MA.

Emmons, L.H. (1987) Comparative feeding ecology of fields in a neotropical forest. *Behav. Ecol. Sociobiol.* **20**: 271–283.

Enders, F. (1975) The influence of hunting manner on prey size, particularly in spiders with long attack distances (Araneidae, Linyphiidae and Salticidae). *Am. Nat.* **109**: 737–763.

Enders, F. (1976) Effects of prey capture, web destruction and habitat physiognomy on web-site tenacity of argiope spiders (Araneidae). *J. Arachnol.* **3**: 75–84.

Enderson, J.H. (1964) A study of the prairie falcon in the central Rocky Mountain region. *Auk* **81**: 332–352.

Endler, J.A. (1978) A predator's view of animal color patterns. *Evol. Biol.* **11**: 319–364.

Endler, J.A. (1980) Natural selection on color patterns in *Poecilia reticulata. Evolution* **34**: 76–91.

Endler, J.A. (1982) Convergent and divergent effects of natural selection on color patterns in two fish faunas. *Evolution,* **36**: 178–188.

Endler, J.A. (1984) Progressive background matching in moths, and a quantitative measure of crypsis. *Biol. J. Linnean Soc.* **22**: 187–231.

Endler, J.A. (1986a) *Natural Selection in the Wild*. Princeton University Press, Princeton.

Endler, J.A. (1986b) Defense against predators. In: Feder, M.E. & Lauder, G.V. (eds) *Predator–Prey Relationships: Perspectives and Approaches from the Study of Lower Vertebrates*, pp. 109–134. University of Chicago Press, Chicago.

Endler, J.A. (1988) Frequency-dependent predation, crypsis and aposematic coloration. *Phil. Trans. R. Soc. Lond. B* **319**: 505–523.

Engen, J.A., Jarvi, T. & Wiklund, C. (1986) The evolution of aposematic coloration by individual selection: a life-span survival model. *Oikos* **46**: 397–403.

Erkert, H.G. (1982) Ecological aspects of bat activity rhythms. In: Kunz, T.H. (ed.) *Ecology of Bats*, pp. 201–242. Plenum Press, New York.

Erlinge, S. (1975) Feeding habits of the weasel *Mustela nivalis* in relation to prey abundance. *Oikos* **28**: 378–384.

Erlinge, S. (1977) Agonistic behaviour and dominance in stoats (*Mustela erminea* L.). *Z. Tierpsychol.* **44**: 375–388.

Erlinge, S., Goransson, G., Hansson, L. et al. (1983) Predation as a regulating factor on small rodent populations in southern Sweden. *Oikos* **40**: 36–52.

Erlinge, S., Goransson, G., Hogstedt, G. et al. (1984) Can vertebrate predators regulate their prey? *Am. Nat.* **123**: 125–133.

Erlinge, S. & Sandell, M. (1988) Coexistence of stoat, *Mustela erminea*, and weasel, *M. nivalis*: social dominance, scent communication, and reciprocal distribution. *Oikos* **53**: 242–246.

Ernsting, G. & Isaacs, J.A. (1988) Reproduction, metabolic rate and survival in a carabid beetle. *Netherlands J. Zool.* **38**: 46–60.

Ernsting, G., Jager, J.C., Van der Meer, J. & Slob, W. (1985) Locomotory activity of a visually hunting carabid beetle in response to non-visual prey stimuli. *Entomologia Experimentalis et Applicata,* **38**: 41–47.

Ernsting, G. & Van der Werf, D.C. (1988) Hunger, partial consumption of prey and prey-size preference in a carabid beetle. *Ecol. Entomol.* **13**: 155–164.

Errington, P.L. (1946) Predation and vertebrate populations. *Q. Rev. Biol.*

21: 144–77, 221–245.

Eshel, I. & Akin, E. (1983) Coevolutionary instability of mixed Nash solutions. *J. Math. Biol.* **18**: 123–133.

Estes, J.A. (1979) Exploitation of marine mammals: r-selection of K-strategists? *J. Fish. Res. Board Can.* **36**: 1009–1017.

Estes, J.A., Smith, N.S. & Palmisano, J.F. (1978) Sea otter predation and community organisation in the western Aleutian Islands, Alaska. *Ecology* **59**: 822–833.

Estes, R.D. (1976) The significance of breeding synchrony in the wildebeest. *E. Afr. Wild. J.* **14**: 135–152.

Estes, R.D. & Estes, R.K. (1979) The birth and survival of wildebeest calves. *Z. Tierpsychol.* **50**: 45–95.

Estes, R.D. & Goddard, J. (1967) Prey selection and hunting behaviour of the African wild dog. *J. Wildlife Manage.* **31**: 52–70.

Eutermoser, A. (1961) Schlagen Beizfalken bevorzügt kranke Krahen? *Vogelwelt* **82**: 101–104.

Evans, A.S. (ed.) (1989) *Viral Infections of Humans — Epidemiology and Control*, 3rd edn. Plenum Medical Book Company, New York.

Evans, H.F. (1976a) The role of predator–prey size ratio in determining the efficiency of capture of *Anthocoris nemorum* and the escape reactions of its prey, *Acyrtosiphon pisum. Ecol. Entomol.* **1**: 85–90.

Evans, H.F. (1976b) The searching behaviour of *Anthocoris confusus* (Reuter) in relation to prey density and plant surface topography. *Ecol. Entomol.* **1**: 163–169.

Evans, P.G.H., Harding, S., Tyler, G. & Hall, S. (1986) Analysis of cetacean sightings in the British Isles, 1958–1985. Unpublished report to the Nature Conservancy Council.

Evans, R.M. (1982) Foraging-flock recruitment at a black-billed gull colony: implications for the information centre hypothesis. *Auk* **99**: 24–30.

Eveleigh, E.S. & Chant, D.A. (1982a) Experimental studies on acarine predator–prey interactions: effects of predator density on prey consumption, predator searching efficiency, and the functional response to prey density. *Can. J. Zool.* **60**: 611–629.

Eveleigh, E.S. & Chant, D.A. (1982b)

Experimental studies on acarine predator–prey interactions: the effects of predator density on immature survival, adult fecundity and emigration rates, and the numerical response to prey density. *Can. J. Zool.* **60**: 630–638.

Ewald, P.W. (1983) Host–parasite relations, vectors, and the evolution of disease severity. *Ann. Rev. Ecol. Systemat.* **14**: 465–485.

Ewer, R.F. (1973) *The Carnivores.* Cornell University Press, Ithaca.

Faaborg, J. (1986) Reproductive success and survivorship of the Galapagos Hawk *Buteo galapagoensis*: potential costs and benefits of co-operative polyandry. *Ibis* **128**: 337–347.

Faeth, S.H. (1985) Host leaf selection by leaf miners: interactions among three trophic levels. *Ecology* **66**: 870–875.

Fairweather, P.G. (1985) Differential predation on alternative prey, and the survival of rocky intertidal organisms in New South Wales. *J. Exp. Marine Biol. Ecol.* **89**: 135–136.

Fairweather, P.G. (1988a) Consequences of supply-side ecology: manipulating the recruitment of intertidal barnacles affects the intensity of predation upon them. *Biol. Bull.* **175**: 349–354.

Fairweather, P.G. (1988b) Predation creates haloes of bare space among prey on rocky shores in New South Wales: *Aust. J. Ecol.* **13**: 401–409.

Fairweather, P.G. & Underwood, A.J. (1983) The apparent diet of predators and biases due to different handling times of their prey. *Oecologia* **56**: 169–179.

Fairweather, P.G., Underwood, A.J. & Moran, M.J. (1984) Preliminary investigations of predation by the whelk *Morula marginalba. Marine Ecol. Prog. Ser.* **17**: 156–184.

Fantino, E. (1969) Choice and rate of inforcement. *J. Exp. Anal. Behav.* **12**: 723–730.

Fantino, E. & Abarca, N. (1985) Choice, optimal foraging and the delay-reduction hypothesis. *Behav. Brain Sci.* **8**: 315–330.

Farmer, J.D. & Sidorowich, J.J. (1989) Exploiting chaos to predict the future and reduce noise. In: Lee, Y.C. (ed.) *Evolution, Learning and Cognition*, pp. 277–304. World Scientific Press, New York.

Fay, F.H., Feder, H.M. & Stoker, S.W. (1977) An estimation of the impact of the Pacific walrus population on its food resources in the Bering Sea. Marine Mammal Commission, Washington, DC. (Natl. Tech. Inf. Serv., PB-273505.)

Fay, F.H. & Lowry, L.F. (1981) Seasonal use and feeding habits of walruses in the proposed Bristol Bay clam fishery area. Final Report Contract No 80–3. N. Pac. Fish. Manage. Council, Anchorage, Alaska.

Fay, F.H. & Stoker, S.W. (1982) Analysis of reproductive organs and stomach contents from walruses taken in the Alaskan Native harvest, Spring 1980. Final report for contract 14–16–0007–81–5216 to US Fish and Wildlife Service, Anchorage, Alaska.

Feare, C.J. (1970) Aspects of the ecology of an exposed shore population of dog-whelks *Nucella lapillus* (L.), *Oecologia* **5**: 1–18.

Feder, M.E. & Lauder G.V. (eds) (1986) *Predator–Prey Relationships: Perspectives and Approaches from the Study of Lower Vertebrates.* University of Chicago Press, Chicago.

Feeny, P. (1976) Plant apparency and chemical defense. In: Wallace, J.W. & Mansell, R.L. (eds) *Biochemical Interactions Between Plants and Insects*, pp. 1–40. *Recent Advances in Phytochemistry* **10**. Plenum Press, New York.

Fenner, F. (1983) Biological control as exemplified by smallpox eradication and myxomatosis. *Proc. R. Soc. Lond. B* **218**: 259–285.

Fenner, F. & Ratcliffe, F.N. (1965) *Myxomatosis.* Cambridge University Press, Cambridge.

Fenton, M.B. (1982) Echolocation, insect hearing, and feeding ecology of insectivorous bats. In: Kunz, T.H. (ed.) *Ecology of Bats*, pp. 261–285. Plenum Press, New York.

Findley, J.S. (1976) The structure of bat communities. *Am. Nat.* **110**: 129–139.

Findley, J.S. & Black, H.L. (1979) Ecological and morphological aspects of community structure in bats. *Am. Zool.* **19**: 989.

Findley, J.S. & Wilson, D.E. (1982) Ecological significance of chiropteran morphology. In: Kunz, T.H. (ed.) *Ecology of Bats*, pp. 243–260. Plenum Press, New York.

Fine, P.E.M. & Clarkson, J. (1982) Measles in England and Wales. 1. An analysis of factors underlying seasonal patterns. *Int. J. Epidemiol.* **11**: 5–14.

Fink, L.S. & Brower, L.P. (1981) Birds can overcome the cardenolide defence of monarch butterflies in Mexico. *Nature* **291**: 67–70.

Fink, L.S., Brower, L.P., Waide, R.B. & Spitzer, P.R. (1983) Overwintering monarch butterflies as food for insectivorous birds in Mexico. *Biotropica*, **15**: 151–153.

Fischlin, A. & Baltensweiler, W. (1979) Systems analysis of the larch budmoth system. Part I. The larch–larch budmoth relationship. In: Delucchi, V. & Baltensweiler, W. (eds) *Dispersal of Forest Insects*, pp. 273–289. IUFRO, Zuoz, Switzerland.

Fisher, R.A. (1930) *The Genetical Theory of Natural Selection.* Clarendon Press, Oxford. (2nd revised edn, Dover, New York, 1958.)

FitzGibbon, C.D. (1988) The antipredator behaviour of Thomson's gazelles, PhD thesis, University of Cambridge.

FitzGibbon, C.D. (1989) A cost to individuals with reduced vigilance levels to groups of Thomson's gazelles hunted by cheetahs. *Anim. Behav.* **37**: 508–510.

FitzGibbon, C.D. (1990a) Anti-predator strategies of immature Thomson's gazelles: hiding and the prone response. *Anim. Behav.* **40**: 846–855.

FitzGibbon, C.D. (1990b) Why do hunting cheetahs prefer male gazelles? *Anim. Behav.* **40**, 837–845.

FitzGibbon, C.D. (1990c) Mixed-species grouping in Thomson's gazelles: the antipredator benefits. *Anim. Behav.* **39**: 1116–1126.

FitzGibbon, C.D. & Fanshawe, J. (1988) Stotting in Thomson's gazelles: an honest signal of condition. *Behav. Ecol. Sociobiol.* **23**: 69–74.

FitzGibbon, C.D. & Fanshawe, J. (1989) The condition and age of Thomson's gazelles killed by cheetahs and wild dogs. *J. Zool.* **218**: 99–107.

Fiuczynski, D. (1978) Zur Populationsökologie des Baumfalken (*Falco subbuteo* L., 1758). *Zool. Jb. Syst. Bd.* **105**: 193–257.

Flor, H.H. (1956) The complementary genic systems in flax and flax rust. *Adv. Genet.* **8**: 29–54.

Foelix, R.F. (1982) *Biology of Spiders*, pp. 239–244. Harvard University Press, Cambridge, MA.

Force, D.C. (1974) Ecology of insect host–parasitoid communities. *Science* **184**: 624–632.

Force, D.C. (1980) Do parasites of Lepidoptera larvae compete for hosts? Probably! *Am. Nat.* **116**: 873–875.

Ford, M.J. (1977) Energy costs of the predation strategy of the web spinning spider *Lepthyphantes zimmermanni* Berkau (Linyphiidae). *Oecologia* **28**: 341–349.

Formanowicz, D.R. Jr (1984) Foraging tactics of one aquatic insect: partial consumption of prey. *Anim. Behav.* **32**: 774–781.

Fosbrooke, H.A. (1963) The stomoxys plague in Ngorongoro, 1962. *E. Afr. Wildlife J.* **1**: 124–126.

Fox, B.J. (1982a) A review of dasyurid ecology and speculation on the role of limiting similarity in community organization. In: Archer, M. (ed.) *Carnivorous Marsupials*, pp. 97–116. Royal Zool. Soc. New South Wales, Sydney.

Fox, B.J. (1982b) Ecological separation and coexistence of *Sminthopsis murina* and *Antechinus stuartii* (Dasyuridae, Marsupialia): a regeneration niche? In: Archer, M. (ed.) *Carnivorous Marsupials*, pp. 187–197. Royal Zool. Soc. New South Wales, Sydney.

Fox, B.J. (1987) Species assembly and the evolution of community structure. *Evol. Ecol.* **1**: 201–213.

Fox, L.R. (1975) Cannibalism in natural populations. *Ann. Rev. Ecol. Systemat.* **6**: 87–106.

Fox, L.R. (1981) Defense and dynamics in plant–herbivore systems. *Am. Zool.* **21**: 853–864.

Fraenkel, G. (1959) The raison d'être of secondary plant substances. *Science* **129**: 1466–1470.

Frank, S.A. (1985) Hierarchical selection theory and sex ratios. II. On applying the theory and a test with fig wasps. *Evolution* **39**: 949–964.

Frank, S.A. (1986) Hierarchical selection theory and sex ratios. I. General solutions for structured populations. *Theor. Pop. Biol.* **29**: 312–342.

Frank, S.A. (1991) Spatial variation in coevolutionary dynamics. *Evol. Ecol.* **5**: 193–217.

Frankel, O.H. & Soulé, M.E. (1981) *Conservation and Evolution.* Cambridge University Press, Cambridge.

Frazer, B.D. & Gilbert, N. (1976) Coccinellids and aphids. A quantitative study of the impact of adult ladybirds (Coleoptera: Coccinellidae) preying on field populations of pea aphids (Homoptera: Aphididae). *J. Entomol. Soc. British Columbia* **73**: 33–56.

Free, C.A., Beddington, J.R. & Lawton, J.H. (1977) On the inadequacy of simple models of mutual interference for parasitism and predation. *J. Anim. Ecol.* **46**: 543–554.

Freeland, W.J. (1983) Parasites and the coexistence of animal host species. *Am. Nat.* **121**: 223–236.

Freeland, W.J. & Janzen, D.H. (1974) Strategies in herbivory by mammals: the role of plant secondary compounds. *Am. Nat.* **108**: 269–289.

Freeman, P.W. (1979) Specialized insectivory: beetle-eating and moth-eating molossid bats. *J. Mammal.* **60**: 467–479.

Freeman, P.W. (1981a) A multivariate study of the family Molossidae (Mammalia, Chiroptera): morphology, ecology, evolution. *Field. Zool., New Ser. No. 7*, Field Mus. Nat. Hist., Chicago.

Freeman, P.W. (1981b) Correspondence of food habits and morphology in insectivorous bats. *J. Mammal.* **62**: 166–173.

Fretwell, S.D. & Lucas, H.L. (1970) On territorial behaviour and other factors influencing habitat distribution in birds. *Acta Biotheor.* **19**: 16–36.

Friedman, M.B. (1975) How birds use their eyes. In: Wright, P., Caryl, P.G. & Vowles, D.M. (eds) *Neural and Endocrine Aspects of Behaviour in Birds*, pp. 181–204. Elsevier, Amsterdam.

Fryxell, J.M., Greever, J. & Sinclair, A.R.E. (1988) Why are migratory ungulates so abundant? *Am. Nat.* **131**: 781–798.

Fujii, Y. (1972) Study of energy utility efficiency in *Agelena opulenta* L. Koch (Araneae: Agelenidae). *Bull. Nippon Dental Coll. Gen. Educ.* No. 1, 79–95.

Fuller, T.K. & Keith, L.B. (1980) Wolf population dynamics and prey relationships in northeastern Alberta. *J. Wildlife Manage.* **44**: 583–602.

Furness, R.W. (1984) Seabird–fisheries relationships in the Northeast Atlantic and North Sea. In: Nettleship, D.N., Sanger, G.A. & Springer, P.F. (eds) *Marine Birds: their Feeding, Ecology and Commercial Fisheries Relationships.* Proceedings of the Pacific Seabird Group

Symposium.

Futuyma, D.J. (1976) Food plant specialization and environmental predictability in Lepidoptera. *Am. Nat.* **110**: 285–292.

Futuyma, D.J. & Slatkin, M. (eds) (1983) *Coevolution*. Sinauer, Sunderland, MA.

Galushin, V.M. (1971) A huge urban population of birds of prey in Delhi, India. *Ibis* **115**: 522.

Gardner, S.M. & Dixon, A.F.G. (1985) Plant structure and foraging success of *Aphidius rhopalosiphi* (Hymenoptera: Aphidiidae). *Ecol. Entomol.* **10**: 171–179.

Gargett, V. (1970) Black eagle experiment II. *Bokmakierie* **22**: 32–35.

Garrett-Jones, C. (1964) Prognosis for interruption of malaria transmission through assessment of the mosquito's vectorial capacity. *Nature* **204**: 1173–1175.

Garrett-Jones, C., Boreham, P.F.L. & Pant, C.P. (1980) Feeding habits of anophelines (Diptera: Culicidae) in 1971–78, with reference to the human blood index: a review. *Bull. Entomol. Res.* **70**: 165–185.

Garrett-Jones, C. & Grab, B. (1964) The assessment of insecticidal impact on the malaria mosquito's vectorial capacity, from data on the proportion of parous females. *Bull. WHO* **31**: 71–86.

Garrett-Jones, C. & Shidrawi, G.R. (1969) Malaria vectorial capacity of a population of *Anopheles gambiae*. An exercise in epidemiological entomology. *Bull. WHO* **40**: 531–545.

Garritty, S.D. & Levings, S.C. (1981) A predator–prey interaction between two physically and biologically constrained tropical rocky shore gastropods: direct, indirect and community effects. *Ecol. Monogr.* **51**: 267–286.

Garson, P.J. & Hunter, M.L. (1979) Effects of temperature and time of year on the singing behaviour of wrens (*Troglodytes troglodytes*) and great tits (*Parus major*). *Ibis* **121**: 481–487.

Gasaway, W.C., Boertje, R.D., Grangaard, D.V., Kelleyhouse, D.G., Stephenson, R.O. & Larsen, D.G. (1991) Predation limiting moose at low densities. *Wildlife Monogr.* (in press).

Gashwiler, J.S. & Robinette, W.L. (1957) Accidental fatalities of the Utah cougar. *J. Mammal.* **38**: 123–126.

Gaskin, D.E. (1972) *Whales, Dolphins and Seals*. Heinemann Educational Books, Oxford.

Gaston, K.J. (1988) The intrinsic rates of increase of insects of different sizes. *Ecol. Entomol.* **14**: 399–409.

Gaston, K.J. & Lawton, J.H. (1988) Patterns in the distribution and abundance of insect populations. *Nature* **331**: 709–712.

Gauld, I. & Bolton, B. (1988) *The Hymenoptera*. Oxford University Press, Oxford.

Gause, G.F. (1934) *The Struggle for Existence*. Macmillan, New York.

Geer, T.A. (1978) Effects of nesting sparrowhawks on nesting tits. *Condor* **80**: 419–422.

Gendron, R.P. (1986) Searching for cryptic prey: evidence for optimal search rates and the formation of search images in quail. *Anim. Behav.* **34**: 898–912.

Gendron, R.P. & Staddon, J.E.R. (1983) Searching for cryptic prey: the effect of search rate. *Am. Nat.* **121**: 172–186.

Gendron, R.P. & Staddon, J.E.R. (1984) A laboratory simulation of foraging behavior: the effect of search rate on the probability of detecting prey. *Am. Nat.* **124**: 407–415.

Gerson, U. & Smiley, R.J. (1990) *Acarine Biocontrol Agents: An Illustrated Key and Manual*. Chapman and Hall, London.

Gibbon, J. (1977) Scalar expectancy theory and Weber's Law in animal timing. *Psychol. Rev.* **84**: 279–325.

Gibbon, J., Church, R.M., Fairhurst, S. & Kacelnik, A. (1988) Scalar expectancy theory and choice between delayed rewards. *Psychol. Rev.* **95**: 102–114.

Gibson, D.I. (1981) Evolution of digeneans. *Parasitology* **82**: 161–163.

Gibson, D.I. (1987) Questions in digenean systematics and evolution. *Parasitology* **95**: 429–460.

Gibson, D.O. (1974) Batesian mimicry without distastefulness. *Nature* **250**: 77–79.

Gibson, D.O. (1980) The role of escape in mimicry and polymorphism: I, the response of captive birds to artificial prey. *Biol. J. Linnean Soc.* **14**: 201–214.

Gilbert, N., Gutierrez, A.P., Frazer, B.D. & Jones, R.E. (1976) *Ecological*

Relationships. W.H. Freeman, San Francisco.

Giller, P.S. (1980) The control of handling time and its effects on the foraging strategy of a heteropteran predator, Notonecta. J. Anim. Ecol. **49**: 699–712.

Gillespie, R.G. (1981) The quest for prey by the web building spider Amaurobius similis (Blackwall). Anim. Behav. **29**: 953–954.

Gillespie, R.G. & Caraco, T. (1987) Risk-sensitive foraging strategies of two spider populations. Ecology **68**: 887–899.

Gillett, J.D. (1971) Mosquitos. Weidenfeld & Nicolson, London.

Gilliam, J.F. & Fraser, D.F. (1987) Habitat selection when foraging under predation hazard: test of a model with foraging minnows. Ecology **68**: 1856–1862.

Gilwicz, M.Z. (1990) Food thresholds and body size in cladocerans. Nature **343**: 638–640.

Giovanardi, A., Fara, G.M. & Giorini, S. (1974) The Role of Environmental Factors in the Epidemiology of Infectious Diseases. Piccin Medical Books, Italy.

Gittleman, J.L. (1985) Carnivore body size: ecological and taxonomic correlates. Oecologia **67**: 540–554.

Gittleman, J.L. (1989) Carnivore group living: comparative trends. In: Gittleman, J.L. (ed.) Carnivore Behavior, Ecology, and Evolution, pp. 183–207. Chapman & Hall, London.

Gittleman, J.L. & Harvey, P.H. (1980) Why are distasteful prey not cryptic? Nature **286**: 149–150.

Gittleman, J.L. & Harvey, P.H. (1982) Carnivore home-range size, metabolic needs and ecology. Behav. Ecol. Sociobiol. **10**: 57–64.

Gittleman, J.L. & Oftedal, O.T. (1987) Comparative growth and lactation energetics of carnivores. Symp. Zool. Soc. Lond. **57**: 41–77.

Givens, R.P. (1978) Dimorphic foraging strategies of a salticid spider (P. audax). Ecology **59**: 309–321.

Glatz, L. (1967) Zur Biologie und Morphologie von Oecobius annulipes Lucas (Araneae, Oecobiidae). Z. Morphol. Tiere. **61**: 185–214.

Glazier, D.S. (1986) Temporal variability of abundance and the distribution of species. Oikos **47**: 309–314.

Gleick, (1987) Chaos: Making a New Science. Viking, New York.

Glen, D.M. (1973) The food requirements of Blepharidopterus angulatus (Heteroptera: Miridae) as a predator of the lime aphid, Eucallipterus tiliae. Entomologia Experimentalis et Applicata **16**: 255–267.

Glendinning, J.I. (1992) Comparative feeding responses of the mice Peromyscus melanotis, P. aztecus hylocetes, Reithrodontomys sumichrasti, and Microtus mexicanus salvus, to overwintering monarch butterflies in Mexico. In: Malcolm, S.B. & Zalucki, M.P. (eds) Biology and Conservation of the Monarch Butterfly. Los Angeles County Natural History Museum, Los Angeles (in press).

Glendinning, J.I., Alonso Mejia, A. & Brower, L.P. (1988) Behavioral and ecological interactions of foraging mice (Peromyscus melanotis) with overwintering monarch butterflies (Danaus plexippus) in Mexico. Oecologia **75**: 222–227.

Glendinning, J.I. & Brower, L.P. (1990) Feeding and breeding responses of five mice species to overwintering aggregations of the monarch butterfly. J. Anim. Ecol. **59**: 1091–1112.

Godfray, H.C.J. (1987a) The evolution of clutch size in parasitic wasps. Am. Nat. **129**: 221–233.

Godfray, H.C.J. (1987b) The evolution of clutch size in invertebrates. Oxf. Surv. Evol. Biol. **4**: 117–154.

Godfray, H.C.J. & Blythe, S.P. (1990) Complex dynamics in multispecies communities. Phil. Trans. R. Soc. Lond. B **330**: 221–233.

Godfray, H.C.J. & Chan, M.A. (1990) How insecticides trigger single-stage outbreaks in tropical pests. Func. Ecol. **4**: 329–337.

Godfray, H.C.J. & Hassell, M.P. (1987) Natural enemies can cause discrete generations in tropical insects. Nature **327**: 144–147.

Godfray, H.C.J. & Hassell, M.P. (1989) Discrete and continuous insect populations in tropical environments. J. Anim. Ecol. **58**: 153–174.

Godfray, H.C.J. & Hassell, M.P. (1990) Ecapsulation and host–parasitoid population dynamics. In: Toft, C. (ed) Parasitism: Coexistence or Conflict?, pp. 131–147. Oxford University Press, Oxford.

Godfray, H.C.J. & Waage, J.K. (1991) Predictive modelling in biological control: the mango mealybug (*Rastrococcus invadens*) and its parasitoids. *J. Appl. Ecol.* **28**: 434–453.

Godfrey, D., Lythgoe, J.N. & Rumball, D.A. (1987) Zebra stripes and tiger stripes: the spatial frequency distribution of the pattern compared to that of the background is significant in display and crypsis. *Biol. J. Linnean Soc.* **32**: 427–433.

Godsell, J. (1982) Aspects of the population ecology of the Eastern Quoll, *Dasyurus viverrinus* (Dasyuridae, Marsupialia), in southern Tasmania. In: Archer, M. (ed.) *Carnivorous Marsupials*, pp. 199–207. Royal Zool. Soc. New South Wales, Sydney.

Goeden, R.D. & Louda, S.M. (1986) Biotic Interference with insects imported for weed control. *Ann. Rev. Entomol.* **21**: 325–342.

Gooch, S., Baillie, S.R. & Birkhead, T.R. (1991) Magpie *Pica pica* and songbird populations. Retrospective investigation of trends in population density and breeding success. *J. Appl. Ecol.* **28**: 1068–1086.

Gordon, D.M., Nisbet, R.M., deRoos, A., Gurney, W.S.C. & Stewart, R.K. (1991) Discrete generations in host–parasitoid models with contrasting life cycles. *J. Anim. Ecol.* **60**: 295–308.

Gosling, L.M. (1980) Defence guilds of savannah ungulates as a context for scent communication. *Symp. Zool. Soc. Lond.* **45**: 195–212.

Goss-Custard, J.D. (1977) Optimal foraging and the size-selection of worms by redshank *Tringa tolarius* in the field. *Anim. Behav.* **25**: 10–29.

Gould, F. (1988) Genetics of pairwise and multispecies plant–herbivore coevolution. In: Spencer, K.C. (ed) *Chemical Mediation of Coevolution*, pp. 13–55. Academic Press, San Diego.

Gould, S.J. & Lewontin, R.C. (1979) The spandrels of San Marco and the Panglossian paradigm: a critique of the adaptationist programme. *Proc. R. Soc. Lond. B* **205**: 581–598.

Gould, S.J. & Vrba, E. (1982) Exaptation — a missing term in the science of form. *Paleobiology* **8**: 4–15.

Gradwohl, J. & Greenberg, R. (1982) The effect of a single species of avian predator on the arthropods of aerial leaf litter. *Ecology* **63**: 581–583.

Grafen, A. (1990) Biological signals as handicaps. *J. Theor. Biol.* **144**: 517–546.

Graham, G.L. (1988) Interspecific associations among Peruvian bats at diurnal roosts and roost sites. *J. Mammal.* **69**: 711–720.

Grant, J.P. (1989) *The State of the World's Children 1989*. UNICEF/Oxford University Press, Oxford.

Grant, P.R. (1986) *Ecology and Evolution of Darwin's Finches*. Princeton University Press, Princeton.

Gray, R.D. (1987) Faith and foraging: a critique of the 'paradigm argument from design'. In: Kamil, A.C., Krebs, J.R. & Pulliam, H.R. (eds) *Foraging Behavior*, pp. 69–140. Plenum Press, New York.

Greany, P.D. & Hagen, K.S. (1981) Prey selection. In: Nordlund, D.A. & Lewis, W.J. (eds) *Semiochemicals, their Role in Pest Control*, pp. 121–135. John Wiley & Sons, New York.

Greathead, D.J. (1989) Biological control as an introduction phenomenon: a preliminary examination of programmes against the Homoptera. *The Entomologist* **108**: 28–37.

Greathead, D.J. & Waage, J.K. (1983) *Opportunities for Biological Control of Agricultural Pests in Developing Countries*. World Bank Technical Paper No. 11, The World Bank, Washington.

Green, R.F. (1980) Bayesian birds: a simple example of Oaten's stochastic model of optimal foraging. *Theor. Pop. Biol.* **18**: 244–256.

Green, R.F. (1984) Stopping rules for optimal foragers. *Am. Nat.* **123**: 30–40.

Greene, H.W. (1988) Anti-predator mechanisms in reptiles. In: Gans, C. & Huey, R.B. (eds) *The Biology of the Reptilia*, pp. 1–152. Alan Liss, New York.

Greenstone, M.H. (1979) Spider feeding behaviour optimises dietary essential amino acid composition. *Nature* **282**: 501–503.

Greenwood, J.J.D. (1984) The functional basis of frequency-dependent food selection. *Biol. J. Linnean Soc.* **23**: 177–199.

Greenwood, J.J.D. & Elton, R.A. (1979) Analysing experiments on frequency-dependent selection by predators.

J. Anim. Ecol. **48**: 721–737.

Grenfell, B.T. (1988) Gastrointestinal nematode parasites and the stability and productivity of intensive ruminant grazing systems. *Phil. Trans. R. Soc. Lond. B* **321**: 541–563.

Grenfell, B.T. & Anderson, R.M. (1985) The estimation of age-related rates of infection from case notifications and serological data. *J. Hygiene (Camb.)* **95**: 419–436.

Griffiths, D. (1980a) Foraging costs and relative prey size. *Am. Nat.* **116**: 743–752.

Griffiths, D. (1980b) The feeding biology of ant-lion larvae: prey capture, handling and utilization. *J. Anim. Ecol.* **49**: 99–125.

Griffiths, K.J. (1959) Observations on the European pine sawfly, *Neodiprion sertifer* (Geoff.) and its parasites in southern Ontario. *Can. Entomol.* **91**: 501–512.

Griffiths, K.L. (1969) Development and diapause in *Pleolophus basizonus* (Hymenoptera: Icheumonidae). *Can. Entomol.* **101**: 907–914.

Grimstad, P.R., Craig, G.B., Ross, Q.E. & Yuill, T.M. (1977) *Aedes triseriatus* and La Crosse virus: geographic variation in vector susceptibility and ability to transmit. *Am. J. Trop. Med. Hyg.* **26**: 990–996.

Grimstad, P.R., Ross, Q.E. & Craig, G.B. (1980) *Aedes triseriatus* (Diptera: Culicidae) and La Crosse virus. II. Modification of mosquito feeding behaviour by virus infection. *J. Med. Entomol.* **17**: 1–7.

Guilford, T. (1985) Is kin selection involved in the evolution of warning coloration? *Oikos* **45**: 31–36.

Guilford, T.C. (1986) How do 'warning colours' work? Conspicuousness may reduce recognition errors in experienced predators. *Anim. Behav.* **34**: 286–288.

Guilford, T.C. (1988) The evolution of conspicuous coloration. *Am. Nat.* **131**: S7–S21.

Guilford, T.C. (1989a) Evolutionary stability and the laboratory assessment of behavior: the role of the go-between. In: Blanchard, R.J., Brain, P., Blanchard, D.C. & Parmigiani, S. (eds) *Ethoexperimental Approaches to the Study of Behavior*, pp. 644–656. Kluwer Academic Publishers, Dordrecht.

Guilford, T.C. (1989b) Studying warning signals in the laboratory. In:

Blanchard, R.J., Brain, P., Blanchard, D.C. & Parmigiani, S. (eds) *Ethoexperimental Approaches to the Study of Behavior*, pp. 87–103. Kluwer Academic Publishers, Dordrecht.

Guilford, T.C. (1990a) The evolution of aposematism. In: Evans, D.L. & Schmidt, J.O. (eds) *Insect Defenses: Adaptive Mechanisms and Strategies of Prey and Predators*, pp. 23–61. State University of New York Press, New York.

Guilford, T.C. (1990b) Evolutionary pathways to aposematism. *Acta Oecologica* **11**: 835–841.

Guilford, T.C. & Dawkins, M.S. (1987) Search images not proven: a reappraisal of recent evidence. *Anim. Behav.* **35**: 1838–1845.

Guilford, T.C. & Dawkins, M.S. (1989a) Search image versus search rate: a reply to Lawrence. *Anim. Behav.* **37**: 160–162.

Guilford, T.C. & Dawkins, M.S. (1989b) Search image versus search rate: two different ways to enhance prey capture. *Anim. Behav.* **37**: 163–165.

Guilford, T.C. & Dawkins, M.S. (1991) Receiver psychology and the evolution of animal signals. *Anim. Behav.* **42**: 1–14.

Guilford, T.C., Nicol, C.J., Rothschild, M. & Moore, B.P. (1987) The biological roles of pyrazines: evidence for a warning odour function. *Biol. J. Linnean Soc.* **31**: 113–128.

Guiler, E.R. (1970) Observations on the Tasmanian Devil, *Sarcophilus harrisii* (Marsupialia: Dasyuridae). I. Numbers, home range, movements and food in two populations. *Aust. J. Zool.* **18**: 49–62.

Gurney, W.S.C. & Nisbet, R.M. (1985) Fluctuation periodicity, generation separation, and the expression of larval competition. *Theor. Pop. Biol.* **28**: 150–180.

Gutierrez, A.P. & Baumgaertner, J.U. (1984a) Multitrophic models of predator–prey energetics. I. Age specific energetics — pea aphid *Acyrtosyphon pisum* (Homoptera: Aphididae) as an example. *Can. Entomol.* **116**: 924–932.

Gutierrez, A.P. & Baumgaertner, J.U. (1984b) Multitrophic models of predator–prey energetics. II. *Can. Entomol.* **116**: 933–949.

Gutierrez, A.P., Baumgaertner, J.U. & Hagen, K.S. (1981) A conceptual model for growth, development and reproduction in the ladybird beetle, *Hippodamia convergens* (Coleoptera: Coccinellidae). *Can. Entomol.* **113**: 21–33.

Gutierrez, A.P., Baumgaertner, J.U. & Summers, C.G. (1984) Multitrophic models of predator–prey energetics. III. A case study in an alfalfa ecosystem. *Can. Entomol.* **116**: 950–963.

Gutierrez, A.P., Neuenschwander, P., Schultess, F., *et al.* (1988) Analysis of biological control of cassava pests in Africa. II. Cassava mealybug, *Phenacoccus manihoti. J. Appl. Ecol.* **25**: 921–929.

Ha, J.C., Lehner, P.N., & Farley, S.D. (1990) Risk-prone foraging behaviour in captive grey jays *Perisoreus canadensis. Anim. Behav.* **39**: 91–96.

Hadfield, M.G. (1978) Metamorphosis in marine molluscan larvae: an analysis of stimulus and response. In: Chia, F.-S. & Rice, M.E. (eds) *Settlement and Metamorphosis of Marine Invertebrate Larvae*, pp. 165–175. Elsevier, New York.

Hafner, M.S. & Nadler, S.A. (1988) Phylogenetic trees support the coevolution of parasites and their hosts. *Nature* **332**: 258–259.

Hagen, Y. (1969) Norwegian studies on the reproduction of birds of prey and owls in relation to micro-rodent population fluctuations. *Fauna* **22**: 73–126.

Hagmeier, A. (1930) Eine Fluktuatik von *Mactra (Spisula) subtruncata* da Costa an der ostfreisische Kuste. *Ber. Deutsch. wiss. komm. Meeresf. N.F.* **5**: 126–155.

Hails, R.S. & Crawley, M.J. (1991) The population dynamics of an alien insect: *Andricus quercuscalicis* (Hymenoptera: Cynipidae). *J. Anim. Ecol.* **60**: 545–562.

Hairston, N.G. (1987) Diapause as a predator avoidance adaptation. In: Kerfoot, W.C. & Sih, A. (eds) *Predation, Direct and Indirect Impacts on Aquatic Communities*, pp. 281–290. University Press of New England, Hanover.

Hairston, N.G. (1989) *Ecological Experiments: Purpose, Design and Execution.* Cambridge University Press, Cambridge.

Hairston, N.G., Smith, F.E. & Slobodkin, L.B. (1960) Community structure, population control, and competition. *Am. Nat.* **94**: 421–425.

Haldane, J.B.S. (1949) Disease and evolution. *La Ricerca Scientifica* (suppl.) **19**: 68–76.

Haldane, J.B.S. (1954) The statics of evolution. In: Huxley, J., Hardy, A.C. & Ford, E.B. (eds) *Evolution as a Process*, pp. 109–121. Allen & Unwin, London.

Hale, R.L. & Shorey, H.H. (1971) Effect of foliar sprays on the green peach aphid on peppers in southern California. *J. Econ. Entomol.* **64**: 547–579.

Hall, R.W. & Ehler, L.E. (1979) Rate of establishment of natural enemies in classical biological control. *Bull. Entomol. Soc. Am.* **25**: 280–282.

Hall, R.W., Ehler, L.E. & Bisabri-Ershadi, B. (1980) Rate of success in classical biological control of arthropods. *Bull. Entomol. Soc. Am.* **26**: 111–114.

Hall, S. (1980) The diets of two coexisting species of *Antechinus* (Marsupialia: Dasyuridae). *Aust. Wildlife Res.* **7**: 365–378.

Hamer, W.H. (1906) Epidemic disease in England. *Lancet* **1**: 733–739.

Hamerstrom, F. (1969) A harrier population study. In: Hickey, J.J. (ed.) *Peregrine Falcon Populations: their Biology and Decline*, pp. 367–385. University of Wisconsin Press, Madison.

Hamerstrom, F., Hamerstom, F.N. & Hart, J. (1973) Nest boxes: an effective management tool for kestrels: *J. Wildlife Manage.* **57**: 400–403.

Hamilton, P.V. (1976) Predation on *Littorina irrorata* (Mollusca: Gastropoda) by *Callinectes sapidus* (Crustacea: Portunidae). *Bull. Marine Sci.* **263**: 403–409.

Hamilton, W.D. (1967) Extraordinary sex ratios. *Science* **156**: 477–488.

Hamilton, W.D. (1971) Geometry for the selfish herd. *J. Theor. Biol.* **31**: 295–311.

Hamilton, W.D. (1980) Sex versus non-sex versus parasite. *Oikos* **35**: 282–290.

Hamilton, W.D. (1982) Pathogens as causes of genetic diversity in their host populations. In: Anderson, R.M. & May, R.M. (eds) *Population Biology of Infectious Diseases*, pp. 269–296. Springer-Verlag, Berlin.

Hammerstein, P. & Riechert, S.E. (1988) Payoffs and strategies in territorial contests: ESS analyses of two ecotypes of the spider *Agelenopsis aperta*. *Evol. Ecol.* **2**: 115–138.

Hammond, P.S. & Harwood, J. (eds) (1985) The impact of grey and common seals on North Sea Resources. Contract Report ENV 665 UK(H) to the Commission of the European Communities.

Hanski, I. (1981) Coexistence of competitors in patchy environments with and without predation. *Oikos* **37**: 306–312.

Hanski, I. (1984) Food consumption, assimilation and metabolic rate in six species of shrew (*Sorex* and *Neomys*). *Ann. Zool. Fenn.* **21**: 157–165.

Hanski, I. (1985a) What does a shrew do in an energy crisis? In: Sibly, R.M. & Smith, R.H. (eds) *Behavioural Ecology*, pp. 247–252. Blackwell Scientific Publications, Oxford.

Hanski, I. (1985b) Single-species spatial dynamics may contribute to long-term rarity and commonness. *Ecology* **66**: 335–343.

Hanski, I. (1987a) Populations of small mammals cycle — unless they don't. *Trends Ecol. Evol.* **2**: 55–56.

Hanski, I. (1987b) Pine sawfly population dynamics: patterns, processes, problems. *Oikos* **50**: 327–335.

Hanski, I. (1988) Four kinds of extra long diapause in insects: a review of theory and observations. *Ann. Zool. Fenn.* **25**: 37–54.

Hanski, I. (1990) Small mammal predation and the population dynamics of *Neodiprion sertifer*. In: Watt, A.D., Leather, S.R., Hunter, M.D. & Kidd, N.A.C. (eds) *Population Dynamics of Forest Insects*, pp. 253–264. Intercept, Andover, Hampshire.

Hanski, I., Hansson, L. & Henttonen, H. (1991) Specialist predators, generalist predators and the microtine rodent cycle. *J. Anim. Ecol.* **60**: 353–367.

Hanski, I. & Otronen, M. (1985) Food quality induced variance in larval performance: comparison between rare and common pine-feeding sawflies (Diprionidae). *Oikos* **44**: 165–174.

Hanski, I. & Parviainen, P. (1985) Cocoon predation by small mammals, and pine sawfly population dynamics. *Oikos* **45**: 125–136.

Hansson, L. (1984) Predation as a factor causing extended low densities in microtine cycles. *Oikos* **43**: 255–256.

Hansson, L. (1987) An interpretation of rodent dynamics as due to trophic interactions. *Oikos* **50**: 308–338.

Hansson, L. & Henttonen, H. (1985) Gradients in density variations of small rodents: the importance of latitude and snow cover. *Oecologia* **67**: 394–402.

Hansson, L. & Hettonen, H. (1988) Rodent dynamics as a community process. *Trends Ecol. Evol.* **3**: 195–200.

Harder, L.D. & Real, L.A. (1987) Why are bumble bees risk averse? *Ecology* **68**: 1104–1108.

Hardman, J.M. & Turnbull, A.L. (1980) Functional response of the wolf spider, *Pardosa vancouveri*, to changes in the density of vestigial-winged fruit flies. *Res. Pop. Ecol.* **21**: 233–259.

Hardy, I.C.W., Griffith, N.T. & Godfray, H.C.J. (1992) Clutch size in a parasitoid wasp: a manipulation experiment. *J. Anim. Ecol.* (in press).

Harkness, R.D. (1977) Further observations on the relation between an ant, *Cataglyphis bicolor* (F.) (Hym., Formicidae) and a spider, *Zodarium frenatum* (Simon) (Araneae, Zodariidae). *Entomol. Mon. Mag.* **112**: 111–121.

Harrison, R.G. (1980) Dispersal polymorphisms in insects. *Ann. Rev. Ecol. Systemat.* **11**: 95–118.

Harvey, P.H., Bull, J.J., Pemberton, M. & Paxton, R.J. (1982) The evolution of aposematic coloration: a family model. *Am. Nat.* **119**: 710–719.

Harvey, P.H. & Clutton-Brock, T.H. (1981) Primate home-range size, metabolic needs and ecology. *Behav. Ecol. Sociobiol.* **8**: 151–155.

Harvey, P.H. & Greene, P.J. (1981) Group composition, an evolutionary perspective. In: Kellerman, H. (ed.) *Group Cohesion: Theoretical and Clinical Perspectives*, pp. 149–169. Grune and Stratton, New York.

Harvey, P.H. & Greenwood, P.J. (1978) Anti-predator defence strategies: some evolutionary problems. In: Krebs, J.R. & Davies, N.B. (eds) *Behavioural Ecology: An Evolutionary Approach*, pp. 129–151. Blackwell Scientific Publications, Oxford.

Harvey, P.H. & Pagel, M.D. (1991) *The Comparative Method in Evolutionary Biology*. Oxford University Press, Oxford.

Harvey, P.H., Pagel, M.D. & Rees, J.A. (1991) Mammalian metabolism and life histories. *Am. Nat.* **137**: 556–566.

Harvey, P.H. & Paxton, R.J. (1981) The evolution of aposematic coloration. *Oikos* **37**: 391–393.

Harwood, J. & Croxall, J.P. (1988) The assessment of competition between seals and commercial fisheries in the North Sea and the Antarctic. *Marine Mammal Sci.* **4**: 13–33.

Hasibeder, G. & Dye, C. (1988) Population dynamics of mosquito-borne disease: persistence in a completely heterogeneous environment. *Theor. Pop. Biol.* **33**: 31–53.

Hassell, M.P. (1978) *The Dynamics of Arthropod Predator–Prey Systems.* Princeton University Press, Princeton.

Hassell, M.P. (1980) Foraging strategies, population models and biological control: a case study. *J. Anim. Ecol.* **49**: 603–628.

Hassell, M.P. (1984) Parasitism in patchy environments: inverse density dependence can be destabilizing. *IMA Journal of Mathematics Applied to Medicine and Biology* **1**: 123–133.

Hassell, M.P. (1985) Parasitoids and population regulation. In: Waage, J.K. & Greathead, D.J. (eds) *Insect Parasitoids*, pp. 201–224. Academic Press, London.

Hassell, M.P. (1986) Parasitoids and population regulation. In: Waage, J. & Greathead, D. (eds) *Insect Parasitoids*, pp. 201–224. Academic Press, London.

Hassell, M.P. (1987) Detecting regulation in patchily distributed animal populations. *J. Anim. Ecol.* **56**: 705–713.

Hassell, M.P. & Anderson, R.M. (1984) Host susceptibility as a component in host–parasitoid systems. *J. Anim. Ecol.* **53**: 611–621.

Hassell, M.P. & Anderson, R.M. (1989) Predator–prey and host–pathogen interactions. In: Cherrett, J.M. (ed.) *Ecological Concepts. The Contribution of Ecology to an Understanding of the Natural World*, pp. 147–196. Blackwell Scientific Publications, Oxford.

Hassell, M.P., Latto, J. & May, R.M. (1989) Seeing the wood for the trees: detecting density dependence from existing life-table studies. *J. Anim. Ecol.* **58**: 883–892.

Hassell, M.P., Lawton, J.H. &

Beddington, J.R. (1976) The components of arthropod predation. I. The prey death rate. *J. Anim. Ecol.* **45**: 135–164.

Hassell, M.P., Lawton, J.H. & Beddington, J.R. (1977) Sigmoid functional responses by invertebrate predators and parasitoids. *J. Anim. Ecol.* **46**: 249–262.

Hassell, M.P. & May, R.M. (1973) Stability in insect host–parasite models. *J. Anim. Ecol.* **42**: 693–726.

Hassell, M.P. & May, R.M. (1974) Aggregation in predators and insect parasites and its effect on stability. *J. Anim. Ecol.* **43**: 567–594.

Hassell, M.P. & May, R.M. (1985) From individual behaviour to population dynamics. In: Sibley, R.M. & Smith, R.H. (eds) *Behavioural Ecology. Ecological Consequences of Adaptive Behaviour*, pp. 3–32. Blackwell Scientific Publications, Oxford.

Hassell, M.P. & May, R.M. (1986) Generalist and specialist natural enemies in insect predator–prey interactions. *J. Anim. Ecol.* **55**: 923–940.

Hassell, M.P. & May, R.M. (1988) Spatial heterogeneity and the dynamics of parasitoid–host systems. *Ann. Zool. Fenn.* **25**: 55–61.

Hassell, M.P., May, R.M., Pacala, S.W. & Chesson, P. (1991) The persistence of host–parasitoid associations in patchy environments. I. A general criterion. *Am. Nat.* **138**: 568–583.

Hassell, M.P. & Pacala, S.W. (1990) Heterogeneity and the dynamics of host–parasitoid interactions. *Phil. Trans. R. Soc. Lond. B* **330**: 203–220.

Hassell, M.P. & Southwood, T.R.E. (1978) Foraging strategies of insects. *Ann. Rev. Ecol. System.* **9**: 75–98.

Hassell, M.P. & Varley, G.C. (1969) New inductive population model for insect parasites and its bearing on biological control. *Nature* **223**: 1133–1137.

Hastings, A. (1977) Spatial heterogeneity and the stability of predator–prey systems. *Theor. Pop. Biol.* **12**: 37–48.

Hausser, J., Catzeflis, F., Meylan, A. & Vogel, P. (1985) Speciation in the *Sorex araneus* complex (Mammalia: Insectivora). *Acta Zool. Fenn.* **170**: 125–130.

Hawes, M.L. (1977) Home range, territoriality and ecological separation in sympatric shrews *Sorex vagrans* and *Sorex obscurus. J. Mammal.* **58**: 354–367.

Hawkins, B. (1988) Species diversity in the third and fourth trophic levels: patterns and mechanisms. *J. Anim. Ecol.* **57**: 137–162.

Hawkins, B. (1990) Global patterns of parasitoid assemblage size. *J. Anim. Ecol.* **59**: 57–72.

Hawkins, B. & Lawton, J.H. (1987) The determinants of species richness for the parasitoids of British phytophagous insects. *Nature* **326**: 788–790.

Hayes, A.J. (1988) A laboratory study on the predatory mite *Typhlodromus pyri* (Acarina: Phytoseiidae). II. The effect of temperature and prey consumption on the numerical response of adult females. *Res. Pop. Ecol.* **30**: 13–24.

Hayes, A.J. & McArdle, B.H. (1987) A laboratory study on the predatory mite *Typhlodromus pyri* (Acarina: Phytoseiidae). I. The effect of temperature and food consumption on the rate of development of the eggs and immature stages. *Res. Pop. Ecol.* **29**: 73–83.

Haynes, D.L. & Sisojevic, P. (1966) Predatory behavior of *Philodromus rufus* Walckenaer (Araneae: Thomisidae). *Can. Entomol.* **98**: 113–133.

Heads, P.A. & Lawton, J.H. (1984) Bracken, ants and extrafloral nectaries. II. The effect of ants on the insect herbivores of bracken. *J. Anim. Ecol.* **53**: 1015–1031.

Heath, R.B. (1987) Varicella–Zoster. In: Zuckerman, A.J., Bonatrala, J.E. & Pattison, J.R. (eds) *Principles and Practice of Clinical Virology*, pp. 51–74. John Wiley, London.

Heessen, H.J.L. & Brunsting, A.M.H. (1981) Mortality of larvae of *Pterostichus oblongopunctatus* (Fabricius) (Col., Carabidae) and *Philonthus decorus* (Gravenhorst) (Col., Staphylinidae). *Netherlands J. Zool.* **31**: 729–745.

Heglund, N.C., Taylor, C.R. & McMahon, T.A. (1974) Scaling stride frequency and gait to animal size: mice to horses. *Science* **186**: 1112–1113.

Helle, T. & Aspi, J. (1983) Does herd formation reduce insect harassment among deer? *Acta Zool. Fenn.* **175**: 129–131.

Helle, W. & Sabelis, M.W. (eds) (1985) *Spider Mites: Their Biology, Natural Enemies and Control.* Elsevier Science Publishers, Amsterdam.

Heller, J. (1975) Visual selection of shell colour in two littoral prosobranchs. *Zool. J. Linnean Soc. Lond.* **56**: 153–170.

Henwood, K. & Fabrik, A. (1979) A quantitative analysis of the dawn chorus: temporal selection for communicatory optimization. *Am. Nat.* **114**: 260–274.

Herrnstein, R.J. (1961) Relative and absolute response as a function of reinforcement. *J. Exp. Anal. Behav.* **4**: 267–272.

Herrnstein, R.J. (1964) Aperiodicity as a factor in choice. *J. Exp. Anal. Behav.* **7**: 178–182.

Herrnstein, R.J. & Vaughan, W. (1980) Melioration and behavioral allocation. In: Staddon, J.E.R. (ed.) *Limits to Action*, pp. 143–176. Academic Press, New York.

Hespenheide, H.A. (1973) Ecological inferences from morphological data. *Ann. Rev. Systemat. Ecol.* **4**: 213–229.

Hethcote, H.W. & Yorke, J.A. (1984) Gonorrhea transmission dynamics and control. *Lecture Notes in Biomathematics* **56**: 1–105.

Heyman, G.M. & Luce, R.D. (1979) Operant matching is not a logical consequence of maximizing reinforcement rate. *Anim. Learning Behav.* **7**: 133–140.

Hickey, J.J. (1942) Eastern populations of the duck hawk. *Auk* **59**: 176–204.

Hildrew, A.G. & Townsend, C.R. (1980) Aggregation, interference and foraging by larvae of *Plectrocnemia conspersa* (Trichoptera, Polycentropodidae). *Anim. Behav.* **28**: 553–560.

Hill, D.E. (1979) Orientation by jumping spiders of the genus *Phidippus* (Araneae: Salticidae) during the pursuit of prey. *Behav. Ecol. Sociobiol.* **5**: 301–322.

Hill, M.G. (1988) Analysis of the biological control of *Mythimna separata* (Lepidoptera: Noctuidae) in New Zealand. *J. Appl. Ecol.* **25**: 197–208.

Hill, W.C.O. & Rewell, R.E. (1954) The caecum of Monotremata and Marsupialia. *Trans. Zool. Soc. Lond.* **28**: 185–240.

Hindsbo, O. (1972) Effects of *Polymorphus* (Acanthocephala) on colour and behaviour of *Gammarus lacustris. Nature* **238**: 333.

Hinson, J.M. & Staddon, J.E.R. (1983) Hill-climbing by pigeons. *J. Exp. Anal. Behav.* **39**: 25–47.

Hinton, H.E. (1974) Lycaenid pupae that mimic anthropoid heads. *J. Entomol. (A)* **49**: 65–9.

Hochberg, M.E., Hassell, M.P. & May, R.M. (1989) The dynamics of host–parasitoid–pathogen interactions. *Am. Nat.* **135**: 74–94.

Hochberg, M.E. & Holt, R.D. (1990) The coexistence of competing parasites. I. The role of cross-specific infection. *Am. Nat.* **136**: 517–541.

Hochberg, M.E. & Lawton, J.H. (1990) Competition between kingdoms. *Trends Ecol. Evol.* **5**: 367–371.

Hodek, I. (1973) *Biology of Coccinellidae*, pp. 296–310. Junk, The Hague.

Höglund, N. (1964) Der Habicht *Accipiter gentilis* Linné in Fennoscandia. *Viltrevy* **2**: 195–270.

Hokkanen, H. & Pimentel, D. (1984) New approach for selecting biological control agents. *Can. Entomol.* **116**: 1109–1121.

Hölldobler, B. (1979) *Steatoda fulva* (Theridiidae), a spider that feeds on harvester ants. *Psyche* **77**: 202–208.

Hölldobler, B. & Lumsden, C.J. (1980) Territorial strategies in ants. *Science* **210**: 732–739.

Hölldobler, B. & Wilson, E.O. (1990) *The Ants.* Springer-Verlag, Berlin.

Holling, C.J. (1959a) The components of predation as revealed by a study of small mammal predation of the European pine sawfly. *Can. Entomol.* **91**: 293–320.

Holling, C.S. (1959b) Some characteristics of simple types of predation and parasitism. *Can. Entomol.* **91**: 395–398.

Holling, C.S. (1961) Principles of insect predation. *Ann. Rev. Entomol.* **6**: 163–182.

Holling, C.S. (1965) The functional response of predators to prey density and its role in mimicry and population regulation. *Mem. Entomol. Soc. Canada* **45**: 1–60.

Holling, C.S. (1966) The functional response of invertebrate predators to prey density. *Mem. Entomol. Soc. Canada* **47**: 3–86.

Holling, C.S. (1968) The tactics of a predator. In: Southwood, T.R.E. (ed.) *Insect Abundance*, pp. 47–58. *Symp. R. Entomol. Soc. Lond. 4.* Blackwell

Scientific Publications, Oxford.

Holling, C.S., Dunbrack, R.L. & Dill, L.M. (1976) Predator size and prey size: presumed relationship in the mantid *Hierodula coarctata* Saussure. *Can. J. Zool.* **54**: 1760–1764.

Hollis, K.L. (1989) In search of a hypothetical construct: a reply to Guilford & Dawkins. *Anim. Behav.* **37**: 162–163.

Holmes, J.C. (1982) Impact of infectious disease agents on the population growth and geographical distribution of animals. In: Anderson, R.M. & Mar, R.M. (eds) *Population Biology of Infectious Diseases*, pp. 37–51. Springer-Verlag, Berlin.

Holmes, J.C. (1983) Evolutionary relationships between parasitic helminths and their hosts. In: Futuyma, D.J. & Slatkin, M. (eds) *Coevolution*, pp. 161–185. Sinauer, Sunderland, MA.

Holt, R.D. (1977) Predation, apparent competition, and the structure of prey communities. *Theor. Pop. Biol.* **12**: 197–229.

Holt, R.D. (1985) Population dynamics in two-patch environments: some anomalous consequences of an optimal habitat distribution. *Theor. Pop. Biol.* **28**: 181–208.

Holt, R.D. & Kotler, B.P. (1987) Short-term apparent competition. *Am. Nat.* **130**: 412–430.

Hope-Simpson, R.E. & Golubev, D.B. (1987) A new concept in the epidemic process of influenza A virus. *Epidemiol. Infect.* **99**: 5–54.

Hornocker, M.G. (1970) An analysis of mountain lion predation upon mule deer and elk in the Idaho Primitive Area. *Wildlife Monogr.* **21**: 1–39.

Houston, A.I. (1983) Optimality theory and matching. *Behav. Anal. Lett.* **3**: 1–15.

Houston, A.I. (1987) Optimal foraging by birds feeding dependent young. *J. Theor. Biol.* **124**: 251–274.

Houston, A.I., Clarke, C.W., McNamara, J.M. & Mangel, M. (1988) Dynamic models in behavioural and evolutionary ecology. *Nature* **332**: 29–34.

Houston, A.I. & Krebs, J.R. (1989) Optimisation in ecology. In: Cherrett, J.M. (ed.) *Ecological Concepts: the Contribution of Ecology to an Understanding of the Natural World*, 29th Symp. British Ecol. Soc., pp. 309–338. Blackwell Scientific

Publications, Oxford.

Houston, A.I., McNamara, J.M. (1982) A sequential approach to risk-taking. *Anim. Behav.* **30**: 1260–1261.

Houston, A.I. & McNamara, J.M. (1985) The choice of two prey types that minimises the probability of starvation. *Behav. Ecol. Sociobiol.* **17**: 135–141.

Howarth, F.G. (1983) Classical biological control: panacea or Pandora's box? *Proc. Hawaiian Entomol. Soc.* **24**: 239–244.

Hoy, M.A. (1985) Improving establishment of arthropod natural enemies. In: Hoy, M.A. & Herzog, D.C. (eds) *Biological Control in Agricultural IPM Systems*, pp. 151–166. Academic Press, London.

Hoy, M.A., Cunningham, G.L. & Knutson, L. (1983) *Biological Control of Pests by Mites*. University of California Special Publication no. 3304, Berkeley.

Hudson, P.J. (1986) *Red Grouse: the Biology and Management of a Wild Gamebird*. The Game Conservancy Trust, Fordingbridge.

Hudson, P.J., Dobson, A.P. & Newborn, D. (1992) The population dynamics of *Trichostrongylus tenuis* in red grouse. *J. Anim. Ecol.* **61** (in press).

Huey, R.B. & Pianka, E.R. (1981) Ecological consequences of foraging mode. *Ecology* **62**: 991–999.

Huffaker, C.B. (1958) Experimental studies on predation: dispersion factors and predator–prey oscillations. *Hilgardia* **27**: 343–383.

Huffaker, C.B. & Kennet, C.E. (1959) A ten year study of vegetational changes associated with biological control of klamath weed. *J. Range Manage.* **12**: 69–82.

Huffaker, C.B. & Kennett, C.E. (1966) Studies of two parasites of olive scale, *Parlatoria oleae* (Colvee): IV. Biological control of *Parlatoria oleae* (Colvee) through the compensatory action of two introduced parasites. *Hilgardia* **37**: 283–335.

Huffaker, C.B., Shea, K.P. & Herman, S.G. (1963) Experimental studies on predation: Complex dispersion and levels of food in an acarine predator–prey interaction. *Hilgardia* **34**: 305–330.

Huffaker, C.B., Kennett, C.E. & Tassan, R.L. (1986) Comparisons of parasitism and densities of *Parlatoria oleae*

(1952–1982) in relation to ecological theory. *Am. Nat.* **128**: 379–393.

Huffaker, C.B. & Rabb, R.L. (eds) (1984) *Ecological Entomology*. John Wiley & Sons, New York.

Hughes, R.D., Duncan, P. & Dawson, J. (1981) Interactions between Camargue horses and horseflies (Diptera: Tabanidae). *Bull. Entomol. Res.* **71**: 227–242.

Hughes, R.N. (1980a) Predation and community structure. In: Price, J.H., Irvine, D.E.G. & Farnham, W.F. (eds) *The Shore Environment, Volume 2, Ecosystems*, pp. 699–728. Systematic Association and Academic Press, London.

Hughes, R.N. (1980b) Optimal foraging theory in the marine context. *Ann. Rev. Oceanography Marine Biol.* **18**: 423–481.

Hughes, R.N. (ed.) (1990) Behavioural mechanisms of food selection. *NATO ASI Series G. Ecological Sciences, Vol. 20*. Springer-Verlag, Berlin.

Hughes, R.N. & Dunkin, S. de B. (1984a) Behavioural components of prey selection by dogwhelks, *Nucella lapillus* (L.), feeding on mussels, *Mytilus edulis* (L.), in the laboratory. *J. Exp. Marine Biol. Ecol.* **79**: 159–172.

Hughes, R.N. & Dunkin, S. de B. (1984b) Effects of dietary history on selection of prey, foraging behaviour among patches of prey by the dogwhelk *Nucella lapillus* (L.) *J. Exp. Marine Biol. Ecol.* **77**: 45–68.

Humphrey, S.R., Courtney, C.H. & Forrester, D.J. (1978) Community ecology of helminth parasites of the brown pelican. *The Wilson Bulletin* **90**: 587–598.

Humphreys, W.F. (1987) Behavioral temperature regulation. In: Nentwig, W. (ed.) *Ecophysiology of Spiders*, pp. 56–65. Springer-Verlag, New York.

Hutchinson, G.E. (1961) The paradox of the plankton. *Am. Nat.* **95**: 137–145.

Hutchinson, G.E. & MacArthur, R.H. (1959) A theoretical ecological model of size distributions among species of animals. *Am. Nat.* **93**: 117–125.

ICES (International Council for the Exploration of the Sea) (1975). Report of the Herring Assessment Working Group for the Area South of 62 Degrees North. ICES-CM-1975/H:2.

ICES (International Council for the Exploration of the Sea) (1981).

Interaction between grey seal populations and fish species. ICES Cooperative Research report No. 101.

Ims, R.A. (1990) On the adaptive value of reproductive synchrony as a predator-swamping strategy. *Am. Nat.* **136**: 485–498.

Ims, R.A. & Steen, H. (1990) Geographical synchrony in microtine population cycles: a theoretical evaluation of the role of nomadic avian predators. *Oikos* **57**: 381–387.

Inglis, I.R., Huson, L.W., Marshall, M.B. & Neville, P.A. (1983) The feeding behaviour of starlings (*Sturnus vulgaris*) in the presence of 'eyes'. *Z. Tierpsychol.* **62**: 181–208.

Inman, A. (1990) Studies of Group Foraging. D. Phil. thesis, Oxford University.

Iperti, G. (1966) Specificity of aphidophagous coccinellids in south-eastern France. In: Hodek, I. (ed.) *Ecology of Aphidophagous Insects*, pp. 31–34. Dr. W. Junk, The Hague.

Itamies, J., Valtonen, E.T. & Feagerholm, H.P. (1980) *Polymorphus minutus* infestation in eiders and its role as a possible cause of death. *Ann. Zool. Fenn.* **17**: 285–289.

Ito, Y. (1964) Preliminary studies on the respiratory energy loss of a spider, *Lycosa pseudoannulata. Res. Popul. Ecol.* **6**: 13–21.

Ives, A.R. (1988) Aggregation and the coexistence of competitors. *Ann. Zool. Fenn.* **25**: 75–88.

Ives, A.R. & Dobson, A.P. (1987) Antipredator behavior and the population dynamics of simple predator–prey systems. *Am. Nat.* **130**: 431–447.

Ivlev, V.S. (1961) *Experimental Ecology of the Feeding of Fishes.* (Translated from Russian by D. Scott.) Yale University Press, New Haven.

Iwasa, Y. (1981) Role of sex ratio in the evolution of eusociality in haplodiploid social insects. *J. Theor. Biol.* **93**: 125–142.

Iwasa, Y, Suzuki, Y. & Matsuda, H. (1984) Theory of oviposition strategy of parasitoids I. Effect of mortality and limited egg number. *Theor. Pop. Biol.* **26**: 205–227.

Jackson, R.R. & Blest, A.D. (1982) The biology of *Portia fimbriata*, a web-building jumping spider (Araneae, Salticidae) (London) *J. Zool.* **196**: 295–305.

Jaenike, J. (1978) A hypothesis to account for the maintenance of sex within populations. *Evol. Theory* **3**: 191–194.

Jagsch, A. (1972) Populationsdynamik und Parasitenkomplex der Laerchenminiermotte, *Coleophora laricella* Hbn. im Naturlichen Verbreitungsgebiet der Eurpaeischen Larche, *Larix decidua* Mill. *Z. Angew. Entomol.* **73**: 1–42.

Janetos, A.C. (1982a) Active foragers vs. sit and wait predators: a simple model. *J. Theor. Biol.* **95**: 381–385.

Janetos, A.C. (1982b) Foraging tactics of two guilds of web-spinning spiders. *Behav. Ecol. Sociobiol.* **10**: 19–27.

Janssen, A. (1989) Optimal host selection by *Drosophila* parasitoids in the field. *Funct. Ecol.* **3**: 469–479.

Janzen, D.H. (1974) Tropical blackwater rivers, animals, and mast fruiting of the Dipterocarpaceae. *Biotropica* **6**: 69–113.

Janzen, D.H. (1981) Evolutionary physiology of personal defence. In: Townsend, C.R. & Calaw, P. (eds) *Physiological Ecology: An Evolutionary Approach to Resource Use*, pp. 145–164. Blackwell Scientific Publications, Oxford.

Jarman, P.J. (1974) The social organisation of antelope in relation to their ecology. *Behaviour* **48**: 215–267.

Jayakar, S.D. (1970) A mathematical model for interaction of gene frequencies in a parasite and its host. *Theor. Pop. Biol.* **1**: 140–164.

Jeffords, M.R., Sternburg, J.G. & Waldbauer, G.P. (1979) Batesian mimicry: field demonstration of the survival value of pipevine swallowtail and monarch color patterns. *Evolution* **33**: 275–286.

Jeffries, M.J. & Lawton, J.H. (1984) Enemy free space and the structure of ecological communities. *Biol. J. Linnean Soc.* **23**: 269–286.

Jenkins, D., Watson, A. & Miller, G.R. (1964) Predation and red grouse populations. *J. Appl. Ecol.* **1**: 183–195.

Jenner, C.E. (1958) An attempted analysis of schooling behavior in the marine snail *Nassarius obsoletus. Biol. Bul.* **115**: 337–338.

Jermy, T. (1984) Evolution of insect/host plant relationships. *Am. Nat.* **124**: 609–630.

Johnson, D.M., Akre, B.C. & Crowley, P.H. (1975) Modeling arthropod predation, wasteful killing by damselfly naiads. *Ecology* **36**: 1081–1093.

Jones, J.S., Leith, B.H. & Rawlings, P. (1977) Polymorphism in *Cepea*: a problem with too many solutions? *Ann. Rev. Ecol. System.* **8**: 109–143.

Jones, R.B. (1980) Reactions of male domestic chicks to two-dimensional eye-like shapes. *Anim. Behav.* **28**: 212–218.

Jones, R.L. (1981) Report of the USDA Biological Control of Stem Borers Study Team's Visit to the People's Republic of China. University of Minnesota, Minneapolis.

Jones, T.H. & Hassell, M.P. (1988) Patterns of parasitism by *Trybliographa rapae*, a cynipid parasitoid of the cabbage root fly, under laboratory and field conditions. *Ecol. Ent.* **13**: 309–317.

Jonsingh, A.J.T. (1983) Large mammalian prey — predators in Bandipur. *J. Bombay Nat. Hist. Soc.* **80**: 1–57.

Jordan, P. & Webbe, G. (1969) *Human Schistomiasis*. William Heinemann, London.

Jordon, P.A. Shelton, P.C. & Allen, D.L. (1967) Numbers, turnover, and social structure of the Isle Royale wolf population. *Am. Zool.* **7**: 233–252.

Judin, B.S. (1962) *Ecology of Shrews (genus* Sorex*) of Western Siberia.* Trudy Biol. Inst. Akad. Nauk SSSR, Sib. Otd., Novosibirsk.

Juliano, S.A. & Lawton, J.H. (1990) The relationship between competition and morphology. I. Morphological patterns among co-occurring dytiscid beetles. *J. Anim. Ecol.* **59**: 403–419.

Juutinen, P. (1967) Zur Bionomie und zum Vorkommen der roten Kiefernbuschhornblattwespe (*Neodiprion sertifer* Geoffr.) in Finnland in den Jahren 1959–65. *Comm. Inst. For. Feen.* **63**: 1–129.

Kacelnik, A. (1984) Central place foraging in starlings (*Sturnus vulgaris*). I. Patch residence time. *J. Anim. Ecol.* **53**: 283–299.

Kacelnik, A. (1988) Short-term adjustments of parental effort in starlings. In: Ovellet, H. (ed.) *Acta XIX Congressus Internationalis Ornithologici*, Vol. II, pp. 1843–1856. University of Ottawa Press, Ottawa.

Kacelnik, A., Bruner, D. & Gibbon, J. (1990) Timing mechanisms in optimal foraging: some applications of scalar expectancy theory. In: Hughes, R.N. (ed.) *Behavioural Mechanisms of Food Selection*, pp. 61–82. Springer-Verlag, Berlin.

Kacelnik, A. & Cuthill, I.C. (1987) *Foraging Behaviour*. Plenum Press, New York.

Kacelnik, A. & Cuthill, I.C. (1990) Central place foraging in starlings (*Sturnus vulgaris*). II. Food allocation to chicks. *J. Anim. Ecol.* **59**: 655–674.

Kacelnik, A. & Houston, A.I. (1984) Some effects of energy costs on foraging strategies. *Anim. Behav.* **32**: 609–614.

Kacelnik, A., Houston, A.I. & Schmid-Hempel, P. (1986) Central-place foraging in honey bees: the effect of travel time and nectar flow on crop filling. *Behav. Ecol. Sociobiol.* **19**: 19–24.

Kacelnik, A. & Krebs, J.R. (1983) The dawn chorus of the great tit: proximate and ultimate causes. *Behaviour* **83**: 287–308.

Kagel, J.H., Green, L. & Caraco, T. (1986a) When foragers discount the future: constraint or adaptation? *Anim. Behav.* **34**: 271–283.

Kagel, J.H., MacDonald, D.N., Battalio, R.C., White, S. & Green, L. (1986b) Risk aversion in rats *Rattus norvegicus* under varying levels of resource availability. *J. Comp. Psychol.* **100**: 95–100.

Kaikusalo, A. & Hanski, I. (1985) Population dynamics of *Sorex araneus* and *S. caecutiens* in Finnish Lapland. *Acta Zool. Fenn.* **173**: 283–285.

Kajak, A. (1967) Productivity of some populations of web spiders. In: Petrusewicz, K. (ed.) *Secondary Productivity of Terrestrial Ecosystems*, pp. 807–820. Warsaw.

Kajak, A. (1978) Analysis of consumption by spiders under laboratory and field conditions. *Ekol. Pol.* **6**: 409–427.

Kakehashi, N., Suzuki, Y. & Iwasa, Y. (1984) Niche overlap of parasitoids in host–parasitoid systems: its consequence to single versus multiple introduction controversy in biological control. *J. Appl. Ecol.* **21**: 115–131.

Kalinowski, J. & Witek, Z. (1981) The physical parameters of krill aggregations in the Western Antarctic. International Council for the Exploration of the Sea. C.M. 1981/1:19. Biological Oceanographic Committee.

Kamil, A.C. (1989) Studies of learning and memory in natural contexts: integrating functional and mechanistic approaches to behavior. In: Blanchard, R.J., Brain, P.F., Blanchard, D.C. & Parmigiani, S. (eds) *Ethoexperimental Approaches to the Study of Behavior*, pp. 30–50. Kluwer Academic Publishers, Dordrecht.

Kamil, A.C., Krebs, J.R. & Pulliam, H.R. (eds) (1987) *Foraging Behavior*. Plenum Press, New York.

Kamil, A.C. & Roitblat, H.L. (1985) The ecology of foraging behavior: implications for animal learning and memory. *Ann. Rev. Psychol.* **36**: 141–169.

Kamil, A.C. & Sargent, T.D. (eds) (1981) *Foraging Behavior: Ecological, Ethological and Psychological Approaches*. Garland, New York.

Kanwisher, J. & Sundnes, G. (1966) Thermal regulation in cetaceans. In: Norris, K.S. (ed.) *Whales, Dolphins and Porpoises*, pp. 397–409. University of California Press, Los Angeles.

Karban, R. & Ricklefs, R.E. (1983) Host characteristics, sampling intensity, and species richness of lepidoptera larvae on broad-leaved trees in southern Ontario. *Ecology* **64**: 636–641.

Kareiva, P. (1986) Habitat fragmentation and the stability of predator–prey interactions. *Nature* **326**: 388–390.

Kareiva, P. (1989) Renewing the dialogue between theory and experiments in population ecology. In: Roughgarden, J., May, R.M. & Levin, S.A. (eds) *Perspectives in Ecological Theory*, pp. 68–88. Princeton University Press, Princeton.

Kareiva, P. & Odell, G. (1987) Swarms of predators exhibit 'preytaxis' if individual predators use area-restricted search. *Am. Nat.* **130**: 233–270.

Kayashima, I. (1961) Study of the lynx spider *Oxyopes sertatus* L. Koch, for the biological control of the crytomerian leaf fly *Contarina inouyei* Mani. *Rev. Appl. Entomol.* **51**: 413.

Kaye, H., Mackintosh, N.J., Rothschild, M. & Moore, B.P. (1989) Odour of pyrazine potentiates an association between environmental cues and unpalatable taste. *Anim. Behav.* **37**: 563–568.

Kayes, R.J. (1985) The decline of porpoises and dolphins in the southern North Sea: a current status report. Political Ecology Research Group Report RR-14.

Keeney, R.C. & Raiffa, H. (1976) *Decisions With Multiple Objectives: Preferences and Value Trade-offs*. John Wiley & Sons, New York.

Keith, L.B. (1963) *Wildlife's Ten-year Cycle*. University of Wisconsin Press, Madison.

Keith, L.B. (1974) Some features of population dynamics in mammals. In: *XI International Congress of Game Biologists*, Stockholm, 3–7 September 1973, pp. 17–58. National Swedish Environmental Protection Board.

Keith, L.B., Cary, J.R., Rongstad, O.J. & Brittingham, M.C. (1984) Demography and ecology of a declining snowshoe hare population. *Wildlife Monogr.* **90**: 1–43.

Keith, L.B. & Rusch, D.H. (1988) Predation's role in the cyclic fluctuations of ruffed grouse. *Proc. Int. Orn. Congr.* **19**: 699–732.

Keith, L.B. & Windberg, L.A. (1978) A demographic analysis of the snowshoe hare cycle. *Wildlife Monogr.* **58**: 4–70.

Keller, M.A. (1984) Reassessing evidence for competitive exclusion of introduced natural enemies. *Environ. Entomol.* **13**: 192–195.

Kelley, S.E., Antonovics, J. & Schmitt, J. (1988) A test of the short-term advantage of sexual reproduction. *Nature* **331**: 714–716.

Kenny, R.D., Hyman, M.A.M., Owen, R.E., Scott, G.P. & Winn, H.E. (1986) Estimation of prey densities required by western North Atlantic right whales. *Marine Mammal Sci.* **2**: 1–13.

Kenward, R.E. (1977) Predation on released Pheasants (*Phasianus colchicus*) by Goshawks (*Accipiter gentilis*) in central Sweden. *Viltrevy* **10**: 79–112.

Kenward, R.E. (1978) Hawks and doves: factors affecting success and selection in goshawk attacks on woodpigeons. *J. Anim. Ecol.* **47**: 449–460.

Kenward, R.E. (1985) Problems of goshawk predation on pigeons and other game. *Proc. Int. Orn. Congr.* **18**: 666–678.

Kenward, R.E., Marcstrom, V. & Karlbolm, M. (1981) Goshawk winter ecology in Swedish pheasant habitats.

J. Wildlife Manage. **45**: 397–408.

Kerfoot, W.C. & Sih, A. (eds) (1987) *Predation: Direct and Indirect Impacts on Aquatic Communities.* University Press of New England, Hanover, NH.

Kermack, W.O. & McKendrick, A.G. (1927) A contribution to the mathematical theory of epidemics. *Proc. R. Soc. Lond. A* **115**: 700–721.

Kettlewell, B. (1973) *The Evolution of Melanism.* Clarendon Press, Oxford.

Kettlewell, H.B.D. (1955) Recognition of appropriate backgrounds by the pale and black phases of lepidoptera. *Nature* **175**: 943–944.

Keymer, A.E. (1980). The influence of *Hymenolepis diminuta* on the survival and fecundity of the intermediate host, *Tribolium confusum. Parasitology* **81**: 405–421.

Keymer, A.E. (1982) Density-dependent mechanisms in the regulation of intestinal helminth populations. *Parasitology* **84**: 537–587.

Keymer, A.E. & Read, A.F. (eds) (1990) The Evolutionary Biology of Parasitism. Symposia of the British Society of Parasitology, 27. *Parasitology* **100** (suppl.).

Kidd, N.A.C. (1982) Predator avoidance as a result of aggregation in the grey pine aphid, *Schizolachnus pineti. J. Anim. Ecol.* **51**: 397–412.

Kidd, N.A.C. & Mayer, A.D. (1983) The effect of escape responses on the stability of insect host–parasite models. *J. Theor. Biol.* **104**: 275–287.

King, C.M. (1989) The advantages and disadvantages of small size to weasels, *Mustela.* In: Gittleman, J.L. (ed.) *Carnivore Behavior, Ecology, and Evolution,* pp. 302–334. Chapman and Hall, London.

King, J.R. (1972) Adaptive periodic fat storage by birds. In: Voous, K.H. (ed.) *Proceedings of the XVth International Ornithological Congress,* pp. 200–217. Brill, Leiden.

Kingsland, S. (1985) *Modeling Nature.* University of Chicago Press, Chicago.

Kingsolver, J.G. (1988a) Thermoregulation, flight, and the evolution of wing pattern in Pierid butterflies: the topography of adaptive landscapes. *Am. Zool.* **28**: 899–912.

Kingsolver, J.G. (1988b) Mosquito host choice and the epidemiology of malaria. *Am. Nat.* **130**: 811–827.

Kiritani, K. & Kakiya, K. (1975) An analysis of the predator–prey system in the paddy field. *Res. Pop. Ecol.* **17**: 29–38.

Kirkwood, J.K. (1981) Maintenance energy requirements and rate of weight loss during starvation in birds of prey. In: Cooper, J.E. & Greenwood, A.G. (eds) *Recent Advances in the Study of Raptor Diseases,* pp. 153–157. Chiron Publications, Keighly.

Kitching, J.A., Muntz, L. & Ebling, F.J. (1966) The ecology of Lough Ine: XV. The ecological significance of shell and body form in *Nucella. J. Anim. Ecol.* **35**: 113–126.

Kitching, R.L. & Pearson, J. (1981) Prey location by sound in a predatory intertidal gastropod. *Marine Biol. Lett.* **2**: 313–333.

Kleiber, M. (1932) Body size and metabolism. *Hilgardia* **6**: 315–353.

Kleiber, M. (1961) *The Fire of Life: An Introduction to Animal Energetics.* Wiley & Sons, New York.

Kleiber, M. (1975) *The Fire of Life: An Introduction to Animal Energetics.* 2nd edn. John Wiley & Sons, New York.

Kohn, A.J. & Waters, V. (1966) Escape responses of three herbivorous gastropods to the predatory gastropod *Conus textile. Anim. Behav.* **14**: 340–345.

Kolmogorov, A.N. (1936) Sulla Teoria di Volterra della Lotta per l'Esisttenza. *Giorn. Inst. Ital. Attuari* **7**: 74–80.

Kolomiets, N.G., Stadnitskii, G.V. & Vorontzov, A.I. (1979). *The European Pine Sawfly.* Amerind Publishing, New Delhi.

Konings, E., Anderson, R.M., Morley, D., O'Riordan, T. & Meegan, M. (1988) Rates of sexual partner change among 2 pastoralist niolotic groups in East Africa. *AIDS* **2**: 245–247.

Kooyman, S.A.L.M. & Metz, J.A.J. (1984) On the dynamics of chemically stressed populations: the deduction of population consequences from effects on individuals. *Ecotoxicol. Environ. Safety* **8**: 254–274.

Korpimaki, E. (1985) Rapid tracking of microtine populations by their avian predators: possible evidence for stabilising predation. *Oikos* **45**: 281–284.

Korpimaki, E. (1988) Factors promoting polygyny in European birds of prey — a hypothesis. *Oecologia* **77**: 278–285.

Korpimaki, E. & Norrdahl, K. (1988) Predation of Tengmalm's owls:

numerical responses, functional responses and dampening impact on population fluctuations of microtines. *Oikos* **54**: 154–164.

Kotler, B.P. & Brown, J.S. (1988) Environmental heterogeneity and the coexistence of desert rodents. *Ann. Rev. Ecol. Systemat.* **19**: 281–307.

Kotler, B.P., Brown, J.S., Smith, R.J. & Wirtz, W.O. II (1988) The effects of morphology and body size on rates of owl predation on desert rodents. *Oikos* **53**: 145–152.

Kraan van der, C. & Strien van, N.J. (1969) Polygamie bij de Blauwe Kiekendief (*Circus cyaneus*). *Limosa* **42**: 34–35.

Krakauer, T. (1972) Thermal responses of the orb weaving spider, *Nephila clavipes* (Araneae: Argiopidae). *Am. Midl. Nat.* **88**: 245–250.

Krantz, G.W. (1983) Mites as biological control agents of dung-breeding flies, with special reference to the Macrochelidae. In: Hoy, M.A., Cunningham, G.L. & Knutson, L. (eds) *Biological Control of Pests by Mites*, pp. 91–98. University of California Special Publication no. 3304, Berkeley.

Krebs, C.J., Gilbert, B.S., Boutin, S., Sinclair, A.R.E. & Smith, J.N.M. (1986) Population biology of snowshoe hares. I. Demography of food-supplemented populations in the southern Yukon, 1976–84 *J. Anim. Ecol.* **55**: 963–982.

Krebs, C.J. & Myers, J.H. (1974) Population cycles in small mammals. *Adv. Ecol. Res.* **8**: 267–399.

Krebs, J.R. (1980) Optimal foraging, predation risk and territory defence. *Ardea* **68**: 83–90.

Krebs, J.R. & Avery, M.I. (1985) Central place foraging theory in the European bee-eater (*Merops apiaster*). *J. Anim. Ecol.* **54**: 459–472.

Krebs, J.R. & Davies, N.B. (1987) *An Introduction to Behavioural Ecology*, 2nd edn. Blackwell Scientific Publications, Oxford.

Krebs, J.R., Erichsen, J.T., Webber, M.I. & Charnov, E.L. (1977) Optimal prey selection in the great tit (*Parus major*). *Anim. Behav.* **25**: 30–38.

Krebs, J.R. & Houston, A.I. (1989) Optimization in ecology. In: Cherrett, J.M. (ed.) *Ecological Concepts*, pp. 309–338. Blackwell Scientific Publications, Oxford.

Krebs, J.R. & May, R.M. (1990) The moorland owner's grouse. *Nature* **343**: 310–311.

Krebs, J.R. & McCleery, R.H. (1984) Optimisation in behavioural ecology. In: Krebs, J.R. & Davies, N.B. (eds) *Behavioural Ecology: An Evolutionary Approach*, 2nd edn, pp. 91–121. Blackwell Scientific Publications, Oxford.

Krebs, J.R. & Kacelnik, A. (1991) Decision-making. In: Krebs, J.R. & Davies, N.B. (eds) *Behavioural Ecology: An Evolutionary Approach*, 3rd edn, pp. 105–136. Blackwell Scientific Publications. Oxford.

Kronk, A.E. & Riechert, S.E. (1979) Parameters affecting the habitat choice of a desert wolf spider, *Lycosa santrita* Chamberlin & Ivie. *J. Arachnol.* **7**: 155–166.

Krugman, S. (ed.) (1965) Rubella symposium. *Am. J. Dis. Child.* **110**: 345–476.

Kruuk, H. (1964) Predators and anti-predator behaviour of the black-headed gull *Larus ridibundus*. *Behaviour* (suppl.) **11**: 1–129.

Kruuk, H. (1972) *The Spotted Hyaena: A Study of Predation and Social Behavior*. Chicago University Press, Chicago.

Kuchlein, J.H. (1966) Mutual interference among the predacious mite, *Typhlodromus longipilus* Nesbitt (Acari, Phytoseiidae). I. Effects of predator density on oviposition rate and migration tendency. *Mededelingen van de Rijksfaculteit der Landbouwwetenschappen in Gent* **31**: 740–746.

Kunz, T.H. (1974) Feeding ecology of a temperate insectivorous bat (*Myotis velifer*). *Ecology* **55**: 693–711.

Kunz, T.H. (ed.) (1982) *Ecology of Bats*. Plenum Press, New York.

Kuyt, E. (1972) Food habits and ecology of wolves on barren-ground caribou range in the Northwest Territories. Canadian Wildlife Service Report No. 21.

Lack, D. (1947) The significance of clutch size. *Ibis* **89** 309–352.

Lack, D. (1954) *The Natural Regulation of Animal Numbers*. Oxford University Press, Oxford.

Laevastu, T. & Favorite, F. (1977) Preliminary Report on Dynamic Numerical Marine Ecosystem Model (DYNUMES II) for Eastern Bering Sea. Processed Report. NOAA Natl. Mar.

Fish. Serv. Northwest and Alaska Fisheries Center.

Lamb, M.R. (1988) Selective attention — effects of cueing on the processing of different types of compound stimuli. *J. Exp. Psychol: Anim. Behav. Proc.* **14**: 96–104.

Lamprecht, J. (1978) On diet, foraging behaviour and interspecific food competition of jackals in the Serengeti National Park, East Africa. *Z. Saeugetierkd.* **43**: 210–223.

LaMunyon, C.W. & Adams, P.A. (1987) Use and effect of an anal defensive secretion in larval Chrysopidae (Neuroptera). *Ann. Entomol. Soc. Am.* **80**: 804–808.

Lande, R. (1987) Extinction thresholds in demographic models of territorial populations. *Am. Nat.* **130**: 624–635.

Lauenstein, G. (1980) Zum Suchverhalten von *Anthocoris nemorum* L. (Heteroptera: Anthocoridae). *Z. Angewandte Entomol.* **89**: 428–442.

Lavigne, D.M., Innes, S., Worthy, G.A.J., Kovacs, K.M., Schmitz, O.J. & Hickie, J.P. (1986) Metabolic rates of seals and whales. *Can. J. Zool.* **64**: 279–284.

Lawrence, E.S. (1985a) Evidence for search image in blackbirds (*Turdus merula* L): long-term learning. *Anim. Behav.* **33**: 1301–1309.

Lawrence, E.S. (1985b) Evidence for search image in blackbirds (*Turdus merula* L): short-term learning. *Anim. Behav.* **33**: 929–937.

Lawrence, E.S. (1986) Can great tits (*Parus major*) acquire search images? *Oikos* **47**: 3–12.

Lawrence, E.S. (1989) Why blackbirds overlook cryptic prey: search rate or search image? *Anim. Behav.* **37**: 157–160.

Lawrence, E.S. & Allen, J.A. (1983) On the term 'search image'. *Oikos* **40**: 313–314.

Laws, R.M. (1960) Problems of whale coservation. *Trans. N. Am. Wildlife Conf.* **25**: 304–319.

Laws, R.M. (1977) The significance of vertebrates in the Antarctic marine ecosystem. In: Llano, G.A. (ed.) *Adaptations within Antarctic Ecosystems*. Third Symposium on Antarctic Ecology. Scientific Committee for Antarctic Research, Smithsonian Institute, Washington.

Lawton, J.H. & Hassell, M.P. (1981) Asymmetrical competition in insects. *Nature* **289**: 793–795.

Lawton, J.H. & Hassell, M.P. (1984) Interspecific competition in insects. In: Huffaker, C.B. & Rabb, R.L. (eds) *Ecological Entomology*, pp. 451–495. John Wiley & Sons, New York.

Lawton, J.H. & McNeill, S. (1979) Between the devil and the deep blue sea: on the problem of being a herbivore. In: Anderson, R.M., Turner, B.D. & Taylor, L.R. (eds) *Population Dynamics*, pp. 223–244. Blackwell Scientific Publications, Oxford.

Lawton, J.H. & Strong, D.R. Jr (1981) Community patterns and competition in folivorous insects. *Am. Nat.* **118**: 317–383.

Lazarus, J. (1979) The early warning function of flocking in birds: an experimental study with captive *Ouelea. Anim. Behav.* **27**: 855–865.

Lee, A.K., Bradley, A.J. & Braithwaite, R.W. (1977) Corticosteroid levels and male mortality in *Antechinus stuartii*. In: Stonehouse, B. & Gilmore, D. (eds) *The Biology of Marsupials*, pp. 209–220. Macmillan, London.

Lee, A.K., Woolley, P. & Braithwaite, R.W. (1982) Life history strategies of dasyurid marsupials. In: Archer, M. (ed.) *Carnivorous Marsupials*, pp. 1–11. Royal Zool. Soc. New South Wales, Sydney.

Lee, K.L. & Cockburn, A. (1985) *Evolutionary Ecology of Marsupials*. Cambridge University Press, Cambridge.

Legg, G. & Jones, R.E. (1988) *Pseudoscorpions*, p. 37. Brill/ Backhuys, Leiden.

Leimar, O., Enquist, M. & Sillén-Tullberg, B. (1986) Evolutionary stability of aposematic coloration and prey unprofitability: a theoretical analysis. *Am. Nat.* **128**: 469–490.

Le Masurier, A.D. (1987) A comparative ¨tudy of the relationship between host size and brood size in *Apanteles* spp. (Hymenoptera: Braconidae). *Ecol. Entomol.* **12**: 383–393.

Lenski, R.E. (1984a) Coevolution of bacteria and phage: are there endless cycles of bacterial defenses and phage counterdefenses? *J. Theor. Biol.* **108**: 319–325.

Lenski, R.E. (1984b) Food limitation and competition: A field experiment with two *Carabus* species. *J. Anim. Ecol.* **53**: 203–216.

Lenski, R.E. (1988a) Experimental

studies of pleiotropy and epistasis in *Escherichia coli*. I. Variation in competitive fitness among mutants resistant to virus T4. *Evolution* **42**: 425–432.

Lenski, R.E. (1988b) Experimental studies of pleiotropy and epistasis in *Escherichia coli*. II. Compensation for maladaptive effects associated with resistance to virus T4. *Evolution* **42**: 425–432.

Lenski, R.E. & Levin, B.R. (1985) Constraints on the coevolution of bacteria and virulent phage: a model, some experiments, and predictions for natural communities. *Am. Nat.* **125**: 585–602.

Lent, P.C. (1974) Mother–infant relationships in ungulates. In: Geist, V. & Walther, F. (eds) *The Behaviour of Ungulates and its Relation to Management. IUCN Publ. New Ser.* **24**: 14–55.

Leonard, K.J. (1969) Genetic equilibria in host–pathogen systems. *Phytopathology* **59**: 1858–1863.

Leonard, K.J. (1977) Selection pressures and plant pathogens. In: Day, P.R. (ed.) *The Genetic Basis of Epidemics in Agriculture. Ann. NY Acad. Sci.* **287**: 207–222.

LeSar, C.D. & Unzicker, J.D. (1978) Life history, habits and prey preferences of *Tetragnatha laboriosa. Environ. Entomol.* **7**: 879–884.

Lessells, C.M. (1985) Parasitoid foraging: should parasitism be density-dependent? *J. Anim. Ecol.* **54**: 27–41.

Letourneau, D.K. (1987) The enemies hypothesis: tritrophic interactions and vegetational diversity in tropical ecosystems. *Ecology* **68**: 1616–1622.

Lett, B.T. (1980) Taste potentiates colour-sickness associations in pigeons and quail. *Anim. Learning Behav.* **8**: 193–198.

Letts, G.A. (1979) (ed.) *Feral Animals in the Northern Territory.* Report of Board of Inquiry, Government Printer, Darwin.

Leuthold, W. (1977) *African Ungulates: A Comparative Review of their Ethology and Behavioural Ecology.* Springer-Verlag, New York.

Levin, B.R. & Lenski, R.E. (1983) Coevolution in bacteria and their viruses and plasmids. In: Futuyma, D.J. & Slatkin, M. (eds) *Coevolution,* pp. 99–127. Sinauer, Sunderland, MA.

Levin, B.R. & Lenski, R.E. (1985)

Bacteria and phage: a model system for the study of the ecology and co-evolution of host and parasites. In: Rollinson, D. & Anderson, R.M. (eds) *Ecology and Genetics of Host–Parasite Interactions,* pp. 227–242. Linnean Society, London.

Levin, S.A. (1972) A mathematical analysis of the genetic feedback mechanism. *Am. Nat.* **106**: 145–164.

Levin, S. & Pimentel, D. (1981) Selection of intermediate rates of increase in parasite–host systems. *Am. Nat.* **117**: 308–315.

Levin, S.A. & Segel, L. (1982) Models of the influence of predation on aspect diversity in prey populations. *J. Math. Biol.* **14**: 253–284.

Levin, S.A. & Udovic, J.D. (1977) A mathematical model of coevolving populations. *Am. Nat.* **111**: 657–675.

Lewis, J.W. (1981a) On the coevolution of pathogen and host: I. General theory of discrete time coevolution. *J. Theor. Biol.* **93**: 927–951.

Lewis, J.W. (1981b) On the coevolution of pathogen and host: II. Selfing hosts and haploid pathogens. *J. Theor. Biol.* **93**: 953–985.

Lewis, W.J., Sparks, A.N. & Redlinger, L.M. (1971) Moth odour: a method of host-finding by *Trichogramma evanescens. J. Econ. Entomol.* **64**: 557–558.

Lewontin, R.C. (1979) Fitness survival and optimality. In: Horn, R.D., Stairs, E.R. & Mitchell, R.D. (eds) *Analysis of Ecological Systems,* pp. 3–21. Ohio State University Press, Columbus.

Lewontin, R.C. (1984) Adaptation. In: Sober, E. (ed.) *Conceptual Issues in Evolutionary Biology,* pp. 237–251. MIT Press, Cambridge, MA.

Lewontin, R.C. (1987) The shape of optimality. In: Dupre, J. (ed.) *The Latest on the Best,* pp. 151–159. MIT Press, Cambridge, MA.

Lima, S. (1983) Downy woodpeckers' foraging behaviour: efficient sampling in simple stochastic environments. *Ecology* **65**: 166–174.

Lima, S.L. (1986) Predation risk and unpredictable feeding conditions: determinants of body mass in birds. *Ecology* **67**: 377–385.

Lima, S.L. & Dill, L.M. (1990) Behavioral decisions made under the risk of predation: a review and prospectus. *Can. J. Zool.* **68**: 619–640.

Linden, H. & Wikman, M. (1983)

Goshawk predation on tetraonids: availability of prey and diet of the predator in the breeding season. *J. Anim. Ecol.* **52**: 953–968.

Lindstedt, S.L., Miller, B.J. & Buskirk, S.W. (1986) Home range, time and body size in mammals. *Ecology* **67**: 413–418.

Litvaitis, J.A., Stevens, C.L. & Mautz, W.W. (1984) Age, sex, and weight of bobcats in relation to winter diet. *J. Wildlife Manage.* **48**: 632–635.

Lively, C.M. (1989) Adaptation by a parasitic trematode to local populations of its snail host. *Evolution* **43**: 1663–1671.

Lockyer, C.H. (1976) Growth and energy budget of large baleen whales from the southern hemisphere. FAO, Advisory Committee on Marine Resources Research. ACMRR/MM/SC/41.

Lockyer, C.H. (1981) Growth and energy budgets of large baleen whales from the southern hemisphere. In: *Mammals in the Seas*, FAO Fish. Ser. V, Vol. 3, pp. 379–488. FAO, Rome.

Lockyer, C.H. (1987) Evaluation of the rate of fat reserves in relation to the ecology of north Atlantic fin and sei whales. In: Huntley, A.C, Costa, D.P., Worlhy, G.A.J. & Castellini, M.A. (eds) *Approaches to Marine Mammal Energetics*, pp. 183–204. Society for Marine Mammalogy Special Publication No. 1.

Logue, A.W. (1988) Research on self-control: an integrating framework. *Behav. Brain Sci.* **11**: 665–709.

London, W.P & Yorke, J.A. (1973) Recurrent outbreaks of measles, chickenpox and mumps, I. Seasonal variation in contact rates. *Am. J. Epidemiol.* **98**: 453–468.

Lotka, A.J. (1925) *Elements of Physical Biology*. Williams & Wilkins, Baltimore.

Lounibos, L.P., Rey, J.R. & Frank, J.H. (eds) (1985) *Ecology of Mosquitos: Proceedings of a Workshop*. Florida Medical Entomology Laboratory, Vero Beach.

Lowry, L.F. (1984) A conceptual assessment of biological interactions among marine mammals and fisheries in the Bering Sea. Proceedings of the workshop on biological interactions among marine mammals and commercial fisheries in the southeastern Bering Sea, 18–21 October, 1983. Alaska Sea Grant

Report 84–1, University of Alaska.

Lowry, L.F. & Frost, K.J. (1985) Biological interactions between marine mammals and commercial fisheries in the Bering Sea. In: Beddington, J.R., Beverton, R.J.H. & Lavigne, D.M. (eds) *Marine Mammals and Fisheries*, pp. 39–61. George Allen & Unwin, London.

Lucas, J.R. (1985a) Metabolic rates and pit-construction costs of two antlion species. *J. Anim. Ecol.* **54**: 295–309.

Lucas, J.R. (1985b) Partial prey consumption by antlion larvae. *Anim. Behav.* **33**: 945–958.

Lucas, J.R. (1987) The influence of time constraints on diet choice of the great tit *Parus major. Anim. Behav.* **35**: 1538–1548.

Lucas, J.R. (1990) Time scale and diet choice decisions. In: Hughes, R.N. (ed.) *Behavioural Mechanisms of Food Selection*, pp. 165–185. Springer-Verlag, Berlin.

Lucas, J.R. & Grafen, A. (1985) Partial prey consumption by ambush predators. *J. Theor. Biol.* **113**: 455–473.

Luck, R.F., Shepard, B.M. & Kenmore, P.E. (1988) Experimental methods for evaluation arthropod natural enemies. *Ann. Rev. Entomol.* **33**: 367–392.

Luckinbill, L.S. (1973) Coexistence in laboratory populations of *Paramecium aurelia* and its predator *Didinium nasitum. Ecology* **54**: 1320–1327.

Lydekker, R. (1895) *Mammals*. John F. Shaw, London.

Lyell, C. (1832) *Principles of Geology*, Vol. 2. John Murray, London.

Mabelis, A.A. (1979) Wood ant wars. The relationship between aggression and predation in the red wood ant (*Formica polyctena* Forst.). *Netherlands J. Zool.* **29**: 451–620.

MacArthur, J.W. (1975) Environmental fluctuations and species diversity. In: Cody, M.L. & Diamond, J.M. (eds) *Ecology and Evolution of Communities*, pp. 74–80. The Belknap Press of Harvard University Press, Cambridge, MA.

MacArthur, R.H. (1955) Fluctuations of animal populations and a measure of community stability. *Ecology* **36**: 533–536.

MacArthur, R.H. (1972) *Geographical Ecology: Patterns in the Distribution of Species*. Harper & Row, New York.

MacArthur, R.H. & Pianka, E.R. (1966)

On optimal use of a patchy environment. *Am. Nat.* **100**: 603–609.

Macdonald, G. (1957) *The Epidemiology and Control of Malaria*. Oxford University Press, London.

Mace, G.M. & Harvey, P.H. (1983) Energetic constraints on home range size. *Am. Nat.* **121**: 120–132.

Mackay, W.P. (1982) The effect of predation of western widow spiders (Araneae: Theridiidae) on harvester ants (Hymenoptera: Formicidae). *Oecologia* **53**: 406–411.

Mackintosh, N.A. (1965) *The Stocks of Whales*. Fishing News Books, London.

Mackintosh, N.J. (1974) *The Psychology of Animal Learning*. Academic Press, London.

Mackintosh, N.J. (1983) *Conditioning and Associative Learning*. Oxford University Press, Oxford.

MacLuick, D.A. (1937) Fluctuations in the numbers of the varying hare (*Lepus americanus*). *University of Toronto Studies in Biology Series*, **43**: 1–136.

Mader, W.J. (1976) Biology of Harris' hawk in southern Arizona. *Living Bird* **14**: 59–85.

Mader, W.J. (1979) Breeding behaviour of a polyandrous trio of Harris' hawks in southern Arizona. *Auk* **96**: 776–788.

Madsen, H.F., Westigard, P.H. & Falcon, L.A. (1961) Evaluation of insecticides and sampling methods against the apple aphid, *Aphis pomi*, *J. Econ. Entomol.* **54**: 892–894.

Mahon, R. & Gibbs, A. (1982) Arbovirus-infected hens attract more mosquitoes. In: Mackenzie, J.D. (ed) *Viral Diseases in Southeast Asia and the Western Pacific*, pp. 502–504. Academic Press, Sydney.

Major, P. (1978) Predator–prey interactions in two schooling fishes, *Caranx ignobilis* and *Stolephorus purpureus*. *Anim. Behav.* **26**: 760–777.

Malcolm, S.B. (1986) Aposematism in a soft-bodied insect: a case for kin selection. *Behav. Ecol. Sociobiol.* **18**: 387–393.

Malcolm, S.B. (1989) Disruption of web structure and predatory behavior of a spider by plant-derived chemical defenses of an aposematic aphid. *J. Chem. Ecol.* **15**: 1699–1716.

Malcolm, S.B. (1990a) Mimicry: status of a classical evolutionary paradigm.

Trends Ecol. Evol. **5**: 57–62.

Malcolm, S.B. (1990b) Chemical defence in chewing and sucking insect herbivores: plant-derived cardenolides in the monarch butterfly and oleander aphid. *Chemoecology* **1**: 12–21.

Malcolm, S.B. & Brower, L.P. (1989) Evolutionary and ecological implications of cardenolide sequestration in the monarch butterfly. *Experientia* **45**: 284–295.

Malcolm, S.B., Cockrell, B.J. & Brower, L.P. (1989) Cardenolide fingerprint of monarch butterflies reared on common milkweed, *Asclepias syriaca* L. *J. Chem. Ecol.* **15**: 819–53.

Malcolm, S.B., Cockrell, B.J. & Brower, L.P. (1992) Spring recolonization of eastern North American by the monarch butterfly: successive brood or single sweep migration? In: Malcolm, S.B. & Zalucki, M.P. (eds) *Biology and Conservation of the Monarch Butterfly*. Los Angeles County Natural History Museum, Los Angeles (in press).

Malherbe, A.P. (1963) Notes on birds of prey and some others at Boshoek north of Rustenburg during a rodent plague. *Ostrich* **34**: 95–96.

Malicky, H. (1976) Trichopteren-emergenz in zwei lunzer bachen 1972–74. *Arch. Hydrobiol.* **77**: 51–65.

Mallet, J. & Singer, M. (1987) Individual selection, kin selection, and the shifting balance in the evolution of warning colours: the evidence from butterflies. *Biol. J. Linnean Soc.* **32**: 337–350.

Mangel, M. (1987) Modelling behavioural decisions of insects. *Lecture Notes on Biomathematics* **73**: 1–18.

Mangel, M. (1988) Analysis and modelling of the Soviet Southern Ocean krill fleet. Presented to the Scientific Committee of CCAMLR, 1988, Hobart.

Mangel, M. (1989) Evolution of host selection in parasitoids: does the state of the parasitoid matter? *Am. Nat.* **133**: 688–705.

Mangel, M. & Clark, C.W. (1988) *Dynamic Modelling in Behavioral Ecology*. Princeton University Press, Princeton.

Mansour-Bek, J.J. (1954) The digestive enzymes of Invertebrata and Protochordata. *Tabulae Biologicae* **21**: 75–382.

Mansour, F., Rosen, D. & Shulov, A. (1980a) Functional response of the spider *Chiricanthium mildei* (Arachnida: Clubionidae) to prey density. *Entomophaga* **25**: 313–316.

Mansour, F., Rosen, D. & Shulov, A. (1981) Disturbing effect of a spider on larval aggregations of *Spodoptera littoralis*. *Entomologica Experimentalis et Applicata* **29**: 234–237.

Mansour, F., Rosen, D., Shulov, A. & Plaut, H.N. (1980b) Evaluation of spiders as biological control agents of *Spodoptera littoralis* larvae on apple in Israel. *Acta Oecologica Applied* **1**: 225–232.

Marcström, V., Kenward, R.E. & Engren, E. (1988) The impact of predation on boreal tetraonids during vole cycles: an experimental study. *J. Anim. Ecol.* **57**: 895–872.

Marinat, P.J. (1987) The role of climatic variation and weather in forest insect outbreaks. In: Barbosa, P. & Schultz, J.C. (eds) *Insect Outbreaks*, pp. 241–268. Academic Press, New York.

Marples, N.M., Brakefield, P.M. & Cowie, R.J. (1989) Differences between the 7-spot and 2-spot ladybird beetles (Coccinellidae) in their toxic effects on a bird predator. *Ecol. Entomol.* **14**: 79–84.

Marr, J.W.S. (1962) The natural history and geography of Antarctic krill (*Euphasia superba* Dana.). *Discovery Rep.* **32**: 33–464.

Martin, A. (1981) *The Ring Net Fishermen*. John Donald, Edinburgh.

Martin, R.D., Chivers, D.J., MacClarnon, A.M. & Hladik, C.M. (1985) Gastrointestinal allometry in primates and other mammals. In: Jungers, W.J. (ed.) *Size and Scaling in Primate Biology*, pp. 61–89. Plenum Press, New York.

Marshall, W.H., Guillion, G.W. & Scwab, R.G. (1962) Early summer activities of porcupines as determined by radio-positioning techniques. *J. Wildlife Manage.* **26**: 75–80.

Mason, J.R. & Reidinger, R.F. (1982) Observational learning of food aversion in red-winged blackbirds (*Agelaius phoeniceus*). *Auk* **99**: 548–554.

Mason, J.R. & Silver, W.L. (1983) Tregeminally mediated odor aversion in starlings. *Brain Res.* **269**: 196–199.

Matsura, T. & Takano, H. (1989) Pit-relocation of antlion larvae in relation to their density. *Res. Pop. Ecol.* **31**: 225–234.

Mattingly, P.F. (1969) *The Biology of Mosquito-Borne Disease*. George Allen & Unwin, London.

May, R.M. (1973) Time-delay versus stability in population models with two and three trophic levels. *Ecology* **54**: 315–325.

May, R.M. (1974) *Stability and Complexity in Model Ecosystems*, 2nd edn. Princeton University Press, Princeton.

May, R.M. (1976) Simple mathematical models with very complicated dynamics. *Nature* **261**: 459–467.

May, R.M. (1978) Host–parasitoid systems in patchy environments: a phenomenological model. *J. Anim. Ecol.* **47**: 833–843.

May, R.M. (ed.) (1981) *Theoretical Ecology: Principles and Applications*, 2nd edn. Blackwell Scientific Publications, Oxford and Sinauer, Sunderland, MA.

May, R.M. (1985) Regulation of populations with non-overlapping generations by microparasites: a purely chaotic system. *Am. Nat.* **125**: 573–584.

May, R.M. (1986a) Population biology of microparasitic infections. *Biomathematics* **17**: 405–442.

May, R.M. (1986b) Population biology of microparasite infections. In: Hallam, T.G. & Levin, S.A. (eds) *Mathematical Ecology: an Introduction*, pp. 405–442. Springer-Verlag, New York.

May, R.M. & Anderson, R.M. (1978) Regulation and stability of host–parasite population interactions. II. Destabilizing processes. *J. Anim. Ecol.* **47**: 249–267.

May, R.M. & Anderson, R.M. (1983a) Epidemiology and genetics in the coevolution of parasites and hosts. *Proc. R. Soc. Lond. B* **219**: 281–313.

May, R.M. & Anderson, R.M. (1983b) Parasite–host coevolution. In: Futuyma, D.J. & Slatkin, M. (eds) *Coevolution*, pp. 186–206. Sinauer, Sunderland, MA.

May, R.M. & Anderson, R.M. (1985) Endemic infections in growing populations. *Math. Biosci.* **77**: 141–156.

May, R.M. & Anderson, R.M. (1987) Transmission dynamics of HIV infection. *Nature* **326**: 137–142.

May, R.M., Beddington, J.R., Horwood, J.W. & Shepherd, J.F. (1978) Exploiting natural populations in an uncertain world. *Math. Biosci.* **42**: 219–522.

May, R.M. & Hassell, M.P. (1981) The dynamics of multiparasitoid–host interactions. *Am. Nat.* **117**: 234–261.

May, R.M., Hassell, M.P., Anderson, R.M. & Tonkyn, D.W. (1981). Density dependence in host–parasitoid models. *J. Anim. Ecol.* **50**: 855–865.

May, R.M. & Leonard, W.J. (1975) Nonlinear aspects of competition between three species. *SIAM J. Appl. Math.* **29**: 243–253.

Maynard Smith, J. (1974) *Models in Ecology*. Cambridge University Press, Cambridge.

Maynard Smith, J. (1978a) Optimization theory in evolution. *Ann. Rev. Ecol. Syst.* **9**: 13–56.

Maynard Smith, J. (1978b) *The Evolution of Sex*. Cambridge University Press, Cambridge.

Maynard Smith, J. (1982) *Evolution and the Theory of Games*. Cambridge University Press, Cambridge.

Maynard Smith, J. & Brown, R.L.W. (1986) Competition and body size. *Theor. Pop. Biol.* **30**: 166–179.

Maynard Smith, J. & Hofbauer, J. (1987) The 'battle of the sexes': a genetic model with limit cycle behavior. *Theor. Pop. Biol.* **32**: 1–14.

Mayr, E. (1963) *Animal Species and Evolution*. Harvard University Press, Cambridge, MA.

Mazak, V. (1981) *Panthera tigris*. *Mammalian Species*, No. 152.

Mazur, J.F. (1984) Tests of an equivalence rule for fixed and variable reinforcer delays. *J. Exp. Psychol. Anim. Behav. Proc.* **10**: 426–436.

Mazur, J.E., Snyderman, M. & Coe, D. (1985) Influences of delay and rate of reinforcement on discrete-trial choice. *J. Exp. Psychol: Anim. Behav. Proc.* **11**: 565–575.

McAlister, W.B. & Perez, M.A. (1976) Ecosystem dynamics of birds and marine mammals. Processed Report Northwest and Alaska Fisheries Center. NMFS NOAA Seattle.

McAtee, W.L. (1932) Effectiveness in nature of the so-called protective adaptations in the animal kingdom, chiefly illustrated by the food habits of nearctic birds. *Smithsonian Miscellaneous Collections* **85(7)**: 1–201.

McCombie Young, T.C. (1924) *Kala azar in Assam*. H.K. Lewis & Co., London.

McCaughley, E. & Murdoch, W.W., (1990) Predator–prey dynamics in environments rich and poor in nutrients. *Nature* **343**: 455–457.

McCleery, R.H. & Perrins, C.M. (1991) Effects of predation on the numbers of great tits *Parus major*. In: Perrins, C.M., Lebreton, J.-D. & Hirons, G.J.M. (eds) *Bird Population Studies*, pp. 129–147. Oxford University Press, Oxford.

McClure, M.S. (1977) Parasitism of the scale insect, *Fiorinia externa* (Homoptera: Diaspididae) by *Aspidiotiphagus citrinus* (Hymenoptera: Eulophidae), in a hemlock forest: density dependence. *Environ. Entomol.* **6**: 551–555.

McClure, M.S. (1980) Competition between exotic species: scale insects on hemlock. *Ecology* **61**: 1391–1401.

McClure, M.S. (1986) Population dynamics of Japanese hemlock scales: a comparison of endemic and exotic communities. *Ecology*, **67**: 1411–1421.

McFarland, D.J. (1977) Decision making in animals. *Nature* **269**: 15–21.

McGowan, J.A. & Walker, P.W. (1985) Dominance and diversity maintenance in an oceanic ecosystem. *Ecol. Monogr.* **55**: 103–118.

McGowan, J.D. (1975) Distribution, density and productivity of goshawks in interior Alaska. Alaska Department of Fish & Game Report.

McGugan, B.M. (1958) Forest Lepidoptera of Canada recorded by the Forest Insect Survey. Vol. 1. Publ. 1034. Canada Department of Agriculture, Division of Forest Biology, Ottawa, Ontario.

McGugan, B.M. & Coppel, H.C. (1962) Biological control of forest pests: 1910–1958. In: *A Review of the Biological Control Attempts Against Insects and Weeds in Canada. Technical Communication of the Commonwealth Institute of Biological Control* **2**: 35–127.

McKey, D. (1984) Interaction of the ant-plant *Leonardoxa africana* (Caesalpiniaceae) with its obligate inhabitants in a rainforest in Cameroon. *Biotropica* **16**: 81–99.

McKillup, S.C. (1983) A behavioural polymorphism in the marine snail *Nassarius pauperatus*: geographic variation correlated with food availability and differences in competitive ability between morphs. *Oecologia* **56**: 58–66.

McLean, A.R. (1986) Dynamics of childhood infectious diseases in high birthrate countries. *Lecture Notes in Biomathematics* **65**: 171–197.

McLean, A.R. & Anderson, R.M. (1988) Measles in developing countries. Part I. Epidemiological parameters and patterns. *Epidemiol. Infect.* **100**: 111–133.

McLeod, J.M. (1966) The spatial distribution of cocoons of *Neodiprion swainei* Middleton in a jack pine stand. I. Cartographic analysis of cocoon distribution, with special reference to predation by small mammals. *Can. Entomol.* **98**: 430–447.

McNab, B.K. (1963) Bioenergetics and the determination of home range size. *Am. Nat.* **97**: 133–140.

McNab, B.K. (1982) Evolutionary alternatives in the physiological ecology of bats. In: Kunz, T.H. (ed.) *Ecology of Bats*, pp. 151–200. Plenum Press, New York.

McNab, B.K. (1986a) Food habits, energetics and reproduction of marsupials. *J. Zool.* **208**: 595–614.

McNab, B.K. (1986b) The influence of food habits on the energetics of eutherian mammals. *Ecol. Monogr.* **56**: 1–19.

McNamara, J.M. (1985) An optimal sequential policy for controlling a Markov neural process. *J. Appl. Prob.* **22**: 324–335.

McNamara, J.M. & Houston, A.I. (1985) Optimal foraging and learning. *J. Theor. Biol.* **117**: 231–249.

McNamara, J.M. & Houston, A.I. (1987a) Partial preferences and foraging. *Anim. Behav.* **35**: 1084–1099.

McNamara, J.M. & Houston, A.I. (1987b) Starvation and predation as factors limiting population size. *Ecology* **68**: 1515–1519.

McNamara, J.M. & Houston, A.I. (1987c) A general framework for understanding the effects of variability and interruptions on foraging behaviour. *Acta Biotheor.* **36**: 3–22.

McNamara, J.M. & Houston, A.I. (1990a) Starvation and predation in a patchy environment. In: Swingland, I., Stenseth, N.C. & Shorrocks, B.

(eds) *Living in a Patchy Environment* Oxford University Press, Oxford.

McNamara, J.M. & Houston, A.I. (1990b) The value of fat reserves and the tradeoff between starvation and predation. *Acta Biotheor.* **38**: 37–61.

McNamara, J.M. & Houston, A.I. (1990c) State dependent ideal free distributions *Evol. Ecol.* **4**: 928–311.

McNamara, J.M., Houston, A.I. & Krebs, J.R. (1990) Why hoard? The economics of food storing in tits. *Behav. Ecol.* (in press).

McNamara, J.M., Mace, R.H. & Houston, A.I. (1987) Optimal daily routines of singing and foraging in a bird singing to attract a mate. *Behav. Ecol. Sociobiol.* **20**: 399–405.

McNeil, W.H. (1976) *Plagues and Peoples*. Blackwell Scientific Publications, Oxford.

Mech, L.D. (1966) The wolves of Isle Royale. Fauna of National Parks of the United States. *Fauna Series* **7**: 1–210.

Mech, L.D. (1977) Wolf-pack buffer zones as prey reservoirs. *Science* **198**: 320–321.

Mendelsohn, J.M. (1983) Social behaviour and dispersion of the black-shouldered kite. *Ostrich* **54**: 1–18.

Meng, H. (1951) The Coopers hawk. PhD thesis, Cornell University, Ithaca, New York.

Menge, B.L. (1976) Organization of the New England rocky intertidal community: role of predation, competition, and environmental heterogeneity. *Ecol. Monogr.* **46**: 335–393.

Menge, B.L. (1978) Predation intensity in a rocky intertidal community: effect of an algal canopy, wave action and desiccation on predator feeding rates. *Oecologia* **34**: 17–35.

Menge, B.L. (1982) Reply to a comment by Edwards, Conover, and Sutter. *Ecology* **63**: 1180–1184.

Menge, B.L. & Lubchenco, J. (1981) Community organization in temperate and tropical rocky intertidal habitats: prey refuges in relation to consumer pressure gradients. *Ecol. Monogr.* **51**: 429–450.

Messenger, P.S. (1975) Parasites, predators and population dynamics. In: Pimentel, D. (ed.) *Insects, Science and Society*, pp. 201–223. Academic Press, New York.

Messier, F. (1985) Social organization, spatial distribution and population density of wolves in relation to moose density. *Can. J. Zool.* **63**: 1068–1077.

Messier, F. & Crete, M. (1985) Moose–wolf dynamics and the natural regulation of moose populations. *Oecologia* **65**: 503–512.

Metz, J.A.J., De Roos, A.M. & Van de Bosch, F. (1988b). Population models incorporating physiological structure: a quick survey of the basic concepts and an application to size-structured population dynamics of water fleas. In: Ebenman, B. & Persson, L. (eds) *Size-Structured Populations*, pp. 106–126. Springer Verlag, Berlin.

Metz, J.A.J., Sabelis, M.W. & Kuchlein, J.H. (1988a) Sources of variation in predation rates at high prey densities: an analytic model and a mite example. *Exp. Appl. Acarol.* **5**: 187–206.

Metz, J.A.J. & Van Batenburg, F.H.D. (1985a) Holling's 'hungry mantid' model for the invertebrate functional response: I. The full model and some of its limits *J. Math. Biol.* **22**: 209–238.

Metz, J.A.J. & Van Batenburg, F.H.D. (1985b) Holling's 'hungry mantid' model for the invertebrate functional response: II. Negligible handling time. *J. Math. Biol.* **22**: 239–257.

Meyburg, B.-U. (1974) Sibling aggression and mortality among nestling eagles. *Ibis* **116**: 224–228.

Michod, R.E. & Levin, B.R. (eds) (1988) *The Evolution of Sex*. Sinauer, Sunderland, MA.

Milinski, M. (1979) An evolutionary stable feeding strategy for sticklebacks. *Z. Tierpsychol.* **51**: 36–40.

Milinski, M. (1984a) A predator's cost of overcoming the confusion effect of swarming prey. *Anim. Behav.* **32**: 233–242.

Milinski, M. (1984b) Parasites determine a predator's optimal feeding strategy. *Behav. Ecol. Sociobiol.* **15**: 35–37.

Milinski, M. & Heller, R. (1978) Influence of a predator on the optimal foraging behaviour of sticklebacks *Gasterosteus aculeatus*. *Nature* **273**: 642–644.

Milinski, M. & Parker, G.A. (1991) Competing for resources. In: Krebs, J.R. & Davies, N.B. *Behavioural Ecology: an Evolutionary Approach*, 3rd edn. Blackwell Scientific Publications, Oxford.

Miller, F.L., Gunn, A. & Broughton, E. (1985) Surplus killing as exemplified by wolf predation on newborn caribou. *Can. J. Zool.* **63**: 295–300.

Milligan, P.J.M. & Baker, R.D. (1988) A model of tsetse-transmitted animal trypanosomiasis. *Parasitology* **96**: 211–239.

Mills, M.G.L. (1989) The comparative behavioural ecology of hyenas: the importance of diet and food dispersion. In: Gittleman, J.L. (ed.) *Carnivore Behavior, Ecology, and Evolution*, pp. 125–142. Cornell University Press, New York.

Mills, M.G.L. (1991) A comparison of methods used to study the feeding habits of large African Carnivores. In: McCullough, D.R. (ed.) *Wildlife 2001: Populations*. Elsevier, Amsterdam. (In press.)

Mills, N.J. (1990) Are parasitoids of significance in endemic populations of forest defoliators? Some experimental observations from gypsy moth, *Lymantria dispar* (Lepidoptera: Lymantriidae). In: Watt, A.D., Leather, S.R., Kidd, N.A.C. & Hunter, M. (eds) *Population Dynamics of Forest Pests*, pp. 265–274. Intercept Ltd, Andover.

Millot, J. (1949) Ordre des Araneides (Araneae). In: Grassé, P.G. (ed.) *Traité de Zoologie*. Masson, Paris.

Milton, K. & May, M.L. (1976) Body weight, diet and home range area in primates. *Nature* **259**: 459–462.

Mims, C.A. (1987) *The Pathogenesis of Infectious Disease*, 3rd edn. Academic Press, London.

Minchella, D.J. (1985) Host life history variation in response to parasitism. *Parasitology* **90**: 205–216.

Minchella, D.J. & LoVerde, P.T. (1981) A cost of increased early reproductive effort in the snail *Biomphalaria glabrata*. *Am. Nat.* **118**: 876–881.

Mindell, D.P., Albuquerque, J.L.B. & White, C.M. (1987) Breeding population fluctuations in some raptors. *Oecologia* **72**: 382–388.

Minks, A.K. & Harrewijn, P. (eds) (1988) *Aphids: their Biology, Natural Enemies and Control*. Elsevier, Amsterdam.

Mitter, C. & Brooks, D.R. (1983) Phylogenetic aspects of coevolution. In: Futuyma, D.J. & Slatkin, M. (eds) *Coevolution*, pp. 65–98. Sinauer, Sunderland, MA.

Miyashita, K. (1968) Growth and

development of *Lycosa T-insignata* Boes et Str. (Araneae Lycosidae) under different feeding conditions. *Appl. Entomol. Zool.* **3**: 81–88.

Miyashita, K. (1969). Effects of locomotory activity, temperature and hunger on the respiratory rate of *Lycosa T-insignata* Boes. et Str. (Araneae: Lycosidae). *Appl. Entomol. Zool.* **4**: 105–113.

Mode, C.J. (1958) A mathematical model for the co-evolution of obligate parasites and their hosts. *Evolution* **12**: 158–165.

Mode, C.J. (1961) A generalized model of a host–pathogen system. *Biometrics* **17**: 386–404.

Molineaux, L. (1985) The impact of parasitic diseases and their control on mortality, with emphasis on malaria and Africa. In: Vallin, J., Lopez, A. (eds) *Health Policy, Social Policy and Mortality Prospects*, pp. 13–44. Ordina editions, Liège.

Mols, P.J.M. (1987) Hunger in relation to searching behaviour, predation and egg production of the carabid beetle, *Pterostichus coerulescens* L.: results of simulations. *Acta Phytopathol. Entomol. Hung.* **22**: 187–205.

Mols, P.J.M. (1988) *Simulation of Hunger, Feeding and Egg Production in the Carabid Beetle*, Pterostichus coerulescens. Agricultural University Wageningen Papers 88(3).

Molyneux, D.H. & Jefferies, D. (1986) Feeding behaviour of pathogen-infected vectors. *Parasitology* **92**: 721–736.

Mook, L.J. (1963) Birds and the spruce budworm. In: Morris, R.F. (ed.) *The Dynamics of Epidemic Spruce Budworm Populations*, pp. 268–271. *Mem. Ent. Soc. Can.* **31**: 268–271.

Moore, D. (1988) Agents used for biological control of mealybugs (Pseudococcidae). *Biocontrol News and Information* **9**: 209–225.

Moore, F.E. & Simm, P.A. (1986) Risk-sensitive foraging by a migratory bird *Dendroica coronata*. *Experientia* **42**: 1054–1056.

Moore, J. (1984) Parasites that change the behaviour of their host. *Sci. Am.* **250**: 108–115.

Moors, P.J. & Atkinson, I.A.E. (1984) Predation on seabirds by introduced mammals, and factors affecting its severity. In: Croxall, J.P., Evans, P.G.H. & Schreiber, R.W. (eds). *Status and Conservation of the World's Seabirds*, pp. 667–690. International Council for Bird Preservation, Cambridge, MA.

Moraleva, N.V. (1987) Interspecific interactions of closely related shrews (Insectivora, *Sorex*). In: Syroechkovsky, E.E. (ed.) *Fauna and Ecology of Birds and Mammals of Central Siberia*, pp. 213–228. Nauka, Moscow. (In Russian.)

Moran, M.J. (1985) The timing and significance of sheltering and foraging behaviour of the predatory intertidal gastropod *Morula marginalba* Blainville (Muricidae). *J. Exp. Marine Biol. Ecol.* **93**: 103–114.

Moran, M.J., Fairweather, P.G. & Underwood, A.J. (1984) Growth and mortality of the predatory intertidal whelk *Morula marginalba*. Blainville (Muricidae): the effects of different species of prey. *J. Exp. Marine Biol. Ecol.* **84**: 1–17.

Morley, D. (1969) Severe measles in the tropics I. *Br. Med. J.* **1**: 297–300.

Morris, R.F., Cheshire, W.F., Miller, C.A. & Mott, D.G. (1958) The numerical response of avian and mammalian predators during a gradation of the spruce budworm. *Ecology* **39**: 487–494.

Morse, D.H. (1980) *Behavioural Mechanisms in Ecology*. Harvard University, Cambridge, MA.

Morse, D.H. & Fritz, R.S. (1982) Experiential and observational studies of patch choice at different scales by the crab spider *Misumena vatia*. *Ecology* **63**: 172–182.

Mountford, M.D. (1988) Population regulation, density dependence and heterogeneity. *J. Anim. Ecol.* **57**: 845–858.

Mueller, H.C. (1974) Factors influencing prey selection in the American kestrel. *Auk* **91**: 705–721.

Mueller, H.C. (1990) The evolution of reversed sexual dimorphism in size in monogamous species of birds. *Biol. Rev.* **65**: 553–585.

Mueller, H.C. & Meyer, J. (1985) The evolution of reversed dimorphism in size: a comparative analysis of the falconiforms of the western Palearctic. *Curr. Ornithol.* **2**: 65–99.

Mulvey, M. & Vrijenhoek, R.C. (1982) Population structure in *Biomphalaria glabrata*: examination of an hypothesis for the patchy distribution of susceptibility to schistosomes. *Am. J. Trop. Med. Hyg.* **31**: 1195–1200.

Murdoch, W.W. (1969) Switching in general predators: experiments on predator and stability of prey populations. *Ecol. Monogr.* **39**: 355–354.

Murdoch, W.W. (1975) Diversity, complexity, stability and pest control. *J. Appl. Ecol.* **12**: 795–807.

Murdoch, W.W. (1990) The relevance of pest-enemy models to biological control. In: Mackauer, M., Ehler, L.E. & Roland, J. (eds) *Critical Issues in Biological Control*, pp. 1–24. Intercept Ltd, Andover.

Murdoch, W.W., Nisbet, R.M., Blythe, S.P., Gurney, W.S.C. & Reeve, J.D. (1987) An invulnerable age class and stability in delay – differential parasitoid–host models. *Am. Nat.* **129**: 263–282.

Murdoch, W.W. & Oaten, A. (1975) Predation and population stability. *Adv. Ecol. Res.* **9**: 1–131.

Murdoch, W.W., Reeve, J.D., Huffaker, C.B. & Kennett, C.E. (1984) Biological control of olive scale and its relevance to ecological theory. *Am. Nat.* **123**: 371–392.

Murdoch, W.W. & Scott, M.A. (1984) Stability and extinction of laboratory populations of zooplankton preyed upon by the backswimmer, *Notonecta. Ecology* **65**: 1231–1248.

Murdoch, W.W. & Sih, A. (1978) Age-dependent interference in a predatory insect. *J. Anim. Ecol.* **47**: 581–592.

Murdoch, W.W. & Steward-Oaten, A. (1989) Aggregation of parasitoids and predators: effects on equilibrium and stability. *Am. Nat.* **134**: 288–310.

Murie, A. (1944) *The Wolves of Mount McKinley*. United States Government Printing Office, Washington.

Murray, B.G. (1971) The ecological consequences of interspecific territorial behavior in birds. *Ecology* **52**: 414–423.

Murray, B.G. (1982) On the meaning of density dependence. *Oecologia* **53**: 370–373.

Murton, R.K., Westwood, N.J. & Isaacson, A.J. (1974) A study of woodpigeon shooting: the exploitation of a natural animal population. *J. Appl. Ecol.* **11**: 61–81.

Muus, B.J. & Dahlstrom, P. (1974) *Collins Guide to the Sea Fishes of Britain and North-western Europe*. Collins, London.

Myerson, J. & Miezin, F.M. (1980) The kinetics of choice: an operant systems analysis. *Psychol. Rev.* **87**: 160–174.

Myllymäki, A. (1969) Productivity of a free-living population of the field vole *Microtus agrestis* (L.) In: Petrusewicz, K. & Ryszkowski, L. (eds) *Energy Flow through Small Mammal Populations*, pp. 225–265. Warzawa.

Nachman, G. (1987a) Systems analysis of acarine predator–prey interactions: I. A stochastic simulation model of spatial processes. *J. Anim. Ecol.* **56**: 247–265.

Nachman, G. (1987b) Systems analysis of acarine predator–prey interactions: II. The role of spatial processes in system stability. *J. Anim. Ecol.* **56**: 267–281.

Nadeau, J.H., Britton-Davidian, J., Bonhomme, F. & Thaler, L. (1988) H-2 polymorphisms are more uniformly distributed than allozyme polymorphisms in natural populations of house mice. *Genetics* **118**: 131–140.

Najera, J.A. (1974) A critical review of the field application of a mathematical model of malaria eradication. *Bull. World Health Org.* **50**: 449–457.

Nakamura, K. (1974) A model of the functional response of a predator to prey density involving the hunger effect. *Oecologia* **16**: 265–278.

Nakamura, K. (1977) A model for the functional response of a predator to varying prey densities, based on the feeding ecology of wolf spiders. *Bull. Natl. Inst. Agric. Sci. Jpn. Ser. C.* **31**: 28–89.

Nakamura, K. (1987) Hunger and starvation. In: Nentwig, W. (ed.) *Ecophysiology of Spiders*, pp. 287–293. Springer-Verlag, New York.

Nakamura, K. (1985) Behavioral mechanism of switchover in search behavior of the ladybeetle, *Coccinella septempunctata. J. Insect Physiol.* **31**: 849–856.

Nakasuji, F., Yamanaka, K. & Kiritani, K. (1973) The disturbing effect of micryphantid spiders on larval aggregation of the tobacco cutworm. *Kontyu* **41**: 220–227.

Narasimham, A.U. & Chacko, M.J. (1988) *Rastrococcus* spp. (Hemiptera: Pseudococcidae) and their natural enemies in India as potential biocontrol agents for *R. invadens*. Williams. *Bull. Entomol. Res.* **78**:

703–708.

Nedelman, J. (1984) Inoculation and recovery rates in the malaria model of Dietz, Molineaux and Thomas. *Math. Biosci.* **69**: 209–233.

Neill, S.R. & Cullen, J.M. (1974) Experiments on whether schooling by their prey affects the hunting behaviour of cephaloplod and fish predators. *J. Zool. Lond.* **172**: 549–569.

Neill, W.E. (1990) Induced vertical migration in copepods as a defence against predation. *Nature* **345**: 524–526.

Nelson, J.B. (1976) The breeding biology of frigate birds. *Living Bird* **14**: 113–155.

Nemoto, T. (1959) Food of baleen whales with reference to whale movements. *Sci. Rep. Whales Res. Inst. Tokyo* **14**: 149–290.

Nentwig, W. (1985) Prey analysis of four species of tropical orb weaving spiders (Araneae: Araneidae) and a comparison with araneids of the temperate zone. *Oecologia* **66**: 580–594.

Nentwig, W. (1987) The prey of spiders. In: Nentwig, W. (ed.) *Ecophysiology of Spiders*, pp. 249–263. Springer-Verlag, New York.

Nentwig, W. & Heimer, S. (1987) Ecological aspects of spider webs. In: Nentwig, W. (ed.) *Ecophysiology of Spiders*. Springer-Verlag, New York.

Nentwig, W. & Wissel, C. (1986) A comparison of prey lengths among spiders. *Oecologia* **68**: 595–600.

Newman, J.A. & Caraco, T. (1987) Foraging, predation hazard, and patch use in grey squirrels. *Anim. Behav.* **35** 1804–1813.

Newman, J.A., Recer, G.M., Zwicker, S.M. & Caraco, T. (1988) Effects of predation hazard on foraging 'constraints': patch-use strategies in grey squirrels. *Oikos* **53**: 93–97.

Newsome, A. (1990) The control of vertebrate pests by vertebrate predators. *Trends Ecol. Evol.* **5**: 187–191.

Newsome, A.E., Catling, P.C. & Corbett, L.K. (1983) The feeding ecology of the dingo. II. Dietary and numerical relationships with fluctuating prey populations in south-eastern Australia. *Aust. J. Ecol.* **8**: 345–366.

Newsome, A.E., Parer, I. & Catling, P.C. (1989) Prolonged prey suppression by

carnivores — predator-removal experiments. *Oecologia* **78**: 458–467.

Newton, A.C. & Crute, I.R. (1989) A consideration of the genetic control of species specificity in fungal plant pathogens and its relevance to a comprehension of the underlying mechanisms. *Biol. Rev.* **64**: 35–50.

Newton, I. (1979) *Population Ecology of Raptors*. Poyser, Berkhamsted.

Newton, I. (1980) The role of food in limiting bird numbers. *Ardea* **68**: 11–30.

Newton, I. (1986) *The Sparrowhawk*. Poyser, Calton.

Newton, I. & Marquiss, M. (1979) Sex ratio among nestlings of the European sparrowhawk. *Am. Nat.* **113**: 309–315.

Newton, I. & Marquiss, M. (1981) Effect of additional food on laying dates and clutch-sizes of sparrowhawks. *Ornis Scand.* **12**: 224–229.

Newton, I., Meek, E. & Little, B. (1986b) Population and breeding of Northumbrian merlins. *Br. Birds* **79**: 155–170.

Newton, I., Wyllie, I. & Mearns, R.M. (1986a) Spacing of sparrowhawks in relation to food supply. *J. Anim. Ecol.* **55**: 361–370.

Nicholson, A.J. (1927) A new theory of mimicry in insects. *Austr. Zool.* **5**: 10–104.

Nicholson, A.J. & Bailey, V.A. (1935) The balance of animal populations. Part 1. *Proc. Zool. Soc. Lond.* **3**: 551–598.

Noah, N.D. (1987) Epidemiology of bacterial meningitis: UK and USA. In: *Bacterial Meningitis*, pp. 93–115. Academic Press, London.

Noah, N.D. (1989) Cyclical patterns and the predictability in infection *Epidemiol. Infect.* **102**: 175–190.

Nokes, D.J. & Anderson, R.M. (1988) The use of mathematical models in the epidemiological study of infectious diesease and in the design of mass immunization programmes. *Epidemiol. Infect.* **101**: 1–20.

Nonacs, P. & Dill, L.M. (1990) Mortality risk vs. food quality trade-offs in a common currency: ant patch preferences. *Ecology* **71**: 1886–1892.

Nordlund, D.A., Jones, R.L. & Lewis, W.J. (eds) (1981) *Semiochemicals, their Role in Pest Control*. John Wiley & Sons, New York.

Nowack, R.M. & Paradiso, J.L. (1983) *Walker's Mammals of the World*, 4th edn, Vol. 2. Johns Hopkins University Press, Baltimore.

Noy-Meir, I. (1975) Stability of grazing systems: an application of predator–prey graphs. *J. Ecol.* **63**: 459–481.

Obrtel, R., Zejda, J. Holisová, V. (1978) Impact of small rodent predation on an overcrowded population of *Diprion pini* during winter. *Folia Zool.* **27**: 97–110.

O'Dowd, D.J. & Willson, M.F. (1989) Leaf domatia and mites on Australasian plants: ecological and evolutionary implications. *Biol. J. Linnean Soc.* **37**: 191–236.

O'Gara, B.W. & Harris, R.B. (1988) Age and condition of deer killed by predators and automobiles. *J. Wildlife Manage.* **52**: 316–320.

Ogden, V.T. (1975) Nesting density and reproductive success of the prairie falcon in southwest Idaho. *Raptor Research Foundation, Raptor Research Report* **3**: 67–69.

Ohsumi, S., Masaki, Y. & Kawamura, A. (1970) Stock of the Antarctic minke whale. *Sci. Rep. Whales Res. Inst. Tokyo* **22**: 75–125.

Okhotina, M.V. (1974) Morpho-ecological features making possible joint habitation for shrews of the genus *Sorex* (*Sorex*, Insectivora). In: Okhotina, M.V. (ed.) *Fauna and Ecology of the Terrestrial Vertebrates of the Southern Part of the Soviet Far East.* New Series Vol. 17, Vladivostok.

Oksanen, L., Fretwell, S.D., Arruda, J., Niemela, P. (1981) Exploitation ecosystems in gradients of primary productivity. *Am. Nat.* **118**: 240–261.

Oliveira, P. & Sazima, I. (1984) The adaptive bases of anti-mimicry in a neotropical aphantochilid spider (Araneae: Aphantochilidae). *Biol. J. Linnean Soc.* **22**: 145–155.

Oliver, J.S. & Slattery, P.N. (1985) Destruction and opportunity on the sea floor: effects of gray whale feeding. *Ecology* **66**: 1965–1975.

Ollason, J.G. (1980) Learning to forage — optimality? *Theor. Pop. Biol.* **18**: 44–56.

Olofsson, E. (1987) Mortality factors in a population of *Neodiprion sertifer* (Hymenoptera: Diprionidae). *Oikos* **48**: 297–303.

Olsen, P.D. & Cockburn, A. (1991) Female-biased sex allocation in peregrine falcons and other raptors. *Behav. Ecol. Sociobiol.* **28**: 417–423.

O'Meara, G.F. & Edman, J.D. (1975) Autogenous egg production in the salt marsh mosquito, *Aedes taeniorhynchus. Biol. Bull.* **149**: 384–396.

Orians, G.H. & Pearson, N.E. (1979) On the theory of central place foraging. In: Horn, D.J., Mitchell, R.D. & Stairs, C.R. (eds) *Analysis of Ecological Systems*, pp. 154–177. Ohio State University Press, Columbus.

Orr, B.K., Murdoch, W.W. & Bence, J.E. (1990) Population regulation, convergence, and cannibalism in *Notonecta* (Hemiptera). *Ecology* **71**: 68–82.

Oster, G.F. & Takahashi, Y. (1974) Models for age-specific interactions in a periodic environment. *Ecol. Monogr.* **44**: 483–501.

Oster, G.F. & Wilson, E.O. (1978) *Caste and Ecology in the Social Insects.* Princeton University Press, Princeton.

Pacala, S. & Crawley, M.J. (1992) Herbivores and plant density. *Am. Nat.* (in press).

Pacala, S. & Hassell, M.P. (1991) The persistence of host–parasitoid associations in patchy environments. II. Evaluation of field data. *Am. Nat.* **138**: 584–605.

Pacala, S.W., Hassell, M.P. & May, R.M. (1990) Host–parasitoid associations in patchy environments. *Nature* **374**: 150–153.

Packer, C. (1983) Sexual dimorphism: the horns of African antelopes. *Science* **221**: 1191–1193.

Packer, C. (1986) The ecology of felid sociality. In: Rubenstein, D.I. & Wrangham, R.W. (eds) *Ecological Aspects of Social Evolution*, pp. 429–451. Princeton University Press, Princeton.

Packer, C.R., Herbst, L., Pusey, A.E. *et al.* (1988) Reproductive success of lions. In: Clutton-Brock, T.H. (ed.) *Reproductive Success. Studies of Individual Variation in Contrasting Breeding Systems.* University of Chicago Press, Chicago.

Packer, C.R. & Pusey, A.E. (1984) Infanticide in carnivores. In: Hausfater, G. & Hrdy, S.B. (eds) *Infanticide: Comparative and Evolutionary Perspectives*, pp. 31–42. Aldine, New York.

Packer, C. & Ruttan, L. (1988) The evolution of cooperative hunting. *Am. Nat.* **132**: 159–198.

Packer, C., Scheel, D. & Pusey, A.E.

(1990) Why lions form groups: food is not enough. *Am. Nat.* **136**: 1–19.

Pagel, M.D. & Harvey, P.H. (1988) Recent developments in the analysis of comparative data. *Q. Rev. Biol.* **63**: 413–440.

Paine, R.T. (1966) Food web complexity and species diversity. *Am. Nat.* **100**: 65–75.

Paine, R.T. (1971) A short-term experimental investigation of resource partitioning in a New Zealand rocky intertidal habitat. *Ecology* **52**: 1096–1106.

Paine, R.T. (1974) Intertidal community structure: experimental studies on the relationship between a dominant competitor and its principal predator. *Oecologia* **15**: 93–120.

Paine, R.T. (1976) Size-limited predation: an observational and experimental approach with the *Mytilus–Pisaster* interaction. *Ecology* **57**: 858–873.

Paine, R.T. (1980) Food-webs: linkage, interaction strength and community infrastructure. *J. Anim. Ecol.* **49**: 667–685.

Paine, R.T. (1988) Food webs: road maps of interactions or grist for theoretical development? *Ecology* **69**: 1648–1654.

Paley, W. (1828) *Natural Theology*, 2nd edn. I. Vincent, Oxford.

Palmer, R.A. (1979) Fish predation and the evolution of gastropod shell sculpture: experimental and geographic evidence. *Evolution* **33**: 697–713.

Palmerino, C.C., Rusinak, K.W. & Garcia, J. (1980) Flavor-illness aversions: the peculiar roles of odor and taste in memory for poison. *Science* **208**: 753–755.

Pan, C.-T. (1965) Studies on the host–parasite relationship between *Schistosoma mansoni* and the snail *Australorbis glabratus*. *Am. J. Trop. Med. Hyg.* **14**: 931–976.

Pankakoski, E. (1984) Relationships between some meteorological factors and population dynamics of *Sorex araneus* L. in southern Finland. *Acta Zool. Fenn.* **173**: 287–289.

Parker, G.A. (1970) The reproductive behaviour and the nature of sexual selection in *Scatophaga stercoraria* L. II. The fertilization rate and the spatial and temporal relationships of each sex around the site of mating and oviposition. *J. Anim. Ecol.* **39**: 205–228.

Parker, G.A. (1983) Arms races in evolution — an ESS to the opponent-independent costs game. *J. Theor. Biol.* **101**: 619–648.

Parker, G.A. (1985) Population consequences of evolutionary stable strategies. In: Sibly, R.M. & Smith, R.H. (eds) *Behavioural Ecology: Ecological Consequences of Adaptive Behaviour*, pp. 33–58. Blackwell Scientific Publications, Oxford.

Parker, G.A. & Courtney, S.P. (1984) Models of clutch size in insect oviposition. *Theor. Pop. Biol.* **26**: 27–48.

Parker, J.W. (1974) Populations of the Mississippi kite in the Great Plains. *Raptor Research Foundation, Raptor Research Report.* **3**: 159–172.

Parker, M.A. (1985) Local population differentiation for compatibility in an annual legume and its host-specific fungal pathogen. *Evolution* **39**: 713–723.

Parmenter, R.R. & MacMahon, J.A. (1988) Factors influencing species composition and population sizes in a ground beetle community (Carabidae); predation by rodents. *Oikos* **52**: 350–356.

Parrish, J.D. & Saila, S.B. (1970) Interspecific competition, predation and species diversity. *J. Theor. Biol.* **27**: 207–220.

Pasteels, J.M., Grégoire, J.-C. & Rowell-Rahier, M. (1983) The chemical ecology of defense in arthropods. *Ann. Rev. Entomol.* **28**: 263–289.

Pasteels, J.N. (1978) Apterous and brachypterous coccinellids at the end of the food chain, *Cionura erecta* (Asclepiadaceae) — *Aphis nerii*. *Entomologia Experimentalis et Applicata* **24**: 379–384.

Pasteur, G. (1982) A classificatory review of mimicry systems. *Ann. Rev. Ecol. Systemat.* **13**: 169–199.

Pauly, D. (1987) On reason, mythologies of natural resource conservation. *NAGA, the ICLARM Quarterly* **10**(4): 11–12.

Pearse, V., Pearse, J., Buchsbaum, M. & Buchsbaum, R. (1987) *Living Invertebrates*. Blackwell Scientific Publications, Boston.

Pearson, D.L. (1980) Patterns of limiting similarity in tropical forest tiger beetles (Coleoptera: Cicindelidae). *Biotropica* **12**: 195–204.

Pearson, D.L. (1985) The function of

multiple anti-predator mechanisms in adult tiger beetles (Coleoptera: Cicindelidae). *Ecol. Entomol.* **10**: 65–72.

Pearson, D.L. (1988) Biology of tiger beetles. *Ann. Rev. Entomol.* **33**: 123–147.

Pearson, D.L. (1989) What is the adaptive significance of multicomponent defensive repertoires? *Oikos* **54**: 251–253.

Pearson, D.L. & Knisley, C.B. (1985) Evidence of food as a limiting resource in the life cycle of tiger beetles (Coleoptera: Cicindelidae). *Oikos* **45**: 161–168.

Pemberton, R.W. & Turner, C.E. (1989) Occurrence of predatory and fungivorous mites in leaf domatia. *Am. J. Bot.* **76**: 105–112.

Pennycuick, C.J. (1975) On the running of the gnu (*Connochaetes taurinus*) and other animals. *J. Exp. Biol.* **63**: 775–799.

Pernetta, J.C. (1976) Diets of the shrews *Sorex araneus* L. and *Sorex minutus* L. in Wytham grasslands. *J. Anim. Ecol.* **45**: 899–912.

Pernetta, J.C. (1977) Population ecology of British shrews in grasslands. *Acta Theriol.* **22**: 279–296.

Perrins, C.M. & Geer, T.A. (1980) The effect of sparrowhawks on tit populations. *Ardea* **68**: 133–142.

Perry, J. (1987) Host–parasitoid models of intermediate complexity. *Am. Nat.* **130**: 955–957.

Person, C. (1959) Gene-for-gene relationships in host: parasite systems. *Can. J. Bot.* **37**: 1101–1130.

Person, C. (1966) Genetic polymorphism in parasitic systems. *Nature* **212**: 266–267.

Peters, R.H. (1983) *The Ecological Implications of Body Size*. Cambridge University Press, Cambridge.

Peterson, R.O. (1979) The wolves of Isle Royale — new developments. In: Klinghammer, E. (ed.) *The Behavior and Ecology of Wolves* pp. 3–18. Garland STPM Press, New York.

Peterson, R.O. & Page, P. (1983) Wolf–moose fluctuations at Isle Royale National Park, Michigan, USA. *Acta Zool. Fenn.* **174**: 251–253.

Phillips, D.W. (1975) Distance chemoreception-triggered avoidance behaviour of the limpets *Acmaea (Colisella) limatula* and *Acmaea (Notoacmaea) scutum* to the predatory starfish *Pisaster ochraceus.*

J. Exp. Marine Biol. Ecol. **191**: 199–210.

Phillips, D.W. (1976) The effect of species–specific avoidance response to predatory starfish on the intertidal distribution of two gastropods. *Oecologia* **23**: 83–94.

Pianka, E.R. (1970) On r- and K-selection. *Am. Nat.* **104**: 592–597.

Picozzi, N. (1984) Sex ratio, survival and territorial behaviour of polygynous Hen Harriers *Circus c. cyaneus* in Orkney. *Ibis* **126**: 356–365.

Pielowski, Z. (1961) Über der Unifikationseinfluss der selektive Nahrungswahl des Habichts auf Haustauben. *Ekologia Polska Ser A* **9**: 183–194.

Pienaar, U. de (1969) Predator–prey relations amongst the larger mammals of Kruger National Park. *Koedoe* **12**: 108–176.

Pierce, G.J. & Ollason, J.G. (1987) Eight reasons why optimal foraging theory is a complete waste of time. *Oikos* **49**: 111–118.

Pietrewicz, A.T. & Kamil, A.C. (1979) Search image formation in the blue jay (*Cyanocitta cristata*). *Science* **204**: 1332–1333.

Pietrewicz, A.T. & Kamil, A.C. (1981) Search images and the detection of cryptic prey: an operant approach. In: Kamil, A.C. & Sargent, T.D. (eds) *Foraging Behavior. Ecological, Ethological and Psychological Approaches*, pp. 311–331. Garland STPM Press, New York.

Pimentel, D. (1961a) An ecological approach to the insecticide problem. *J. Econ. Entomol.* **54**: 108–114.

Pimentel, D. (1961b) Animal population regulation by the genetic feedback mechanism. *Am. Nat.* **95**: 65–79.

Pimentel, D. (1968) Population regulation and genetic feedback. *Science* **159**: 1432–1437.

Pimentel, D. (1984) Genetic diversity and stability in parasite–host systems. In: Shorrocks, B. (ed.) *Evolutionary Ecology*, pp. 295–311. Blackwell Scientific Publications, Oxford.

Pimm, S.L. (1982) *Food Webs*. Chapman and Hall, London.

Pimm, S.L. (1984a) The complexity and stability of ecosystems. *Nature* **307**: 321–326.

Pimm, S.L. (1984b) Food chains and return times. In: Strong, D.R. Jr, Simberloff, D., Abele, L.G. & Thistle,

A.B. (eds) *Community Ecology: Conceptual Issues and the Evidence*, pp. 397–412. Princeton University Press, Princeton.

Pimm, S.L. (1991) *The Balance of Nature?* Chicago University Press, Chicago.

Pimm, S.L. & Lawton, J.H. (1977) The number of trophic levels in ecological communities. *Nature* **268**: 329–331.

Pimm, S., Lawton, J.H. & Cohen, J.E. (1991) Food webs: patterns, causes and consequences. *Nature* (in press).

Pitcher, T.J. (1986) Functions of shoaling behaviour in teleosts. In: Pitcher, T.J. (ed.) *The Behaviour of Teleost Fishes*, pp. 294–337. Croom Helm, Beckenham.

Platt, J.B. (1977) *The breeding behaviour of wild and captive gyr falcons in relation to their environment and human disturbance.* PhD thesis, Cornell University.

Platt, W.J. & Blackley, N.R. (1973) Short-term effects of shrew predation upon invertebrate prey. *Proc. Iowa Acad. Sci.* **80**: 60–66.

Polis, G.A. (1981) The evolution and dynamics of intraspecific predation. *Ann. Rev. Ecol. Systemat.* **12**: 225–251.

Polis, G.A. & McCormick, S.J. (1986) Patterns of resource use and age structure among species of the desert scorpion. *J. Anim. Ecol.* **55**: 59–73.

Polis, G.A. & McCormick, S.J. (1987) Intraguild predation and competition among desert scorpions. *Ecology* **68**: 332–343.

Polis, G.A., Myers, C.A. & Holt, R.D. (1989) The ecology and evolution of intraguild predation: potential competitors that eat each other. *Ann. Rev. Ecol. Systemat.* **20**: 297–330.

Porter, R.D. & Wiemeyer, S.N. (1972) Reproductive patterns in captive American kestrels (sparrowhawks). *Condor* **74**: 46–53.

Possingham, H.P., Houston, A.I. & McNamara, J.M. (1991) Risk-averse foraging in bees: a comment on the model of Harder and Real. *Ecology* **71**: 1622–1624.

Post, W.M. & Travis, C.C. (1979) Quantitative stability in models of ecological communities. *J. Theor. Biol.* **79**: 547–553.

Potter, C.W. (1987) Influenza. In: Zuckerman, A.J., Banatvala, J.E. & Pattison, J.R. (eds) *Principles and Practice of Clinical Virology*, pp. 199–223. John Wiley & Sons, New York.

Potts, G.R. (1980) The effects of modern agriculture, nest predation and game management on the population ecology of partridges (*Perdix perdix* and *Alectoris rufa*). *Adv. Ecol. Res.* **11**: 2–81.

Pough, F.H. (1988a) Mimicry and related phenomena. In: Gans, C. & Huey, R.B. (eds) *The Biology of the Reptilia*, pp. 153–234. Alan Liss, New York.

Pough, F.H. (1988b) Mimicry of vertebrates: are the rules different? *Am. Nat.* **131**: S67–S102.

Poulton, E.B. (1890) *The Colours of Animals. Their Meaning and Use. Especially Considered in the Case of Insects.* Kegan Paul, London.

Powell, G.V.N. (1974) Experimental analysis of the social value of flocking by starlings (*Sturnus vulgaris*) in relation to predation and foraging. *Anim. Behav.* **22**: 501–505.

Power, G. & Gregoire, J. (1978) Predation by freshwater seals on the fish community of Lower Seal Lake, Quebec. *J. Fish Res. Board Can.* **35**: 844–850.

Pratt, D.M. (1974) Attraction to prey and stimulus to attack in the predatory gastropod *Urosalpinx cinerea*. *Marine Biol.* **27**: 37–46.

Pravosudov, W. (1985) Search for and storage of food by *Parus cinctus japonicus* and *P. montanus borealis* (Paridae). *Zool. Zh.* **64**: 1031–1034.

Prelec, D. (1982) Matching, maximising, and the hyperbolic reinforcement feedback function. *Psychol. Rev.* **89**: 189–230.

Prentice, R.M. (1962) *Forest Lepidoptera of Canada recorded by the Forest Insect Survey*, Vol. 2. Bull. 128. Canada Department of Forestry, Ottawa, Ontario.

Prentice, R.M. (1963) *Forest Lepidoptera of Canada recorded by the Forest Insect Survey*, Vol. 3. Publ. 1013. Canada Department of Forestry, Ottawa, Ontario.

Prentice, R.M. (1965) *Forest Lepidoptera of Canada recorded by the Forest Insect Survey*, Vol. 4. Publ. 1142. Canada Department of Forestry, Ottawa, Ontario.

Prestwich, K.N. (1977) The energetics of web-building in spiders. *Comp. Biochem. Physiol.* **57A**: 321–326.

Price, M.V. (1978) Seed dispersion preferences of coexisting desert rodent species. *J. Mammal.* **59**: 624–626.

Price, P.W. (1973) Parasitoid strategies and community organization. *Environ. Entomol.* **2**: 623–626.

Price, P.W. (1980) *Evolutionary Biology of Parasites.* Princeton University Press, Princeton.

Price, P.W. (1984) *Insect Ecology*, 2nd edn. John Wiley & Sons, New York.

Price, P.W. (1988) Inversely density-dependent parasitism: the role of plant refuges for hosts. *J. Anim. Ecol.* **57**: 89–96.

Prins, H.H.T. & Iason, G.R. (1989) Dangerous lions and nonchalant buffalo. *Behaviour* **108**: 262–296.

Provencher, L. & Vickery, W. (1988) Territoriality, vegetation complexity and biological control: the case for spiders. *Am. Nat.* **132**: 257–266.

Pucek, Z. (1959) Sexual maturation and variability of the reproductive system in young shrews (*Sorex* L.) in the first calendar year of life. *Acta Theriol.* **3**: 269–296.

Pyke, G.H. (1984) Optimal foraging theory: a critical review. *Annual Rev. Ecol. System.* **15**: 523–575.

Pyke, H.G., Pulliam, H.R. & Charnov, E.L. (1977) Optimal foraging: a selective review of theory and tests. *Q. Rev. Biol.* **52**: 137–154.

Rachlin, H., Battalio, R., Kagel, J. & Green, L. (1981) Maximization theory and behavior. *Behav. Brain Sci.* **4**: 371–388.

Radinsky, L. (1978) Evolution of brain size in carnivores and ungulates. *Am. Nat.* **112**: 815–831.

Rae, B.B. (1965) The food of the common porpoise (*Phocoena phocoena*). *Proc. Zool. Soc. Lond.* **146**: 114–122.

Rae, B.B. (1973) Additional notes on the food of the common porpoise (*Phocoena phocoena*). *J. Zool. Lond.* **169**: 127–131.

Raizenne, H. (1952) *Forest Lepidoptera of Southern Ontario and their Parasites.* Canada Department of Agriculture, Division of Forest Biology, Ottawa, Ontario.

Ralls, K.B., Kranz, K. & Lundrigan, B. (1987) Mother–young relationships in captive ungulates: behavioural changes over time. *Ethology* **75**: 1–14.

Randolph, S.E. (1979) Population regulation in ticks: the role of acquired resistance in natural and unnatural hosts. *Parasitology* **79**:

141–156.

Rapport, D.J. & Person, C.O. (1980) Games that genes play: host–parasite interactions in a game-theoretic context. *Evol. Theory* **4**: 275–287.

Ratcliffe, D.A. (1972) The peregrine population of Great Britain in 1971. *Bird Study* **19**: 117–156.

Raymont, J.E.G. (1983) *Plankton and Productivity in the Oceans. Volume 2. Zooplankton*, 2nd edn. Pergamon Press, Oxford.

Read, A.J. & Gaskin, D.E. (1985) Radio tracking the movements and activities of harbour porpoises, *Phocoena phocoena* (L.) in the Bay of Fundy, Canada. *Fish. Bull.* **83** (4).

Real, L.A. (1981) Uncertainty and pollinator–plant interactions: the foraging behavior of bees and wasps on artificial flowers. *Ecology* **62**: 20–26.

Real, L.A. & Caraco, T. (1986) Risk and foraging in stochastic environments: theory and evidence. *Ann. Rev. Ecol. Systemat.* **17**: 371–390.

Real, L.A., Ott, J. & Silverfine, E. (1982) On the tradeoff between the mean and the variance in foraging: effect of spatial distribution and color preference. *Ecology* **63**: 1617–1623.

Recer, G.M. & Caraco, T. (1989) Sequential-encounter prey choice and effects of spatial resource variability. *J. Theor. Biol.* **139**, 239–249.

Rechten, C., Avery, M.I. & Stevens, J.A. (1983) Optimal prey selection: why do great tits show partial preferences? *Anim. Behav.* **31**: 576–584.

Redfearn, A. & Pimm, S.L. (1987) Insect pest outbreaks and community structure. In: Barbosa, P. & Schultz, J.C. (eds) *Insect Outbreaks* pp. 99–134. Academic Press, San Diego.

Redfearn, A. & Pimm, S.L. (1988) Population variability and polyphagy in herbivorous insect communities. *Ecol. Monogr.* **58**: 39–55.

Redford, K.H. (1988) Ants and termites as food. In: Genoways, H.H. (ed.) *Current Mammalogy*, Vol. 1, pp. 349–399.

Redpath, S.M. (1991) The impact of hen harriers on red grouse breeding success. *J. Appl. Ecol.* **28**: 659–671.

Reese, J.G. (1970) Reproduction in a Chesapeake Bay osprey population. *Auk* **87**: 747–759.

Reeve, J.D. (1988) Environmental variability, migration and persistence in host–parasitoid systems. *Am. Nat.* **132**: 810–836.

Reeve, J.D. (1990) Stability, variability, and persistence in host–parasitoid systems. *Ecology* **71**: 422–426.

Reid, M.L. (1987) Costliness and reliability in the singing vigour of Ipswich sparrows. *Anim. Behav.* **35**: 1735–1744.

Rejmanek, M. & Spitzer, K. (1982) Bionomic strategies and long-term fluctuations in abundance of Noctuidae (Lepidoptera). *Acta Entomol. Bohem.* **79**: 81–96.

Rescorla, R.A. (1988) Pavlovian conditioning: is not what you think it is. *Am. Psychol.* **43**: 151–160.

Rescorla, R.A. & Wagner, A.R. (1972) A theory of Pavlovian conditioning: variations in the effectiveness of reinforcement and nonreinforcement. In: Black, A.H. & Prokasy, W.F.G. (eds) *Classical Conditioning II; Current Research and Theory,* pp. 64–99. Appleton-Century-Crofts, New York.

Rettenmeyer, C.W. (1970) Insect mimicry. *Ann. Rev. Entomol.* **15**: 43–75.

Reynolds, J.W. (1977) *The Earthworms (Lumbricidae and Sparganophilidae) of Ontario.* Life Sci. Misc. Publ. R. Ont. Mus., Toronto.

Reynolds, J.C., Angelstam, P. & Redpath, S. (1988) Predators, their ecology and impact on game-bird populations. In: Hudson, P.J. & Rands, M.W. (eds) *Ecology and Management of Gamebirds,* pp. 72–97. Blackwell Scientific Publications, Oxford.

Reynoldson, T.B. (1983) The population biology of Turbellaria with special reference to the freshwater triclads of the British Isles. *Adv. Ecol. Res.* **13**: 235–326.

Reznick, D.A., Bryga, H. & Endler, J.A. (1990). Experimentally induced life-history evolution in a natural population. *Nature* **346**: 357–359.

Rhoades, D.F. (1979) Evolution of plant chemical defense against herbivores. In: Rosenthal, G.A. & Janzen, D.H. (eds) *Herbivores. Their Interaction with Secondary Plant Metabolites,* pp. 3–54. Academic Press, New York.

Rhoades D.F. (1983) Herbivore population dynamics and plant chemistry. In: Denno, R.F. & McClure, M.S. (eds) *Variable Plants and Herbivores in Natural and Managed Systems,* pp. 155–222. Academic Press, New York.

Rhoades, D.F. (1985) Offensive–defensive interactions between herbivores and plants: their relevance in herbivore population dynamics and ecological theory. *Am. Nat.* **125**: 205–238.

Rhoades, D.F. & Cates, R.G. (1976) Toward a general theory of plant antiherbivore chemistry. In: Wallace, J.W. & Mansell, R.L. (eds) *Biochemical Interactions Between Plants and Insects,* pp. 168–213. *Recent Advances in Phytochemistry 10.* Plenum Press, New York.

Rhodes, L.I. (1972) Success of osprey nest structures at Martin National Wildlife Refuge. *J. Wildlife Manage.* **36**: 1296–1299.

Ribeiro, J.M.C., Rossignol, P.A. & Spielman, A. (1985) *Aedes aegypti:* model for blood finding strategy and prediction of parasite manipulation. *Exp. Parasitol.* **60**: 118–132.

Richards, C.S. (1970). Genetics of a molluscan vector of schistosomiasis. *Nature* **227**: 806–810.

Richards, C.S. (1975). Genetic factors in susceptibility of *Biomphalaria glabrata* for different strains of *Schistosoma mansoni. Parasitology* **70**: 231–241.

Richman, D.B., Hemenway, R.C. & Whitcomb, W.H. (1980) Field cage evaluation of predators of the soybean looper *Pseudoplusia includens* (Walker) Lepidoptera: Noctuidae. *Environ. Entomol.* **9**: 315–317.

Richter, C.J.J. (1970) Aerial dispersal in relation to habitat in wolf spider species (*Pardosa,* Araneae, Lycosidae). *Oecologia* **5**: 200–214.

Riechert, S.E. (1976) Web-site selection in the desert spider, *Agelenopsis aperta* (Gertsch) *Oikos* **27**: 311–315.

Riechert, S.E. (1978) Energy-based territoriality in populations of the desert spider *Agelenopsis aperta* (Gertsch). *Symp. Zool. Soc. Lond.* **42**: 211–222.

Riechert, S.E. (1981) The consequences of being territorial: spiders, a case study. *Am. Nat.* **117**: 871–892.

Riechert, S.E. (1982) Spider interaction strategies: communication versus coercion. In: Witt, P.N. & Rovner, J. (eds) *Spider Communication: Mechanisms and Ecological Significance,* pp. 281–313. Princeton University, Princeton.

Riechert, S.E. (1985a) Why do some spiders cooperate? *Agelena consociata*, a case study. *Florida Entomol.* **68**: 5–16.

Riechert, S.E. (1985b) Decisions in multiple goal contexts: habitat selection of the spider, *Agelenopsis aperta* (Gertsch). *Z. Tierpsychol.* **70**: 53–69.

Riechert, S.E. (1986) Spider fights and the role of behavior in ecotypic adaptation. *Am. Sci.* **74**: 604–610.

Riechert, S.E. & Bishop, L. (1990) Prey control by an assemblage of generalist predators in a gradient test system. *Ecology* **71**: 1441–1450.

Riechert, S.E. & Cady, A.B. (1983) Patterns of resource use and tests for competitive release in a spider community. *Ecology* **64**: 899–913.

Riechert, S.E. & Gillespie, R. (1986) Habitat choice and utilization in web building spiders. In: Shear, B. (ed.) *Spiders, Webs, Behavior and Evolution*, pp. 23–49. Stanford University Press, Stanford.

Riechert, S.E. & Harp, J. (1987) Nutritional ecology of spiders. In: Slansky, F. Jr. & Rodriquez, J.G. (eds) *Nutritional Ecology of Insects, Mites and Spiders*, pp. 645–672. John Wiley & Sons, New York.

Riechert, S.E. & Lockley, T. (1984) Spiders as biological control agents. *Ann. Rev. Entomol.* **29**: 299–320.

Riechert, S.E. & Luczak, J. (1982) Spider foraging: behavioral responses to prey. In: Witt, P.N. & Rovner, J. (eds) *Spider Communication: Mechanisms and Ecological Significance*, pp. 353–384. Princeton University Press, Princeton.

Riechert, S.E. & Maynard Smith, J. (1989) Genetic analyses of two behavioral traits linked to individual fitness in the desert spider, *Agelenopsis aperta*. *Anim. Behav.* **37**: 624–637.

Riechert, S.E. & Tracy, C.R. (1975) Thermal balance and prey availability: bases for a model relating web-site characteristics to spider reproductive success. *Ecology* **56**: 265–285.

Ringelberg, J. (1991) Enhancement of the phototactic reaction in *Daphnia hyalina* by a chemical mediated by juvenile perch (*Perca fluviatilis*). *J. Plankton Res.* **12**: 17–25.

Risch, S.J., Andow, D. & Altieri, M. (1983) Agroecosystem diversity and pest control; data, tentative conclusions and new research directions. *Environ. Entomol.* **12**: 625–629.

Roach, S.H. (1980) Arthropod predators on cotton, corn, tobacco, and soybeans in South Carolina. *J. Georgia Entomol. Soc.* **15**: 124–131.

Robinson, M.H., Mirick, H. & Turner, O. (1969) The predatory behavior of some araneid spiders and the origin of immobilization wrapping. *Psyche* **76**: 487–501.

Robinson, M.H. & Olazarri, J. (1971) Units of behavior and complex sequences in the predatory behavior of *Argiope argentata* Fabricius (Araneae: Araneidae). *Smithsonian Contributions to Zoology* **65**: 1–36.

Robinson, M.H. & Robinson, B.C. (1971) The predatory behavior of the ogre-faced spider, *Dinopis longipes* F. Cambridge (Araneae: Dinopidae). *Am. Midl. Nat.* **85**: 85–96.

Robinson, M.H. & Valerio, C.E. (1977) Attacks on large or heavily defended prey by tropical salticid spiders. *Psyche* **84**: 1–10.

Rockenbauch, D. (1968) Zur Brutbiologie des Turmfalken (*Falco tinnunculus* L.). *Anz. orn. Ges. Bayern* **8**, 267–276.

Rocklin, S. & Oster, G. (1976) Competition between phenotypes. *J. Math. Biol.* **3**: 225–261.

Roeske, C.N., Seiber, J.N., Brower, L.P. & Moffitt, C.M. (1976) Milkweed cardenolides and their comparative processing by monarch butterflies. (*Danaus plexippus* L.). In: Wallace, J.W. & Mansell, R.L. (eds) *Biochemical Interactions Between Plants and Insects*, pp. 93–167. *Recent Advances in Phytochemistry* 10. Plenum Press, New York.

Rogers, D. (1972) Random search and insect population models. *J. Anim. Ecol.* **41**: 369–384.

Rogers, D.J. (1988) A general model for the African trypanosomiases. *Parasitology* **97**: 193–212.

Rogers, D.J. & Randolph, S.E. (1984a) From a case study to a theoretical basis for tsetse control. *Insect Science and its Application* **5**: 419–423.

Rogers, D.J. & Randolph, S.E. (1984b) A review of density-dependent processes in tsetse populations. *Insect Science and its Application* **5**: 397–402.

Rogers, D.J. & Randolph, S.E. (1985). Population ecology of tsetse. *Ann. Rev. Entomol.* **30**: 197–216.

Rogers, D.J., Randolph, S.E. & Kuzoe, F.A.S. (1984) Local variation in the population dynamics of *Glossina palpalis palpalis* (Robineau-Desvoidy) (Diptera:Glossinidae). I. Natural population regulation. *Bull. Entomol. Res.* **74**: 403–423.

Roland, J. (1988) Decline in winter moth populations in North America: direct versus indirect effect of introduced parasites. *J. Anim. Ecol.* **57**: 523–531.

Rollinson, D. & Anderson, R.M. (eds) (1985) *Ecology and Genetics of Host–Parasite Interactions*. Academic Press, London.

Rollinson, D. & Southgate, V.R. (1985) Schistosome and snail populations: genetic variability and parasite transmission. In: Rollinson, D. & Anderson, R.M. (eds) *Ecology and Genetics of Host–Parasite Interactions*, pp. 91–109. John Wiley & Sons, New York.

Rollinson, D. & Wright, C.A. (1984) Population studies on *Bulinus cernicus* from Mauritius. *Malacologia* **25**: 447–463.

Root, R.B. (1973) Organization of a plant arthropod association in simple and diverse habitats; the fauna of collards, *Brassica oleracea. Ecol. Monogr.* **43**: 95–124.

Roper, T.J. (1990) Responses of domestic chicks to artificially coloured insect prey: effects of previous experience and background colour. *Anim. Behav.* **39**: 466–473.

Roper, T.J. & Cook, S.E. (1989) Responses of chicks to brightly coloured insect prey. *Behaviour* **110**: 276–293.

Roper, T.J. & Redston, S. (1987) Conspicuousness of distasteful prey affects the strength and durability of one-trial avoidance learning. *Anim. Behav.* **35**: 739–747.

Roper, T.J. & Wistow, R. (1986) Aposematic colouration and avoidance learning in chicks. *Q. J. Exp. Psychol.* **38**: 141–149.

Rosenberg, A.A., Beddington, J.R. & Basson, M. (1986) Growth and longevity of krill during the first decade of pelagic whaling. *Nature* **324**: 152–154.

Rosenzweig, M.L. (1971) Paradox of enrichment: destabilisation of exploitation ecosystems in ecological time. *Science* **171**: 385–387.

Rosenzweig, M.L. & MacArthur, R.H. (1963) Graphical representation and stability conditions of predator–prey interactions. *Am. Nat.* **97**: 209–273.

Rosenzweig, M.L. & Schaffer, W.M. (1978) Homage to the red queen. II. Coevolutionary response to enrichment of exploitation ecosystems. *Theor. Pop. Biol.* **14**: 158–163.

Ross, R. (1911) *The Prevention of Malaria*, 2nd edn. Murray, London.

Rossignol, P.A., Ribeiro, J.M.C., Jungery, M.J., Turell, M.J., Spielman, A. & Bailey, C.L. (1985) Enhanced mosquito blood-finding success on parasitaemic hosts: evidence for vector–parasite mutualism. *Proc. Natl. Acad. Sci. USA* **82**: 7725–7727.

Roth, G. (1986) Neural mechanisms of prey recognition: an example in amphibians. In: Feder, M.E. & Lauder, G.V. (eds) *Predator–Prey Relationships. Perspectives and Approaches from the Study of Lower Vertebrates*, pp. 42–68. Chicago University Press, Chicago.

Rotheray, G.E. (1983) Feeding behaviour of *Syrphus ribesii* and *Melanostoma scalare* on *Aphis fabae. Entomologia Experimentalis et Applicata* **34**: 148–154.

Rotheray, G.E. (1986) Colour, shape and defence in aphidophagous syrphid larvae (Diptera). *Zool. J. Linnean Soc.* **88**: 201–216.

Rothschild, M. (1973) Secondary plant substances and warning coloration in insects. In: van Emden, H.F. (ed.) *Insect/Plant Relationships*, pp. 59–83. *Symp. R. Entomol. Soc. Lond.* **6**. Blackwell Scientific Publications, Oxford.

Rothschild, M. (1984) Aide memoire mimicry. *Ecol. Entomol.* **9**: 311–319.

Rothschild, M. (1985) British aposematic lepidoptera. In: Heath, J. & Maitland Emmet, A. (eds) *The Moths and Butterflies of Great Britain and Ireland*, 2, pp. 9–62. Harley, Colchester.

Rothschild, M. & Moore, B.P. (1986) *Pyrazines as Alerting Signals in Toxic Plants and Insects*, pp. 97–101. Junk, Pau, France.

Rothschild, M., Moore, B.P. & Brown, V. (1984) Pyrazines as warning odour components in the monarch butterfly, *Danaus plexippus*, and in moths of the genera *Zygaena* and *Amata* (Lepidoptera). *Biol. J. Linnean Soc.* **23**: 375–380.

Roughgarden, J. (1979) *Theory of*

Population Genetics and Evolutionary Ecology: An Introduction. Macmillan, New York.

Roughgarden, J. (1983). The theory of coevolution. In: Futuyma, D.J. & Slatkin, M. (eds) *Coevolution*, pp. 33–64. Sinauer, Sunderland, MA.

Roughgarden, J. & Feldman, M. (1975) Species packing and predation pressure. *Ecology* **56**: 459–492.

Roughgarden, J., Iwasa, Y. & Baxter, C. (1985) Demographic theory for an open marine population with space-limited recruitment. *Ecology* **66**: 54–67.

Roughgarden, J., May, R.M. & Levin, S.A. (1989) *Perspectives in Ecological Theory.* Princeton University Press, Princeton.

Royama, T. (1971) A comparative study of models of predation and parasitism. *Res. Pop. Ecol.* **1**: 1–91.

Rudge, M.R. (1968) The food of the common shrew *Sorex araneus* (Insectivora, Soricidae) in Britain. *J. Anim. Ecol.* **37**: 565–581.

Rudnai, J. (1974) The pattern of lion predation in Nairobi Park. *E. Afr. Wild. J.* **12**: 213–225.

Rutberg, A.T. (1987) Adaptive hypotheses of birth synchrony in ruminants: an interspecific test. *Am. Nat.* **130**: 692–710.

Ryan, R.B. (1986) An analysis of life tables for the larch casebearer (Lepidoptera: Coleophoridae) in Oregon. *Can. Entomol.* **118**: 1255–1263.

Ryan, R.B. (1990) Evaluation of biological control: introduced parasites of the larch casebearer (Lepidoptera: Coleophoridae) in Oregon. *Environ. Entomol.* **19**: 1873–1881.

Rypstra, A.L. (1983) The importance of food and space in limiting web-spider densities; a test using field enclosures. *Oecologia* **59**: 312–316.

Saarikko, J. & Hanski, I. (1990) Timing of sleep and rest in foraging shrews. *Anim. Behav.* **40**: 861–869.

Sabelis, M.W. (1981) *Biological Control of Two-Spotted Spider Mites Using Phytoseiid Predators.* Agricultural Research Reports **910**, Pudoc, Wageningen.

Sabelis, M.W. (1985) Predation on spider mites. In: Helle, W. & Sabelis, M.W. (eds) *Spider Mites: Their Biology, Natural Enemies and Control*, pp. 103–129. Elsevier, Amsterdam.

Sabelis, M.W. (1986) The functional response of predatory mites to the density of two-spotted spider mites. In: Metz, J.A.J. & Diekmann, O. (eds) *Dynamics of Physiologically Structured Populations*, pp. 298–321. Springer-Verlag, Berlin.

Sabelis, M.W. (1990) How to analyse prey preference when prey density varies? A new method to discriminate between effects of gut fullness and prey type composition. *Oecologia* **82**: 289–298.

Sabelis, M.W. & De Jong, M.C.M. (1988) Should all plants recruit bodyguards? Conditions for a polymorphic ESS of synomone production in plants. *Oikos* **53**: 247–252.

Sabelis, M.W. & Dicke, M. (1985) Long-range dispersal and searching behaviour. In: Helle, W. & Sabelis, M.W. (eds) *Spider Mites: Their Biology, Natural Enemies and Control*, pp. 141–160. Elsevier, Amsterdam.

Sabelis, M.W., Diekmann, O. & Jansen, V.A.A. (1991) Metapopulation persistence despite local extinction: predator–prey patch models of the Lotka–Volterra type. *Biol. J. Linnean Soc.* **42**: 267–283.

Sabelis, M.W., Janssen, A. & Helle, W. (eds) (1988a) Population dynamics of predatory mites and spider mites. Part I. *Exp. Appl. Acarol.* **4**: 187–318.

Sabelis, M.W., Janssen, A. & Helle, W. (eds) (1988b) Population dynamics of predatory mites and spider mites. Part II. *Exp. Appl. Acarol.* **5**: 186–347.

Sabelis, M.W. & Nagelkerke, C.J. (1988) Evolution of pseudo-arrhenotoky. *Exp. Appl. Acarol.* **4**: 301–318.

Sabelis, M.W. & Van de Baan, H.E. (1983) Location of distant spider-mite colonies by phytoseiid predators. Demonstration of specific kairomones emitted by *Tetranychius urticae* and *Panonychus ulmi* (Acari: Phytoseiidae, Tetranychidae). *Entomologia Experimentalis et Applicata* **33**: 303–314.

Sabelis, M.W. & Van Der Meer, J. (1986) Local dynamics of the interaction between predatory mites and two-spotted spider mites. In: Metz, J.A.J. & Diekmann, O. (eds) *Dynamics of Physiologically Structured Populations*, pp. 322–344. Springer-Verlag, Berlin.

Sabelis, M.W. & Van der Weel, J.J.

(1992) Orientation of a phytoseiid mite to air currents with and without prey kairomones. *Exp. Appl. Acarol.* (in press).

Sabelis, M.W., Vermaat, J.E. & Groeneveld, A. (1984) Arrestment responses of the predatory mite, *Phytoseiulus persimilis*, to steep odour gradients of a kairomone. *Physiol. Entomol.* **9**: 437–446.

Salt, G. (1934) Experimental studies in insect parasitism. II. Superparasitism. *Proc. R. Soc. Lond. B* **144**: 455–476.

Salt, G. (1961) Competition between insect parasitoids. *Symp. Soc. Exp. Biol.* **15**: 96–119.

Samarawickrema, W.A. & Laurence, B.R. (1978) Loss of filarial larvae in a natural mosquito population. *Ann. Trop. Med. Parasitol.* **72**: 561–565.

Sandell, M. (1989) Ecological energetics, optimal body size and sexual size dimorphism: a model applied to the stoat, *Mustela erminea. Funct. Ecol.* **3**: 315–324.

Sandness, J.N. & McMurtry, J.A. (1970) Functional responses of three species of Phytoseiidae to prey density. *Can. Entomol.* **102**: 692–704.

Sandness, J.N. & McMurtry, J.A. (1972) Prey consumption behaviour of *Amblyseius largoensis* in relation to hunger. *Can. Entomol.* **104**: 461–470.

Sargent, T.D. (1966) Background selections of Geometrid and Noctuii moths. *Science* **154**: 1674–1675.

Sargent, T.D. (1968) Cryptic moths: effects on background selections of painting the circumocular scales. *Science* **159**: 100–101.

Sargent, T.D. (1969a) Background selections of the pale and melanic form of the cryptic moth, *Phigalia titea* (Cramer). *Nature* **222**: 585–586.

Sargent, T.D. (1969b) Behavioral adaptation of cryptic moths. iii. Resting attitudes of two bark-like species, *Melanophia canadaria* and *Catocala ultronia. Anim. Behav.* **17**: 670–672.

Sargent, T.D. (1973) Behavioral adaptations of cryptic moths. vi. Further experimental studies on bark-like species. *J. Lepidopterists' Soc.* **27**: 8–12.

Savidge, J.A. (1987) Extinction of an island forest avifauna by an introduced snake. *Ecology* **68**: 660–668.

Saville, A. & Bailey, R.S. (1980) The assessment of management of the herring stocks in the North Sea and to the West of Scotland. *Rapp. P-v. Reun. Cons. Int. Explor. Mer* **177**: 112–142.

Savory, T.H. (1928) *The Biology of Spiders*. Sidgwick and Jackson, London.

Scaife, M. (1976a) The response to eye-like shapes by birds. I. The effect of context: a predator and a strange bird. *Anim. Behav.* **24**: 195–199.

Scaife, M. (1976b) The response to eye-like shapes by birds. II. The importance of staring, pairedness and shape. *Anim. Behav.* **24**: 200–206.

Schad, G.A. & Anderson, R.M. (1985) Predisposition to hookworm infection in humans. *Science* **228**: 1537–1540.

Schaffer, W.M. (1987) Chaos in ecology and epidemiology. In: Degn, H., Holden, A.V. & Olsen, L.F. (eds) *Chaos in Biological Systems*, pp. 233–248. Plenum Press, London.

Schaffer, W.M. & Rosenzweig, M.L. (1978) Homage to the Red Queen. I. Coevolution of predators and their victims. *Theor. Pop. Biol.* **14**: 135–157.

Schaller, G.B. (1972) *The Serengeti Lion: A study of Predator–Prey Relations*. University of Chicago Press, Chicago.

Schaller, G.B., Jinchu, H., Wenshi, P. & Jing, Z. (1985) *The Giant Pandas of Wolong*. University of Chicago Press, Chicago.

Schelde, O. (1960) Danske Spurvhoges (*Accipiter nisus* (L.)) Traekforhold. *Dansk Orn. Foren. Tidsk.* **54**: 88–102.

Scheltema, R.S. (1971) Larval dispersal as a means of genetic exchange between geographically separated populations of shallow-water benthic marine gastropods. *Biol. Bull.* **140**: 284–322.

Schindler, U. (1965) Zur Parasitierung des Kiegernknospen-triebwichlers (*Rhyacionia buoliana* Schiff.) in Norddeutschland. *Z. angew. Entomol.* **55**: 353–364.

Schmid-Hempel, P., Kacelnik, A. & Houston, A.I. (1985) Honeybees maximise efficiency by not filling their crop. *Behav. Ecol. Sociobiol.* **17**: 61–66.

Schmidt-Nielsen, K. (1984) *Scaling, Why is Animal Size so Important?* Cambridge University Press, Cambridge.

Schneider, P. (1971) Ameisenjagende Spinnen (Zodariidae) an *Cataglyphis-*

Nestern in Afghanistan. *Zool. Anz.* **187**: 199–201.

Schoener, T.W. (1968) Sizes of feeding territories among birds. *Ecology* **49**: 704–726.

Schoener, T.W. (1971) Theory of feeding strategies. *Ann. Rev. Ecol. Systemat.* **2**: 369–404.

Schoener, T.W. (1983) Simple models of optimal feeding-territory size: a reconciliation. *Am. Nat.* **121**: 608–629.

Schoener, T.W. (1986) Mechanistic approaches to community ecology: a new reductionism. *Am. Zool.* **26**: 81–106.

Schoener, T.W. (1987) A brief history of optimal foraging theory. In: Kamil, A.C., Krebs, J.R. & Pulliam , H.R. (eds) *Foraging Behavior*, pp. 5–67. Plenum Press, New York.

Schoenfelder, T.W., Housewearth, M.W., Thompson, L.C. Kulman, H.M. & Martin, F.B. (1978) Insect and mammal predation of yellow-headed spruce sawfly cocoons (Hymenoptera, Tenthredinidae). *Environ. Entomol.* **7**: 711–713.

Schofield, C.J. (1985) Population dynamics and control of *Triatoma infestans*. *Ann. Soc. Belge Med. Trop.* **65** (suppl. 1): 149–164.

Schofield, C.J. & Marsden, P.D. (1982) The effect of wall plaster on a domestic population of *Triatoma infestans*. *Bull. Pan Am. Health Organ.* **16**: 356–360.

Scholander, P.F. (1940) Experimental investigations on the respiratory functions in diving mammals and birds. *Hvalrad. Skr.* **22**: 1–133.

Schroeder, D.S. (1974) A study of the interactions between internal larval parasites of *Rhyacionia buoliana* (Lepidoptera: Olethreutidae). *Entomophaga* **19**: 145–171.

Schuler, W. (1989) Ethoexperimental analysis of the response of young chicks to warningly colored prey. In: Blanchard, R.J., Brain, P., Blanchard, D.C. & Parmigiani, S. (eds) *Ethoexperimental Approaches to the Study of Behavior*, pp. 104–111. Kluwer Academic Publishers, Dordrecht.

Schuler, W. & Hesse, E. (1985) On the function of warning coloration: a black and yellow pattern inhibits prey-attack by naive domestic chicks. *Behav. Ecol. Sociobiol.* **16**: 249–255.

Schwartz, I.B. (1985) Multiple recurrent outbreaks and predictability in seasonally forced nonlinear epidemic models. *J. Math. Biol.* **21**: 347–361.

Scott, M.E. (1988) The impact of infection and disease on animal populations: implications for conservation biology. *Conservation Biol.* **2**: 40–56.

Scudo, F.M. & Ziegler, J.R. (eds) (1978) *The Golden Age of Theoretical Ecology: 1923–1940.* Springer-Verlag, Berlin.

Seeley, R.H. (1986) Intense natural selection caused a rapid morphological transition in a living marine snail *Littorina obtusata*. *Proc. Nat. Acad. Sci. USA* **83**: 6897–6901.

Seger, J. (1988) Dynamics of some simple host–parasite models with more than two genotypes in each species. *Phil. Trans. R. Soc. Lond. B* **319**: 541–555.

Sergeant, D.E. (1969) Feeding rates of Cetacea. *Fisker Direktoratets Skrifter Serie Havundersokelser, Norway* **15**: 246–258.

Shaw, J. (1988) Arrested larval development of *Trichostrongylus tenuis* as third stage larvae in red grouse. *Res. Vet. Sci.* **45**: 256–258.

Sheehan, W. (1986) Response by specialist and generalist natural enemies to agroecosystem diversification: a selective review. *Environ. Entomol.* **15**: 456–462.

Sheftel, B.I. (1983) Zonal distribution of insectivorous mammals in the Yenisei taiga and forest tundra. In: Syroechkovsky, E.E. (ed.) *Animals of the Yenisei Taiga and Forest Tundra*, pp. 184–203. Nauka, Moscow. (In Russian.)

Sheftel, B.I. (1989) Long-term and seasonal dynamics of shrews in Central Siberia. *Ann. Zool. Fenn.* **26**: 331–479.

Shettleworth, S.J. (1972) The role of novelty in learned avoidance of unpalatable prey by domestic chicks. *Anim. Behav.* **20**: 29–35.

Shettleworth, S.J. (1988) Foraging as operant behavior and operant behavior as foraging: what have we learned? In: Bower, G. (ed.) *The Psychology of Learning and Motivation: Advances in Research and Theory*, Vol. 22, pp. 1–99. Academic Press, New York.

Shettleworth, S.J., Krebs, J.R., Stephens, D. & Gibbon, J. (1988) Tracking a changing environment: a study of sampling. *Anim. Behav.* **36**: 87–105.

Shipman, P. & Walker, A. (1989) The costs of becoming a predator. *J. Hum. Evol.* **18**: 373–392.

Shoop, W.L. (1988) Trematode transmission patterns. *J. Parasitol.* **74**: 46–59.

Sibly, R.M. & Smith, R.H. (eds) (1985) *Behavioural Ecology: Ecological Consequences of Adaptive Behaviour.* Blackwell Scientific Publications, Oxford.

Sih, A. (1980) Optimal foraging: partial consumption of prey. *Am. Nat.* **116**: 281–290.

Sih, A. (1981) Stability, prey density and age-dependent interference in an aquatic insect predator. *Notonecta hoffmanni. J. Anim. Ecol.* **50**: 625–636.

Sih, A. (1987) Predators and prey lifestyles: an evolutionary and ecological overview. In: Kerfoot, W.C. & Sih, A. (eds) *Predation: Direct and Indirect Impacts on Aquatic Communities*, pp. 203–224. University of New England Press, Hanover, NH.

Silberglied, R.E., Aiello, A. & Windsor, M. (1980) Disruptive coloration in butterflies: lack of support in *Anartia fatima. Science* **209**: 617–619.

Sillén-Tullberg, B. (1985) The significance of coloration *per se*, independent of background for predator avoidance of aposematic prey. *Anim. Behav.* **33**: 1382–1384.

Simms, D.A. (1979) North American weasels: resource utilization and distribution. *Can. J. Zool.* **57**: 504–520.

Sinclair, A.R.E. (1979) The eruption of the ruminants. In: Sinclair, A.R.E. & Norton-Griffiths (eds) *Serengeti: The Dynamics of an Ecosystem*, pp. 82–103. University of Chicago Press, Chicago.

Sinclair, A.R.E. (1989) Population regulation in animals. In: Cherrett, J.M. (ed.) *Ecological Concepts*, pp. 197–241. Blackwell Scientific Publications, Oxford.

Sinclair, A.R.E., Dublin, H. & Borner, M. (1985) Population regulation of Serengeti wildebeest: a test of the food hypothesis. *Oecologia* **65**: 266–268.

Sinclair, A.R.E., Krebs, C.J., Smith, J.N.M. & Boutin, S. (1988) Population biology of snowshoe hares. III. Nutrition, plant secondary compounds and food limitation. *J. Anim. Ecol.* **57**: 787–806.

Sinclair, A.R.E. & Norton-Griffiths, M. (1979) *Serengeti: Dynamics of an Ecosystem.* University of Chicago Press, Chicago.

Sissenwine, M.P. (1977) The effect of random fluctuations on a hypothetical fishery. *Int. Comm. Northw. Atlant. Fish. Sel. Par.* **2**: 137–144.

Skinner, J.W. (1985) Clutch size as an optimal foraging problem for insects. *Behav. Ecol. Sociobiol.* **17**: 231–238.

Skorping, A. (1985) Parasite-induced reduction in host survival and fecundity: the effect of the nematode *Elephostrongylus rangiferi* on the snail intermediate host. *Parasitology* **91**: 555–562.

Slade, N.A. & Balph, D.F. (1974) Population ecology of Uinta ground squirrels. *Ecology* **55**: 989–1003.

Slansky, F. Jr & Rodriguez, J.G. (eds) (1987) *Nutritional Ecology of Insects, Spiders and Related Invertebrates.* John Wiley & Sons, New York.

Slansky, F. Jr & Scriber, J.M. (1985) Food consumption and utilization. In: Kerkut, G.A. & Gilbert, L.I. (eds) *Comprehensive Insect Physiology, Biochemistry and Pharmacology*, pp. 87–163. Pergamon Press, Oxford.

Slatkin, M. & Maynard Smith, J. (1979) Models of coevolution. *Q. Rev. Biol.* **54**: 233–263.

Slobodkin, L.B. (1974) Prudent predation does not require group selection. *Am. Nat.* **108**: 665–678.

Smith, A.D.M. (1982) Epidemiological patterns in directly transmitted human infections. *ICCET (Imperial College), Series E 1* (unpublished).

Smith, D.G., Wilson, C.R. & Frost, H.H. (1972) The biology of the American kestrel in central Utah. *Southwest Naturalist* **17**: 73–83.

Smith, G. (1984a) Density-dependent mechanisms in the regulation of *Fasciola hepatica* populations in sheep. *Parasitology* **88**: 449–461.

Smith, H.R. (1985) Wildlife and the gypsy moth. *Wildlife Soc. Bull.* **13**: 166–176.

Smith, H.S. (1929) Multiple parasitism: its relation to the biological control of insect pests. *Bull. Entomol. Res.* **20**: 141–149.

Smith, R.B. (1984b) *Feeding ecology of Araneus diadematus.* M.S. thesis, University of British Columbia,

Vancouver.

Smith, R.B. & Wellington, W.G. (1986) The functional response of a juvenile orb-weaving spider. Proceedings of the IX International Congress on Arachinology, 1985, Smithsonian Institution.

Smith, R.H. & Mead, R. (1974) Age structure and stability in models of predator–prey systems. *Theor. Pop. Biol.* **6**: 308–322.

Smith, S.M. (1975) Innate recognition of coral snake pattern by a possible avian predator. *Science* **187**: 759–760.

Smith, S.M. (1977) Coral snake pattern recognition and stimulus generalisation by naive great kiskadees (Aves: Tyrannidae). *Nature* **265**: 535–536.

Snow, D.W. (1968) Movements and mortality of British kestrels *Falco tinnunculus*. *Bird Study* **15**: 65–83.

Snyder, N.F.R. & Wiley, J.W. (1976) Sexual size dimorphism in hawks and owls of North America. *Ornithol. Monogr.* **20**: 1–96.

Snydermann, M. (1983) Optimal prey selection: the effects of food deprivation. *Behav. Anal. Lett.* **3**: 359–369.

Solomon, D.M., Glen, D.M., Kendall, D.A. & Milsom, N.F. (1976) Predation of overwintering larvae of codling moth (*Cydia pomonella* (L.)) by birds. *J. Appl. Ecol.* **13**: 341–353.

Solomon, M.E. (1949) The natural control of animal populations. *J. Anim. Ecol.* **18**: 1–35.

Soper, H.C. (1929) Interpretation of periodicity in disease prevalence. *J. R. Stat. Soc.* **92**: 34–73.

Sota, T. (1985) Limiting of reproduction by feeding conditions in a carabid beetle, *Catabus vaconicus. Res. Pop. Ecol.* **27**: 171–184.

Southern, H.N. (1970) The natural control of a population of tawny owls (*Strix aluco*). *J. Zool.* **162**: 197–285.

Southgate, B.R. (1983) Intensity of transmission and development of disease: their relationship in lymphatic filariasis. Mimeographed Document FIL/EC/WP/83.29. WHO, Geneva.

Southwood, T.R.E. (1976) Bionomic strategies and population parameters. In: May, R.M. (ed.) *Theoretical Ecology*, pp. 26–48. Blackwell Scientific Publications, Oxford.

Southwood, T.R.E. (1977) Habitat, the template of ecological strategies? *J. Anim. Ecol.* **46**: 337–365.

Southwood, T.R.E. (1981) Bionomic strategies and population parameters. In: May, R.M. (ed.) *Theoretical Ecology. Principles and Applications,* pp. 30–52. Blackwell Scientific Publications, Oxford.

Southwood, T.R.E. & Comins, H.N. (1976) A synoptic population model. *J. Anim. Ecol.* **45**: 949–965.

Spangler, H.G. (1988) Hearing in tiger beetles (Cicindelidae). *Physiol. Entomol.* **13**: 447–452.

Spencer, K.C. (ed.) (1988) *Chemical Mediation of Coevolution.* Academic Press, San Diego.

Spiller, D.A. (1986) Interspecific competition between spiders and its relevance to biological control by general predators. *Environ. Entomol.* **15**: 177–181.

Spitzer, K. & Leps, J. (1988) Determinants of temporal variation in moth abundance. *Oikos* **53**: 31–36.

Spitzer, K., Rejmanek, M. & Soldan, T. (1984) The fecundity and long-term variability in abundance of noctuid moths (Lepidoptera, Noctuidae). *Oecologia* **62**: 91–93.

Staddon, J.E.R. (1983) *Adaptive Behaviour and Learning.* Cambridge University Press, Cambridge.

Staddon, J.E.R. & Horner, J.M. (1989) Stochastic choice models: a comparison between Bush-Mosteller and a Sonrie-independent reward-following model. *J. Exp. Anal. Behav.* **52**: 57–64.

Stanley, S.M. (1973) An ecological theory for the sudden origin of multi-cellular life in the late precambrian. *Proc. Nat. Acad. Sci. USA* **70**: 1486–1489.

Stearns, S.C. & Schmid-Hempel, P. (1987) Evolutionary insights should not be wasted. *Oikos* **49**: 118–125.

Stephens, D.W. (1981) The logic of risk-sensitive foraging preferences. *Anim. Behav.* **29**: 628–629.

Stephens, D.W. (1985) How important are partial preferences? *Anim. Behav.* **33**: 667–669.

Stephens, D.W. (1991) Risk and uncertainty in behavioral ecology. In: Cashdan, E. (ed.) *Risk and Uncertainty in Tribal and Peasant Economies.* Westview Press, Lawrence.

Stephens, D.W. & Krebs, J.R. (1986) *Foraging Theory.* Princeton

University Press, Princeton.

Stephens, D.W. & Paton, S.R. (1986). How constant is the constant of risk aversion? *Anim. Behav.* **34**: 1659–1667.

Sternburg, J.G., Waldbauer, G.P. & Jeffords, M.R. (1977) Batesian mimicry: selective advantage of color pattern. *Science* **195**: 681–683.

Stewart, F.M. (1971) Evolution of dimorphism in a predator–prey model. *Theor. Pop. Biol.* **2**: 493–506.

Stewart, I. (1989) *Does God Play Dice? The Mathematics of Chaos*. Basil Blackwell, Oxford.

Stewart, L.A. & Dixon, A.F.G. (1989) Why big species of ladybird beetles are not melanic. *Funct. Ecol.* **3**: 165–177.

Stiling, P.D. (1987) The frequency of density dependence in insect host–parasitoid systems. *Ecology* **68**: 844–856.

Stoltz, D.B. & Vinson, S.B. (1979) Viruses and parasitism in insects. *Adv. Virus Res.* **24**: 125–170.

Stonehouse, B. (1960) The king penguin *Aptenodytes patagonica* of South Georgia. I. Breeding behaviour and development. Falkland Islands Dept. Surv. Sci. Report **23**, p. 81.

Stowe, M.K. (1978) Observations of two nocturnal orb weavers that build specialized webs: *Scloloderus cordatus* and *Wixia ectypa* (Araneae: Araneidae). *J. Arachnol.* **6**: 141–146.

Stradling, D.J. (1976) The nature of the mimetic patterns of the brassolid genera, *Caligo* and *Eryphanis*. *Ecol. Entomol.* **1**: 135–138.

Strand, M.R. (1986) The physiological interactions of parasitoids with their hosts and their influence on reproductive strategy. In: Waage, J. & Greathead, D. (eds) *Insect Parasitoids*, pp. 97–136. Academic Press, London.

Strong, D.R., Lawton, J.H. & Southwood, T.R.E. (1984) *Insects on Plants: Community Patterns and Mechanisms*. Blackwell Scientific Publications, Oxford.

Strong, D.R. Jr, Lawton, J.H. & Southwood, T.R.E. (1984) *Insects on Plants*. Blackwell Scientific Publications, Oxford.

Stuart-Harris, C.H., Schild, G.C. & Oxford, J.S. (1985) *Influenza. The Viruses and the Disease*. Edward Arnold, London.

Suarez Miranda, J.A. (1658) *Travels of Praiseworthy Men*. Quoted by J.L. Borges in *A Universal History of Infamy*, Penguin Books, London, 1954.

Sudd, J.H. & Franks, N.R. (1987) *The Behavioural Ecology of Ants*. Chapman and Hall, New York.

Sugihara, G., Grenfell, B.T. & May, R.M. (1990) Distinguishing error from chaos in ecological time series. *Phil. Trans. R. Soc. Lond. B* **330**: 235–251.

Sugihara, G. & May, R.M. (1990) Nonlinear forecasting as a way of distinguishing chaos from measurement error in time series. *Nature* **344**: 734–741.

Sulimski, A. (1959) Pliocene insectivores from Weze. *Acta Palaeont. Polonica* **4**, Warzawa.

Sulkava, S. (1964) Goshawk (*Accipiter gentilis*) and its feeding habits in south and central Finland. *Suomen Riista* **17**: 22–42.

Sunderland, K.D., Chambers, R.J., Stacey, D.L. & Crook, N.E. (1985) Invertebrate polyphagous predators and cereal aphids. *Bulletin Section Regionale Ouest Palearctic/West Palearctic Regional Section* **VIII/3**: 105–114.

Sutherland, W.J. (1983) Aggregation and the 'ideal free' distribution. *J. Anim. Ecol.* **52**: 821–828.

Sutherland, W.J. & Parker, G.A. (1985) Distribution of unequal competitors. In: Sibly, R.M. & Smith, R.H. (eds) *Behavioural Ecology*, pp. 255–273. Blackwell Scientific Publications, Oxford.

Swartz, L.G., Walker, W., Roseneau, D.G. & Springer, A.M. (1974) Populations of gyrfalcons on the Seward Peninsula, Alaska, 1968–1972. *Raptor Research Foundation, Raptor Research Report* **3**: 71–75.

Sweeney, B.W. & Vannote, R.L. (1982) Population synchrony in mayflies: a predator satiation hypothesis. *Evolution* **36**: 810–821.

Sweet, S.S. (1985) Geographic variation, convergent crypsis and mimicry in gopher snakes (*Pituophis melanoleucus*). *J. Herpetol.* **19**: 55–67.

Syme, P.D. (1981) *Rhyacionia buoliana* (Schiff.), European pine shoot moth (Lepidoptera: Tortricidae). In: Kelleher, J.S. & Hulme, M.A. (eds) Biological control programmes against insects and weeds in Canada, 1969–1980, pp. 387–394. Commonwealth

Agricultural Bureau, Farnham Royal, UK.

Takens, F. (1981) Detecting strange attractors in turbulence. *Lecture Notes in Mathematics* **898**: 366–381.

Tanaka, K. & Ito, Y.G. (1982) Decrease in respiratory rate of a wolf spider, *Pardosa astrigera*, under starvation. *Res. Pop. Ecol.* **24**: 360–374.

Tanaka, K., Ito, Y. & Saito, T. (1985) Reduced respiratory quotient by starvation in a wolf spider, *Pardosa astrigera* (L. Koch). *Comp. Biochem. Physiol.* **80A**: 415–418.

Tanner, J.T. (1975) The stability and the intrinsic growth rates of prey and predator populations. *Ecology* **56**: 855–867.

Taylor, A.D. (1988) Large-scale spatial structure and population dynamics in arthropod predator–prey systems. *Ann. Zool. Fenn.* **25**: 63–74.

Taylor, A.D. (1990) Metapopulations, dispersal, and predator–prey dynamics: an overview. *Ecology* **71**: 429–433.

Taylor, C.R., Caldwell, S.L. & Rowntree, V.J. (1972) Running up and down hills: some consequences of size. *Science* **178**: 1096–1097.

Taylor, C.R. & Lyman, C.P. (1973) Heat storage in running antelopes: independence of brain and body temperatures. *Am. J. Physiol.* **222**: 114–117.

Taylor, C.R., Schmidt-Nielsen, K. & Raab, J.L. (1970) Scaling of energetic cost of running to body size in mammals. *Am. J. Physiol.* **219**: 1104–1107.

Taylor, L.R. (1986) Synoptic dynamics, migration and the Rothamsted Insect Survey. *J. Anim. Ecol.* **55**: 1–38.

Taylor, M.E. (1989) Locomotor adaptations by carnivores. In: Gittleman, J.L. (ed.) *Carnivore Behavior, Ecology and Evolution*, pp. 382–409. Cornell University Press, New York.

Taylor, P.D. (1981) Intra-sex and inter-sex sibling interactions as sex ratio determinants. *Nature* **291**: 64–66.

Taylor, P.D. & Bulmer, M.G. (1980) Local mate competition and the sex ratio. *J. Theor. Biol.* **86**: 409–419.

Taylor, R.H. (1979) How the Macquarie Island parakeet became extinct. *N. Z. J. Ecol.* **2**: 42–45.

Taylor, R.J. (1976) Value of clumping to prey and the evolutionary response of ambush predators. *Am. Nat.* **110**: 13–29.

Taylor, R.J. (1984) *Predation*. Chapman & Hall, London.

Templeton, A.R., Lawlor, I.R. (1981) The fallacy of averages in ecological optimization theory. *Am. Nat.* **117**: 390–393.

Thaler, R. (1980) Toward a positive theory of consumer choice. *J. Econ. Behav. Org.* **1**: 39–60.

Thapar, V. (1986) *Tiger: Portrait of a Predator*. Facts on File Publications, New York.

Thiel, H. (1975) The size structure of the deep-sea benthos. *Int. Rev. Ges. Hydrobiol.* **60**: 575–606.

Thiele, H.-U. (1977) *Carabid Beetles in Their Environments*, pp. 90–101. Springer-Verlag, Berlin.

Thompson, D.J. (1975) Towards a predator–prey model incorporating age structure: the effects of predator and prey size on the predation of *Daphnia magna* by larvae of the damselfly, *Ischnura elegans*. *J. Anim. Ecol.* **44**: 907–916.

Thompson, D.J. (1978) Towards a realistic predator–prey model: the effect of temperature on the functional response and life history of larvae of the damselfly, *Ischnura elegans*. *J. Anim. Ecol.* **47**: 757–767.

Thompson, T.E. (1958) The natural history, embryology, larval biology and post-larval development of *Adalaria proxima* (Alder & Hancock) (Gastropoda: Opisthobranchia). *Phil. Trans. R. Soc. Lond. B* **242**: 1–58.

Thompson, V. (1973) Spittlebug polymorphism for warning colour. *Nature* **242**: 126–128.

Thorson, G. (1950) Reproductive and larval ecology of marine bottom invertebrates. *Biol. Rev.* **25**: 1–45.

Tilman, D. (1982) *Resource Competition and Community Structure*. Princeton University Press, Princeton.

Tinbergen, J. (1981) Foraging decisions in starlings. *Ardea* **69**: 1–67.

Tinbergen, L. (1946) Sperver als Roofvijand van Zangvogels. *Ardea* **34**: 1–123.

Tinbergen, L. (1960) The natural control of insects in pine woods. I. Factors influencing the intensity of predation by songbirds. *Arch. Néerland. Zool.* **13**: 265–336.

Tinbergen, N., Impekoven, M. & Franck, D. (1976) An experiment on spacing-out as a defence against predation. *Behaviour* **28**: 307–327.

Tomilin, A.G. (1967) Cetacea. In: Ognev, S.I. (ed.) *Mammals of the USSR and Adjacent Countries*, Vol. 9. Israel Program for Scientific Translation, Jerusalem. IPST Cat. No. 1124.

Torgersen, T.R. & Campbell, R.W. (1982) Some effects of avian predators on the western spruce budworm in north central Washington. *Environ. Entomol.* **11**: 429–431.

Tostowaryk, W. (1972) The effect of prey defense on the functional response of *Podisus modestus* (Hemiptera: Pentatomidae) to densities of the sawflies *Neodiprion swainei* and *N. pratti banksianae* (Hymenoptera: Neodiprionidae). *Can. Entomol.* **104**: 61–69.

Trager, W. (1939) Acquired immunity to ticks. *J. Parasitol.* **25**: 57–81.

Traniello, J.F.A. (1989) Foraging strategies of ants. *Ann. Rev. Entomol.* **34**: 191–210.

Trautman, C.G., Frederickson, L.F. & Carter, A.V. (1973) Relationship of red foxes and other predators to populations of ring-necked pheasants and their prey, 1964–71. South Dakota Dept. Game Fish and Parks, P-R Report, Project W-75-R-9, Job F-8.2-9.

Travis, J., Keen, W.H. & Juilianna, J. (1985) The role of relative body size in a predator–prey relationship between dragonfly naiads and larval anurans. *Oikos*, **45**: 59–65.

Treherne, J.E. & Foster, W.A. (1980) The effects of group size on predator avoidance in a marine insect. *Anim. Behav.* **28**: 1119–1123.

Treherne, J.E. & Foster, W.A. (1981) Group transmission of predator avoidance behaviour in a marine insect: the Trafalgar effect. *Anim. Behav.* **29**: 911–917.

Turell, M.J., Bailey, C.L. & Rossi, C.A. (1985) Increased mosquito feeding on Rift Valley fever virus-infected lambs. *Am. J. Trop. Med. Hyg.* **33**: 1232–1238.

Turnbull, A.L. (1962) Quantitative studies of the food of *Linyphia triangularis* (Clerck) (Araneae, Linyphiidae). *Can. Entomol.* **94**: 1233–1249.

Turnbull, A.L. (1964) The search for prey by the web building spider *Achaearanea tepidariorum*. (C.L. Koch) (Araneae: Theridiidae). *Can. Entomol.* **96**: 568–579.

Turnbull, A.L. (1966) A population of spiders and their potential prey in an overgrazed pasture in eastern Ontario, Canada. *Can. J. Zool.* **44**: 557–583.

Turnbull, A.L. (1967) Population dynamics of exotic insects. *Bull. Entomol. Soc. Am.* **13**: 333–337.

Turnbull, A.L. (1973) Ecology of the true spiders (Araneomorphae). *Ann. Rev. Entomol.* **18**: 305–348.

Turnbull, A.L. & Chant, D.A. (1961) The practice and theory of biological control of insects in Canada. *Can. J. Zool.* **39**: 697–753.

Turner, J.R.G. (1975) A tale of two butterflies. *Nat. Hist.* **84**: 28–37.

Turner, J.R.G. (1984a) Darwin's coffin and Doctor Pangloss — do adaptationist models explain mimicry? In: Shorrocks, B. (ed.) *Evolutionary Ecology*, pp. 313–361. Blackwell Scientific Publications, Oxford.

Turner, J.R.G. (1984b) Mimicry: the palatability spectrum and its consequences. In: Vane-Wright, R.I. & Ackery, P.R. (eds) *The Biology of Butterflies*, pp. 141–161. Academic Press, London.

Turner, M. (1979) Diet and feeding phenology of the green lynx spider, *Peucetia viridans* (Araneae: Oxyopidae). *J. Arachnol.* **7**: 149–154.

Tuttle, M.D. & Stevenson, D. (1982) Growth and survival of bats. In: Kunz, T.H. (ed.) *Ecology of Bats*, pp. 105–150. Plenum Press, New York.

Uetz, G.W., Kane, T.C. & Stratton, G.E. (1982) Variation in the social grouping tendency of a communal web-building spider. *Science* **217**: 547–549.

Uetz, G.W., Kane, T.C., Stratton, G.E. & Benton, M.J. (1986) Environmental and genetic influences on the social grouping tendency of a communal spider. In: Huettel, M.D. (ed.) *Evolutionary Genetics of Invertebrate Behavior*, pp. 43–54. Plenum Press, New York.

Underwood, A.J., Denley, E.J. (1984) Paradigms, explanations and generalizations for the structure of intertidal communities on rocky shores. In: Strong, D.R., Simberloff, D., Abele, L.G. & Thistle, A.B. (eds) *Ecological Communities: Conceptual Issues and the Evidence*, pp. 151–180. Princeton University Press,

Princeton.

Underwood, A.J., Denley, E.J. & Moran, M.J. (1983) Experimental analyses of the structure and dynamics of mid-shore rocky intertidal communities in New South Wales. *Oecologia* **56**: 202–219.

Underwood, A.J. & Fairweather, P.G. (1989) Supply-side ecology and benthic marine assemblages. *Trends Ecol. Evol.* **4**: 16–19.

Utida, S. (1950) On the equilibrium state of the interacting population of an insect and its parasite. *Ecology* **31**: 165–175.

Utida, S. (1957) Population fluctuation, an experimental and theoretical approach. *Cold Spring Harb. Symp. Quant. Biol.* **22**: 139–151.

Vale, G.A. (1977) Feeding responses of tsetse flies (Diptera: Glossinidae) to stationary hosts. *Bull. Entomol. Res.* **67**: 635–649.

Van Alphen, J.J.M. & Visser, M.E. (1990) Superparasitism as an adaptive strategy for insect parasitoids. *Ann. Rev. Entomol.* **35**: 59–79.

Van Alphen, J.J.M. & Vet, L.E.M. (1986) An evolutionary approach to host finding and selection. In: Waage, J. & Greathead, D. (eds) *Insect Parasitoids*, pp. 23–61. Academic Press, London.

Van Ballenberghe, V. (1987) Effects of predation on moose numbers: a review of recent North American studies. *Swedish Wildlife Res. Viltrevy.* Suppl. 1: 431–460.

Van Buskirk, J. (1989) Density-dependent cannibalism in larval dragonflies. *Ecology* **70**: 1442–1449.

van den Berg, H., Nyambo, B.T. & Waage, J.K. (1990) Parasitism of *Helicoverpa armigera* (Lepidoptera: Noctuidae) in Tanzania: analysis of parasitoid–crop associations. *Environ. Entomol.* **19**: 1141–1145.

van Driesche, R.G. (1983) The meaning of 'percent parasitism' in studies of insect parasitoids. *Environ. Entomol.* **12**: 1611–1622.

Van Dyck, S. (1982) The relationship of the Australian marsupials *Antechinus stuartii* and *A. flavipes* (Dasyuridae, Marsupialia) in Queensland. In: Archer, M. (ed.) *Carnivorous Marsupials*, pp. 723–766. Royal Zool. Soc. New South Wales, Sydney.

van Emden, H.F. (1990) Plant diversity and natural enemy efficiency in agroecosystems. In: Mackauer, M.,

Ehler, L.E. & Roland, J. (eds) *Critical Issues in Biological Control*, pp. 63–80. Intercept Ltd, Andover.

van Hamburg, H., & Hassell, M.P. (1984) Density dependence and the augmentative release of egg parasitoids against graminaceous stalkborers. *Ecol. Entomol.* **9**: 101–108.

van Lenteren J.C. (1981) Host discrimination by parasitoids. In: Nordlund, D.A., Jones, R.L. & Lewis, W.J. (eds) *Semiochemicals, their Role in Pest Control*, pp. 153–180. John Wiley & Sons, New York.

Van Orsdol, K.G. (1984) Foraging behaviour and hunting success of lions in Queen Elizabeth National Park, Uganda. *Afr. J. Ecol.* **22**: 79–99.

van Riper III, C. (1991) Parasite communities in wet and dry forest subpopulations of the Hawaii common anakihi. In: Loye, J.E. & Zuk M. (eds) *Bird–Parasite Interactions*, pp. 140–153. Oxford University Press, Oxford.

van Riper III, C., van Riper, S.G., Goff, M.L. & Laird, M. (1986) The epizootiology and ecological significance of malaria in Hawaiian birds. *Ecol. Monogr.* **56**: 327–344.

Van Valen, L. (1973) A new evolutionary law. *Evol. Theory* **1**: 1–30.

Van Valkenburgh, B. (1988) Trophic diversity in past and present guilds of large predatory mammals. *Paleobiology* **14**: 155–173.

Van Wingerden, W.K.R.E. & Vugts, H.F. (1974) Factors influencing aeronautic behaviour of spiders. *Bull. Br. Arachnol. Soc.* **3**: 6–10.

Various authors (1932) Protective adaptations of animals — especially insects. *Proc. Entomol. Soc. Lond.* **7**: 79–105.

Various authors (1934) Replies to Dr W.L. McAtee's rejoinder in *Proc. Roy. Ent. Soc. Lond.*, **8**: 113–126, to papers on protective adaptations published by the Society (loc. cit. 7: 79–105). *Proc. R. Ent. Soc. Lond.* **9**: 21–40.

Varley, G.C. (1959) The biological control of insect pests. *J. Roy. Soc. Arts* **107**: 475–490.

Varley G.C. & Gradwell, G.R. (1968) Population models for the winter moth. In: Southwood, T.R.E. (ed.) *Insect Abundance.* Symposium of the Royal Entomological Society of London, 4, 132–142. Blackwell Scientific Publications, Oxford.

Varley, G.C., Gradwell, G.R. & Hassell, M.P. (1973) *Insect Population Ecology*. Blackwell Scientific Publications, Oxford.

Vaughan, T.A. (1977) Foraging behaviour of the giant leaf-nosed bat (*Hipposideros commersoni*). *E. Afr. Wildlife J.* **15**: 237–249.

Vaughan, W. Jr & Herrnstein, R.J. (1987) Stability, amelioration and natural selection. In: Green, L. & Kagel, J.H. (eds) *Advances in Behavioral Economics*, pp. 185–245. Ablex Publishing Co., Norwood, NJ.

Vermeij, G.J. (1978) *Biogeography and Adaptation*. Harvard University Press, Cambridge, MA.

Vermeij, G.J. (1982) Unsuccessful predation and evolution. *Am. Nat.* **120**: 701–720.

Vermeij, G.J. (1987) *Evolution and Escalation: an Ecological History of Life*. Princeton University Press, Princeton.

Vet, L.E.M., Janse, C.J., van Achterberg, C. & van Alphen, J.J.M. (1984) Microhabitat location and niche segregation in two sibling species of drosophilid parasitoids: *Asobara tabida* (Nees) and *A. rufescens* (Foerster) (Braconidae: Alysiinae). *Oecologia* **61**: 182–188.

Viitala, J. & Hoffmeyer, I. (1985) Social organization in *Clethrionomys* compared with *Microtus* and *Apodemus*: social odours, chemistry and biological effects. *Ann. Zool. Fenn.* **22**: 359–371.

Village, A. (1983) The role of nest-site availability and territorial behaviour in limiting the breeding density of kestrels. *J. Anim. Ecol.* **52**: 635–645.

Village, A. (1989) Factors limiting European kestrel *Falco tinnunculus* numbers in different habitats. In: Meyburg, B.-U. & Chancellor, R.D. (eds) *Raptors in the Modern World*, pp. 193–202. World Working Group on birds of prey and owls, London.

Village, A. (1990) *The Kestrel*. Poyser, Calton.

Vinson, S.B. (1981) Habitat location. In: Norland, D.A., Jones, R.L. & Lewis, W.J. *Semiochemicals: Their Role in Pest Control*, pp. 51–77. John Wiley, New York.

Vinson, S.B. (1986) How parasitoids locate their hosts: a case of insect espionage. In: Lewis, T. (ed.) *Insect Communication*, pp. 325–348. Academic Press, London.

Vogel, P. (1980) Metabolic levels and biological strategies of shrews. In: Schmidt-Nielsen, K., Bolis, L. & Taylor, C.R. (eds) *Comparative Physiology: Primitive Mammals*, pp. 170–180. Cambridge University Press, Cambridge.

Vollrath, F. (1985) Web spiders' dilemma: a risky move or site dependent growth. *Oecologia* **68**: 69–72.

Volterra, V. (1926) Variation and fluctuations of the number of individuals in animal species living together. In: Chapman, R.N. (ed.) *Animal Ecology*, pp. 409–448. McGraw Hill, London.

von Bertalanffy, L. (1938) A quantitative theory of organic growth. *Hum. Biol.* **10**: 181–213.

Waage, J.K. (1978) Arrestment responses of the parasitoid, *Nemeritis canescens*, to a contact chemical produced by its host, *Plodia interpunctella*. *Physiol. Entomol.* **3**: 135–146.

Waage, J.K. (1979a) Foraging for patchily-distributed hosts by the parasitoid *Nemeritis canescens*. *J. Anim. Ecol.* **48**: 353–371.

Waage, J.K. (1979b) The evolution of insect/vertebrate associations. *Biol. J. Linn. Soc.* **12**: 187–224.

Waage, J.K. (1989) The population ecology of pest–pesticide–natural enemy interactions. In: Jepson, P. (ed.) *Pesticides and Non-target Invertebrates*, pp. 81–93. Intercept Ltd, Andover.

Waage, J.K. (1990) Ecological theory and the selection of biological control agents. In: Mackauer, M., Ehler, L.E. & Roland, J. (eds) *Critical Issues in Biological Control*, pp. 135–158. Intercept Ltd, Andover.

Waage, J.K. & Davies, C.R. (1986) Host mediated competition in a bloodsucking insect community. *J. Anim. Ecol.* **55**: 171–180.

Waage, J.K. & Greathead, D.J. (eds) (1986) *Insect Parasitoids*. Academic Press, London.

Waage, J.K. & Greathead, D.J. (1989) Biological control: challenges and opportunities. In: Wood, R.K.S. & Way, M.J. (eds) *Biological Control of Pests, Pathogens and Weeds: Developments and Prospects*, pp. 1–18. Cambridge University Press, Cambridge.

Waage, J.K., Hassell, M.P. & Godfray,

H.C. (1985) The dynamics of pest—parasitoid—insecticide interactions. *J. Appl. Ecol.* **22**: 825–838.

Wainer, J.W. (1976) Studies of an island population of *Antechinus minimus* (Marsupialia: Dasyuridae). *Aust. Zool.* **19**: 1–7.

Wakelin, D. (1978) Genetic control of susceptibility and resistance to parasitic infection. *Adv. Parasitol.* **16**: 219–308.

Wakelin, D. (1984) Evasion of the immune response: survival within low responder individuals of the host population. *Parasitology* **88**: 639–657.

Wakelin, D. (1985) Genetics, immunity and parasite survival. In: Rollinson, D. & Anderson, R.M. (eds) *Ecology and Genetics of Host–Parasite Interactions*, pp. 39–54. John Wiley & Sons, New York.

Waldbauer, G.P. (1988a) Asynchrony between Batesian mimics and their models. *Am. Nat.* **131** (suppl.): S103–S121.

Waldbauer, G.P. (1988b) Aposematism and Batesian mimicry. Measuring mimetic advantage in natural habitats. *Evol. Biol.* **22**: 227–259.

Waldbauer, G.P., Sternburg, J.G. & Maier, C.T. (1977) Phenological relationships of wasps, bumblebees, their mimics, and insectivorous birds in an Illinois sand area. *Ecology* **58**: 583–591.

Walde, S.J. & Murdoch, W.W. (1988) Spatial density dependence in insect parasitoids. *Ann. Rev. Entomol.* **33**: 441–466.

Walker, B.H. & Noy Meir, I. (1982) Aspects of the stability and resilience of savanna ecosystems. In: Huntley, B.J. & Walker, B.H. (eds) *Ecology of Tropical Savannas* pp. 556–590. Springer-Verlag, Berlin.

Wallace, A.R. (1867) *Proc. Entomol. Soc. Lond.* 1867(15): lxxvii.

Wallace, J.B., Webster, J.R. & Woodall, W.R. (1977) The role of filter feeders in flowing water. *Arch Hydrobiol.* **79**: 506–532.

Walsh, J.A. (1983) Selective primary health care: strategies for control of disease in the developing world. IV. Measles. *Rev. Infect. Dis.* **55**: 330–340.

Ward, P. & Zahavi, A. (1975) The importance of certain assemblages of birds as 'information centres' for food finding. *Ibis* **115**: 517–534.

Warner, R.E. (1968) The role of introduced diseases in the extinction of the Hawaiian avifauna. *Condor* **70**: 101–120.

Warren, G.L. (1970) Introduction of the masked shrew to improve control of forest insects in Newfoundland. *Proc. Tall Timbers Conf., Tallahassee, Florida.*

Warren, J.H. (1985) Climbing as an avoidance behaviour in the salt marsh periwinkle *Littorina irrorata* (Say). *J. Exp. Marine Biol. Ecol.* **89**: 11–38.

Warren, P.H. & Lawton, J.H. (1987) Invertebrate predator—prey body size relationships: an explanation for upper triangular food webs and patterns in food web structure? *Oecologia* **74**: 231–235.

Wassom, D.L., DeWitt, C.W. & Grundmann, A.W. (1973) Host resistance in a natural host—parasite system. Resistance to *Hymenolepis citelli* by *Peromyscus maniculatus*. *J. Parasitol.* **59**: 117–121.

Watson, A.P. & Gaskin, D.E. (1983) Observations on the ventilation cycle of the harbour porpoise *Phocoena phocoena* L. in coastal waters of the Bay of Fundy. *Can. J. Zool.* **61**: 126–132.

Watson, D. (1977) *The Hen Harrier*. Poyser, Berkhamsted.

Watson, J. & Langslow, D.R. (1989) Can food supply explain variation in nesting density and breeding success amongst Golden Eagles *Aquila chrysaetos*? In: Meyburg, B.-U. & Chancellor, R.D. (eds) *Raptors in the Modern World*, pp. 181–186. Berlin, London & Paris: World Working Group on Birds of Prey.

Watt, K.E.F. (1964) Comments on fluctuations of animal populations and measures of community stability. *Can. Entomol.* **96**: 1434–1442.

Watt, K.E.F. (1965) Community stability and the strategy of biological control. *Can. Entomol.* **97**: 887–895.

Weems, H.V. & Whitcomb, W.H. (1977) The green-lynx spider *Peucetia viridans* (Hentz) (Araneae: Oxyopidae). *Ent. Circ.* 1811.

Wells, M.J. (1978) *Octopus: Physiology and Behaviour of an Advanced Invertebrate*. Chapman and Hall, London.

Werner, E.E. & Gilliam, J.F. (1984) The ontogenetic niche and species interactions in size-structured populations. *Ann. Rev. Ecol.*

Systemat. **15**: 393–425.

Werren, J.H. (1980) Sex ratio adaptations to local mate competition in a parasitic wasp. *Science* **208**: 1157–1159.

Werren, J.H. (1983) Sex ratio evolution under local mate competition in a parasitic wasp. *Evolution* **37**: 116–124.

Wesloh, R.M. (1976) Behaviour of forest insect parasitoids. In: Anderson, J.F. & Kaya, H.K. (eds) *Perspectives in Forest Entomology*, pp. 99–110. Academic Press, New York.

West, L. (1988) Prey selection by the tropical snail *Thais melonges*: a study of interindividual variation. *Ecology* **69**: 1839–1854.

Westbrook, R.F. & Brookes, N. (1988) Potentiation and blocking of conditioned flavour and context aversions. *Q. J. Exp. Psychol. B* **40**: 3–30.

Wetterer, M. (1989) Central place foraging theory: when load size affects travel time. *Theor. Pop. Biol.* **36**: 267–280.

White, D.O. & Fenner, F. (1986) *Medical Virology*, 3rd edn. Academic Press, London.

Whitfield, D.P. (1988) Sparrowhawks *Accipiter nisus* affect the spacing behaviour of wintering turnstone *Arenaria interpres* and redshank *Tringa totanus. Ibis* **150**: 284–287.

Whitfield, P. (1985) Raptor predation on wintering waders in SE Scotland. *Ibis* **127**: 544–558.

Whitman, D.W., Blum, M.S. & Jones, C.H. (1986) Olfactorily mediated attack suppression in the southern grasshopper mouse towards an unpalatable prey. *Behav. Processes* **13**: 77–83.

Wickler, W. (1968) *Mimicry in Plants and Animals*. World University Library, London.

Wiklund, C. & Järvi, T. (1982) Survival of distasteful insects after being attacked by naive birds: a reappraisal of the theory of aposematic coloration evolving through individual selection. *Evolution* **36**: 998–1002.

Wikman, M. (1976) Sex ratio of Finnish nestling goshawks *Accipiter gentilis* (L.). *Congr. Int. Union Game Biol.* **12**.

Wildt, D.E., Bush, M., Goodrowe, K.L., *et al.* (1987) Reproductive and genetic consequences of founding isolated lion populations. *Nature* **329**: 328–331.

Williams, D.J. (1986) *Rastrococcus invadens* sp. n. (Hemiptera: Pseudococcidae) introduced from the Oriental Region to West Africa and causing damage to mango, citrus and other trees. *Bull. Entomol. Res.* **76**: 695–699.

Williams, G.C. (1975) *Sex and Evolution*. Princeton University Press, Princeton.

Williamson, M.H. (1984) The measurement of population variability. *Ecol. Entomol.* **9**: 239–241.

Willink, E. & Moore, D. (1988) Aspects of the biology of *Rastrococcus invadens* Williams (Hemiptera: Pseudococcidae), a pest of fruit crops in West Africa, and one of its primary parasitoids, *Gyranusoidea tebygi* (Hymenoptera: Encyrtidae). *Bull. Entomol. Res.* **78**: 709–715.

Wilson, D.S. (1973) Size selective predation among copepods. *Ecology* **54**: 909–914.

Wilson, D.S. (1975) The adequacy of body size as a niche difference. *Am. Nat.* **109**: 769–784.

Wilson, D.S. (1978) The adequacy of body size as a niche difference. *Am. Nat.* **109**: 769–784.

Wilson, D.S. (1980) *The Natural Selection of Populations and Communities*, p. 116. Benjamin Cummings, Menlo Park.

Wilson, D.S. & Clark, A.B. (1977) Above ground predator defence in the harvester termite, *Hodotermes mossambicus* (Hagen). *J. Entomol. Soc. S. Africa* **40**: 271–282.

Wilson, R.S. (1970) Some comments on the hydrostatic system of spiders (Chelicerata, Araneae). *Z. Morphol. Tiere.* **68**: 308–322.

Wise, D.H. (1975) Food limitation in the spider *Linyphia marginata*: experimental field studies. *Ecology* **56**: 637–646.

Wise, D.H. (1979) The effects of an experimental increase in prey abundance upon the reproductive rates of two orb weaving spider species (Araneae; Araneidae). *Oecologia* **41**: 289–300.

Wise, D.H. (1984) The role of competition in spider communities: insights from field experiments with a model organism. In: Strong, D.R., Simberloff, D.S., Abele, L.G. & Thistle, A. (eds) *Ecological Communities: Conceptual Issues and* the Evidence, pp. 42–54. Princeton University Press, Princeton.

Woets, J. & van Lenteren, J.C. (1976) The parasite–host relationship between *Encarsia formosa* (Hymenoptera: Aphelinidae) and *Trialurodes vaporarium* (Homoptera: Aleyrodidae). VI. The influence of the host-plant on the greenhouse whitefly and its parasite *Encarsia formosa*. *Proc. 3rd Conf. Biol. Control Greenhouses*. OILB/SROP **76**: 125–137.

Woffinden, N.D. & Murphy, J.R. (1977) Population dynamics of the ferruginous hawk during a prey decline. *Great Basin Nat.* **37**: 411–425.

Wolda, H. (1983) Long-term stability of tropical insect populations. *Res. Pop. Ecol.* Suppl. **3**: 112–126.

Wollkind, D.J., Collings, J.B. & Logan, J.A. (1988) Metastability in a temperature-dependent model system for predator–prey mite outbreak interactions on fruit trees. *Bull. Math. Biol.* **50**: 379–411.

Wood, B.J. (1971) Development of integrated control programs for pests of tropical perennial crops in Malaysia. In: Huffaker, C.B. (ed.) *Biological Control*, pp. 422–457. Plenum Press, New York.

Wood, D.H. (1970) An ecological study of *Antechinus stuartii* (Marsupialia) in a south-east Queensland rainforest. *Aust. J. Zool.* **18**: 185–207.

Woodruff, D.S. (1986) Genetic control of schistosomiasis: a technique based on the manipulation of intermediate host snail populations. In: Cheng, T. (ed.) *Parasitic and Related Disease: Basic Mechanisms, Manifestations and Control*, pp. 41–68. Comparative Pathobiology Vol. 8. Plenum Press, New York.

Wright, C.J. (1966) The pathogenesis of helminths in the mollusca. *Helminthol. Abstr.* **35**: 201–224.

Wunderle, J.M., Santa Castro, M. & Fetcher, N. (1987) Risk-averse foraging by bananaquits on negative energy budgets. *Behav. Ecol. Sociobiol.* **21**: 249–255.

Yamaguchi, M. (1974) Growth of juvenile *Acanthaster planci* (L.) in the laboratory. *Pacific Sci.* **28**: 123–138.

Yamanaka, K., Nakasuji, F., Kivitani, K. (1972) Life tables of the tobacco cutworm *Spodoptera litera* (Lepiduptera: Noctuidae) and the

evaluation of effectiveness of natural enemies. *Jap. J. Appl. Entomol. Zool.* **16**: 205–214.

Yao, D.S. & Chant, D.A. (1989) Population growth and predation interference between two species of predatory phytoseiid mites (Acarina: Phytoseiidae) in interactive systems. *Oecologia* **80**: 443–455.

Yasui, W.Y. & Gaskin, D.E. (1986) Energy budget of a small cetacean, the harbour porpoise, *Phocoena phocoena* (L.). *Ophelia* **25**: 183–197.

Yodzis, P.P. (1984) Energy flow and the vertical structure of real ecosystems. *Oecologia* **65**: 86–88.

Yodzis, P. (1989) *Introduction to Theoretical Ecology*. Harper & Row, New York.

Young, C.M. & Gotelli, N.J. (1988) Larval predation by barnacles: effects on patch colonization in a shallow subtidal community. *Ecology* **69**: 624–634.

Young, M.D., Hardman, N.F., Burgess, R.N., Frohne, W.C. & Sabrosky, C.W. (1948) The infectivity of native malarias in South Carolina to *Anopheles quadrimaculatus*. *Am. J. Trop. Med. Hyg.* **28**: 303–311.

Yu, P. (1972) Some host–parasite genetic interaction models. *Theor. Pop. Biol.* **3**: 347–357.

Zahavi, A. (1987) The theory of signal selection and some of its implications. In: Delfino, V.P. (ed.) *International Symposium of Biological Evolution*, pp. 305–327. Adriatica Editrice, Bari.

Zalom, F.G. (1978) A comparison of predation rates and prey handling times of adult *Notonecta* and *Buenoa* (Hemiptera: Notonectidae). *Ann. Entomol. Soc. Am.* **71**: 143–148.

Zeigler, B.P. (1977) Persistence and patchiness of predator–prey systems induced by discrete event population exchange mechanisms. *J. Theor. Biol.* **67**: 687–713.

Zwölfer, H. (1971) The structure and effect of parasite complexes attacking phytophagous insects. In: den Boer, P.J. & Gradwell, G.R. (eds) *Dynamics of Populations*, pp. 405–418. Centre for Agricultural Publishing and Documentation, Wageningen.

Zwölfer, H. (1963) The structure of the parasite complexes of some Lepidoptera. *Z. Angew. Entomol.* **51**: 346–357.

Index